Electrofocusing and Isotachophoresis

Electrofocusing and Isotachophoresis

Proceedings of the International Symposium
August 2-4, 1976, Hamburg, Germany

Editors

B. J. Radola · D. Graesslin

W
DE
G

Walter de Gruyter · Berlin · New York 1977

Editors

Bertold J. Radola, Dr. rer. nat.
Institut für Lebensmitteltechnologie und
Analytische Chemie
Technische Universität München,
D-8050 Freising-Weihenstephan

Dieter Graesslin, Dr. rer. nat.
Abteilung für klinische und experimentelle
Endokrinologie, Universitäts-Frauenklinik Hamburg,
Martinistraße 52, D-2000 Hamburg 20

With 168 Figures

CIP-Kurztitelaufnahme der Deutschen Bibliothek

Electrofocusing and isotachophoresis: proceedings of the
internat. symposium, August 2–4, 1976,
Hamburg, Germany/ed. B. J. Radola; D. Graesslin.
– 1 Aufl. – Berlin, New York; de Gruyter, 1977.
ISBN 3-11-007026-X
ME: Radola, Bertold J. [Hrsg.]

Library of Congress Cataloging in Publication Data

International Symposium on Electrofocusing and Isotachophoresis.
Hamburg, 1976.
Electrofocusing and isotachophoresis.

Bibliography: p.
Includes index.
1. Isoelectric focusing-Congresses. 2. Isotachophoresis-Congresses.
I. Radola, B.J., 1931 – II. Graesslin, D., 1937 – III. Title.
QP519.9.18158 1976 574.1'9285 77-1446
ISBN 3-11-007026-X

 Printing: Karl Gerike, Berlin. – Binding: Lüderitz & Bauer, Buchgewerbe GmbH, Berlin. Printed in Germany.

Preface

The present volume is based on the lectures and communications presented at the International Symposium on Electrofocusing and Isotachophoresis, held in Hamburg from August 2 to 4, 1976. A decade after the synthesis of carrier ampholytes, this was the fourth conference dealing with these methods, the previous were organized in New York (1972), Glasgow (1973), and Milan (1974). Although these activities and the increasing number of publications both reflect a growing interest, it seems to us that the potential of these methods has only begun to be exploited.

Isoelectric focusing offers an exceedingly attractive dimension for the fractionation and characterization of amphoteric substances, mainly proteins. Isoelectric focusing surpasses other techniques with respect to its resolving power and many substances considered to be homogeneous by other charge-dependent methods have been found to be heterogeneous when analysed by isoelectric focusing. The method defines a fundamental characteristic of amphoteric substances: the isoelectric point. Determination of this important physico-chemical constant provides information on composition and conformation, and helps to select optimum conditions for other separation techniques, for example ion exchange chromatography. Isoelectric focusing thus constitutes part of a generalized strategy for studying amphoteric substances.

In the past few years increasing attention has been paid to preparative isoelectric focusing which is attractive because it covers a range of loads for which high resolution techniques did not exist. Preparative isoelectric focusing offers a straight physico-chemical approach to the simultaneous isolation and purification of several components from complex mixtures, either in a single step or in combination with fractionation methods based on different properties of the separated molecules. Economical considerations are not a seriously limiting factor in preparative isoelectric focusing since at the high load capacity the cost of the carrier ampholytes will in most cases represent only a minor fraction of the value of the already prefractionated material. Application of preparative isoelectric focusing will certainly be promoted by the less expensive "technical" carrier ampholytes which have recently become commercially available.

Sometimes it is argued that the complex patterns obtained on isoelectric focusing are an obstacle to some applications, particularly in the area of clinical chemistry. We should, however, remember that by applying a high resolution technique we are not creating the complexity of nature but only trying to understand it. In view of the undoubtedly great potential of isoelectric focusing it is encouraging to see that a number of promising clinical applications have been described at the Hamburg meeting and that even routine screening has already been envisaged.

Isotachophoresis is a separation technique which holds considerable promise for the separation of all kind of ionic species. A variety of analytical problems can be resolved using isotachophoresis and for a number of applications it could be the method of choice. Preparative isotachophoresis, although still in the development stage is likely to provide a useful and powerful complementation of other charge-dependent methods.

The reports presented at the Symposium cover a broad area of biological and clinical research. It is hoped that these proceedings prove to be a useful summary of an obviously successful conference, and that they will stimulate further applications and developments. Finally, we wish to express our gratitude to all contributors – the success of the conference and the rapid publication of this book must be credited to them.

Freising-Weihenstephan and Hamburg, October 1976

B.J. Radola
D. Graesslin

Acknowledgements

We would like to acknowledge the assistance of our secretarial staff, particularly Mrs. Schedlinski, in organizing the Symposium and in preparing the Proceedings for publication.

Financial and organizational help was provided by: Colora Messtechnik, Lorch; C. Desaga, Heidelberg; LKB Produkter, Bromma, Sweden, and Serva, Heidelberg.

We greatly appreciate the efforts of Water de Gruyter Publishing Company, Berlin, which led to the rapid publication of the Proceedings.

Contents

Part II. Isotachophoresis

Contributors

Allen, R.C., Department of Pathology, Medical University of South Carolina, Charleston, South Carolina 29401, USA

Altland, K., Institut für Humangenetik, Universität Heidelberg, D-6900 Heidelberg

Arlinger, L., Development Department, LKB Produkter AB, S-16125 Bromma

Arnaud, P., Laboratoire de Biochimie, Hôpital Ed. Herriot, F-69 374 Lyon-Cedex

Baumstark, J.S., School of Medicine, Creighton University, Omaha, Nebraska 68108, USA

Bengtsson, G., Department of Physiological Chemistry, University of Umeå, S-90 187 Umeå

Bier, M., Veterans Administration Hospital and University of Arizona, Tucson, Arizona 85723, USA

Bispink, G., Max-Planck-Institut für experimentelle Medizin, Forschungsstelle Neurochemie, D-3400 Göttingen

Bjellqvist, B., Aminkemi AB, S-16 120 Bromma

Bohn, B., Institut für Biochemie II, Universität Heidelberg, D-6900 Heidelberg

Bordi, S., Department of Physical Chemistry, University of Florence, I-50121 Florence

Borja, C., Serva-Feinbiochemica GmbH, D-6900 Heidelberg

Bours, J., Klinisches Institut für experimentelle Ophthalmologie, Universität Bonn, D-5300 Bonn-Venusberg

Brogren, C.H., Immunological Department, Pathological-Anatomical Institute, DK-1399 Copenhagen

Brossmer, R., Institut für Biochemie II, Universität Heidelberg, D-6900 Heidelberg

Brown, R.K., Department of Biochemistry, School of Medicine; Wayne State University, Detroit, Michigan 48201, USA

Caspers, M.L., Department of Biochemistry, School of Medicine, Wayne State University, Detroit, Michigan, 48201, USA

Catsimpoolas, N., Biophysics Laboratory, Department of Nutrition and Food Science, Massachusetts Institute of Technology, Cambridge, Massachusetts 02139, USA

Chapuis-Cellier, C., Laboratoire de Biochimie, Hôpital Ed. Herriot, F-69 374 Lyon-Cedex

Chrambach, A., Reproduction Research Branch, National Institute of Child Health and Human Development, National Institutes of Health, Bethesda, Maryland 20014, USA

Czempiel, W., Embryonalpharmakologie, Freie Universität Berlin, D-1000 Berlin 33

Das, P.D., Institut für Humangenetik, Universitätskrankenhaus Eppendorf, D-2000 Hamburg 20

Delincée, H., Bundesforschungsanstalt für Ernährung, D-7500 Karlsruhe

Delmotte, P., National Centre for Multiple Sclerosis, B-1910 Melsbroek

Denckla, W.D., Thorndike Laboratory, Beth Israel Hospital, Boston, Massachusetts 02111, USA

Drysdale, J.W., Department of Biochemistry and Pharmacology, Tufts University School of Medicine, Boston, Massachusetts 02111, USA

Everaerts, F.M., Department of Instrumental Analysis, Eindhoven University of Technology, NL-Eindhoven

Fawcett, J.S., Department of Experimental Biochemistry, London Hospital Medical College, Queen Mary College, London E1 4NS, U.K.

Fredriksson, S., Department of Physical Chemistry, Chalmers Institute of Technology and University of Gothenburg, S-40 220 Gothenburg

Fudenberg, H.H., Department of Basic and Clinical Immunology and Microbiology, Medical University of South Carolina, Charleston, South Carolina 29401, USA

Galante, E., Laboratory for Plant Biosynthesis, CNR, I-20 133 Milano

Gasparic, V., Aminkemi AB, S-16 120 Bromma

Gelfand, D.H., Department of Biochemistry and Biophysics, University of California Medical Center, San Francisco, California 94143, USA

Gianazza, E., Department of Biochemistry, University of Milano, I-20 133 Milano

Goedde, H.W., Institut für Humangenetik, Universitätskrankenhaus Eppendorf, D-2000 Hamburg 20

Graesslin, D., Abteilung für klinische und experimentelle Endokrinologie, Universitäts-Frauenklinik Hamburg, D-2000 Hamburg 20

Griffith, A.L., Biophysics Laboratory, Department of Nutrition and Food Science, Massachusetts Institute of Technology, Cambridge, Massachusetts 02139, USA

Grubhofer, N., Serva-Feinbiochemica GmbH, D-6900 Heidelberg

Hamann, A., Heinrich-Pette-Institut für experimentelle Virologie und Immunologie, Universität Hamburg, D-2000 Hamburg 20

Hamberg, U., Department of Biochemistry, University of Helsinki, SF-00170 Helsinki 17

Hansén, L., National Board of Occupational Safety and Health, Department of Occupational Health, Chemical Division, Fack, S-10 026 Stockholm

Hees, M., Institut für Humangenetik, Philipps-Universität, D-3550 Marburg

Hjalmarsson, S., Application Laboratory, LKB Produkter AB, S-16 125 Bromma

Holmberg, K., Department of Bacteriology, Statens Bakteriologiska Laboratorium, S-10 521 Stockholm

Ivarie, R.D., Department of Biochemistry and Biophysics, University of California, Medical Center, San Francisco, California 94143, USA

Jeppsson, J.O., Department of Clinical Chemistry, University of Lund, Malmö General Hospital, S-21 401 Malmö

Jones, P.P., Department of Biochemistry and Biophysics, University of California Medical Center, San Francisco, California 94143, USA

Just, W.W., Max-Plack-Institut für Hirnforschung, Arbeitsgruppe Neurochemie, D-6000 Frankfurt/Main

Knüfermann, H., Max-Planck-Institut für Immunbiologie, D-7800 Freiburg

Knuutinen, U., Department of Biochemistry, University of Helsinki, SF-00170 Helsinki

Koistinen, J., Department of Basic and Clinical Immunology and Microbiology, Medical University of South Carolina, Charleston, South Carolina 29401, USA

Kolin, A., Molecular Biology Institute, School of Medicine, University of California, Los Angeles, California 90024, USA

Kopwillem, A., LKB Produkter AB, S-161 25 Bromma

Kubicz, A., Institute of Biochemistry, University of Wroclaw, PL-50-137 Wroclaw,

Leaback, D.H., Department of Biochemistry, Institute of Orthopaedics, Stanmore, Middlesex HA7 4LP, U.K.

Lundin, H., Development Department, LKB Produkter AB, S-161 125 Bromma

Manske, W., Institut für Biochemie II, Universität Heidelberg, D-6900 Heidelberg

Markotai, J., Pharmakognostisches Institut, Universität Wien, A-1090 Wien

Moberg, U., Application Laboratory, LKB Produkter AB, S-161 25 Bromma

Möllby, R., Department of Bacteriology, Karolinska Institutet, S-104 01 Stockholm

Monsher, T.A., Department of Basic and Clinical Immunology and Microbiology, Medical University of South Carolina, Charleston, South Carolina 29401, USA

Narayanan, K.R., Gesellschaft für Strahlen- und Umweltforschung, D-8042 Neuherberg,

Neuhoff, V., Max-Planck-Institut für experimentelle Medizin, Forschungsstelle Neurochemie, D-3400 Göttingen

Nguyen, N.Y., Reproduction Research Branch, National Institute of Child Health and Human Development, National Institutes of Health, Bethesda, Maryland 20014, USA

O'Farrell, P.H., Department of Biochemistry and Biophysics, University of California Medical Center, San Francisco, California 94143, USA

O'Farrell, P.Z., Department of Biochemistry and Biophysics, University of California Medical Center, San Francisco, California 94143, USA

Olivecrona, T., Department of Physiological Chemistry, Umeå University, S-901 87 Umeå

Olsson, B., Department of Bacteriology, Statens Bakteriologiska Laboratorium, S-105 21 Stockholm

Oulla, P.M., Department of Pathology, Medical University of South Carolina, Charleston, South Carolina 29401, USA

Papeschi, G., Department of Physical Chemistry, University of Florence, I-50 121 Florence

Peltre, G., Immunologie Cellulaire, Institut Pasteur, F-75015 Paris

Polisky, B.A., Department of Biochemistry and Biophysics, University of California Medical Center, San Francisco, California 94143, USA

Pronk, J.C., Institute of Human Biology, State University Utrecht, NL-Utrecht

Quast, R., Hygiene Institut, Philipps-Universität, D-3550 Marburg

Radola, B.J., Institut für Lebensmitteltechnologie und Analytische Chemie, Technische Universität München, D-8050 Freising-Weihenstephan

Raj, A.S., Gesellschaft für Strahlen- und Umweltforschung München, D-8042 Neuherberg

Righetti, P.G., Department of Biochemistry, University of Milano, I-20 133 Milano

Rilbe, H., Department of Physical Chemistry, Chalmers Institute of Technology and University of Gothenburg, S-40 220 Gothenburg

Rosengren, A., Aminkemi AB, S-16 120 Bromma 20

Rücker, W., Pharmakognostisches Institut, Universität Wien, A-1090 Wien

Salamini, F., Institute of Cereal Research, I-24100 Bergamo

Schuricht, H., Institut für Lebensmitteltechnologie und Analytische Chemie, Technische Univerität München, D-8050 Freising-Weihenstephan

Sluyterman, L.A.AE., Philips Research Laboratories, NL-Eindhoven

Smyth, C.J., Department of Bacteriology, Veterinärhögskolan, Biomedicum, S-75123 Uppsala

Soave C., Laboratory for Plant Biosynthesis, CNR, I-20 133 Milano

Söderholm, J., Department of Bacteriology, Statens Bakteriologiska Laboratorium, S-10521 Stockholm

Stegemann, H., Institut für Biochemie, Biologische Bundesanstalt, D-3300 Braunschweig

Steinberg, R.A., Cardiovascular Research Institute, University of California Medical Center, San Francisco, California 94143, USA

Tschesche, H., Organisch-Chemisches Institut, Technische Universität München, D-8000 München

Turpeinen, U. Department of Biochemistry, University of Helsinki, SF-00 170 Helsinki 17

Utermann, G., Institut für Humangenetik, Philipps-Universität, D-3550 Marburg

Vesterberg, O., National Board of Occupational Safety and Health, Department of Occupational Health, Chemical Division, Fack, S-10 026 Stockholm

Vinogradov, S.N., Department of Biochemistry, Wayne State University School of Medicine, Detroit, Michigan 48201, USA

Viotti, A., Laboratory for Plant Biosynthesis, CNR, I-20 133 Milano

Vogelberg, K.H., Institut für Humangenetik, Philipps-Universität, D-3550 Marburg

Wadström, T., Department of Bacteriology, Veterinärhögskolan, Biomedicum, S-75123 Uppsala

Werner, G., Max-Planck-Institut für Hirnforschung, Arbeitsgruppe Neurochemie, D-6000 Frankfurt

Wilson, G.B. Department of Basic and Clinical Immunology and Microbiology, Medical University of South Carolina, Charleston, South Carolina 29401, USA

Wielders, J.P.M., Department of Instrumental Analysis, Eindhoven University of Technology, NL-Eindhoven

Wijdenes, J., Philips Research Laboratory, NL-Eindhoven

Winter, A., Application Laboratory, LKB Produkter AB, S-16 125 Bromma

Wolanska, L., Institute of Biochemistry, University of Wroclaw, PL-50 137 Wroclaw

Zeineh, R.A., Medical Laser Unit, Division of Medical Research, Arab Development Institute, Tripoli, Libyan Arab Republic

Part I. Isoelectric Focusing

1. Theoretical and General Aspects

Reminiscences about the Genesis of Isoelectric Focusing and Generalization of the Idea

A. Kolin

INTRODUCTION

Fruitful developments have not always originated in highly competent quarters. They often came from dilettantes whose naive outlook led them in directions ignored by experts. The genesis of isoelectric focusing offers an example of such a course of events.

If one reads accounts of the origin of isoelectric focusing in the most authoritative reviews (1,2,3), one is given the impression that all the basic foundations underlying Svensson's spectacularly successful approach to isoelectric analysis was on hand well over one third of a century ago: 1) The fourteen-compartment apparatus of Williams and Waterman (4) which accumulated ampholytes in one or more of its membrane-bounded chambers was described in 1929. 2) Tiselius (5), unaware of Williams and Waterman, described a similar arrangement with ten compartments in 1941 and clearly outlined the principle of establishing a stable steady-state pH gradient by prolonged "stationary electrolysis" of ampholyte solutions involving equilibrium between diffusion and electromigration. The above reviews give the impression that nothing new and significant has been added beyond these publications until Svensson's paper of 1961 (1). The question arises then: why was isoelectric focusing not introduced and developed in the nineteen forties?

I believe that the reason lies in a missing key insight: The concept of *focusing ions in a continuous pH gradient,* without which the impressive resolution of the Svensson - Vesterberg column is unthinkable. The jump from the stair-case pH distribution in the instruments of Williams and Waterman and of Tiselius to the idea of a continuous pH-distribution was apparently not self-evident since nobody took this step until 1954 (6). In fact, Tiselius, who was in the forefront of electrophoretic developments and well aware of essential innovations in this field, greeted the continuous pH gradient idea and the concept of isoelectric focusing expounded in the 1954 publication as innovative steps offering great possibilities for further development of electrophoretic methods, as can be seen from his letter of September 6, 1955 reproduced in Fig. 1 *).

We shall view in this presentation isoelectric focusing as a physical effect rather than merely an analytical method. In fact, we shall show that we are dealing with a family of three closely related effects and that the idea of isoelectric focusing in an electric field is a special case of a general principle encompassing analogous focusing phenomena obtainable with other types of force fields. Before I go into this I would like to briefly review the events and lines of thought which led to the discovery of isoelectric focusing.

*) The papers to which Tiselius refers in this letter were as follows:

1) Separation and Concentration of Proteins in a pH Field Combined with an Electric Field. J. Chem. Phys. 22, 1628 (1954).
2) Electrophoretic "Line Spectra". J. Chem. Phys. 23, 407 (1955).
3) Isoelectric Spectra and Mobility Spectra: A New Approach to Electrophoretic Separation. Proc. Nat. Ac. Sc. 41, 101 (1955).
4) An Electromagnetokinetic Phenomenon Involving Migration of Neutral Particles. Science 117, 134 (1953).
5) Method for Elimination of Thermal Convection. J. Appl. Physics 25, 1442 (1954).

THE INSTITUTE OF BIOCHEMISTRY
UNIVERSITY OF UPPSALA
UPPSALA, SWEDEN

AT/mr

Uppsala Sept. 6th 1955

Professor
Alexander Kolin
5512 Kimbark Ave
C h i c a g o 37
Ill. U.S.A.

Dear Professor Kolin,

I wish to thank you for your kindness in sending me two sets of reprints in reply to my letter of April 1st, 1955. I have been extremely interested in your highly original work, which seems to me to offer great possibilities for further development of electrophoretic methods. We have discussed your publications at a special seminar at this Institute and my collaborators in this field were extremely interested. I hope you will continue this important work and that you will be good enough to continue to send me reprints.

Sincerely yours,

Arne Tiselius

FIG. 1

ELECTROMAGNETOPHORESIS AS STARTING POINT

In 1952 I discovered an effect (7,8,9) which could be classified as an electrokinetic or just as well as a magnetokinetic phenomenon. In 1952, in the course of a lecture to students of Biology and Medicine at the University of Chicago, I was demonstrating the effect of an electric current passing through a rubber tubing filled with NaCl solution upon an adjacent bar magnet (compass needle). I pointed out that this interaction is not limited to the deviation of the magnet and that the ions experience a force perpendicular to the current and to the transverse magnetic field established across the rubber tube by the magnet. The force upon the ions will in-turn be transmitted to the wall of the tube and would cause, if strong enough, visible displacement of the tube. At this point a question suddenly came into my mind: "Suppose there were a particle (say, a glass sphere) suspended in the solution filling the rubber tube - would it also experience a force like the wall of the tube?"

I thought about it after the lecture and concluded that the suspended sphere would indeed experience a force and that this force would be *opposite* in direction to the force exerted upon the tube wall! Further consideration showed that this electromagnetically engendered force would vanish for a sphere whose conductivity σ' would equal the conductivity σ'' of the solution and that the force upon the sphere will have the same direction as the force upon the rubber tube for $\sigma' > \sigma''$ and be of opposite direction for $\sigma' < \sigma''$. This disappearance and reversal of the force, depending on the value of $(\sigma' - \sigma'')$ is reminescent of Archimedes' principle for bodies submerged in fluids in a gravitational field. I called this effect electromagnetophoresis. *It has no relation to electrophoresis.* In fact, it can exhibit migration of electrically *neutral* particles and can be produced by a combination of alternating electric and magnetic fields as well as by constant, unidirectional ones. It can be roughly regarded as an electromagnetic synthesis of a quasi-gravitational field. I will consider it quickly not only because my work on it led me accidently to the idea of isoelectric focusing but also because it eventually led me to ideas of non-isoelectric focusing which, in-turn, suggested the more general principle of "isoperichoric focusing" described in the last section of this paper.

Fig. 2 shows a configuration in which this effect can be observed. The two coils generate a uniform magnetic field perpendicular to the plane of the coils. Assume a cell with two electrodes E_1, E_2 placed in the field as shown. The top and bottom of the cell are metallic, serving as electrodes, and the cell is filled with an electrolyte which forms a non-polarizable combination with the top and bottom electrodes. The current density $\vec{J}$ is perpendicular to the uniform magnetic field $\vec{B}$ in the optimal configuration. Each particle of the fluid experiences an electromagnetic force $\vec{F}$ (per unit

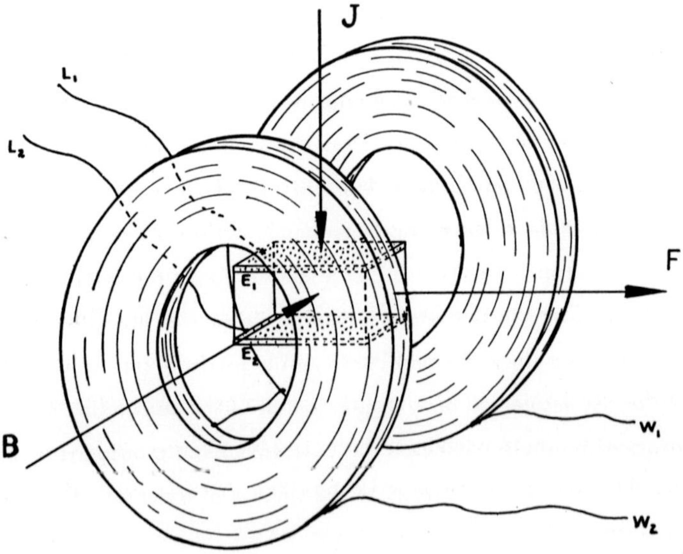

FIG. 2 *Arrangement for demonstration of electromagnetophoresis, showing a migration cell with electrodes $\mathcal{E}_1$ and $\mathcal{E}_2$ in a magnetic field (B) of two Helmholtz coils.*

volume :

(1) $\vec{F}^* = [\vec{J} \times \vec{B}]$

(where $\vec{F}^*$ is measured in dynes/cm^3, $\vec{J}$ in abamperes/cm^2, $\vec{B}$ in gauss). This force reverses in direction if we reverse $\vec{B}$ or $\vec{J}$. On the other hand, the simultaneous reversal of $\vec{B}$ and $\vec{J}$ leaves the direction of the force $\vec{F}$ unaltered. This makes it possible to generate unidirectional migration of particles with an alternating current coupled with an alternating magnetic field, thus obviating the need for non-polarizable electrodes.

If we turn the figure 90° so that $\vec{F}$points down, we have a more accustomed orientation in our analogy to the gravitational field, where $[\vec{J} \times \vec{B}]$ has the meaning of weight density so that a hydrostatic pressure (increasing with depth in the direction of $\vec{F}$) is generated in the cell. The gradient of this pressure is responsible for the force of buoyancy. (i.e., the surface integral of the normal pressure forces on the surface of a submerged body [the glass spheres in our example] has an "upward" resultant directed as $-\vec{F}$.) Thus, if we submerge a nonconductor (e.g., a plastic sphere) into the fluid, it will experience the force of "electromagnetic buoyancy" and will start moving in the direction of $-\vec{F}$. If the body (sphere A of Fig. 3) has the same conductivity as the surrounding fluid, the "electromagnetic buoyancy" is exactly balanced by "electromagnetic gravity". The latter is the $[\vec{J} \times \vec{B}]$ force exerted upon the interior of the sphere. Such a body will remain in stationary suspension in such a $[\vec{J} \times \vec{B}]$-field much like a body submerged into a fluid of equal density in a gravitational field. Finally, if the conductivity of the submerged body surpasses that of the suspending medium, the force of "electromagnetic gravity" exceeds the "electromagnetic buoyancy" and the body moves in the direction of $+\vec{F}$ similarly to the sinking of a denser object in the gravita-

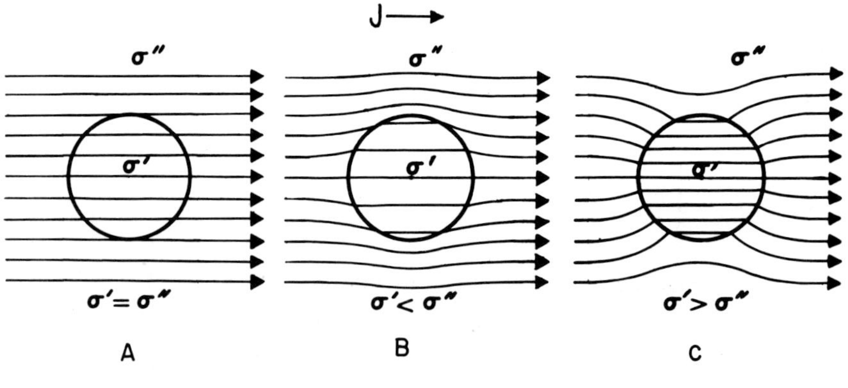

FIG. 3 *Current density ($\vec{J}$) distribution for spheres. A: conductivities of sphere (σ ′) and surrounding fluid (σ ″) are equal. B: the sphere has a smaller conductivity than the environment. C: the sphere is a better conductor than the surrounding fluid. The current density inside the sphere is increased in C, decreased in B and unchanged in A as compared to the current density in the fluid far from the sphere.*

tional analogy. The two cases of main general interest are those of a body of slightly lesser conductivity σ' than the surrounding fluid: $\sigma' < \sigma''$ (sphere B of Fig. 3) and of an object of higher conductivity σ' than the surrounding fluid $\sigma' > \sigma''$ (sphere C of Fig.3). As we see on sphere B, the current lines diverge near it, making the interior $[\vec{J} \times \vec{B}]$ *volume force* smaller than the exterior *surface force* of "electromagnetic buoyancy" which thus predominates. The sphere thus moves in the direction of $-\vec{F}$.

On the other hand, the current lines converge near sphere C, making the $[\vec{J} \times \vec{B}]$ *volume force* in it larger than the *surface force* of "electromagnetic buoyancy". This sphere will move in the direction of $+\vec{F}$.

We see, thus, that if we choose the electrical conductivity of the fluid σ'' intermediate between the conductivities σ_a' and σ_b' of two species of particles, they would move in opposite directions and be separated out of a mixed suspension.

In general the forces on the particles and their velocities are governed by the following equations:

$$(2) \quad F = 3/2\,(JBV)\left(\frac{\sigma' - \sigma''}{\sigma' + 2\sigma''}\right) = 2\pi a^3 [J \times B]\left(\frac{\sigma' - \sigma''}{\sigma' + 2\sigma''}\right)$$

$$(3) \quad v = JB\left(\frac{\sigma' - \sigma''}{\sigma' + 2\sigma''}\right)\frac{a^2}{3\eta}$$

(where V is the volume of the sphere, v its velocity in cm/sec, a its radius in cm and η the viscosity of the suspension medium in poise). Both equations contain the factor $(\sigma' - \sigma'')/(\sigma' + 2\sigma'')$. We see that the particle velocity becomes $v = 0$ for $\sigma' = \sigma''$.

Imagine now that we have a conductivity gradient in the direction of $\vec{F} = \vec{J} \times \vec{B}$. We can arrange for a zone in this gradient to be equiconductive at point A with a particle species of conductivity $\sigma_a' = [\sigma'']_A$. The particles of this species will be at rest in this zone and will move in opposite directions on either side of zone A. Eventually they will come to a standstill upon reaching this zone in which they are thus condensed or "focused". This will happen when the directions of $[\vec{J} \times \vec{B}]$ and grad σ'' coincide. In this case the forces on the $\sigma' < \sigma''$ side of zone A and those on the $\sigma' > \sigma''$ side of the zone

A are opposite to each other and both pointing toward zone A. On the other hand, when $[\vec{J} \times \vec{B}]$ and grad σ'' point in opposite directions, the forces on the particles of conductivity σ_a' will point away from the zone A on either side of it and will cause particle evacuation from the vicinity of this zone.

If we have now a second particle species characterized by conductivity σ_b', it will come to rest in a different equiconductive zone B $(\sigma_b' = [\sigma'']_B)$. We could thus spread a heterogeneous population of biological cells into a sorting pattern according to their electrical conductivities or, in practice, perhaps according to differences in cell membrane permeabilities (depending on the frequency of the alternating electric and magnetic fields employed).

TURNING TO ELECTROPHORESIS

Demonstration of electromagnetophoresis looked quite simple in theory; but as soon as I turned to experiments I encountered great difficulties mainly due to two types of convection: 1) thermal convection and 2) electromagnetic convection. I decided to look into the literature on electrophoresis in the hope of finding some useful information and ideas on how to suppress disturbances due to convection. Dr. Harold Abramson once presented me with a volume of his "Electrokinetic Phenomona". I took it along on a train ride from Chicago to Fort Wayne where I went on a consulting mission. As I read about the amphoteric properties of proteins, with which I was totally unfamiliar, and about the modification of their charge in the process of titration, I visualized a "gedanken-experiment". Instead of changing the pH of the environment of the protein ions in a conventional manner by titration, I imagined a fluid column in which the pH was changing monotonically from point to point. The pH distribution was viewed by me as a "chemical field", i.e. a region of space in which a certain property of the embedded molecular species (the protein charge) was modified from point to point by physico-chemical interaction. I singled out a positive protein ion in a region of low pH and started dragging it as it were by a leash in the direction of increasing pH, watching its charge diminish as the ion approached the zone whose pH was equal to the pI of the protein. Having reached electric neutrality in the isoelectric zone, the charge of the protein reversed its sign as it was dragged further toward zones of increasing pH.

The next step in this imaginary experiment was to release the protein ion from its leash and let an electric field drag it through the pH gradient column in the direction of increasing pH. The result was a little different. The ion came to rest in its isoelectric zone where it lost its charge. The interesting aspect of this arrest was the stability of the particle's equilibrium. No matter in which direction it would be pushed away from its equilibrium position by brownian movement, thermal convection or any other method of displacement, it would acquire in its new, non-isoelectric environment a charge of such a sign as to be returned to its point of origin.

The next step was to direct attention to the millions of other protein ions that acted in the same fashion and were filling the entire pH gradient column at equal concentration. This step introduced a new surprise. Unlike a conventional electrophoresis column, the pH gradient column did not exhibit unidirectional migration of the protein ions, but rather, the protein ions performed so-to-speak a "centripetal" motion, the center of their convergence being their isoelectric zone in the pH gradient, where they kept condensing into a sharp zone until all of them were swept into it from the ambient solution. [I referred to this condensation also as "focusing" in my first papers (6, 10) .]

This appeared to be a reversible phenomenon. Reversing the current created a "centrifugal" movement of the ions away from the isoelectric zone dispersing them in the solution. (This requires, of course, some disturbances due to diffusion or convection to remove the ions from the ideal zone of electroneutrality in the isoelectric "focus".) Starting with ampholyte ions at uniform concentration in the pH gradient column, the reversed current caused an evacuation of the ampholyte ions from their isoelectric zone instead of focusing. For this direction of current, the isoelectric zone proved to be a zone of unstable equilibrium. The slightest displacement of the neutral ampholyte particle from the isoelectric zone caused them to acquire a charge in the pH gradient away from their point of origin.

It was thus obvious that the effect of isoelectric condensation or focusing of ampholytes in a pH gradient column traversed by a current in proper direction offered a new approach to protein analysis and preparative separation. A mixture of N proteins would yield a sorting pattern of proteins condensed in N discrete focal zones corresponding to their isoelectric pH values in the pH gradient column. Measurement of the pH in the condensation foci would characterize each protein by its pI value. I called such sorting patterns "isoelectric spectra" and likened the process of sweeping the ions into an isoelectric band to the process of "*focusing*" (6, 10).

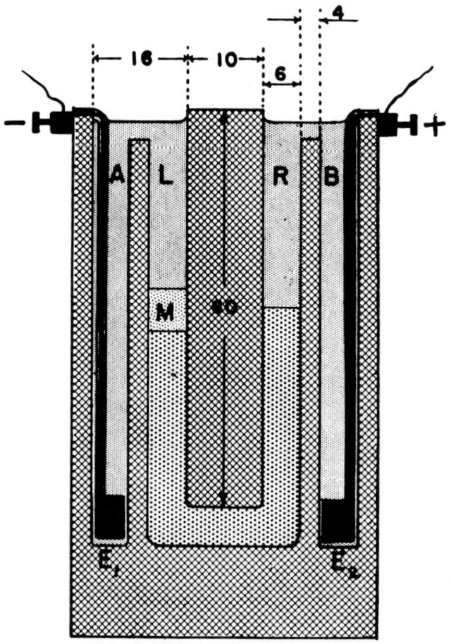

FIG. 4 *Electrofocusing cell. The dimensions are given in mm. (From A. Kolin. J. Chem. Phys. 22, 1628, 1954).*

THE INITIAL DEMONSTRATION OF ISOELECTRIC FOCUSING, EVACUATION AND STABILIZATION EFFECTS

The idea of isoelectric focusing made obvious the existence of the converse effect of isoelectric *defocusing* or *evacuation*. Since the ampholyte ions are positively charged on the low pH side of the isoelectric zone and negatively above it, reversing the focusing current i.e., directing it toward decreasing pH in the pH gradient should sweep the ions on either side of the isoelectric zone away from it, thus creating an ampholyte ion vacuum around it.

The demonstration of the condensation and evacuation effects was performed simultaneously in a simple U-tube apparatus in 1954 (6). Fig. 4 shows such a simple electrofocusing cell. The legs L and R of the central U-tube communicate with lateral vertical electrode tubes A and B. E_1 and E_2 are Pt electrodes. The bottom fluid in the U-tube is an acid buffer of pH 2.6 made very dense by a high concentration in sucrose. The upper parts of the U-tube (L and R) as well as the electrode tubes A and B are filled with a low-density basic buffer of pH 9.6. The specimen containing the ampholyte to be focused into an isoelectric zone is introduced in the left leg L as a short column ("M-layer"). The density of the M-layer is made intermediate between the dense bottom buffer and the upper low-density buffer by dissolving a suitable amount of sucrose or glycerol in it. A similar M-layer can also be injected into the leg R (for demonstration of "defocusing" or "evacuation"). Michaelis buffers of the following composition were used: (a) *basic buffer:* 4.85 g of sodium acetate plus 7.35 g of sodium barbital/liter of solution; (b) *acid buffer:* prepared by adding to 285 ml of water 100 ml of the basic buffer plus 75 ml of N/10 HCl.

A pH gradient is established in the M-layer by diffusion and turbulence engendered during the filling process. The buffer and ampholyte solution in the M-layer is very dilute and of much lower conductivity than the buffers below and above the M-layer. Thus, a large proportion of the voltage applied to the electrodes is placed across the M-layer which results in very rapid ion migration within the M-layer.

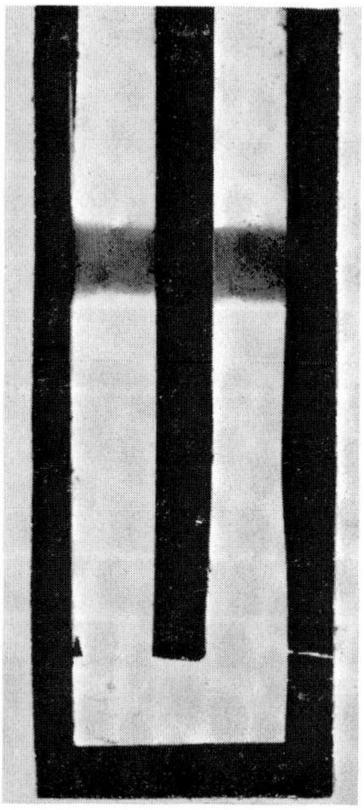

FIG.5

FIG.6

FIG. 5 "M-layers" in electrofocusing cell prior to passage of current. (From A. Kolin. Proc. Nat. Ac. Sc. 41, 101, 1955).

FIG. 6 Formation of a condensation zone (focusing) of hemoglobin in the left "M-layer" and of a dilution zone (dispersion) in the right "M-layer" after passage of a current of 9mA for 40 seconds in the direction of increasing pH in the left leg of the U-tube and in the direction of decreasing pH in the right leg. The polarity is as indicated in Fig. 4. The pH changes from 2.6 below the "M-layers" to 9.6 above them. (From A. Kolin. Proc. Nat. Ac. Sc. 41, 1955).

Fig. 5 shows how two M-layers were introduced in both U-tube legs. The current flow was in the direction of increasing pH in the left M-layer and of decreasing pH in the right M-layer. This resulted in rapid (within 40 sec.) condensation of hemoglobin into a sharp isoelectric zone on the left side and evacuation of hemoglobin toward the boundaries of the M-layer on the right side exhibiting the two converse effects in the same experiment, as shown in Fig. 6.

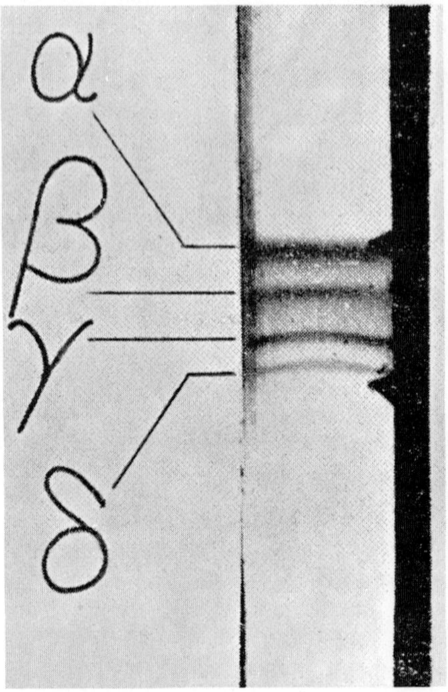

FIG. 7 *Protein separation pattern obtained in 4 minutes in a preformed pH gradient. α: Cytochrome C, β: hemoglobin, γ: catalase, δ: "Azocoll". pH range : 4.8 - 7.7. The zone α is not formed by isoelectric focusing, but rather by ion retardation in a conductivity gradient at the upper boundary of the M-layer. The zones β, γ and δ are formed by isoelectric focusing. (From A. Kolin. Proc. Nat. Ac. Sc. 41, 101, (1955).*

These focusing and evacuation effects are reversible. By reversing the direction of the current after formation of the focusing and evacuation patterns shown in Fig. 6, the original appearance of the M-layers, as seen in Fig. 5 can be restored. By continuation of the reversed current, one can obtain a similar condensation and evacuation pattern as seen in Fig. 6, but now the condensation will be in the right leg of the U-tube and the evacuation in the left leg.

The third of the group of effects manifested near an isoelectric zone of an ampholyte in a pH gradient in the presence of an electric current is the stability of the isoelectric condensation zone. Removal of ampholyte molecules from the zone by diffusion or convection imparts to them an electric charge which results in a force in the imposed electric field returning them to the condensation zone. Thus, a condensation zone rendered diffuse by stirring can be refocused by passing a current through the M-layer in appropriate direction.

Fig. 7 demonstrates the simultaneous focusing of four different proteins into four condensation zones from a solution containing four protein components in the M-layer: α: cytochrome C, β: hemoglobin, γ: catalase, and δ: "Azocoll". The pH difference across the M-layer was pH 4.8 - 7.7 (10). (Zone α is formed by condensation in a conductivity gradient rather than by isoelectric focusing.)

In a somewhat modified apparatus where the M-layer was supported by a cellophane membrane, Tuttle (11) succeeded, using multiple tubes in parallel, to separate simultaneously different types of hemoglobins by isoelectric focusing. Fig. 8 shows separation of hemoglobins C and A; S and A; C, S and A; as well as C and S.

LIMITATIONS OF THE PREFORMED pH GRADIENT APPROACH

The preceding separations in *preformed pH gradients* are primitive as compared to the high-resolution fractionations achievable in the Svensson-Vesterberg ampholine column based on so-called *natural pH gradients* established by prolonged electrolysis.

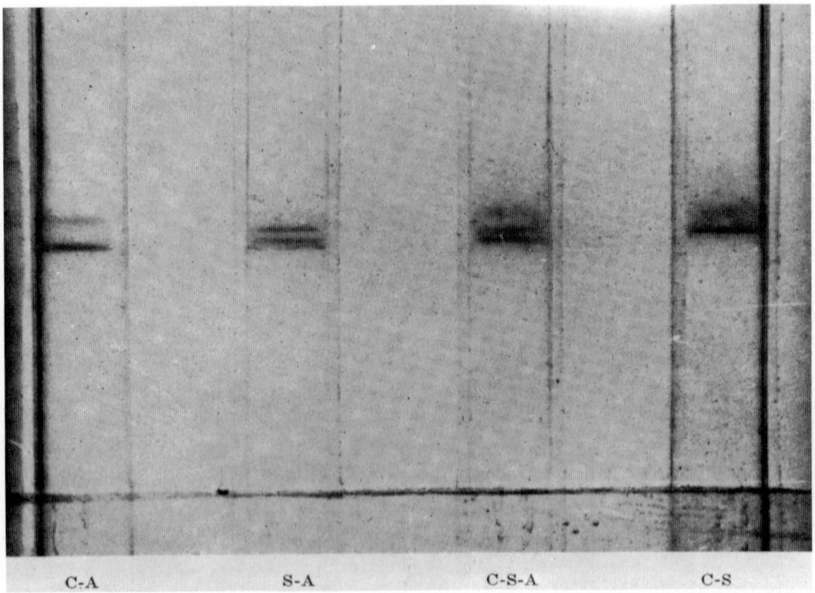

FIG. 8 *Tuttle's separations of hemoglobins A, C and S by isoelectric focusing in preformed pH gradients.(From A. H. Tuttle. J. Lab. Clin. Med. 47, 811, (1956).*

Nevertheless, they served a useful function by stimulating men of vision as evidenced by Tiselius' 1955 letter and Svensson's accomplishments which greatly enriched the present diversity of powerful analytical tools.

The virtue of preformed pH gradient columns was the speed of their preparation and the short time in which isoelectric condensation and separation could be achieved. However, the pH gradient was not stable, it was subject to drift due to passage of current. The zones were not very sharp and did not persist for long. The protein zones were not localized exactly in regions of their isoelectric pH and became diffuse as their positions drifted slowly.

In 1958 I visited Sweden and met Svensson (Rilbe) who became acquainted with my "isoelectric spectra" at Tiselius' special seminar in 1955. We had a long evening walk during which we touched upon experimental difficulties in isoelectric focusing. I was, frankly, pessimistic about the future of isoelectric focusing and was actually shifting my activities to blood flow studies. I mentioned in passing a vague idea which occurred to me of using mixtures of amino acids and polypeptides of a wide spectrum of isoelectric points instead of conventional buffer mixtures to obtain pH gradient columns stable in the presence of a current. My idea was superficial and certainly far from the insightful analysis of electrofocusing in ampholyte columns presented later by Svensson (Rilbe). With remarkable insight and effectiveness he combined the pre-existing ideas of isoelectric focusing in continuous pH gradients, the formation of stable pH gradients by prolonged stationary electrolysis pioneered by Tiselius, and Vesterberg's synthesis of "ampholines" to create the powerful analytical method which stimulated these international symposia.

EXTENSION OF THE IDEA OF ISOELECTRIC FOCUSING

The title of my first paper: " Separation and Concentration of Proteins in a pH Field Combined with an Electric Field" (6) reflects a way of thinking that led me to a generalization of the concept of isoelectric focusing. I considered the environment of the solute molecules (or suspended particles) as a "chemical field" in which a force- or velocity-determining parameter (in this case the pH which determines the particle charge and thus its mobility) was changing from point to point in space and varying the force exerted upon the particles of interest by the imposed electric field. The scalar quantity characterizing the chemical field in this example was the pH of the solution.

We shall recapitulate now the description of isoelectric focusing and dispersion in a pH-gradient column in more general terms which will facilitate the transition to phenomena of non-isoelectric focusing and dispersion described below.

We have a chemical field in a vertical column of fluid characterized by the environmental parameter $p'' = pH = \varphi(z)$ (where z is the axial coordinate of the column and $\varphi(z)$ a monotonic function of z). A dispersion of ampholyte particles in the fluid is characterized by the particle parameter $p' = pI$. (pI need not be constant. It could for instance vary with the temperature, which, in turn, could be a function of z.) We shall call the difference $p^* = p' - p'' = (pI - pH)$ the "force-determining parameter". This quantity will be a function of z and there will be a plane perpendicular to z in which $p^* = pI - pH = 0$. This is the isoelectric plane in which the neutral ampholyte particles experience no force in the presence of an electric field.

In a gradient of the environmental pH parameter p'', the sign of $p^* = (pI - pH)$ will depend on the direction of grad (pH). In the case where pH increases with z, $p^* < 0$ above the isoelectric plane $p^* = pI - pH = 0$. All ampholyte ions above this plane are negatively charged and migrate *downward* toward the $p^* = 0$ plane in an upward electric field. Conversely, the ampholyte ions below the $p^* = 0$ plane ($p^* > 0$) are positively charged. They too migrate toward the $p^* = 0$ plane but in the *upward* direction under these conditions which bring about isoelectric focusing.

Inversion of either the direction of the electric field or of the gradient of the environmental parameter (pH) leads to inversion of the electrical forces upon the ampholyte particles. These forces point toward the zone of $p^* = 0$ above and below it in the case of isoelectric focusing and away from it in the case of isoelectric dispersion.

We shall now apply the same terminology and mode of description to a familiar case of non-isoelectric focusing: *isopyknic condensation in a density gradient* effected by the gravitational field. The particle parameter p' is its density d', the environmental parameter p'' is the density of the fluid d''. The force-determining parameter is $p^* = d' - d''$. The chemical field is characterized by a concentration gradient which creates a density gradient grad (d''). The electrical field of the previous example is replaced by the gravitational field.

The density gradient is chosen so that there is an *isopyknic zone* $p^* = p' - p'' = 0$. In a density gradient where p'' increases in the downward direction, the suspended particles will experience resultant forces due to gravity and buoyancy which will point toward the plane of p * = 0 from above and below. These converging forces account for the focusing of the particles in the isopyknic zone. In principle, one could reverse the direction of these forces, thus causing isopyknic dispersion. This is, however, practically not feasible because of the instability of the inverse density gradient and our inability to reverse the gravitational field.

The case of *centrifugation in a preformed density gradient* is analogous to the above case of isopyknic focusing by gravitational sedimentation. The main distinction is the non-uniformity of the field in a centrifuge.

The *density gradient centrifugation method of Meselson and Stahl (12)* is a particularly interesting case because of its *analogy to Rilbe's pH gradient electrophoresis in a "natural" pH gradient (1-3)*. While Rilbe established an *equilibrium distribution of pH* under combined action of electrical forces and diffusion ("natural" pH gradient) by prolonged electrolysis of a micromolecular solution of "carrier ampholytes" in order to achieve isoelectric focusing of macromolecular ampholytes (proteins), Meselson and Stahl established an *equilibrium distributionof density* under combined action of inertial centrifugal forces and diffusion ("natural" density gradient) by prolonged centrifugation of a micromolecular solution (CsCl) in order to achieve isopyknic focusing of a macromolecular solute (nucleic acids).

ISOPERICHORIC FOCUSING — A GENERALIZATION OF ISOELECTRIC FOCUSING

The diagrams A and B of Fig. 9 describe in general terms phenomena in unspecified force fields and parameter gradients which follow the same pattern as isoelectric focusing and dispersion. A chemical environment is generated in which an environmental property described by parameter p" is changing from point to point along the z-axis [$p'' = \varphi(z)$]. A species of particles characterized by the parameter value p', corresponding to the parameter p" of the suspension fluid, is dispersed with an arbitrary

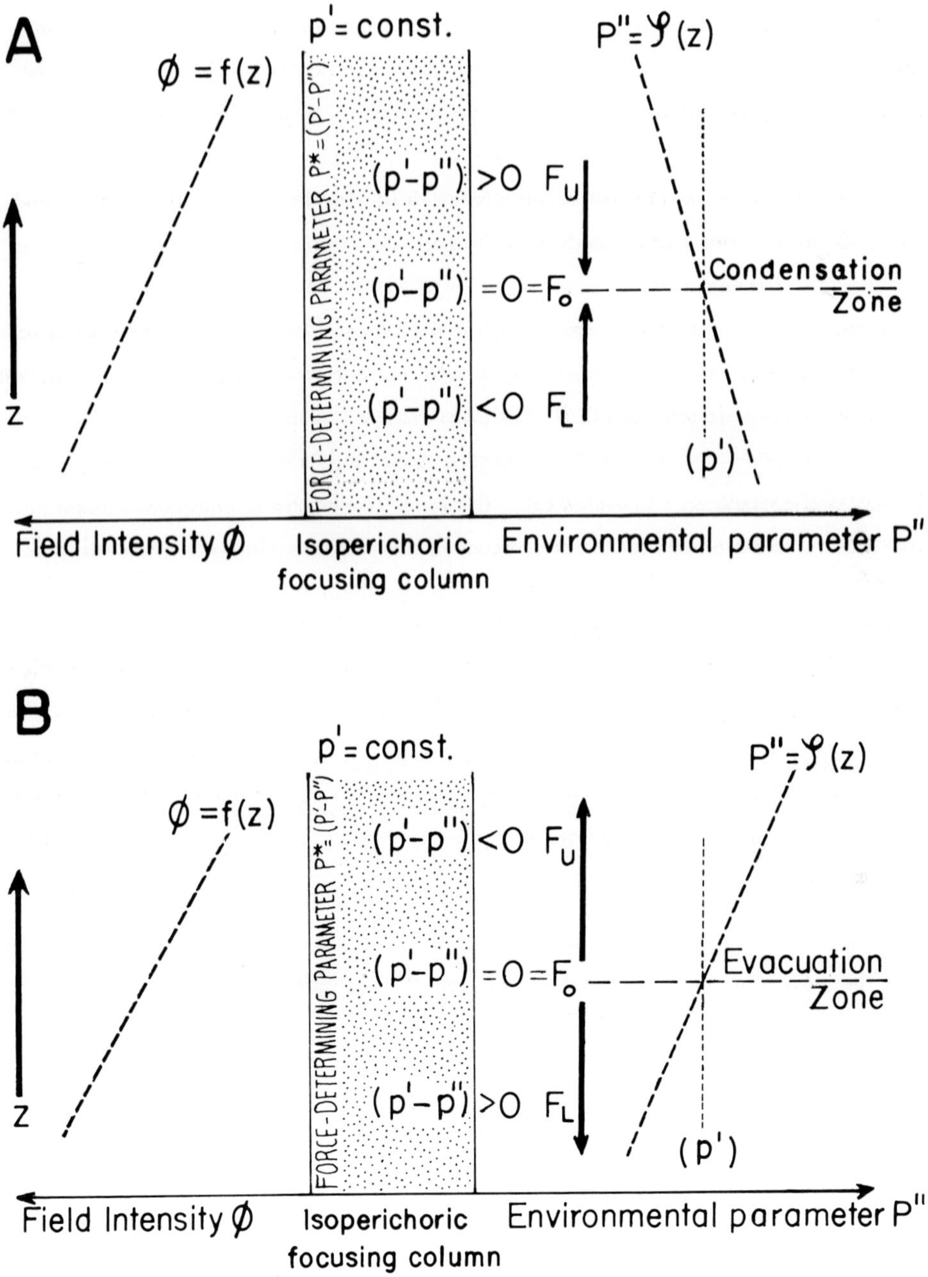

ISOPERICHORIC FOCUSING (A) AND DISPERSION (B).
FIG 9

distribution in the fluid column. We assume that there is a point within the gradient of the p" distribution at which p' = p", i.e., where the particle parameter equals the environmental parameter.

A further assumption is that we impose a force field which acts upon the particles and upon the fluid. The force F exerted on the particles is proportional to the difference (p' - p"). It reverses direction when (p' - p") < 0 changes to (p' - p") > 0. Depending on the slope of the gradient of p", the imposed forces will sweep the particles toward the zone where (p' - p") = 0 (and hence F = 0), as illustrated in Fig. 9A (*"isoperichoric focusing"*) or away from it, as shown in Fig. 9B (*"isoperichoric dispersion"*). The term *"isoperichoric"* is derived from Greek: isos = equal and perichoron = environment. The term "isoperichoric" designates a condition where a particle parameter becomes equal to the corresponding environmental parameter. Instead of reversing the slope of the p" gradient to generate dispersion in place of focusing, we can reverse the direction of the imposed force field ϕ and/or of its gradient.

The particle parameter p' may be constant or it may, like p", be a function of z. The force field may be uniform or be a function of z. The parameter p" must vary monotonically with z but not necessarily in linear fashion. We shall show in the following section how non-isoelectric focusing (and defocusing) effects could be obtained with different force fields offering analytical separation and characterization possibilities on the basis of diverse physicochemical properties of particles.

GENERALIZATION OF ARCHIMEDES' PRINCIPLE

Before we procede to examples of predicted isoperichoric focusing effects, it will be helpful to formulate Archimedes' principle, (which has been originally stated for interactions in a gravitational field and later extended to a centrifugal inertial force field) in general terms. We imagine a vertical cylindrical column of liquid in the field of gravity and consider a small volume element of fluid of arbitrary shape which rests in equilibrium in the ambient element of fluid despite the fact that each particle within the volume element experiences a downward gravitational force. This equilibrium is due to the fact that the force of gravity, acting on every particle within the cylindrical fluid column, generates a gradient of hydrostatic pressure which increases in downward direction. The pressure forces acting on the lower portion of the surface of our floating volume element are thus greater than those acting on its upper surface elements. The integral of the *surface forces* exerted upon the boundary of the volume element within the hydrostatic pressure gradient yields an upward resultant force (the force of buoyancy). This force is exactly equal to the gravitational *volume force* acting on the enclosed volume of fluid, which accounts for its perfect equilibrium.

We can disturb this equilibrium by changing the content of our volume element. This does not change the *surface force* of buoyancy, but the *volume force* experienced by the volume element, its weight, can be changed by filling the volume element with a material differing in specific gravity from the ambient fluid. If the force per unit volume *(the "force density")* experienced by the new content of the volume element is greater than the corresponding parameter of the ambient fluid, the volume force of gravity surpasses the surface force of buoyancy which acts on our volume element and the resultant force points in the direction of the gravitational field causing the volume element to sink. Conversely, in the case of replacement of our volume element by a specifically lighter substance, the unchanged surface force of buoyancy surpasses the gravitational force upon the volume element and the resultant force upon it is directed upward, as is the force of buoyancy.

We shall see below that there is a variety of force fields which can act upon fluids so as to generate a pressure gradient within them. A particle differing from the fluid in the value of a force-determining parameter, within such a force field and pressure gradient, will experience a force exerted upon its volume and a quasi-buoyant force upon its surface which forces will, in general, not be equal. The result will be particle migration in a direction which will be determined by the relative magnitude of the volume force and the opposing surface force.

We shall show below that establishing a gradient in a force-determining parameter in such a body of fluid can, in a suitable force field, give rise to effects of focusing of suspended particles according to the pattern of isoelectric focusing in pH gradients or isopyknic focusing in density gradients.

EXAMPLES OF PREDICTED NON-ISOELECTRIC FOCUSING EFFECTS

Isoconductivity Focusing

In the second section we became acquainted with electromagnetophoresis which is based on generation of a force upon a fluid traversed by a current in a transverse magnetic field. The force-density $\vec{F}^* = [\vec{J} \times \vec{B}]$ was compared to the weight-density in the analogous gravitational case. It was shown that objects differing in electrical conductivity σ' from the ambient solution conductivity (σ'') were "rising" or "sinking" depending on the sign of $\Delta\sigma = \sigma' - \sigma''$ and remain at rest at $\sigma' = \sigma''$ in analogy to the behavior of bodies of density d' submerged in a fluid of density d'' in a field of gravity.

At the end of the section, we mentioned the idea of maintaining a conductivity gradient in the fluid column, as shown in Fig. 9 when we replace p by σ. This led to focusing effects analogous to isopyknic focusing in the field of gravity.

Assume, for instance, the electromagnetic force $[\vec{J} \times \vec{B}]$ to be pointing downward in Fig. 9A. The electrical conductivity σ'' is increasing in the downward direction. The conductivity σ' of the suspended particles exceeds that of the ambient solution above the "iso-conductive" zone ($\sigma' = \sigma''$) where the particles rest at $F = 0$. Above this zone $\sigma' > \sigma''$. "Electromagnetic gravity" exceeds the "electromagnetic buoyancy" so that the particles move downward in the direction of $\vec{F}^* = [\vec{J} \times \vec{B}]$. Below this zone ($\sigma' < \sigma''$), the "electromagnetic buoyancy" is predominant and the particles are swept upward toward the isoconductive zone. (The force F* need not vary with z.)

Reversal of either the current $\vec{J}$ or the magnetic field $\vec{B}$ reverses both force vectors $\vec{F}_U$ and $\vec{F}_L$ converting the "isoconductive zone" from a focal plane of *particle condensation* to a zone of *particle evacuation*. Simultaneous reversal of $\vec{J}$ and $\vec{B}$ leaves the vectors $\vec{F}_U$ and $\vec{F}_L$ unchanged. These vectors can, however, be reversed by inversion of the conductivity gradient by making σ'' decrease in the downward direction as shown in Fig. 9B. [Stability of such conductivity gradients has been discussed elsewhere (13)].

While the above process of electromagnetophoresis in a conductivity gradient is analogous to gravitational sedimentation in a density gradient, it is also possible to conduct electromagnetophoresis so as to obtain, *without rotational motion*, a centrifugal or centripetal force-field analogous to the inertial field in a centrifuge (14). Electromagnetophoretic focusing effects in such a force field would be similar to isopyknic condensation patterns in density-gradient centrifugation.

Isomagnetic Focusing

When Faraday described his discovery of panmagnetism of matter he distinguished between two classes of non-ferromagnetic materials: paramagnetic substances with a magnetic permeability $\mu > 1$ and dimagnetic substances with $\mu < 1$. The properties of these materials are most simply visualized in an idealized configuration in a vacuum comprising a diamagnetic or a paramagnetic test object, say a sphere, in a radial magnetic field of a magnetic point source (say a magnetic north pole). The paramagnetic test sphere is attracted toward the magnetic north pole (i.e., toward regions of increasing field strength), whereas the diamagnetic sphere is repelled from it (toward decreasing field strength) . Such forces are not limited to a radial force field. The paramagnetic sphere will be pulled in the direction of increasing field intensity and the diamagnetic sphere in the opposite direction in arbitrary non-uniform magnetic fields. This difference in behavior is due to opposite magnetization induced in the test objects in the two cases. The paramagnetic sphere is magnetized so that its induced south pole is proximal to the north-magnetic point source and its induced north pole is distal to it. Thus, the mutual attraction between the less distant points of opposite magnetic polarity surpasses the mutual repulsion between the magnetic north pole and the north magnetic distal side of the test sphere and the resultant force is one of attraction between the magnetic point source and the magnetic dipole represented by the magnetized paramagnetic test sphere.

The magnetization of the diamagnetic test sphere is such that the equivalent dipole is of opposite polarity as compared to the paramagnetic sphere. This results in mutual repulsion between the magnetic point source and the magnetic dipole (i.e., the diamagnetic test sphere).

Faraday made on this occasion a surprising discovery. He found, that under certain conditions the *paramagnetic* test sphere could be *repelled* from the magnetic point source and the *diamagnetic* test sphere could be *attracted* toward it, acting in converse fashion to their behavior in a vacuum.

This modification of the magnetic interaction can be achieved by submerging the test sphere in a fluid of appropriate magnetic permeability. The magnetic force upon the test object will vanish in a fluid of equal magnetic permeability. The force upon a paramagnetic sphere is one of attraction toward maximum field intensity as long as its magnetic permeability is larger than that of the ambient fluid. It becomes a force of repulsion when the magnetic permeability of the fluid exceeded that of the test sphere. — An analogous transformation of the repulsive force upon a diamagnetic sphere surrounded by a less diamagnetic fluid into a force of attraction is achieved by replacement of the ambient fluid by a diamagnetic fluid of higher absolute value of permeability than that of the test object. These phenomena bear a close resemblance to manifestations of Archimedes' principle in a field of gravity.

To understand the mechanism of this modification of forces, we shall limit ourselves to the idealized example of a paramagnetic sphere immersed in a paramagnetic fluid in the radial field of our isolated magnetic north pole. The magnetized paramagnetic test sphere will not be the only object attracted toward the region of maximal magnetic field strength. The same will be true of everyone of the contiguous volume elements of which we can imagine the ambient paramagnetic fluid to consist. Like in the case of a fluid in the field of gravity, such forces upon the fluid elements will give rise to a pressure gradient and the pressure will increase in the direction of the center of attraction. Thus, an immersed hollow object (e.g., a bubble) will experience pressure forces upon its surface which will be maximal on the side facing the region of maximal field strength. The surface integral of the pressure forces will yield a resultant quasi buoyant surface force directed *away* from the region of maximal magnetic field strength. In close analogy to the considered gravitational case, this surface force will be superimposed upon the volume force exerted upon any material that we could introduce into our hollow test volume. For objects of higher magnetic permeability than the ambient fluid the magnetic volume force exerted in the non-uniform magnetic field will surpass the quasi buoyant surface force and our paramagnetic test object will move in the same direction as it would in a vacuum. However, in a fluid of higher magnetic permeability than that of the submerged test object, the quasi buoyant surface force will be predominant and the object would move in the opposite direction. In a similar fashion the direction of motion of a diamagnetic test object could be inverted by submersion in a fluid of suitable magnetic permeability.

<u>Isoparamagnetic Focusing</u> Our final step will now be to imagine a gradient of the magnetic permeability in the paramagnetic fluid and to assume that the magnetic permeability μ' of small particles suspended in the permeability gradient is the same as that of the fluid at a distance R^* from the magnetic point source. In a gradient of the magnetic permeability μ'' in which its value increases with decreasing R (the distance from the magnetic point source), "paramagnetic buoyancy" exceeds the magnetic volume force for distances $R < R^*$ so that the resultant force will sweep the particles centrifugally toward the spherical zone $R = R^*$ where the resultant force becomes zero and the particles accumulate in an *isoparamagnetic zone*. On the other hand, at distances $R > R^*$, the magnetic volume force upon the particles exceeds the surface force of "magnetic buoyancy" and the particles are swept centripetally toward the

isoparamagnetic focusing zone of $R = R^*$.

In a converse gradient with μ'' increasing with increasing radial distance R, the isoparamagnetic zone of $R = R^*$ is not a focal zone of condensation, but, rather, a zone of dispersion from which suspended particles would recede. This case is of lesser interest because the permeability distribution will be unstable, whereas the converse permeability gradient will be a stable one.

Isodiamagnetic Focusing The same considerations as were detailed above for a paramagnetic fluid can be applied to a diamagnetic particle suspension in a diamagnetic permeability gradient exposed to a non-homogeneous magnetic field. We must only remember that the force upon a diamagnetic object in a radial magnetic field will be centrifugal and the "diamagnetic buoyancy" centripetal. To obtain *isodiamagnetic focusing* of particles in a zone $R = R^*$ where the diamagnetic particle permeability μ' is equal to the permeability of the ambient fluid, one needs a magnetic permeability gradient where the value of μ'' decreases with R.

The Magnetic Force upon a Submerged Body in a Non-homogenous Magnetic Field

Whereas gravitational and buoyancy forces as well as isopyknic focusing are possible in an homogeneous gravitational field, isomagnetic focusing presupposes a *non-homogeneous* magnetic field which is a prerequisite for exertion of a magnetic force to produce translation of a magnetically polarized object.

The expression for the magnetic force F_m upon a particle of volume V and magnetic polarizability α^* in a non-homogeneous magnetic field H is (15):

$$(4) \quad \vec{F}_m = \frac{V\alpha^*}{2} \text{ grad } (H)^2 .$$

For a small spherical particle of radius a and magnetic permeability μ' suspended in a fluid of permeability μ'' the force is:

(5) $\vec{F}_m = 2\pi a^3 \mu'' \operatorname{grad}(H)^2 \left(\frac{\mu' - \mu''}{\mu' + 2\mu''}\right)$.

The factor in the parenthesis is analogous to the expression for the electrical conductivities in equation (2) for electromagnetophoresis and to the expression for dielectric constants in below equation (7) for the electric force upon a dielectric sphere in a heterogeneous electric field.

The magnetic force vanishes for $\mu' = \mu''$ and its direction depends on whether $\mu' > \mu''$ or $\mu'' > \mu'$. The direction of the force is independent of the direction of the magnetic field. For paramagnetic materials $\mu > 1$ and for diamagnetic materials $\mu < 1$.

Isodielectric Focusing

Dielectric materials submerged in dielectric fluids of different dielectric constants experience in a heterogeneous electric field forces analogous to those exerted in a non-homogeneous magnetic field upon paramagnetic objects in a paramagnetic fluid of different magnetic permeability. The forces are expressed by the same equations except for replacement of the magnetic permeability μ by the dielectric constant K. Designating by K' the dielectric constant of the suspended particle and by K'' that of the suspension medium we can write for the force upon a particle of volume V and dielectric polarizability α in a (non-homogeneous) electric field E:

(6) $\vec{F} = \frac{V\alpha}{2} \operatorname{grad}(E)^2$.

For a small spherical particle of radius a the force is

(7) $F = 2\pi a^3 K'' \operatorname{grad}(E)^2 \left(\frac{K' - K''}{K' + 2K''}\right)$

in close analogy to the force upon a paramagnetic sphere (equation 5) .

We see that $F = 0$ for $K' = K''$ and that its direction is opposite for the two cases $K' > K''$ and $K' < K''$. *The force is independent of the direction of the electric field E* and , hence, unidirectional particle migration and focusing effects are obtainable in alternating non-homogeneous electric fields.

Dielectrophoretic focusing phenomena are predicted on the basis of considerations similar to those which lead to prediction of paramagnetic focusing. A surface force analogous to buoyancy would be generated by the hydrostatic pressure gradient set up in the fluid by the non-homogeneous electric field. The superposition of this force upon the volume force experienced by the electric dipole resulting from dielectric polarization of the suspended dielectric particle accounts for the changes in the direction of the resultant force upon the particle depending on the sign of $\Delta K = K' - K''$ and for its disappearance at $K' = K''$.

We would thus expect to obtain in a non-uniform electric field combined with a gradient of the dielectric constant K'' (the latter increasing in the gradient with the field intensity $\vec{E}$) phenomena of *isodielectric focusing* i.e., sweeping of suspended particles of dielectric constant K' into a stable zone where $K' = K''$. This would yield an *isodielectric spectrum*, i.e., a sorting pattern of particle species differing in the value of K' accumulated in sharp isodielectric zones. Measurement of the value of K in a given zone will characterize the particle species by its value of dielectric constant.

The above considerations were greatly simplified by their limitations to ideal dielectrics. Actually, suspended particles and the suspension medium seldom conform closely enough to this idealization. This is particularly true of biologically interesting particles, such as cells and cell organelles. The medium will, in general, have a finite electrical conductivity and the particles will be characterized by an overall electrical conductivity as well as a dielectric constant and the two may be frequency-dependent, which is of special importance when alternating electric fields are used and a further characteristic parameter, the "dielectric relaxation time" is introduced (16, 17). The equation describing the forces exerted upon biological particles in alternating non-uniform fields of varying frequency are quite complex (17), but the complexity of these phenomena offers additional possibilities of resolution in potential applications of isodielectric focusing to preparative and analytical problems of cellular and molecular biology.

CONCLUSION

Although these reminescences are primarily concerned with paths of discovery of certain physicochemical effects, we also got a glimpse of a sociological phenomenon which could be designated as "defocusing of the origins of ideas". It consists of blurring of historical perspectives and usually takes the form of shifting the responsibility for discoveries and ideas to (preferably multiple) sources in the remote past. This account presents references to correspondence and publications that may bring the historical background of isoelectric focusing into somewhat sharper focus.

We have seen that isoelectric focusing is a physical effect and that it, in fact, belongs in a group of three related phenomena: isoelectric focusing, - dispersion and - zone stability. The idea and initial demonstration of isoelectric focusing and of its analytical potential did not result from a directed effort of analytical chemists to create a new form of protein analysis. This idea was a byproduct of research devoted to an unrelated phenomenon. However, the practical creation of the analytical method of isoelectric protein analysis in "natural" pH gradients is primarily due to a brilliant synthesis of pre-existing ideas by one man (Rilbe). The ultimate effective implementation and widespread use of this method rest on the contributions of Rilbe and Vesterberg.

In physical measurements, it is well known that great accuracy and sensitivity can be achieved by using "null methods". The potentiometer and bridge circuits are the best known examples of this family of methods. In a certain sense, isoelectric focusing can be considered as a null method in which particles reach equilibrium in a focal zone of vanishing force.

Null methods analogous to isoelectric focusing may be devised based on other physical effects. It was shown that isoelectric focusing may be considered as a member of a family of phenomena — *isoperichoric focusing effects*—in which particles migrating in a force- field along a chemical concentration gradient are swept into sharp, stable stationary zones of vanishing force in which a physical parameter of the suspension medium matches that of the particles. Examples of such condensation phenomena were isopyknic - , isoconductivity - , isomagnetic - and isodielectric - focusing. The generalization of isoelectric focusing leading to the wider class of isoperichoric focusing phenomena may add new resources to biochemical and microbiological analysis.

BIBLIOGRAPHY

1. Svensson, H.: Isoelectric Fractionation, Analysis and Characterization of Amlytes in Natural pH Gradients I. Acta Chem. Skand. 15, 325 - 341 (1961).

1a. Svensson, H.: Fokuserande Electrofores. Svensk Kem. Tidskr. 79, 654 - 659 (1967).

2. Rilbe, H.: Historical and Theoretical Aspects of Isoelectric Focusing. Ann. N. Y. Ac. Sc. 209, 11 - 22 (1973).

3. Rilbe, H.: Isoelectric Focusing - Development from Notion to Practically Working Tool. Science Tools 23, 18 - 22 (1976).

4. Williams, R. R. and Waterman, R. E.: Electrodialysis as a Means of Characterizing Ampholytes. Proc. Soc. Exp. Biol. & Med. 27, 56 - 65 (1929).

5. Tiselius, A.: Stationary Electrolysis of Ampholyte Solutions. Svensk Kem. Tidskr. 53, 305 - (1941).

6. Kolin, A.: Separation and Concentration of Proteins in a pH Field Combined with an Electric Field. J. Chem. Phys. 22, 1628 - 1629 (1954).

7. Kolin, A.: An Electromagnetokinetic Phenomenon Involving Migration of Neutral Particles. Science 117, 134 - 137 (1953).

8. Kolin, A.: Electromagnetophoresis in Medical Physics vol. 3, pp 268 - 274 (Yearbook Publ. Co., Chicago 1960) O. Glasser, Ed.

9. Leenov, D. and Kolin, A.: Theory of Electromagnetophoresis. J. Chem. Phys. 22, 683 - 688 (1954).

10. Kolin, A.: Isoelectric Spectra and Mobility Spectra: A New Appraach to Electrophoretic Separation. Proc. Nat. Ac. Sc. (U.S.) 41, 101 - 110 (1955).

11. Tuttle, A. H.: The Separation and Identification of Human Hemoglobin by "Isoelectric Line Spectra" Formation. J. Lab. & Clin. Med. 47, 811 - 816 (1956).

12. Meselson, M. and Stahl, F. W.: The Replication of DNA in Escherichia coli. Proc. Nat. Ac. Sc. (U.S.) 44, 671 - 682 (1958).

13. Kolin, A.: Suppression of Thermal Convection and of Similar Types of Fluid Streaming by an Electromagnetic Force Field. Biochim. Biophys. Acta 32, 538 - 543 (1959).

14. Kolin, A.: Centripetal and Centrifugal Electromagnetophoresis. J. Appl. Phys. 25, 1065 - 1066 (1954).

15. Bhatnagar, S. S. and Mathur, K. N. Physical Principles and Applications of Magnetochemistry. (Macmillan & Co. Ltd., London 1935).

16 Pohl, H. A.: in Electrostatics and Its Applications (John Wiley & Sons, N. Y. 1973) A. D. Moore, Ed.

17. Pohl, H. A. and Crane, J. S. in Electrostatics and Its Applications (John Wiley & Sons, N. Y. 1973) A. D. Moore, Ed.

Stable pH Gradients – A Key Problem in Isoelectric Focusing

H. Rilbe

SUMMARY

The possible causes of pH gradient instability in density gradient columns and in gels are reviewed. Arguments are presented favouring the view that all observed forms of pH gradient instability can be explained by electroösmotic water flow and hydration water transport together with secondary phenomena, but a quantitative theory accompanied by affirmative experiments is still lacking.

INTRODUCTION

First of all I wish to remind of the elegant separation of proteins in a few minutes in combined pH and density gradients reported by Kolin in 1954-55. In his first paper (1) on the subject, he wrote: "The present process is distinct from the conventional electrophoretic separation in that the latter sorts ions according to differences in their mobilities, whereas the former separates them according to their isoelectric pH values. It remains to be seen whether or not a general method can be developed for obtaining an "isoelectric spectrum" of a protein mixture whose components could be identified by their isoelectric pH rather than electrophoretic mobility." Kolin was thus the initiator of isoelectric focusing and the creator of the term isoelectric spectrum, and he could have added that the isoelectric point of a protein is a much more valuable characteristic than its mobility at a certain pH. Kolin made his pH gradients by allowing diffusion between two layers with different pH.

He described his method in more detail in subsequent papers (2, 3). He noted difficulties with the instability of the pH gradient and concluded that its stabilisation against modification by the current was the main problem that still remained to be worked out.

In the year 1961, natural versus artificial pH gradients were discussed (4), the former being defined as gradients created by the current itself by its electrolytic action. Preference was given to natural pH gradients due to their expected unlimited stability in spite of the great draw-back of non-constant conductivity. From this paper, it may be cited: "Hitherto available experimental evidence supports the view that it is possible to reach a steady state by prolonged electrolysis under the conditions defined earlier (freedom from convection; absence of redox reactions at the electrodes except for the water ions). A steady state includes the demand of a stationary solvent, that is, absence of electroosmosis by the use of a density gradient as stabilising agent, or an electrically inert capillary system, or possibly a hydrodynamic back flow of solvent balancing the electroosmotic flow. All these alternatives are experimentally possible. If somebody is worried about the continuous consumption of water in the form of hydrogen and oxygen at the electrodes, he can imagine a hydrogen electrode as the anode and a device conducting the hydrogen formed at the cathode to this anode. In such an idealised stationary state, which is most suited for a theoretical attack, the only thing that happens is a continuous circulatory flow of hydrogen. All other chemical components are stationary, they do not change their concentrations anywhere in the system." According to this citation, I looked upon isoelectric focusing as a true steady state process. I also judged electro-osmosis to be impossible in a density gradient column. This is probably in error, and this question will be taken up later.

When I made my first experiments with isoelectric focusing in a density gradient column of bigger cross-section than generally used to-day, I noticed a slow drift towards the cathode. However, since this drift only amounted to a small fraction of the column length even in an experiment with 2 weeks' duration, I regarded it as insignificant and did not mention it in the report (5) of these experiments. I devoted no time to speculation about the cause of this drift. On the contrary, I was satisfied with the close experimental realisation of the ideal steady state described in the above citation.

The situation became drastically changed when isoelectric focusing began to be carried out in gels. It was then found that a real steady state was never reached, that a pH plateau developed in the centre of the gel, that it was necessary to adopt an optimum focusing time, etc. The notion of "pH gradient decay" became an often discussed theme. The instability of pH gradients in gels has been studied especially by Finlayson and Chrambach (6), Miles et al. (7), and Baumann and Chrambach (8). As density gradient columns of smaller cross-section were introduced, the cathodic drift of the pH gradient became more accentuated. The factors influencing this drift have been studied by Fawcett (9).

The instability of pH gradients, especially in gels, must be regarded as the most severe problem in isoelectric focusing at the present time. Although it is possible to perform very fine protein analyses by adhering to an optimum time of focusing, it is a nuisance to be afflicted with this phenomenon not covered by the elementary theory and very difficult to explain and master. Several attempts at explanation have been made, but none has been generally accepted as true. I too am unable to offer a complete explanation. I can only give some new points of view and express some opinions. Results at different laboratories are partly conflicting, and it does not seem possible to account for all observed phenomena on the basis of one single physico-chemical mechanism.

LIQUID FLOW DUE TO EVOLUTION OF HYDROGEN AND OXYGEN AT THE ELECTRODES

Liquid flow due to this cause can occur in density gradient columns only. One mole of water disappears per faraday electricity passed, thus the volume loss is $18 \cdot 10^{-5}$ cm^3/As. At a current of $\underline{i}$ A, this corresponds to a volume loss of 0.65 $\underline{i}$ cm^3/h. One fraction thereof occurs at the anode, the remaining part at the cathode, and only the difference between these two volumes will give rise to a flow through the column. By neglect of this, an upper limit to the flow due to this cause is of course obtained. On the other hand, the escaping gases also drag along saturated water vapour, which will also be neglected here. Division by the cross-sectional area q gives an upper limit of 6.5 i/q mm/h for the linear rate of flow. Since the current density is less than 1 mA/cm^2, it can be concluded that the flow due to electrolytic consumption of water is far too small to be of any interest.

CONDENSATION OF WATER AT ITS ISOELECTRIC POINT

Ornstein (10, Postscript, p. 59) pointed out that, since water is an ampholyte isoelectric at pH 7, it will tend to collect there, pushing away the bulk of water together with solutes toward both ends. This effect may be operative in density gradient columns, but in gels, with their great hydrodynamic resistance, it can only be expected to cause a certain swelling of the gel in the neutral region. An attempt at quantifying this effect will be made as follows through a consideration of the mass transport of water ions.

The concentration of free hydrogen ions is about 10^{-pH} moles per dm^3, and with a mobility $\underline{u}$, its mass transport at a field strength $\underline{E}$ becomes $E \cdot u \cdot 10^{-pH}$. Since pH is bound to dm^3 as the volume unit, both $\underline{E}$ and $\underline{u}$ have to be given with dm as the length unit. The unit of mass transport then becomes moles per second and dm^2 of the cross-section of the column. If we also multiply by the molar volume $\underline{v}$ (in dm^3/mole) of hydrated hydrogen ions, we get the following expression for the volume effect of the migration of hydrogen ions: $E \cdot u \cdot v \cdot 10^{-pH}$. It has the dimension of linear velocity (dm/s) and represents the magnitude of the anodic back flow of the bulk of solvent.

As is well known, hydrogen ions are about 10 times as mobile as other small ions. Its excessively high mobility, however, is due to proton jumps from hydronium ions to water molecules, and such jumps have virtually no volume effect at all. It is therefore justified to insert the mobility of an ordinary small ion, say Na^+, in the above expression. On the other hand, we should make allowance for a hydration shell of about 5 molecules of water, giving $\underline{v}$ the value 0.09 dm^3/mole. With a field strength of 500 V/dm, we arrive at the following linear velocity: $0.09 \cdot 10^{-3} \cdot 10^{-pH}$ dm/s = $32 \cdot 10^{-pH}$ mm/h. On the alkaline side, pH has to be replaced by 14-pH, otherwise there will not be much difference. The conclusion is that this effect can cause a drift of the order of magnitude of 3 mm/h at pH 1 or 13, and that this drift diminishes with one power of 10 for every pH unit closer to neutrality. The isoelectric condensation of water to the neutral region can thus be ruled out as a contributing cause of a drifting pH gradient. The same conclusion was also drawn by Fawcett (9) on purely experimental grounds, and any swelling of the neutral part of the gel has never been reported.

ELECTROCHEMICAL MODIFICATION OF CARRIER AMPHOLYTES

Special electrode solutions containing acid and base are in general used in order to protect the carrier ampholytes from anodic oxidation and cathodic reduction. If the protection is not entirely successful, it might be suspected that the ampholytes undergo modification making them more acidic than before. The resulting successive acidification would of course give rise to a cathodic drift of the pH gradient, whereas the plateau formation in gels would be more difficult to explain in this way. Fawcett (9) has described special experiments for testing this hypothesis and reported negative results, but Denckla (11) is convinced of the opposite. Continued study must be devoted to this problem.

WATER TRANSPORT DUE TO UNEQUAL HYDRATION OF THE CATIONIC AND ANIONIC FORMS OF CARRIER AMPHOLYTES

Good carrier ampholytes in the isoelectric state have appreciable fractions of their molecules in cationic and anionic forms. These ions move in the electric field, and they are the sole conductors of the current in the neutral region. At the isoelectric point the mass transports of cations and anions balance each other, whereas there is a net cathodic migration on the acid side and a net anodic migration on the alkaline side. Each individual ampholyte particle incessantly exchanges protons with the solvent and thus performs an irregular back and forth movement across the isoelectric point, taking up protons on the acid side and leaving them on the alkaline side.

It is generally believed that deprotonated carboxyl groups are heavily hydrated, whereas protonated amino groups don't carry much water along. Thus, if the isoelectric point of an ampholyte lies between two pK's belonging to carboxyl groups, which occurs most frequently in acid solution, there will be an appreciable difference in hydration between the two ions and a comparatively great transport of hydration water toward the anode. If the isoelectric point lies between the pK's of two amino groups, which occurs most frequently in alkaline solution, there will be a small difference in hydration and a comparatively small transport of hydration water toward the cathode. Ampholytes being isoelectric between the pK's of one carboxyl and one amino group, probably predominating in the neutral range, will be expected to have intermediate water transport properties.

As to the consequences of hydration water transport, one must again distinguish between density gradient columns and gels. In the former case, there cannot be any net water flow for hydrostatic reasons, and consequently there must be a bulk flow of water together with solutes in the direction opposite to that of the hydration water flow. Since the latter is predominantly anodic in the acidic and neutral regions, the bulk flow will be cathodic, in agreement with observations. However, the bulk flow must be expected to decrease with increasing pH, whereas Fawcett (9) has given conclusive evidence of a pH drift which increases with pH. This discrepancy may be explained as a secondary phenomenon. The bulk flow transports the more acidic ampholytes in the direction of the current all the time. As a result, pH goes down in the region of less acidic ampholytes, which are then forced to migrate away from the advancing acidity, that is, toward the cathode. Thus there is a qualitative agreement between experimental evidence in density gradient columns and the theory of hydration water transport. However, there seems to be a quantitative discrepancy in the experimental fact that the cathodic pH drift decreases with an increasing ampholyte concentration, whereas the theory of hydration water transport requires a water transport proportional to the current and thus to the ampholyte concentration at a constant voltage. Thus hydration water transport cannot be the only cause of the pH drift in density gradient columns.

In the case of gels there cannot be any back flow of bulk water since the hydration water flow does not build up any hydrostatic pressure and since the flow resistance of gels is too great. In gels the hydration water flow shall thus be expected to go toward the anode in acidic and neutral solution, whereas a certain cathodic water flow should be expected in alkaline solution. The gel should thus be expected to shrink in the central part and to swell at the ends, especially the acidic end. This is in excellent agreement with what Davies (12) wrote about LKB's gel slab apparatus: "...the gel shrinks and becomes very thin in the middle. At the same time, the gel swells considerably at the electrode strips. The swelling is greater at the anodic than the cathodic isoelectric focusing strip." Felgenhauer et al. (13) wrote: "The plateau gel shrinks, the acid end is pulled into the glass tube and the gel may even protrude into the alkaline solution." Fawcett (9) wrote: "However, when gels containing higher concentrations of Ampholine were used, a significant liquid flow into the anode occurred. When the Ampholine concentration was 4% the volume of liquid transferred into the anode was equivalent to 8% of the gel volume." Fawcett also reported occasional cathodic water flow when the normal ampholyte concentration of 1% was used. Baumann and Chrambach (8) have studied the water transport in gels with the aid of tritiated water, but since they placed this radioactive water in one of the electrode vessels and not in the gel, their results are difficult to interpret because radioactivity of course enters the gel owing to the electric migration of H^+ and OH^- ions, whereas hydration water transport cannot be indicated. However, they reported quite small water movements, only 0.15 ml cathodically in a 1.6 ml gel during 100-150 hours.

In the absence of bulk water flow no pH changes are to be expected. Hydration water transport thus cannot explain the development of a pH plateau, nor the cathodic drift. Nevertheless, it probably plays some role in electrofocusing in gels.

ELECTROOSMOSIS

Although it is generally agreed that polyacrylamide gel is virtually free from electroosmotic effects when used in disc electrophoresis, one cannot be quite safe of this in isoelectric focusing due to the extremely low ionic strength developing in this technique. Most authors in the field (13-19) believe that electroosmosis is at least an important contributing factor to the plateau phenomenon and the cathodic drift, whereas others (7, 9, 10), although admitting that electroosmosis exists in gels, mean that it is too small to account for the observations and thus prefer other mechanisms to explain observed phenomena. The effects to be expected from electroosmosis in gels and in density gradient columns will now be qualitatively discussed and compared with observations.

Electroosmosis in Gels

All writers in the field agree that electroosmosis is due to fixed charges on the gel matrix and that the main source of these charges is a certain low concentration of carboxyl groups. If so, these charges must be expected to disappear at pH values lower than about 3.5, whereas they acquire their full value at pH'es higher than about 5.5. Thus, if carboxyl groups are the only source of electroosmosis, a cathodic flow of water together with solutes should be expected in the neutral and alkaline parts of the gel, whereas no water flow should occur in the acidic part. As a consequence, gel shrinkage must be expected round the pK of the carboxyl groups and a corresponding flooding at the alkaline edge of the gel.

Gel shrinkage in the region of the plateau has been reported by several authors (6, 12, 13, 15, 20, 21). In gel tubes this shrinkage manifests itself either in that the acid end is pulled into the glass tube (13) or in a deficient wall adherence (6), which can be improved by coating the wall with high molecular weight linear polyacrylamide. A deficient wall adherence is fatal since a film of free liquid tends to develop between wall and gel. Gel slabs can of course shrink and swell without such side effects.

Gel shrinkage, to be expected in the slightly acidic region of the gel, must give rise to two important secondary phenomena. First, the decreasing cross-sectional area in a gel slab causes a correspondingly increasing field strength since the electroösmotic flow does not change the composition and conductivity of the flowing liquid. The increased field strength of course enhances the linear velocity of the electroosmotic flow. Second, the decreasing cross-sectional area also has the consequence that a fixed amount of a focused ampholyte will require more space in the direction of the current since its concentration remains unchanged.

These two secondary phenomena greatly amplify the effects of electroösmosis and thus explain that very extensive changes in the isoelectric pattern may be accompanied by very small displacements of water. They appear to be quite sufficient to explain the plateau phenomenon and the cathodic drift of the pH gradient, at least for gel slabs. The theory is also in agreement with the experimental evidence that the plateau phenomenon and the cathodic drift increase with voltage and time and that they decrease with increasing viscosity and increasing concentration of carrier ampholytes (7).

The theory of electroosmosis cannot explain all results that have been reported for gel tubes. No attempt will be made here to explain the reported fact (7, 10) that focusing in a gel tube with a pH 3-10 gradient gave a bidirectional pH drift, whereas a unidirectional cathodic drift was obtained in a pH 3-6 tube, other conditions being the same. The theory of electroosmosis, as presented so far, is also unable to explain the gel swelling at the acidic end reported by Davies (12). An anodic electroosmotic flow in the acidic region of the gel must depend on fixed positive charges in a concentration increasing towards the anode. Amino groups of Ampholine cannot be responsible for such an electroosmotic effect since Miles et al. (7) as well as Bates and Deyoe (14) state that Ampholine, even if present during gel polymerisation, has nothing to do with the plateau phenomenon.

Whereas Bates and Deyoe (14) believe that the carboxyl oxygen, being the most electronegative element of the amide group, imparts a small negative charge to the fixed gel support matrix, Edward and Wang (22) have measured the basicity of the nitrogen atom in propionamide and found a pK value of -0.9. This means that one out of 10,000 amide nitrogen atoms should have a positive charge at pH 3.1. In view of the abundance of amide groups in the gel matrix, the concentration of positively charged nitrogen atoms could possibly be great enough to give a noticeable electroosmotic flow towards the anode.

Electroosmosis in Density Gradient Columns

Only very little of what has been said about electroosmosis in gels has any application to electroosmosis in density gradient columns. For hydrostatic reasons, any existing electroösmotic flow close to the walls is exactly balanced by a counterflow at a distance from the walls. In the presence of a density gradient, it seems inconceivable to imagine a continuous, unbroken liquid flow from the highest to the lowest density or vice versa. The author therefore believed (4) that electroosmotic flow could not occur in density gradient columns. Recent studies by Jonsson (unpublished) have shown this assumption to be incorrect. There is electroosmotic flow even in density gradient columns, but this flow and the corresponding counterflow are restricted to very small density intervals. Thus, everywhere at the walls there are small circulatory liquid movements, probably not penetrating very far from the walls. According to Jonsson's observations, these whirlpools with a vividness increasing with the voltage have a tendency to blur and deteriorate the zones and are, rather than heat convection, the limiting factor for the voltage that can be applied. Electroosmosis is thus a severe limitation even in density gradient electrofocusing, and the rapidity and resolving power of the latter could probably be much improved in columns completely devoid of wall charges.

The cross-sectional area of a density gradient column cannot change, and neither the electroosmostic, nor the counterflow, change the conductivity. Thus the field strength is constant, which is consistent with the fact that pH plateaus have never been observed in density gradient columns.

Let us now proceed to the question whether electroosmosis is responsible for the observed cathodic drift of the pH gradient. The following points of view are a result of vivid discussions with Dr. Jonsson in connexion with his extensive methodical experiments. Since the counterflow is as great as the electroosmotic flow, it is not immediately understandable why these flows should cause a drift of the pH gradient in any direction. However, since all ampholytes are carried small distances to pH'es somewhat higher than their isoelectric points, they revert to the latter not only by counterflow, but also by electric migration. Consequently, there is everywhere a net anodic migration of all ampholytes, proportional to the electroosmotic flow. This anodic migration is accompanied by an anodic transport of hydration water which, again for hydrostatic reasons, gives rise to a cathodic counterflow of bulk water. According to this hypothesis, the cathodic drift in density gradient columns should be a combined effect of electroosmosis and hydration water transport. Since electroosmosis is the primary cause, the pH drift should be expected to decrease with an increasing ampholyte concentration, in agreement with observations.

OTHER PROPOSED MECHANISMS

Those authors who don't believe in electroosmosis as the cause of the instability of the pH gradient have suggested other mechanisms to explain the observed phenomena. These proposals will now be critically discussed.

Rammler's Hypothesis

A theory originally conceived by Dr. D. H. Rammler was presented by Miles et al. (7). Undoubtedly this theory is based on physico-chemically unsound ideas, as for instance an idea of electrophoretic migration of Ampholine components away from their isoelectric points. This is in conflict with the very definition of an isoelectric point as the pH where there is no net migration and where the transport number vanishes. Another misinterpretation of a physico-chemical concept is done on p. 115, where it is stated that a protein with pI = 6 must possess more carboxyl than amino groups and that a transient net charge at equilibrium therefore would tend to be negative. The Falseness of this statement is most easily understood through the fact that glycine and alanine, both with pI = 6.0, have one carboxyl and one amino group and have the same tendency of acquiring a net positive and a net negative transient charge.

Fawcett's Hypothesis

Fawcett (9) pointed out that the cationic, anionic, and uncharged forms of an ampholyte may have unequal volumes and conformation, that only the ions contribute to the isoelectric condensation, whereas even the uncharged forms contribute to the balancing diffusional flow, and that this may give rise to an asymmetric concentration distribution. All this is correct, but when Fawcett concludes that there will be a slow net migration of the more mobile ion (which probably means migration in the direction of the more mobile ion), it is no longer possible to follow his thinking. Isoelectric focusing is a balance between electric condensation and diffusional spreading irrespective of a possible skewness of the concentration distribution. At a steady state with reference to the solvent, the ampholyte is isoelectric at the peak concentration whether the concentration curve is symmetric or not, and it is difficult to see how an asymmetric distribution could cause a migration away from the isoelectric point. Nevertheless Fawcett's proposal is worthy of attention in attempts at a deeper understanding of non-ideal electrofocusing.

CONCLUDING REMARKS

As is evident from the preceding discussion, the author considers it possible to explain the plateau phenomenon and the cathodic drift of the pH gradient as effects of electroosmosis and hydration water transport. What has been brought forward here, however, is by no means a proof of this hypothesis, it is only some evidence. Much additional research is required in order to clarify these problems completely. A mathematical theory is desirable, using only measurable quantities and leading to several equations which can be tested by suitably designed experiments.

If it is true that electroosmosis is the main cause of the pH gradient instability, the practical question arises whether it is possible to make gels and column walls completely devoid of electric charges. Hydrophobisation of glass walls by siliconisation has been suggested from time to time, but it has not given any definite improvement in the author's laboratory. Radola (18) reported good results by treatment of Sephadex with propylene oxide in order to neutralise carboxyl groups. It should be remembered, however, that even if it were possible to prepare a completely uncharged solid phase, there would be a permanent risk that this complete neutrality could get lost by adsorption of solutes with ionogenic groups. It is therefore doubtful whether it really pays to remove all electric charges from a solid phase in contact with a solution. Most probably, one has to resort to an extensive reduction of the effects by already known methods: increase in viscosity, increase of carrier ampholyte concentration.

PROSPECTS FOR THE FUTURE

Despite the great difficulties in getting rid of electroosmosis in isoelectric focusing, this lecture can be closed with some optimistic prospects for the future. My own research group has initiated a project for creation of stable pH gradients without use of Ampholine or other ampholytes. Since ordinary buffers are used, the ionic strength can be chosen arbitrarily high, and consequently electroosmotic effects are quite negligible. This technique, which will be described in a forthcoming paper, operates in multi-membrane types of apparatus for preparative purposes, but does not seem to be adaptable to analytical isoelectric focusing.

Quite certainly there is place for many new inventions in this field, and I am looking forward with great interest against what will be presented at this symposium. Possibly doors will be opened to a new era of electrofocusing with wide possibilities of further development and important biochemical applications.

REFERENCES

1. Kolin, A. J. Chem. Phys. 22, 1628 (1954).
2. Kolin, A. Proc. Natl. Acad. Sci. U. S. 41, 101 (1955).
3. Kolin, A. Methods of Biochemical Analysis (D. Glick, Ed.), Interscience, New York 1958, p. 259.
4. Svensson, H. Acta Chem. Scand. 15, 325 (1961).
5. Svensson, H. Arch. Biochem. Biophys., Suppl. 1, 132 (1962).
6. Finlayson, G. R., Chrambach, A. Anal. Biochem. 40, 292 (1971).
7. Miles, L. E. M., Simmons, J. E., Chrambach, A. Anal. Biochem. 49, 109 (1972).
8. Baumann, G., Chrambach, A. Progress in Isoelectric Focusing and Isotachophoresis (P. G. Righetti, Ed.), North-Holland, Amsterdam 1975, p. 13.
9. Fawcett, J. S. Ibid., p. 25.
10. Chrambach, A., Doerr, P., Finlayson, G. R., Miles, L. E. M., Sherins, R., Rodbard, D. Ann. N. Y. Acad. Sci. 209, 44 (1973).
11. Denckla, W. D., This Volume, p. (1977).
12. Davies, H. Isoelectric Focusing (J. P. Arbuthnott and J. A. Beeley, Eds.), Butterworths, London 1975, p. 97
13. Felgenhauer, K. Graesslin, D., Huismans, B. D. Protides of the Biological Fluids 19, 575 (1972).
14. Bates, L. S., Deyoe, C. W. J. J. Chromatog. 73, 296 (1972).
15. Bours, J., van Doorenmaalen, W. J. Science Tools 17, 36 (1970).
16. Karlsson, Ch., Davies, H., Öhman, J., Andersson, U.-B. LKB Application Note No. 75 (1973).
17. Leaback, D. H., Rutter, A. C. Biochim. Biophys. Res. Comm. 32, 447 (1968).
18. Radola, B. J. Biochim. Biophys. Acta 295, 412 (1973).
19. Righetti, P. G., Drysdale, J. W. J. Chromatog. 98, 271 (1974).
20. Frater, A. Anal. Biochem. 38, 536 (1970).
21. King, T. P. Biochemistry 11, 367 (1972).
22. Edward, J. T., Wang, I. C. Can. J. Chem. 40, 966 (1962).

Electrofocusing in Buffers: Formation of Natural pH Gradients, Flexibility, Gradient Stability, Relation to Isotachophoresis and Preparative Potential

A. Chrambach and Nga Y. Nguyen

SUMMARY

Natural pH gradients can be formed and Isoelectric Focusing (IF) can be carried out when Ampholine is replaced by a mixture of common amphoteric buffers (Buffer Electro-Focusing=BEF). The same is true when a mixture of non-amphoteric buffers of varying $pK(NH_2)$ is applied, suggesting that in electrofocusing of these buffers, and possibly of amphoteric compounds as well, the buffers are aligned in order of mobility, i.e. isotachophoretically. BEF provides increased freedom in the choice of pH and the relative flatness or steepness of the pH gradient.

Isotachophoresis (ITP) itself gives rise to a pH gradient within the stack which can be made to span the entire gel length if the ordinarily used single trailing (terminator) constituent is replaced by multiple such constituents (ITP-B or "cascade stacking"). Such a pH gradient formed in ITP-B can be stabilized if the cascade stacking gel is transposed between strong acid and basic electrolytes, i.e. into an electrofocusing system.

Similarly, the reverse transposition of a gel electrofocused in ITP-B buffers ("cascade focused") between the Upper and Lower Buffers of the particular cascade stacking system results in the mobilization of isoelectric, stationary protein zones. This method appears to be potentially valuable as a high load-capacity preparative method for proteins.

INTRODUCTION

This report is both a sad and a jubilant postscript to a recent review on gel isoelectric focusing (IFPA) [1]. It seems sad that this review, which has not even seen the light of day, is already obsolete.

The jubilant note derives from the fact that recent and rapid new developments in electrofocusing have provided it a) with a broadened applicability and flexibility, b) with the possibility to mobilize isoelectric protein zones and c) with the beginnings of an understanding of its relation to isotachophoresis, which might aid in a better theoretical understanding of electrofocusing. To some degree these new developments in electrofocusing derive from an idea of T. M. Jovin [2] to apply "cascades" of multiple trailing constituents in multiphasic zone electrophoresis (MZE) to form pH gradients for use in electrofocusing.

MATERIALS AND METHODS

1) Buffers were selected from the kit of multiphasic zone electrophoresis (MZE) constituent buffers provided by Serva (Heidelberg, G.F.R.), from Calbiochem (La Jolla, CA, USA) or from Sigma (St. Louis, MO, USA). An analytical, thermostated all-Pyrex polyacrylamide gel electrophoresis (PAGE) apparatus recently described [3] was used. Electrophoresis was carried out at 0-4°C. Polacrylamide gels "non-restrictive" to molecular weights of up to 0.5 million (5 %T, 15 $\%C_{DATD}$) were used as described [4] unless stated otherwise.

2) Buffer electrofocusing (BEF): BEF was carried out in either amphoteric or non-amphoteric buffers by the procedures previously described [5].

3) Cascade focusing: BEF was carried out in the buffer mixture used for isotachophoresis with multiple trailing constituents (cascade stacking) (see section 4). Concentrations of the various buffer constituents were obtained from the program of Routs as described [6]. Gels were polymerized in the presence of the mixture of all buffers selected and defined by the output of the Routs program. Conditions of cascade focusing were otherwise those of BEF, as detailed previously [6].

4) Cascade stacking (ITP-B): ITP-B used the identical buffers employed for cascade focusing. However, only the leading constituent (MES in the selected mixture of buffers) was incorporated as a gel buffer into the polymerization mixture. The remaining buffers, in the concentrations stated by the Routs output, were applied with the protein sample as a single mixture [6]. Conditions of ITP-B were those previously described [6].

5) pH gradient measurement and staining: pH gradients were determined either manually [7] or by an automated device recently described [8]. A gel scanning time of 12 minutes was selected. After scanning, gels from BEF were stained by the procedure of Diezel [9], those from IFPA by procedure B of Vesterberg [10].

6) Two-dimensional BEF--SDS-PAGE: First dimensional BEF gels were dissected longitudinally [11], and the two halves were applied on top of a Stacking Gel (MZE system 4192.VI.) of SDS-PAGE, using a vertical slab apparatus (Hoefer Scientific, No. DE-102, San Francisco, CA, USA). Conditions of SDS-PAGE were those described [12]. Conditions of the 2-D fractionation are given elsewhere [13].

RESULTS AND DISCUSSION

1) Buffer electrofocusing (BEF): A random selection of either amphoteric or non-amphoteric buffers differing from another in the pK's of their amino-groups (Fig. 1) when mixed together and subjected to the ordinary conditions of gel electrofocusing give rise to pH gradients across the length of the gel (Fig. 2). These pH gradients arise, stabilize and decay as a function of the time of electrofocusing [5] in a

BUFFER ELECTROFOCUSING

AMPHOTERIC—BEF—NON-AMPHOTERIC

BUFFER	pK_2	BUFFER	pK_2
MES	6.4	PYRIDINE	5.5
MOPS	7.2	4-PICOLINE	6.2
ACES	7.3	BISTRIS	6.9
TES	8.0	LUTIDINE	7.0
TRICINE	8.6	IMIDAZOLE	7.5
BICINE	8.7	HEM	7.2
TAURINE	9.7	NEM	8.0
GLYCINE	10.5	TEA	8.4
GABA	11.3	TRIS	8.8
		MEA	10.4

Fig. 1. Composition of buffer mixtures used in BEF. Buffer selection has not been optimized.

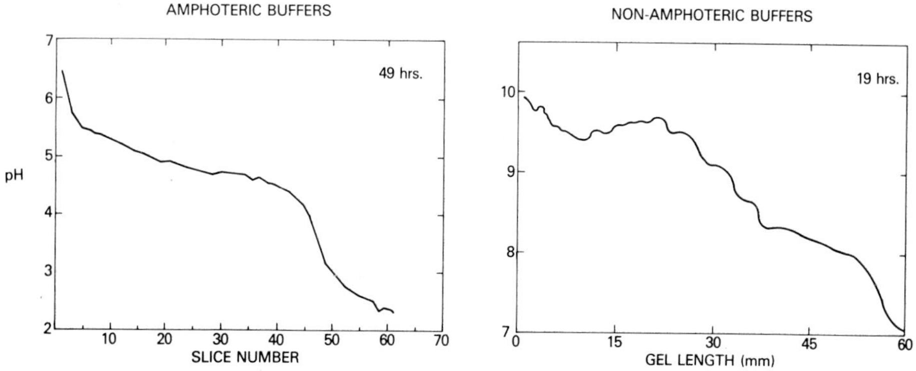

Fig. 2. Representative pH gradients formed in BEF with amphoteric and non-amphoteric buffer mixtures defined in Fig. 1.

fashion qualitatively similar to the natural pH gradients formed in Ampholine [7]. Since relatively high molarities of the buffers (0.1 M) provide the most stable pH gradients, gradient stability can surpass that obtainable with Ampholine. Depending on the buffer mixture used, pH gradients made in BEF can be flattened or steepened at will [14]. Equally, there appears to be no limit to the choice of desired pH-range in BEF. The fact that non-amphoteric buffer mixtures form natural pH gradients appears inconsistent with the notion [15] that pH gradients form in electrofocusing by an isoelectric mechanism. BEF also has advantages over IFPA in 2-dimensional fractionations where it avoids a) the extended stacks of Ampholine precluding PAGE or SDS-PAGE fractionation of proteins with high R_f; b) the difficulties in protein detection due to Ampholine staining with protein stains [13].

2) Mobilization of isoelectric protein zones in cascade focusing by cascade stacking (ITP-B): Isoelectric zones in IFPA can be mobilized and eluted from the gel by substituting the anodic acid with a buffer of suitable pI or pH [16]. However, this procedure does not seem to provide assurance against an interchange of zone positions when the zones ordered in sequence of isoelectric points are eluted in order of constituent mobility.

BEF can be carried out in a set of particular buffers computed to provide steady-state stacking with multiple trailing (terminator) constituents (see below). A linear, stable pH gradient is obtained [6], and

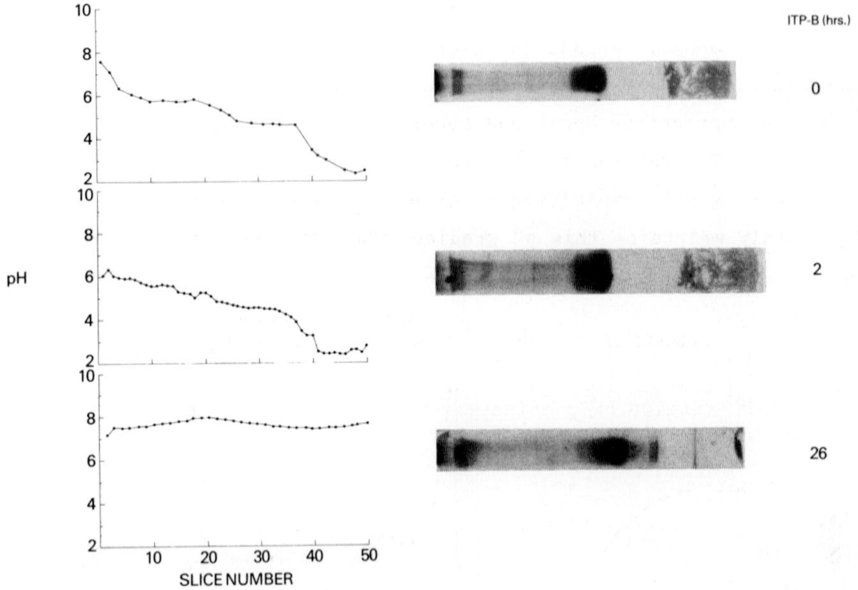

Fig. 3. Mobilization of a pH gradient formed in **cascade focusing and of** protein zones after transposition into a cascade stacking system.

proteins migrate in this gradient to their respective pI-positions [6]. When one then replaces the acid and basic electrolytes by the proper Upper and Lower Buffers of the steady-state stacking system used, the protein zones, as well as the pH gradient, are displaced progressively from the gel (Fig. 3). It appears that this mobilization of isoelectric zones formed in Stacking buffers (i.e. in "cascade focusing") by transformation of the system into one of steady-state stacking (ITP-B or "cascade stacking") does not lead to a change in the order of eluting proteins. Such zone mobilization appears of great preparative promise in protein fractionation in particular since the protein load capacity of both electrofocusing and steady-state stacking is 2-3 orders of magnitude higher than that of PAGE [3,17].

The sequential application of cascade focusing and cascade stacking to protein preparation appears more promising than the sole preparative application of ITP-B when and if the resolubilization of isoelectrically precipitated protein zones is easy and rapid. If this is not so, ITP-B by itself appears as the more promising preparative approach, using the high loads and the elution techniques previously described for steady-state stacking without buffer spacers, i.e. ITP-A [17].

3) Conversion of steady-state stacking to electrofocusing systems: As pointed out above, cascade focusing systems can be converted to cascade stacking systems by substitution of the acid and basic electrolytes by the appropriate Upper and Lower Buffers of the multiphasic buffer systems used. The reverse is also the case. Figure 4 shows that a cascade stacking gel, exhibiting a certain pH gradient within the migrating stack, roughly maintains this pH gradient when the Upper and Lower Buffers are replaced by strong acid and basic electrolytes.

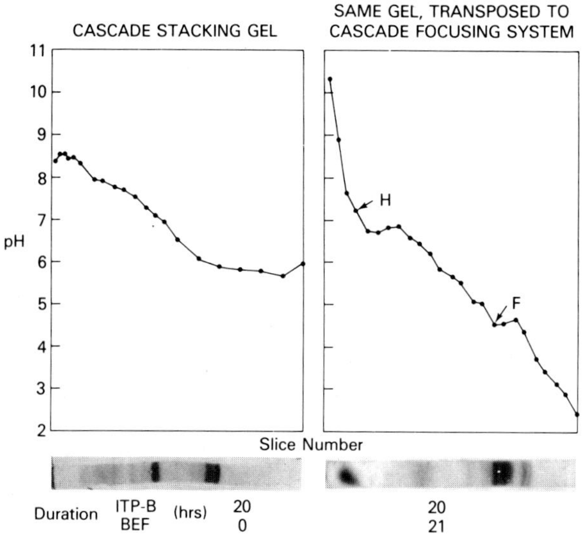

Fig. 4. Stabilization of pH gradient formed in cascade stacking after transposition into a cascade focusing system. The position on the pH gradients of hemoglobin (H) and ferritin (F) are indicated.

However, two changes occur: first, the protein which previously had aligned in order of constituent mobility and was found in the stack at a pH characteristic for its segment of the stack-pH-gradient, will move into its proper isoelectric position after transposition of the gel into an electrofocusing system; second, the anodic electrolyte "connects" by a steep new pH gradient to the "stack-pH-gradient". To a less pronounced degree, a "connection" is also formed to the cathodic electrolyte.

The fact that the stack-pH-gradient appears similar to the cascade focusing gradient formed from it would indicate that the order of buffer constituents, an order of constituent mobility, which prevails in the stack is maintained after the exchange of anodic and cathodic buffers for strongly acid and basic electrolytes. Thus, it is possible that the buffers in cascade focusing are aligned in order of constituent mobility, not of isoelectric point. It should be possible to verify this point experimentally. However, even if this point can be verified, we cannot assume that all electrofocusing is non-isoelectric. The fact of the protein shift towards its pI would argue the opposite. Conceivably, the "caging" of a stack-pH-gradient between strong acid and base generates an exceptional case in which a pH gradient is maintained as it exists within the stack. Irrespective, it appears that the existence of such an exceptional case must lead to a broadening in our physical-chemical analysis of electrofocusing.

REFERENCES

1. Chrambach, A., Baumann, G.: Isoelectric focusing on polyacrylamide gel. In Isoelectric Focusing (N. Catsimpoolas, ed.), Academic Press, New York, pp. 77-91 (1976).

2. Jovin, T. M.: Multiphasic zone electrophoresis. IV. Design and analysis of discontinuous buffer systems with a digital computer. Ann. N. Y. Acad. Sci. 209, 477-496 (1973).

3. Chrambach, A., Jovin, T. M., Svendsen, P. J., Rodbard, D.: Analytical and preparative polyacrylamide gel electrophoresis: An objectively defined fractionation route, apparatus, and procedures. In Methods of Protein Separation (N. Catsimpoolas, ed.), Vol. 2, Plenum Press, New York, pp. 27-144 (1976).

4. Baumann, G., Chrambach, A.: A highly crosslinked, transparent polyacrylamide gel with improved mechanical stability for use in isoelectric focusing and isotachophoresis. Anal. Biochem. 69, 649-651 (1975).

5. Nguyen, N. Y., Chrambach, A.: Non-isoelectric focusing in buffers. Anal. Biochem. 74, 145-153 (1976).

6. Nguyen, N. Y., Rodbard, D., Svendsen, P. J., Chrambach, A.: Cascade stacking, cascade electrofocusing and their basic identity. Anal. Biochem., in press.

7. Baumann, G., Chrambach, A.: Mechanisms of pH-gradient instability in isoelectric focusing. In Progress in Isoelectric Focusing and Isotachophoresis (P. G. Righetti, ed.), North-Holland Publ. Co., pp. 13-23 (1975).

8. Chidakel, B. E., Nguyen, N. Y., Chrambach, A.: A device for the automated measurement of pH-gradients in gel electrofocusing. Anal. Biochem., submitted.

9. Diezel, W., Kopperschlaeger, G., Hofmann, E.: An improved procedure for protein staining in polyacrylamide gels with a new type of Commassie Brilliant Blue. Anal. Biochem. 48, 617-662 (1972).

10. Vesterberg, O., Hansen, L.: This Volume.

11. Kohler, P. O., Bridson, W. E., Chrambach, A.: Human growth hormone produced in tissue culture: Characterization by polyacrylamide gel electrophoresis. J. Clin. Endocrinol. Metab. 32, 70-76 (1971).

12. Wyckoff, M., Rodbard, D., Chrambach, A.: Polyacrylamide gel electrophoresis in sodium dodecylsulfate containing buffers, using multiphasic buffer systems: Properties of the stack, valid R_f measurement and optimized procedure. Anal. Biochem., submitted.

13. Nguyen, N. Y., Chrambach, A.: Characterization of the protein components of a plasmin preparation capable of catalyzing the transformation of human growth hormone to a biologically activated species.

14. Nguyen, N. Y., Sakokangas, A., Chrambach, A.: Electrofocusing in natural pH gradients formed by buffers: Gradient modification and stability with time. Anal. Biochem., submitted.

15. Svensson, H.: Isoelectric fractionation, analysis and characterization of ampholytes in natural pH gradients. Acta Chem. Scand. 15, 325 (1961) and 16, 456 (1962).

16. McCormick, A., Miles, L.E.M., Chrambach, A.: Selective elution of zones from preparative isoelectric focusing gels by ampholytes of buffers. Anal. Biochem. 75, XXX-XXX (1976).

17. Baumann, G., Chrambach, A.: Gram-preparative protein fractionation by isotachophoresis: Isolation of human growth hormone isohormones. Proc. Natl. Acad. Sci. USA 73, 732-736 (1976).

Abbreviations: %C, crosslinking agent (g)/monomer (g) + crosslinking agent (g); DATD, N, N'-diallyltartardiamide; BEF, buffer electrofocusing; IF, isoelectric focusing; IFPA, isoelectric focusing on polyacrylamide gel; ITP, isotachophoresis; ITP-A, ITP without spacers; ITP-B, ITP with buffer spacers; MZE, multiphasic zone electrophoresis; PAGE, polyacrylamide gel electrophoresis; SDS-PAGE, PAGE in sodium dodecylsulfate containing buffers; %T, total gel concentration (monomer (g) + crosslinking agent (g)/100 ml).

Comparison of pH Gradients Used for Isoelectric Focusing

J.S. Fawcett

ABSTRACT

A comparison has been made between the pH gradients formed when Ampholine, Servalyt and Bio-Lyte are used in isoelectric focusing with sucrose density gradient and polyacrylamide gel techniques. Differences in resolution of marker proteins and in the conductance profile have been observed. In addition pH gradients obtained from the peptide mixture - Bactopeptone Difco - have been studied. The extent of the pH drift has been determined for each of these gradients.

INTRODUCTION

A number of different methods have been proposed for the formation of pH gradients suitable for isoelectric focusing. Some of these methods are restricted to relatively narrow pH regions while others, particularly the amphoteric mixtures, are applicable to both wide and narrow pH ranges. Indubitably, during the past ten years Ampholines (mixtures of polyamino-polycarboxylic acids) have proved most popular.

Recently, new carrier ampholytes - Servalyt and Bio-Lyte - have been introduced. They contain sulphonic and phosphonic acids in addition to amino and carboxylic acid groups (1). It is of interest, therefore, to compare the pH gradients obtained from these compounds together with pH gradients obtained from peptide mixtures. The latter have been suggested as suitable substitutes for the commercial carrier ampholyte. For example, Blanický and Bihar (2) have reported that satisfactory pH gradients are obtained from certain commercial samples of peptones. We have investigated one grade of Bactopeptone Difco which, however, was the least satisfactory of those tested by Blanický and Pihar and we have fractionated this mixture by zone convection isoelectric focusing (3) in an attempt to obtain improved pH gradients from this material.

We have also measured the rate of pH drift during prolonged isoelectric focusing with these different ampholytes. In a previous report (4) we concluded that electroosmosis was not the major cause of the pH drift, nor was the uptake of carbon dioxide by the cathode solution nor electrochemical changes to the ampholytes. It was suggested that the drift could be due to an imbalance between the electrophoretic transport towards the isoelectric point and the diffusional transport away from that point. Difference in configuration and hydration between the cationic and anionic forms of the ampholyte could cause this non-equilibrium condition. Any variation in the rate of pH drift with the newly introduced carrier ampholytes (where the hydration of the sulphonic and phosphonic groups is likely to be different from that of the carboxylic) would be of considerable interest.

MATERIALS AND METHODS

Ampholine(LKB, Bromma, Sweden), Servalyt(Serva Feinbiochemica Heidelberg, G.F.R.) and Bio-Lyte(Bio-Rad Laboratories, Richmond, California, U.S.A.) were obtained from the manufacturers. Horse and sperm whale myoglobin(Koch-Light Laboratories, Colnbrook, U.K.) and horse spleen ferritin (MilesResearch Products Division, Kankakee, U.S.A.) were used as coloured markers to follow the isoelectric separations and to measure the rate of pH drift.

Bactopeptone Difco was fractionated in zone convection isoelectric focusing apparatus based on the design of Kalous and Vacik (5). One unit with 38 compartments had a total

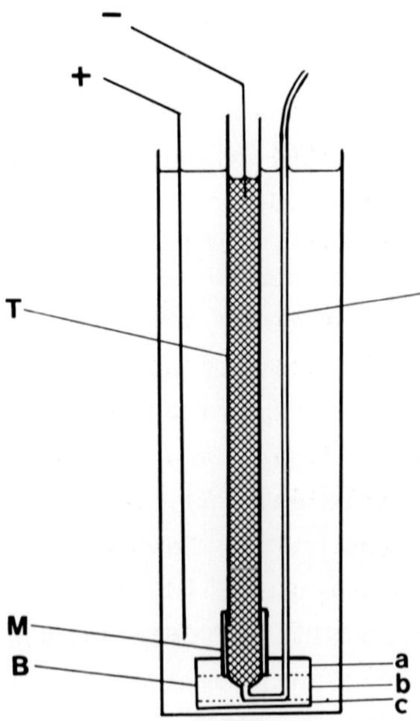

Fig. 1. Diagram of small density gradient column. The glass tube (T) is cemented to a porous polyethylene-poly-acrylamide gel membrane (M) which is fixed to a plastic block (B) fitted with capillary tubing (P). Unit is positioned in a tank containing dense electrode solution.

working volume of 30 ml, while a second model of a wider design had 11 compartments and could fractionate 100 ml of solution. Both units were cooled externally by partial immersion in a cooling bath.

Density gradient isoelectric focusing was performed in LKB 8100-1 column (110 ml) or, for economy in ampholytes, in small columns (12 ml) of our own design. The LKB column was used when precise pH and conductance profiles were required, as the larger fraction volume allowed a more accurate determination. Columns were filled from the lower end with sucrose gradients (0-50%). They were run at 4° C and the position of the coloured zones determined twice daily. By pumping water into the top compartment the columns were emptied at the rate of 1 ml per minute. To minimise mixing the fraction collector was positioned under the column and the elute dripped directly from the column into the collector tube. All pH and conductance measurements were made at 25°C.

For other comparative experiments small columns (Fig. 1) were used. They consisted of a glass tube (T) 9 mmID - 22 cm long which was cemented with epoxy adhesive to a rigid cylindrical membrane (M). Membranes were prepared from porous polyethylene tubing, the pores of which had been plugged with polyacrylamide gel (6). When dry they were easily glued into position. The lower end of the membrane was cemented into a Perspex block (B) which was machined with a channel to connect with polyethylene capillary tubing (P). The block is conveniently made by gluing together three sections (a, b and c). Using the capillary tubing the columns were filled from the lower end with density gradient (10 ml) followed by dense electrode solution (1.5 ml) - usually anode solution. The latter ensures that the membrane is immersed in solution of high conductance and also minimises the possibility of precipitation within the membrane. Cooling was effected by immersing the units in a narrow tank fitted with electrode, stirrer and cooling coil and filled with dense electrode solution. For convenience six of these units were placed side by side in a tank 5 x 15 x 24 cm. Fractions were collected by pumping dense solution into the capillary tubing, displacing the gradient upwards through a silicon rubber adapter attached to the top of the tube.

The movement of zones of coloured proteins during isoelectric focusing in polyacrylamide gels was followed by time lapse photography as previously described (4).

RESULTS AND DISCUSSION

Fractionation of Bactopeptones

The results of the zone convection isoelectric focusing of Bactopeptone-Difco are given in Fig. 2. A 38-compartment apparatus was filled with 30 ml of Bactopeptone solution (4%) and a potential of 600v was applied for 4 days. The pH gradient and conductance were measured by interrupting the experiment after 48, 72 and 96 hours.

Although the system readily focused human haemoglobin into one compartment (due to the steep pH gradient in that position) the irregular pH gradient is clearly unsatisfactory. It was considered that preliminary fractionation of a more concentrated Bactopeptone mixture would allow selective blending of fractions in order to give a more linear pH gradient. In addition, this procedure would effectively remove the components insoluble at their isoelectric point.

The short 11-compartment apparatus was filled with 100 ml of Bactopeptone (10%) and a potential of 550v applied for two days. The pH gradient was similar to that obtained with the

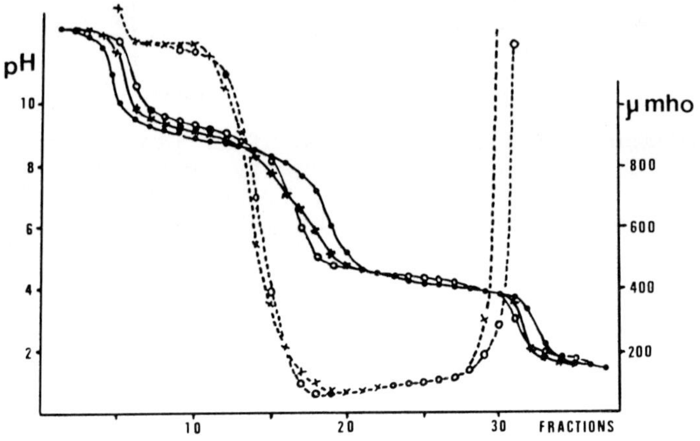

Fig.2. Zone convection isoelectric focusing of Bactopeptone Difco in 38 compartment apparatus. Continuous line pH, broken line conductance, -●- 48h, -x- 72h and -o- 96h.

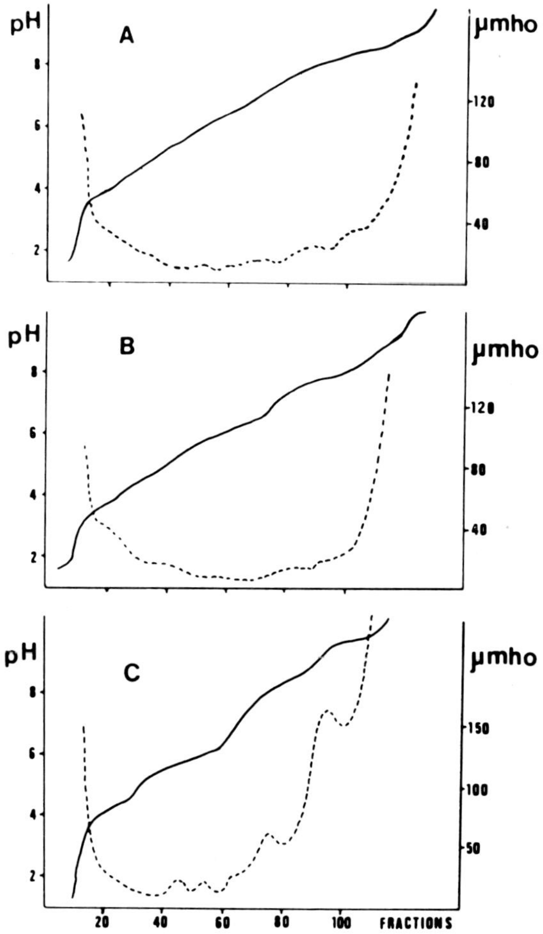

Fig.3. Conductance and pH curves after isoelectric focusing in a LKB column using sucrose density gradient at 4°C with cathode at top. Solid line pH, broken line conductance, measured at 25°C. (A) Ampholine pH 3.5-10, (B) Servalyt pH 2-11, (C) Bactopeptone Difco reconstituted mixture from zone convection fractionation. (A & B) run at 1200v for 4 days. (C) 800v for 2 days. Carrier ampholyte concentration 1%.

38-compartment unit. By making a number of successive experiments sufficient material was accumulated to recombine fractions in various proportions so as to eliminate much of the acidic fraction and to enrich the neutral region. This blended mixture was used for subsequent experiments.

Density Gradient Experiments

The pH and conductance profiles for density gradient isoelectric focusing using the different carrier ampholytes are given in Fig. 3. The results obtained from Servalyt and Bio-Lyte were almost identical, both having very low conductance in the pH range 5.75-6.5 and an inflection in the pH profile at pH 7.8. It is interesting to note that all three commercial carrier ampholytes had a minimum conductance at pH 6.2 - possibly the same polyamino compound is used in the synthesis of all three carrier ampholytes.

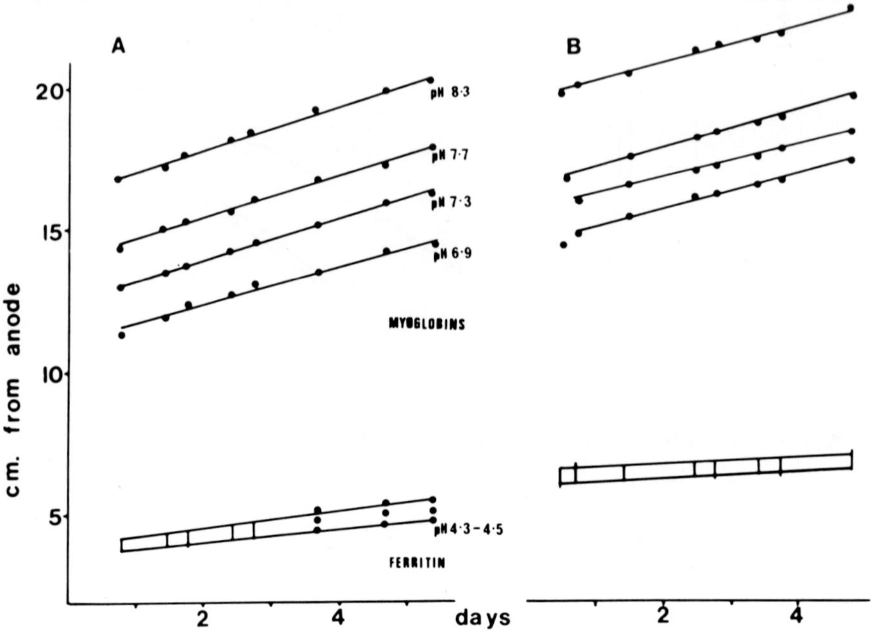

Fig.4. Drift of the pH gradient as shown by the movement of coloured proteins in LKB sucrose density gradient column using 1% solution of carrier ampholyte. (A) Ampholine pH 3.5 -10, (B) Servalyt pH 2-11. Electrofocusing carried out at 4°C with applied potential of 600v. Cathode at upper end.

Gradients prepared from Servalyt and Bio-Lyte appeared to focus proteins more rapidly and the current decreased to a lower value than with gradients prepared from Ampholine.

The pH gradient obtained from the blended Bactopeptone mixture is shown in Fig. 3c. Compared with the original Bactopeptone this mixture gave considerable improvement in the pH profile but failed to give high resolution of horse and whale myoglobins and remains unsatisfactory as a wide pH range carrier ampholyte.

The rate of pH drift in these density gradient experiments is given in Fig. 4. The pH drift was slightly greater with gradients prepared from Ampholine. With pH gradients prepared from Servalyt and Bio-Lyte myoglobins focused

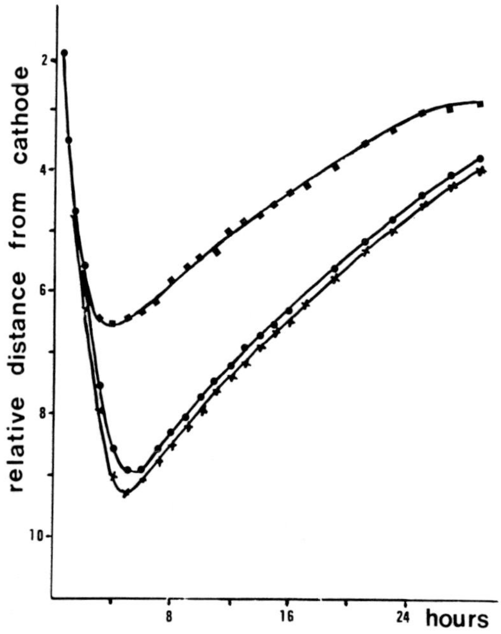

Fig.5. Movement of sperm whale myoglobin during isoelectric focusing in polyacrylamide gels containing 1% solution of carrier ampholyte. -■- Servalyt pH 2-11, -●- Ampholine pH 3.5-10, -x- Ampholine pH 3.5-10 enriched with Ampholine pH 4-6 (10%) and Ampholine pH 5-7 (10%). Gels were run at 20°C with an average potential gradient of 25v/cm.

nearer to the cathode and the resolution of these proteins deteriorated more rapidly during prolonged isoelectric focusing in these gradients.

Polyacrylamide Gels

The drift of the focused zone of sperm whale myoglobin during isoelectric focusing in polyacrylamide gel is given in Fig. 5. The myoglobin appears to focus sooner in the gradient prepared from Servalyt, although it must be emphasised that the samples were applied to the cathode end and due to the distribution of the pH gradient, the protein migrates a shorter distance in this ampholyte. During prolonged isoelectric focusing in polyacrylamide gel the resolution between the myoglobin zones rapidly deteriorates with gels prepared from Servalyt and Bio-Lyte. On the other hand, these ampholytes gave a better resolution of ferritin components at the anode end of the gradient.

In an attempt to see the effect of increasing the concentration of ampholyte in the neutral region Ampholines pH 4-6 and pH 5-7 were added to Ampholine pH 3.5-10 as suggested by the manufacturers (LKB 2117 Multiphor Application Note 1973). Although it is stated that this mixture gives an improved pH gradient it has no effect on the rate of the pH drift (Fig. 5).

The sample of Bio-Lyte was supplied with a graph showing an almost linear pH gradient obtained after isoelectric focusing for 20h in polyacrylamide gel containing glycerol. Using the conditions specified by Bio-Rad, we were unable to obtain this linear gradient. Bio-Rad recommend adding calcium hydroxide to the cathode solution, presumably to remove carbon dioxide as calcium carbonate. We find that this procedure does not reduce the rate of pH drift towards the cathode.

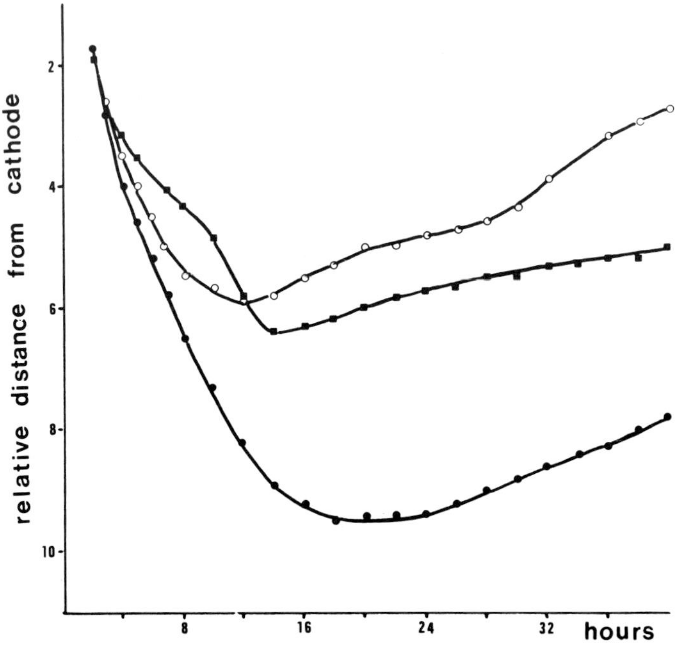

Fig.6. Movement of sperm whale myoglobin during gel isoelectric focusing in polyacrylamide gel containing Bactopeptone mixture obtained from zone convection fractionation. -o- Peptone mixture (2%), -■- Peptone mixture (4%) and -●- Ampholine pH 3.5-10 (1%). Gels run at 20°C with average potential gradient of 5v/cm.

The drift of the pH gradient in polyacrylamide gel when using Bactopeptone mixture is given in Fig. 6 together with that obtained from Ampholine for comparison. The myoglobin mixture was only partly resolved and it was necessary to apply a lower voltage to eliminate excessive heating in the regions of low conductance. As with Ampholine gradients (7) a reduced rate of pH drift is observed with higher concentrations of peptones.

Blanický and Pihar fractionated their Bactopeptones by ethanol precipitation before use in isoelectric focusing. When we applied this procedure to Bactopeptone Difco we found it did not remove all the less soluble components and turbid zones were always formed during isoelectric focusing. On the other hand, after centrifuging the fractions obtained from zone convection isoelectric focusing, we obtained material which did not give further precipitation when used for subsequent isoelectric focusing experiments. Of the peptones tested by Blanický and Pihar, Bactopeptone-Spofa gave the most satisfactory results and it is possible that further improvements could be obtained by additional isoelectric fractionation of this material.

Results presented in this report show that the various ampholytes produce only small differences in the rate of pH drift towards the cathode. The drift is concentration dependant (7) and these small variations in rate may be due partly to slight differences in concentration of the commercial samples. The results do not provide any further understanding towards the mechanism of the pH drift and further work is required to elucidate this phenomenon.

REFERENCES

1. Grubhofer, N., Borja, C.: Carrier ampholytes with sulphonic and phosphonic acid groups. This volume.
2. Blanický, P., Pihar, O.: Isoelectric focusing of proteins on bactopeptone. Collect.Czech. Commun. 37, 319-325 (1972).
3. Valmet, E.: Zone convection electrofocusing. Science Tools 16, 8-13 (1969).
4. Fawcett, J.S.: The pH drift during isoelectric focusing. Progress in Isoelectric Focusing and Isotachophoresis, Ed. P.G. Righetti, North-Holland, Amsterdam, 25-37 (1975).
5. Kalous, V., Vacik, J.: Horizontal preparative electrophoresis chamber without carrier. Chem. Listy 53, 35-37 (1959).
6. Fawcett, J.S.: Some recent developments in preparative isoelectric focusing, Isoelectric Focusing, Ed. J.P. Arbuthnott and J.A. Beeley, Butterworths, London, 23-43 (1975).
7. Righetti, P.G., Drysdale, J.W.: Isoelectric focusing in polyacrylamide gels. Biochem. Biophys. Acta 236, 17-28 (1971).

On the Temperature Dependence of Isoelectric Points of Proteins with Special Reference to Isoelectric Focusing

S. Fredriksson

SUMMARY

The pI shifts accompanying the increase of temperature from 4 to 25 $^{\circ}$C have been calculated for a number of model proteins by means of the Linderstrøm-Lang equation. On comparison of these shifts with the corresponding shifts of pH for Ampholine carrier ampholyte fractions being isoelectric at 4 $^{\circ}$C, it is evident that for the great majority of proteins, one should not expect to get accurate pI values, if the isoelectric focusing is performed at 4 $^{\circ}$C, whereas the pH gradient is measured at 25 $^{\circ}$C. For certain weakly acid proteins having no titratable imidazole groups, the difference between the apparent pI thus obtained and the true pI at 25 $^{\circ}$C may be as much as 0.2 pH units. For strongly acid proteins, on the other hand, almost correct pI values will be obtained regardless of the focusing temperature owing to the small standard heats of ionization of carboxyl groups. The consequence of the results to the future evaluation and reporting of pI values of proteins are discussed. A way of converting the apparent pI values obtained at 25 $^{\circ}$C to true pI values at 4 $^{\circ}$C is suggested.

INTRODUCTION

Under carefully controlled conditions, the method of isoelectric focusing enables absolute isoelectric points to be determined with an accuracy usually better than $\pm$ 0.02 pH units. This makes the pI value a very attractive parameter for characterization and identification of a protein (1), but also for examination of possible charge heterogeneity (2, 3) or for revealing conformational changes brought about by specific reagents (4 - 6).

One experimental variable that has to be carefully controlled to enable the above accuracy is temperature. As, apart from the carboxyls, the protolytic groups in proteins have rather large standard heats of ionization [6 - 14 kcal/mole (7)], pI values of neutral and basic proteins can be expected to be sensitive to temperature changes. According to the van't Hoff equation, a ΔH^o equal to 10 kcal/mole, for example, means a temperature coefficient for the corresponding acid dissociation constant equal to -0.03 pK units per degree at 4°C. To allow a meaningful comparison of pI values obtained for a given protein in different laboratories, it is thus essential that the data really refer to the same temperature.

An examination of the vast literature on isoelectric focusing shows that the great majority of runs have been made at 4°C. Although a higher temperature should give a more rapid focusing due to increased mobilities, there are strong arguments for keeping the focusing temperature at this low value. Many proteins run the risk of being partially denatured during a run already at room temperature. The diffusion of proteins is slower at lower temperatures, which is important during the fractionation. The stability of a density gradient is higher at lower temperatures.

As regards the subsequent pH measurements, however, there are strong arguments for performing them at 25°C. An isoelectric point referring to 25°C is of special interest as most pK values in the literature have been measured at this temperature (8). The response of the glass electrode is considerably faster at 25°C than at 4°C, especially if the electrode is of the capillary type (9, 10).

On the basis of the above arguments, it were thus desirable that true isoelectric points referring to 25°C could be evaluated from pH measurements at this temperature despite a focusing temperature of 4°C. However, due to the large differences in the ΔH^{o} values of different protolytic groups in a protein (see above), the net charge of the latter should a priori be expected to change with temperature. Nevertheless, some practical results indicate that the difference between the true pI at 25°C and the pI obtained by focusing at 4°C and mapping the pH course at 25°C may be negligible for many proteins as compared with the normal experimental uncertainty. For two myoglobin species, Vesterberg and Svensson (8) thus found the pI values at 25°C (6.92 and 6.97) to be independent of whether the focusing temperature was 4 or 25°C. According to Jonsson and Pettersson (11), this seems to be the case also for fungal laccase A (pI 3.07, 3.27) and for bovine carbonic anhydrase B (pI 5.89).

These findings made the present author presume that the isoelectric points of the β-lactoglobulins A and B (5.13 and 5.23) were independent of the focusing temperature, too (10). Recent control runs at 25°C have revealed, however, that the true pI's at 25°C are about 0.1 pH unit higher than the apparent ones reported in reference 10.

In this situation, and with regard to the fact that also other workers are reporting pI values at 25°C from runs at 4°C, it was felt worth-while to make a theoretical investigation of the temperature dependence of pI for some model proteins. However, before presenting the resulting pI shifts, it is essential to comment upon the role of the Ampholine in this connection.

BUFFER CAPACITIES OF AMPHOLINE AND PROTEINS

According to the general theory of isoelectric focusing, the carrier ampholytes must fulfil certain criteria to be useful (12). Inter alia, they should have a certain buffer capacity in the isoelectric state to be able to dictate the pH gradient also in presence of proteins.

As regards Ampholine, it is generally assumed that at an average Ampholine concentration of 1% (w/v), the buffer capacity of the focused carrier ampholytes will be sufficient to permit them to dictate the pH even in the presence of as much as 1% of protein [cf. Vesterberg (13) and Davies (14)]. Although this assumption is probably valid for most proteins, the data of Table 1 indicate that there may be several exceptions. The buffer capacities reported for the proteins have been calculated from the slopes of the titration curves (as given in the literature) at the isoelectric points. The corresponding buffer capacities of 1% Ampholine have been evaluated from the data of Davies (14). If it is assumed that a focused protein has a negligible effect on the pH gradient as long as its buffer capacity is less than 10% of that of the surrounding Ampholine, it is evident from the data given that this requires a protein concentration less than 1.1% for sperm-whale myoglobin component I, but less than 0.2% for β-lactoglobulin A.

For the future discussion, it is presupposed, however, that the total protein concentration is kept as low as to permit the Ampholine molecules to dictate the pH in the focused zone, and thereby in the corresponding fractions. Then the pH value assigned to the concentration maximum of the focused protein at a temperature different from the focusing temperature will refer to the corresponding carrier ampholyte fraction rather than to the protein. As a consequence, there is a theoretical chance that the true isoelectric point at the temperature of the pH measurement will be obtained regardless of the focusing temperature, viz. if dpH/dT for this carrier ampholyte fraction equals dpI/dT for the protein in the actual temperature interval.

CALCULATION OF pI SHIFTS FOR MODEL PROTEINS

To get an idea of to what extent the condition dpH/dT = dpI/dT is met in practice, the pI shift accompanying the increase of the focusing temperature from 4 to 25°C was calculated for a number of model proteins. The resulting shifts then could be compared with the corresponding pI shifts reported for Ampholine (14).

Table 1. Buffer Capacities of Ampholine and Various Proteins at the Isoelectric Points of the Latter

Protein *	pI(25°C)	Buffer capacity of protein µequiv/mg pH	Buffer capacity of 1% Ampholine µequiv/ml pH **
β-Lactoglobulin A (15)	5.1	0.34	6.4
Bovine serum albumin (16)	5.4	0.11	6.0
Horseradish apoperoxidase (17)	6.8	0.03	6.6
Sperm-whale myoglobin, MbI (18)	8.2	0.09	10
Horseradish peroxidase (17)	8.7	0.02	25
Ribonuclease (19)	9.6	0.25	>30

* Reference number refers to the source of the titration curve from which the buffer capacity of the protein was calculated.

** Calculated from data of Davies (14).

The calculations were based on Linderstrøm-Lang's equation:

$$(1)\quad pH = pK^{i}_{int} + \log [x_i/(1-x_i)] - 0.868\ wZ$$

which correlates the dissociation of ionizable groups on proteins with the prevailing pH value. [x_i refers to the average degree of dissociation at any pH of groups of type <u>i</u> having an intrinsic dissociation constant K^{i}_{int}, Z means the net charge (in units of proton charge) of the protein, and w is a parameter accounting for the electrostatic interaction between charged groups.] At the isoelectric pH, Z is zero, and Eqn. (1) is simplified to

$$(2)\quad pI = pK^{i}_{int} + \log [x_i/(1-x_i)]$$

The isoelectric condition of Z = 0 is also expressed by the equation:

$$(3)\quad \sum n_{ic}(1-x_{ic}) + \sum n_{ia}(-x_{ia}) = 0$$

n_{ic} and n_{ia} are the total numbers of cationic and anionic groups of type <u>i</u>, respectively, whereas $(1-x_{ic})$ and $(-x_{ia})$ are the corresponding average charges on each of these groups. By solving Eqn. (2) for x_i and inserting the result into Eqn. (3), the isoelectric condition was obtained in a form suitable for the present purpose:

$$(4)\quad \sum n_{ic} \left[1 - 10^{pI-pK^{ic}_{int}} \left(1 + 10^{pI-pK^{ic}_{int}}\right)^{-1}\right] -$$

$$- \sum n_{ia}\, 10^{pI-pK^{ia}_{int}} \left(1 + 10^{pI-pK^{ia}_{int}}\right)^{-1} = 0$$

In principle, the calculations were continued as follows. Reasonable model values of pK^{i}_{int} and ΔH^{o}_{i} for the various types of ionizable groups at 25°C were chosen from the literature. Neglecting the temperature dependence of ΔH^{o}_{i}, the corresponding values of pK^{i}_{int} at 4°C were calculated by means of the van't Hoff equation. The next step could have been to specify the numbers (n_i) of <u>all</u> the types of ionizable groups present in the actual model protein and calculate the corresponding pI at $4^{o}C$ by solving Eqn. (4) with the method of trial and error. However, to get an even distribution of the pI values of the model proteins along the pH scale, the number of ε-amino groups in each model protein was not specified in advance, but was adjusted to the (generally non-integer) value that made the protein satisfy Eqn. (4) at a prechosen pI value (4.0, 4.5, 5.0, etc; cf. Fig. 1) at $4^{o}C$.

With a knowledge of the protolytic composition giving the model protein a specified pI value at $4^{o}C$, and the pK^{i}_{int} values at $25^{o}C$, it was finally possible to calculate from Eqn. (4) the corresponding pI value at $25^{o}C$.

To simplify the calculations, the α-carboxyl of each model protein was treated as a side-chain carboxyl. Since the pK_{int}, and probably also the ΔH^{o}, of an α-COOH are close to those of a β- or γ-COOH, this simplification will have negligible influence on the pI shifts obtained. For similar reasons, guanidyl groups were treated as ε-amino groups. Thiol groups, and in most cases also phenolic groups, were omitted in the model proteins.

The number of carboxyl groups per molecule of model protein was constantly chosen to be 20. Further, the N-terminal α-amino group was presupposed to be present in all the model proteins. From preliminary calculations, it was evident that the number of imidazole groups strongly influences the pI shifts for proteins being isoelectric in the neutral region. Therefore, two series of model proteins were studied: one having 20 carboxyls and 5 imidazoles, and one having 20 carboxyls but no imidazoles. In some calculations, two phenolic groups were added to the latter type of proteins.

The values of $pK^{i}_{int}(25^{o}C)$ and ΔH^{o}_{i} used in the calculations are given in Table 2. These data were chosen on the basis of those reported by Tanford (7) and Bull (20). Table 2 also contains

Table 2. Values of pK^i_{int} and ΔH^o_i Used in the Calculation of pI Shifts for Model Proteins

Ionizable group	$pK^i_{int}(25^oC)$	ΔH^o_i/kcal	$pK^i_{int}(4^oC)$*	ΔpK^i_{int}
Carboxyl	4.50	1	4.56	0.06
Imidazole	6.50	7	6.89	0.39
α-Amino	7.80	10	8.36	0.56
Phenolic	9.80	6	10.13	0.33
ε-Amino	10.00	12	10.67	0.67

* Calculated from the van't Hoff equation and the ΔH^o_i given.

the corresponding values of $pK^i_{int}(4^oC)$ and ΔpK^i_{int} as calculated from the van't Hoff equation.

RESULTS AND DISCUSSION

The results of the pI calculations are presented in Fig. 1. The abscissa represents the isoelectric point of the specified protein at 4^oC, whereas the ordinate represents the negative shift of this isoelectric point accompanying the increase of temperature from 4 to 25^oC. For comparison, the figure also contains the corresponding shifts of pH for Ampholine fractions being isoelectric at 4^oC.

To judge from the data of Fig. 1, it was a mere coincidence that the condition dpH/dT = dpI/dT mentioned above proved to be valid (within the experimental uncertainty) for the two horse myoglobins studied by Vesterberg and Svensson (8). For the great majority of proteins, one obviously should not expect to get accurate pI values if the focusing is performed at 4^oC, whereas the pH gradient is measured at 25^oC. As is evident from the figure, the difference between the apparent pI thus obtained and the true pI at 25°C may be as much as 0.2 pH units for certain weakly acid proteins. For strongly acid proteins, on the other hand, almost correct pI values will be obtained regardless of the focusing temperature owing to the small standard heats of ionization of carboxyl groups (≈ 1 kcal). Cf. the results for fungal laccase A mentioned in the introduction.

It is also evident from Fig. 1 that the number of titratable imidazole groups in a protein strongly influences the temperature dependence of its isoelectric point, especially if the latter at 4°C is situated in the region 6 - 7. As might have been anticipated, the effect on dpI/dT of adding the first imidazole to an imidazole-free protein was found to be larger than that of adding the second imidazole, and so on (cf. the points 1 - 3 in the figure). For proteins isoelectric above pH 8 at 4°C, dpI/dT is reduced by the presence of phenolic groups.

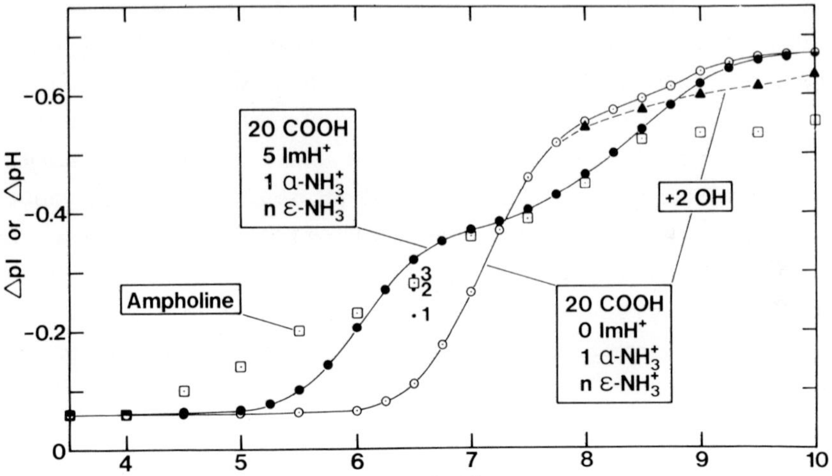

Fig. 1. The pI shifts (ΔpI) accompanying the increase of temperature from 4 to 25°C vs. pI at 4°C for a number of model proteins. The assumed protolytic compositions of the model proteins are given within the squares. The number (n) of ε-amino groups in each model protein was not specified in advance but was adjusted to the (generally non-integer) value that made the protein isoelectric at a prechosen pH (4.0, 4.5, etc.) at 4°C. The points numbered 1 - 3 refer to model proteins having pI = 6.5 at 4°C and containing 1, 2, and 3 imidazole groups per molecule, respectively. The figure also contains the pH shifts (ΔpH) accompanying the warming up from 4 to 25°C of Ampholine carrier ampholyte fractions being isoelectric at 4°C [calculated from data of Davies (14)]. For further explanation, see text.

The present calculations thus fully verify the apprehensions expressed by several authors (13, 21, 22, 9, 1) about the risk of artefacts when the temperatures of the focusing and the pH measurement do not coincide. An isoelectric point evaluated at a temperature different from the zone focusing temperature is per se ambiguous and lacks an exact physico-chemical significance. In routine work, where the species are well-known and absolute pI values are of minor interest, this will generally be unimportant since such a "mixed" pI may still be used for identification. On the other hand, a mixed pI should never be used as a measure of the relative acidity of a protein.

With regard to the above results, it is essential that a policy for evaluation and reporting of pI data is established as soon as possible. Since the pK_{int} values of protolytic groups given in the literature generally refer to 25 °C, it would be most practical if future pI values were evaluated at this temperature, too, rather than at 4 °C. However, as it is of primary importance that the proteins studied are not damaged by the focusing procedure, it seems more reasonable that workers within the field of isoelectric focusing agree to use 4 °C as the common temperature for focusing and pH measurement. Of course, it would also be conceivable to compromise by letting the main focusing at 4 °C be followed by a final focusing at 25 °C. However, besides being time-consuming, this technique may cause problems at the fractionation especially when microcolumns are used, due to the increased diffusion and the decreased viscosity at 25 °C as compared with 4 °C.

It is thus recommendable that the pI values henceforth reported refer to 4 °C. To circumvent the disadvantage of mapping the pH course at this temperature, it would be desirable to accurately determine the shifts of pH which accompany the warming up to 25 °C of different Ampholine fractions being focused at 4 °C. If expressed as a function of the resulting pH values at 25 °C, these shifts would enable the apparent pI value of a protein at 25°C to be transformed into a true value at 4°C.

In principle, the desired pH data are already obtainable from the paper of Davies (14; cf. Fig. 1). However, to avoid new artefacts, it is necessary to investigate carefully the repeatability of the pH shifts for different Ampholine batches. It is also important to get a knowledge of the influence of added sucrose.

An additional source of error that deserves to be considered at evaluation of pI data is the difference between the cooling-water temperature and the temperature in the focused protein zone. In most reports so far published, this difference has been neglected. This is probably warranted for the 110-ml LKB column as long as the wattage is kept below the recommended value of 0.01 W/ml. At higher wattages or in apparatuses with less efficient cooling, however, the zone temperature may rise several degrees above that of the coolant. [Cf., for instance, the papers of Righetti and Righetti (23), and of Söderholm and Wadström (24).] As the average value of dpH/dT is about -0.015 pH units per degree (cf. Fig. 1), it is obvious that a pI value evaluated from such a run may be seriously incorrect. To get as accurate pI values as possible, it is thus important to keep the wattage at the end of a run as low as is reasonable with respect to the resolution. In addition, some knowledge of the temperature distribution in the actual density-gradient column or gel plate for different wattages and pH intervals would be desirable.

ACKNOWLEDGEMENTS

I am indebted to Professor Harry Rilbe for helpful criticism during the preparation of this paper. My thanks are also due to Dr. Mats Jonsson for valuable discussions.

REFERENCES

1. Righetti, P.G., Caravaggio, T. J. Chromatogr. 127, 1 (1976).
2. Vesterberg, O. Acta Chem. Scand. 21, 206 (1967).
3. Beeley, J.G. Biochim. Biophys. Acta 230, 595 (1971).
4. Ui, N. Biochim. Biophys. Acta 229, 567 (1971).
5. Ui, N. Ann. N.Y. Acad. Sci. 209, 198 (1973).
6. Jonsson, M. Acta Chem. Scand. 26, 3435 (1972).
7. Tanford, C. Advan. Protein Chem. 17, 69 (1962).
8. Vesterberg, O., Svensson, H. Acta Chem. Scand. 20, 820 (1966).
9. Fredriksson, S. Isoelectric Focusing in Small Density-Gradient Columns. Thesis, Chalmers Institute of Technology, Gothenburg 1975.
10. Fredriksson, S. Anal. Biochem. 50, 575 (1972).
11. Jonsson, M., Pettersson, E. Personal communication.
12. Svensson, H. Acta Chem. Scand. 16, 456 (1962).
13. Vesterberg, O. Ann. N.Y. Acad. Sci. 209, 23 (1973).
14. Davies, H. Protides of the Biological Fluids 17, 389 (1970).
15. Basch, J.J., Timasheff, S.N. Arch. Biochem. Biophys. 118, 37 (1967).
16. Tanford, C., Swanson, S.A., Shore, W.S. J. Am. Chem. Soc. 77, 6414 (1955).
17. Phelps, C., Forlani, L., Antonini, E. Biochem. J. 124, 605 (1971).
18. Shire, S.J., Hanania, G.I.H., Gurd, F.R.N. Biochem. 13, 2967 (1974).
19. Tanford, C., Hauenstein, J.D. J. Am. Chem. Soc. 78, 5287 (1956).
20. Bull, H.B. An Introduction to Physical Biochemistry, F.A. Davis Co., Philadelphia 1971, p. 127.
21. Beeley, J.A., Stevenson, S.M., Beeley, J.G. Biochim. Biophys. Acta 285, 293 (1972).
22. Bours, J. Science Tools 20, 29 (1973).

23. Righetti, P.G., Righetti, A.B.B. Isoelectric Focusing (J.P. Arbuthnott and J.A. Beeley, Eds.), Butterworths, London 1975, p. 114.
24. Söderholm, J., Wadström, T. Ibid., p. 132.

2. Carrier Ampholytes

Carrier Ampholyte Distribution

R.K. Brown, M.L. Caspers and S.N. Vinogradow

Cation exchange chromatography resolves 12 to 15 ninhydrin positive components from each 1-2 mm slice of polyacrylamide gel after focusing. Each component is distributed over several slices and in some cases is non-Gaussian. This was confirmed by focusing radiolabeled histidine, leucine, or glycine. Although small amounts of histidine have a nearly Gaussian distribution large amounts are quite assymetric as are even small loads of glycine and leucine. When large amounts of histidine or leucine are focused, their concentration rises rapidly, remains almost constant over a considerable distance and then falls sharply. Increasing the duration of focusing or the potential gradient has little effect on the pattern.

Adding amphoteric substances (separators) modifies the pH gradient and provides a simple, controllable means of improving the separation during focusing. These findings have been applied to the separation of human hemoglobin subtypes, barley beta amylase, goat antibodies and to human gamma globulin. Factors affecting the shape of the pH gradient include the nature and concentration of the specific separator added, the amount of ampholyte used and its pH range.

INTRODUCTION

Carrier ampholytes are complex mixtures of undefined composition. Large numbers of components can be identified by various staining procedures [1-3], by scanning photographs of thin-layer polyacrylamide gels [4] and by cation exchange chromatography followed by automated ninhydrin analysis [5]. Many components have a sharp narrow band while others are spread over a considerable distance.

The theoretical treatment of isoelectric focusing predicts a Gaussian distribution for individual components in model systems [6]. In practice the unequal buffering capacity of individual components near their isoelectric point, the non-constancy of the conductivity [4], the departure of the system from ideality and other factors can be expected to alter the distribution. Catsimpoolas and his coworkers have examined the distribution of histidyl-tyrosine during isoelectric focusing and found it to be nearly Gaussian for loads in the 10 to 66 μg range [7]. We have examined the distribution of low molecular weight amphoteric substances, ampholytes and amino acids, using cation exchange chromatography followed by automated ninhydrin analysis. Radiolabeled amino acids have also been used as models. In many cases the distribution was non-Gaussian.

Although much effort has gone into the development of linear pH gradients, non-linear gradients often can provide a better means of altering the gradient in a controlled manner. The modified gradients have been used for improved separation of a variety of proteins. Although the pH gradients are shallow the separated bands are quite discrete.

MATERIALS AND METHODS

Gels were sliced by hand within 6-8 minutes after the current was turned off. Unless otherwise noted ampholyte synthesized from pentaethylenehexamine (PEHA) was used [8]. Automated ninhydrin analyses were performed as described previously [5]. In one set of experiments, proline (6 mg/ml of gel solution) was added prior to polymerization. After focusing, the gel was sliced into 34 slices, eluted in water and the pH profile determined. Eluate from slices 11 (pH 6.35), 13 (pH 6.79) and 15 (pH 6.88) was analyzed. ^{14}C-U-leucine (Lot 949 072) and ^{3}H-2-glycine (Lot 929 042) obtained from New England Nuclear had the specific activities of 302 mCi/mmole for leucine and 11.4 Ci/mmole for glycine. Gels containing these substances were focused, sliced and each slice was weighed prior to treatment with Protosol:water (95%:5%). The slices were counted in a Packard Model 3375 liquid scintillation spectrometer. The pH gradients in gels were determined by slicing the gel and soaking each slice in 0.5 or 1.0 ml of boiled water. Human hemoglobin was prepared from freshly drawn blood and the A_2 component was chromatographed on DEAE Sephadex A50 [9]. Hemoglobin F was obtained from a sample of cord blood. The human gamma globulin was a gift from Dr. George R. Anderson, the barley β-amylase was Lot 106 401 from Calbiochem. Polyacrylamide gel isoelectric focusing was done as previously [10].

EXPERIMENTAL

Our previous studies [10] show that there are a minimum of 12-15 ampholyte species in a 1 mm slice of a polyacrylamide gel. Time studies in which the gel was frozen in liquid nitrogen 45 sec, 2 min, or 5 min after the current was turned off indicate that the large number of components is not due to diffusion from adjacent areas. Automated analysis of eluate from 6 successive slices of a gel shows that the number of components results in large part from the rather broad, 3-6 mm, distribution of individual species.

In the experiment reported here proline was added to the ampholyte (Fig. 1). Proline is easily differentiated from ampholytes by the yellow color it gives in the ninhydrin reaction. In all the slices represent 1 cm of gel length with 2 mm gaps between the slices analyzed. There are a large number of components in each of the slices and many are found in 2 or even 3 of the slices. The species appear as nearly focused as they do in the absence of added proline. Those slices nearer the cathode contain, as expected, more of the basic ampholyte components. The proline concentration is almost constant in each of the slices although the pH of the slices

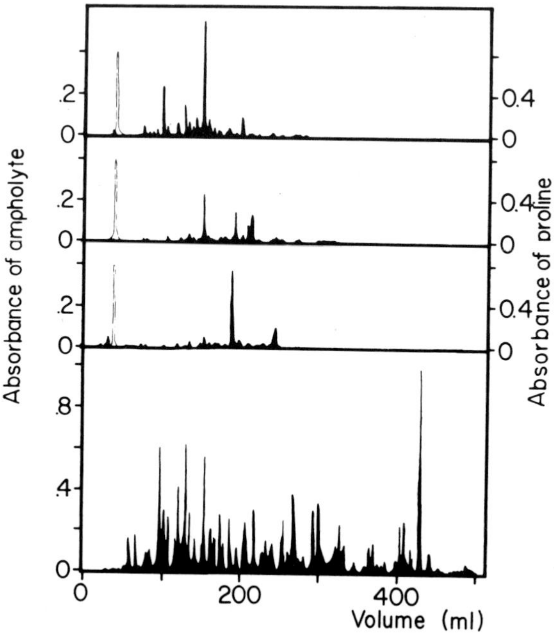

Fig. 1

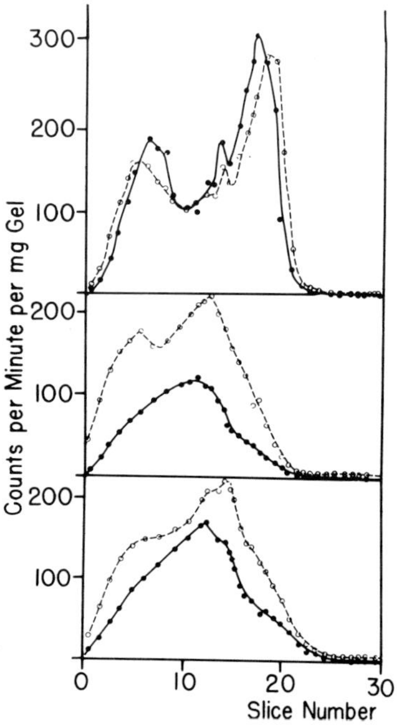

Fig. 2

Fig. 1. The distribution of proline and PEHA ampholyte during isoelectric focusing. Slices 11 (top), 13 (next) and 15 (next to bottom) were analyzed as was the PEHA ampholyte (bottom) used. The proline is the larger open peak near the left.

Fig. 2. Distribution of the ^{14}C-leucine (o—o) and ^{3}H-glycine (●—●) focused for various periods, top 5 hr, middle 20 hr, bottom 30 hr.

ranges from 6.35 to 6.88. The pI of proline is 6.30. Its distribution along the gel is clearly non-Gaussian. It overlaps and engulfs the ampholytes rather than focusing between them.

Studies of the Distribution of Labeled Amino Acids in Isoelectric Focusing.

Histidine was chosen because its pK_2 and pK_3 are relatively close, 1.6 pH units, to its isoelectric point. Small amounts of histidine focus symmetrically but at larger concentrations the pattern becomes assymetric and at 0.6% the pattern is very non-Gaussian, rising sharply remaining level for an appreciable distance and then falling sharply. This band is 23.5 cm wide. This distribution is not appreciably altered between 18 and 42 hr nor is the sharpness improved by increasing the potential gradient from 20 volts/cm to 100 volts/cm. The concentration of ampholyte does have some modest effect upon the distribution. At high concentrations (4%) the distribution is narrower and more symmetric than at low (0.2%) concentrations [10].

These studies were continued using ^{14}C-leucine and ^{3}H-glycine. They were chosen because their mobilities (gly 0.97 and leu 0.67 at pH 1.5) are quite different [11] while their isoelectric points are nearly identical (gly 5.97 and leu 5.98). Both are distributed over a wide portion of the gel and have a non-Gaussian distribution although only small amounts (.18 μg of leu and 2.4 ng of gly) were used (Fig. 2). The major peak lies somewhat alkaline to the pI of each amino acid. The distributions change only slightly between 20 and 30 hr of focusing. When the potential gradient is doubled the amino acid distributions are not narrowed and remain quite non-Gaussian. When unlabeled leucine (10 mg/ml) is added to the system the concentration rises rapidly, remains constant for a considerable distance, then falls fairly quickly.

pH Gradients of Gels Containing Separators

The pH gradient is often flattened in gels which contain added amphoteric substances. The isoelectric point of the separator is probably the most important variable. In some cases buffer ions which have a pK within the range of the ampholyte also affect the separation. In general the nature and concentration of the separator and character and amount of carrier ampholyte all influence the pH gradient. By proper manipulation of these variables the pH gradient can be altered controllably. In some cases the flat portion of the gradient is near the isoelectric point of the separator. For example, histidine (pI=7.59) and threonine (pI=6.53) affect the pH gradient near their pI (Fig. 3). β-alanine (pI=6.90) flattens the gradient at about pH 6.80. On the other hand, tetraglycine (pI=5.20) gives a flattening at pH 5.90; 5-aminovaleric acid (pI=7.51) is good for separations near pH 7.0. The pI is concentration dependent but in the case of 5-aminovaleric acid this factor is not the major one affecting the pH gradient. Titration experiments yielded a pI of 7.58 for a 0.4 M solution at 23° which decreased to 7.43 for a .005 M solution.

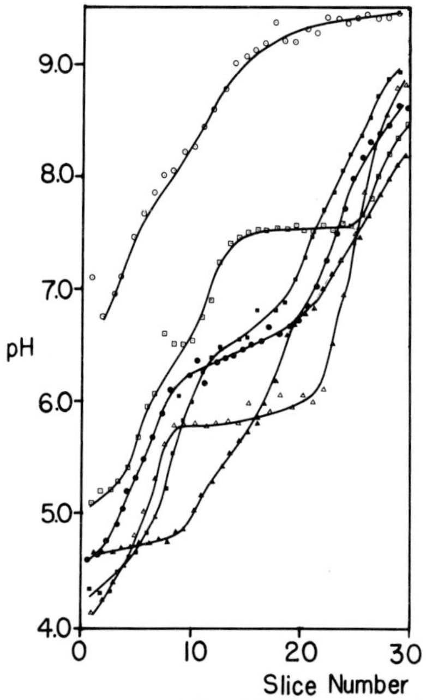

Fig. 3

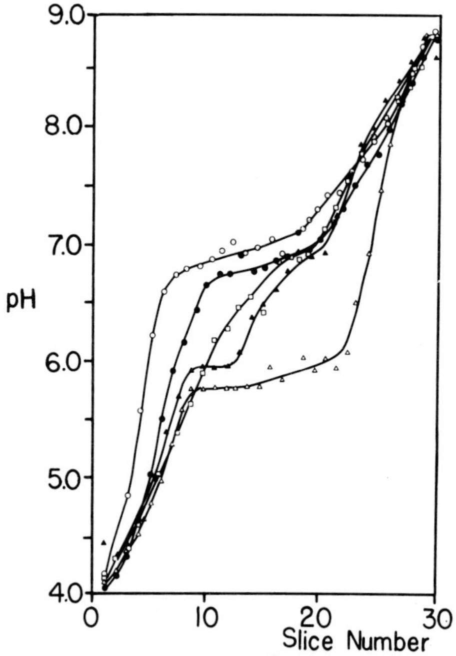

Fig. 4

Fig. 3. pH gradients obtained in the presence of arginine (o—o), 12.0 mg/ml or threonine (●—●), 7.0 mg/ml each in LKB Lot 29, pH 3.5-10 Ampholine, histidine, 6.1 mg/ml in LKB Lot 17, pH 3.5-10 Ampholine (□—□), tetraglycine, 15.0 mg/ml in PEHA ampholyte (△—△), or glutamic acid, 8 mg/ml in PEHA (▲—▲), Control (■—■) in PEHA ampholyte.

Fig. 4. Effect of concentration of 5-aminovaleric acid (5 mg/ml ●—● and 20 mg/ml O—O) and tetraglycine (15 mg/ml ▲—▲ and 30 mg/ml △—△ on the pH gradient. PEHA ampholyte (■—■) was used in all tubes. A band of cloudiness developed in the tube containing 30 mg/ml of tetraglycine during focusing. It seems probable that this compound reached its solubility limit under the conditions of the experiment.

The molarity of 5-aminovaleric acid after focusing is probably of the order of 0.4 M while the major flat portion of the curve lies at pH 7.0. Certain separators, such as taurine and glycine give no appreciable flattening of the curve. Nevertheless, they accentuate protein separations in many cases.

The concentration of separator is the second factor influencing the pH gradient (Fig. 4). Increasing the amount of separator increases the width of the flat portion of the pH gradient. In general, flattening of the gradients has been obtained with concentration of 2 to 20 mg/ml. Smaller amounts have little effect while larger amounts generally produce a long flat gradient with limited utility. The concentration of added substance also has a modest effect upon the pH of the flat region. The pH of the flat portion increases from about 6.7 for 5 mg of 5-aminocaproic acid to about 7.1 at a concentration of 40 mg/ml.

Separation of Proteins in Non-linear pH Gradients

Adding separator is an easy method of improving the separation of proteins during isoelectric focusing [12]. A sample containing hemoglobin F and one of hemoglobin A_2 prepared by chromatography are well separated from other hemoglobin components when focused in a gel containing 10 mg/ml of 5-aminovaleric acid (Fig. 5). That hemoglobin A_2 runs as a single component indicates that the additional bands are not arti-

+

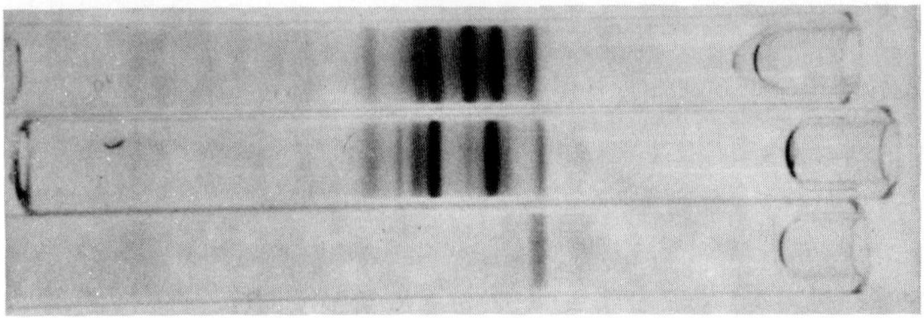

A

B

C

Fig. 5. Isoelectric focusing of hemolysates from human adult (a), cord blood (b) and hemoglobin A_2 (c) in gels containing 10 mg/ml of 5-aminovaleric acid.

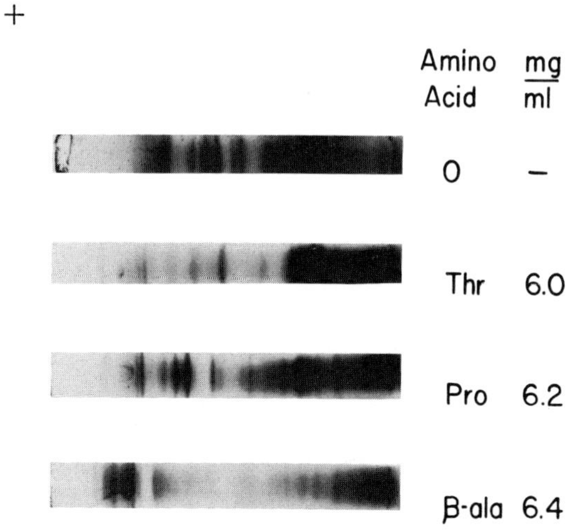

Fig. 6

Fig. 7

Fig. 6. Effect of amino acids on the focusing of goat antibody to oxidized ribonuclease.

Fig. 7. Isoelectric focusing of barley β-amylase in the presence and absence of 10 mg/ml of threonine and of human gamma globulin in the presence of 1 mg/ml (a) or 10 mg/ml (b) of 8-aminocaprylic acid or in its absence.

facts produced by the focusing. The major individual components isolated after focusing occur as single components with the expected pI upon refocusing. In a complex mixture such as goat antibody to oxidized ribonuclease different separators resolve different components (Fig. 6). Even mixtures as complex as human gamma globulin may be partially separated; threonine improves the resolution of the isozymes of barley β-amylase (Fig. 7).

DISCUSSION

The distribution of many of the ampholytes and amino acids studied is clearly non-Gaussian. There are many possible reasons, among them the differences in buffering capacity of the components, conductivity changes and non-ideality of the system. It is possible that the ampholyte components visualized by the ninhydrin reaction are atypical. Ninhydrin presumably reacts with only a portion of the ampholyte components. Amino acids such as proline, leucine and glycine are very poor buffers at their pI and accordingly would be expected to be widely distributed. Histidine, a better buffer, is more sharply distributed at low concentrations. At higher concentrations the effect is less clear. A 0.039 molar solution of histidine in 2% ampholine was distributed over 2.35 cm of a 6.5 cm gel while a 0.076 M solution of leucine extended 4.65 cm under similar conditions. It is also difficult to define a steady state in a system that is continually changing. As major criteria we have used invarience of distribution with time or potential gradient. Because of the small change in mobility with pH found with amphoteric substances which have pK's distant from their pI it is possible that the amino acids, glycine, leucine and proline, are not focused while the ampholytes which have a larger $\delta\mu/\delta$ pH are.

Controlled modification of the pH gradient in isoelectric focusing provides a means of improving protein separations. The isoelectric point, buffering capacity at the isoelectric point, and the molecular weight of the separator are among the properties which influence the pH gradient. Amphoteric substances which are good buffers at their pI and individual ampholyte species should be especially useful. The concentration of separator also plays an important role. Large concentrations give a shallow gradient over a long distance. Solubility may be an additional factor. Effective concentrations of separator are often nearly saturated solutions.

Ampholytes often display an anomolous behavior in the pH range between 6.5 and 7.0. A plateau develops in this area especially as the focusing time increases. Many separators flatten the pH profile in this pH range, especially those which have a low buffering capacity at their pI. Those with pI alkaline to this pH shift the plateau toward the alkaline side and those with acid pI shift it toward acidic pH's.

REFERENCES

1. Frater, R.: Behavior of Ampholines during Isoelectric Focusing. Anal. Biochem. 38, 536-538 (1970).

2. Felgenhauer, K., Pak, S. J.: Detection of Ampholine Patterns. Ann. N. Y. Acad. Sci. 209, 147-153 (1973).

3. Radola, B. J., Tschesche, H., Schuricht, H.: Characterization of Carrier Ampholytes by New Visualization Reactions and Chromatographic Methods. This volume.

4. Righetti, P. G., Pagani, M., Gianazza, E.: Characterization of Synthetic Carrier Ampholytes for Isoelectric Focusing. J. Chromatog. 109, 341-356 (1975).

5. Brown, R. K., Lull, J. M., Lowenkron, S., Bagshaw, J. C., Vinogradov, S. N.: Analysis of Carrier Ampholytes by Ion Exchange Chromatography. Anal. Biochem. 71, 325-332 (1976).

6. Svensson, H.: Isoelectric Fractionation, Analysis, and Characterization of Ampholytes in Natural pH Gradients. Acta Chem. Scand. 15, 325-341 (1961).

7. Catsimpoolas, N., Yotis, W. W., Griffith, A. L., Rodbard, D.: Transient State Isoelectric Focusing. Arch. Biochem. Biophys. 163, 113-121 (1974).

8. Vinogradov, S. N., Lowenkron, S., Andonian, M. R., Bagshaw, J., Felgenhauer, K. and Pak, S. J.: Synthetic Ampholytes for the Isoelectric Focusing of Proteins. Biochem. Biophys. Res. Comm. 54, 501-506 (1973).

9. Williams, R. C., Jr., Tsay, K-Y.: A Convenient Chromatographic Method for the Preparation of Human Hemoglobin. Anal. Biochem.54, 137-145 (1973).

10. Brown, R. K., Caspers, M. L., Lull, J. M., Vinogradov, S. N., Felgenhauer, J., Nekic, M.: Carrier Ampholyte Distribution in Isoelectric Focusing. J. Chromatography in press.

11. Edward, J. T., Waldron-Edward, D.: Mobilities of Amino Acids and Peptides in Paper Electrophoresis. J. Chromatography 24, 125-130 (1966).

12. Caspers, M. L., Posey, Y., Brown, R. K.: Separator Isoelectrofocusing - An Improved Method of Protein Analysis and Purification. Anal. Biochem., submitted.

Characterization of Carrier Ampholytes by New Visualization Reactions and Chromatographic Methods

B.J. Radola, H. Tschesche and H. Schuricht

ABSTRACT

Following thin-layer isoelectric focusing in Sephadex G-75 carrier ampholytes were visualized with formaldehyde, lactose or ninhydrin. In pH 2 - 11 "Servalyt" carrier ampholytes 80 - 100 visible and fluorescent species were detected. Several hundred species were visualized by analyzing commercially available narrow range carrier ampholytes and 0.5 pH cuts obtained by preparative isoelectric focusing. On ion exchange chromatography of the 0.5 pH cuts highly complex patterns were observed while only one or a few components were found when carrier ampholytes with a defined pH-value were analyzed. Thin-layer gel chromatography was employed to study the molecular weight distribution of carrier ampholytes. Neutral and alkaline narrow range "Ampholine" contained a substantial portion of high molecular weight ampholytes while a lower molecular weight and more uniform distribution were characteristic of the corresponding ranges of "Servalyt" carrier ampholytes.

INTRODUCTION

It is surprising that carrier ampholytes for isoelectric focusing have remained insufficiently characterized for many years. Hundreds of isomers and homologues are assumed to be present in carrier ampholytes prepared by current synthetic procedures (1,2) and a large number of ampholyte species has been found to be essential for obtaining an adequately stabilized pH gradient (3). However, these hundreds of postulated ampholyte species have never been demonstrated. A variety of approaches has been used to study the composition and distribution of carrier ampholytes, including physical e.g. refractive index variation (4-6) and chemical methods e.g. staining and glucose caramelization (7-9). Ion exchange chromatography has been recently found to offer a high potential for a study of the composition of carrier ampholytes (10,11). Judged by the number of detected ampholyte species most of the methods so far employed offer rather poor resolution even when narrow range carrier ampholytes are investigated. Some 40 - 80 components have been detected in wide range carrier ampholytes, and the discrepancy of this figure with the hundreds of postulated species is obvious.

In our approach to the characterization of carrier ampholytes we used simultaneously several visualization reactions. Visualization with formaldehyde, lactose, and ninhydrin proved to be particularly useful. In addition, ion-exchange chromatography and thin-layer gel chromatography have been applied. Recent findings of artefactual patterns in analytical isoelectric focusing (12) and difficulties in separating carrier ampholytes from the focused proteins (13), which could impair preparative applications of the method, stress that it is important to characterize carrier ampholytes with respect to an irreversible interaction with dyes and molecular weight distribution.

MATERIAL AND METHODS

"Servalyt" carrier ampholytes were obtained from Serva, Heidelberg, F.R.G., while "Ampholine" was from LKB, Bromma, Sweden. Sephadex G-75 "Superfine" came from Pharmacia, Uppsala, Sweden, and Bio-Gel P-4 and P-6, both "minus 400 mesh" from Bio-Rad Laboratories, Richmond, California.

Visualization: The print technique was applied following thin-layer isoelectric focusing in Sephadex G-75 as described earlier (9). The chromatographic paper Schleicher & Schüll 2043 b Mgl (Dassel, F.R.G.) was employed throughout. Formaldehyde visualization: The paper for printing was first impregnated with an aqueous solution of 0,4 % formaldehyde and 5 % glycerol, and subsequently dried at 60°C for 30 minutes. After focusing the paper was rolled on to the gel layer, removed after contact for 1-2 minutes, and finally dried in an oven at 115-118°C for 10 minutes. Lactose visualization: A 6 % solution of lactose containing 0.3 M citrate buffer pH 2.5 and 10 % glycerol was used to impregnate the paper which was dried at room temperature and could be stored indefinitely. The lactose-impregnated paper was kept in contact with the focusing gel for 1-2 minutes, and dried at 115°C for 5-10 minutes. Ninhydrin visualization: An unbuffered paper was impregnated with a methanolic solution containing 0.2 % ninhydrin, 5 % glycerol and 5 % collidine. After drying at room temperature for no more than 20 minutes the paper was used for printing. Optimum conditions for colour development were 60°C for 15-20 minutes. Alternatively, a paper buffered with 0.3 M citrate buffer, pH 5.5, was dipped into a ninhydrin solution from which collidine was omitted.

Preparative isoelectric focusing, at a 5-10 % concentration of carrier ampholytes, was performed as described previously (14). Narrow cuts of 0.5 - 1 pH were isolated, and used either directly for analytical refocusing in visualization experiments or, after elution with water, in ion exchange chromatography (15). In thin-layer gel chromatography (16) on Bio-Gel P-4 and P-6 it was necessary to use 1 mm gel layers to provide adequate amounts of interstitial liquid for printing.

RESULTS AND DISCUSSION

Visualization: On formaldehyde visualization a yellow-brownish colour developed on heating, the intensity and colour depending both on temperature and heating time. In addition to the visible zones many fluorescent zones could be seen when viewed under UV at 366 nm, the fluorescence being quenched with increasing temperature and heating time. Formaldehyde visualization affords the advantage that the print did not change on prolonged storage and that colour development was only little influenced by pH. Lactose visualization has proved superior to the previously described glucose caramelization (8). Also with this visualization reaction, in addition to the visible zones, fluorescent zones could be seen in UV at 366 nm. Lactose visualization depends strongly on pH and tends to give a heavy colour particularly in the alkaline pH range, an effect which is only partly compensated for by buffering the paper for printing. For the majority of visualized zones the patterns were very similar on formaldehyde and lactose visualization but some distinct differences were noted both for the visible and fluorescent zones. Ninhydrin visualization supplements the formaldehyde and lactose reaction by detecting many species which are not revealed by the latter reagents. A minor drawback of ninhydrin visualization is colour instability, evaluation of the ninhydrin prints must therefore be carried out briefly after printing.

Figure 1 demonstrates the results obtained when different visualization reactions were applied simultaneously to detect the patterns of TETA carrier ampholytes (prepared from triethylene tetramine and kindly supplied by Dr. Brown, Wayne State University). These carrier ampholytes contained only relatively few species some of which reacted preferentially with only one reagent while other species gave a reaction with all visualization reagents. Some of the TETA ampholytes occurred as rather broad zones of a type never observed in

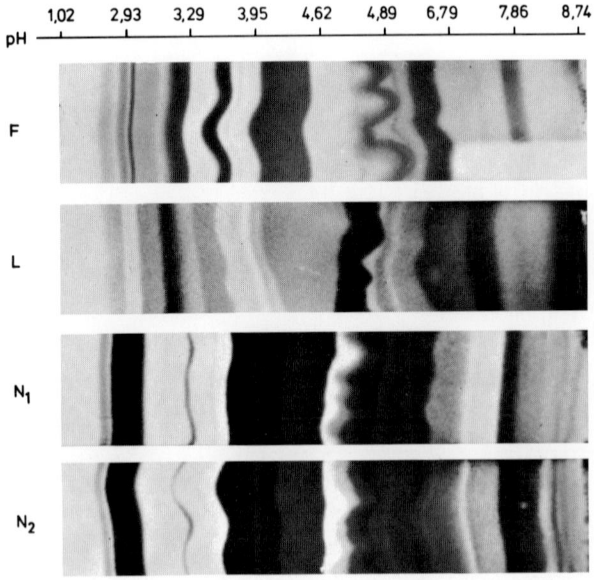

Fig. 1. Visualization of TETA carrier ampholytes after thin-layer isoelectric focusing in Sephadex G-75. Plate: 20 x 20 cm. Ampholyte concentration - 1.5 %. F - formaldehyde, L - lactose, N_1 - ninhydrin with collidine, N_2 - ninhydrin pH 5.5 buffered paper. pH values indicated for 2 cm intervals.

commercial carrier ampholytes. The pH gradient of the TETA ampholytes covering the pH range 3 - 9 was non-linear, and poor results were obtained on thin-layer isoelectric focusing of pH marker proteins in compact polyacrylamide and in granulated gels.

In commercial carrier ampholytes a large number of ampholyte species has been consistently detected as shown in Figure 2 for the pH 2 - 11 "Servalytes". The visualization patterns of commercial carrier ampholytes is composed of relatively narrow zones arranged characteristically depending on their pH range, origin and batch. The visualized ampholyte species are distributed virtually over the whole pH gradient, and there were no gaps between the zones nor broad zones of the type found in the TETA ampholytes.

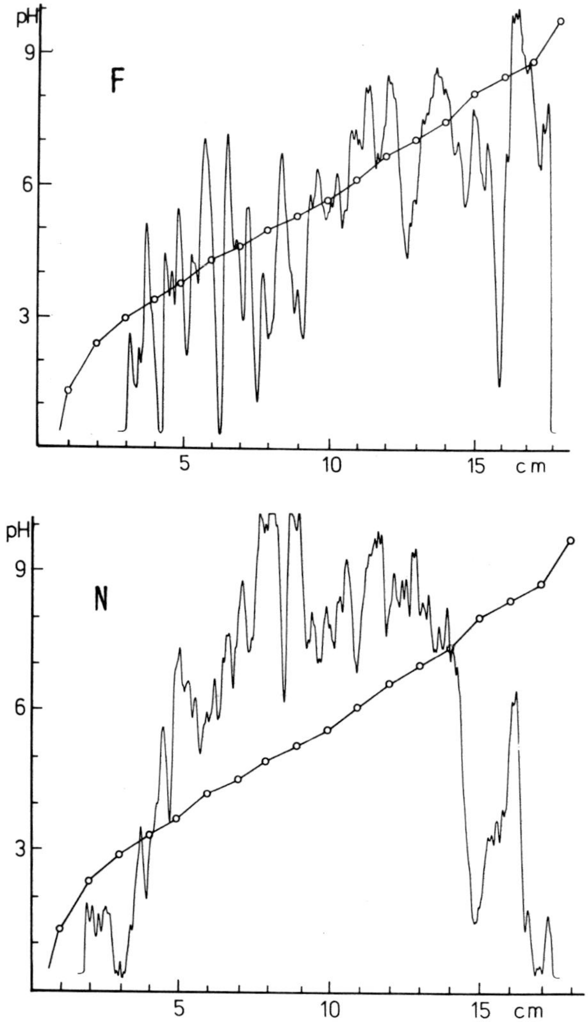

Fig. 2. Densitograms of pH 2-11 "Servalyt" carrier ampholytes after thin-layer isoelectric focusing. Visualization: F - formaldehyde, densitometry at 435 nm, N - ninhydrin, densitometry at 619 nm.-o-o-o- pH gradient determined at 25°C.

Considering the different response, visible and fluorescent, to the various reagents, it must be concluded that the individual zones detected by a single visualization reaction actually correspond to clusters of ampholyte species. A total of 80 - 100 ampholyte species could be detected in pH 2-11 "Servalyt" carrier ampholytes by summation of all visible and UV detectable zones. The same number of species was revealed in pH 3.5 - 10 "Ampholine" carrier ampholytes.

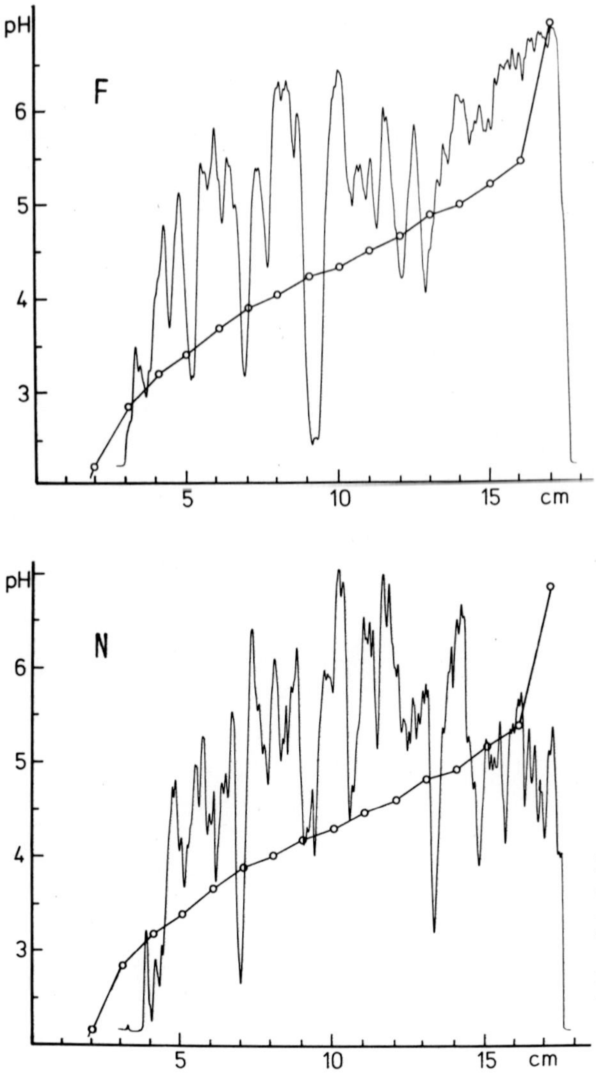

Fig. 3. Densitograms of pH 3 - 5 "Servalyt" carrier ampholytes. Visualization: F - formaldehyde, N - ninhydrin.
For details see Fig.2.

Further analysis of narrow range carrier ampholytes, namely commercially available 2 pH ranges and 0.5 - 1 pH cuts obtained by preparative isoelectric focusing, has shown that the actual number of species is much higher. Visualization of pH 3 - 5 "Servalyt" carrier ampholytes revealed in the nominal range some 40 - 50 species (Fig. 3). Here again it must be assumed that some of the individual zones represent clusters which could be further resolved by an analysis of the 0.5 - 1 pH cuts. By this approach we arrived at approximately 150 major and minor components in the pH 3 - 5 range. While satisfactory results were obtained by the visualization reactions in the acidic and neutral pH range, the ampholyte species of the alkaline range could be only poorly differentiated. Particularly when studying narrow pH ranges we observed instead of well defined zones a rather uniform background staining with only a few broad zones. The composition and distribution of carrier ampholytes in the alkaline pH range could be studied more efficiently by ion exchange chromatography.

Ion exchange chromatography: Analysis of wide range pH 2 - 11 "Servalyt" carrier ampholytes, using a single column system with automated gradient mixing (15), revealed some 50 - 80 peaks which is in good agreement with what has been recently demonstrated (10). When fifteen 0.5 pH cuts, isolated by preparative isoelectric focusing from pH 2 - 11 "Servalyt" carrier ampholytes, were analyzed, it was found that with increasing pH value of the cuts the peaks of the elution patterns shifted towards more alkaline ranges of the eluting buffer. Surprisingly, most of the cuts showed highly complex elution patterns with considerable overlapping in terms of the retention values for the adjacent cuts. In ion exchange chromatography of carrier ampholytes identical retention values do not necessarily prove identity of the components from different pH ranges because in addition to the charge properties other factors appear to effect the chromatographic behaviour of carrier ampholytes. The elution profiles of pH 8.5 - 9 range carrier ampholytes in Figure 4 demonstrate that the ion exchange chromatography system is capable of very high resolution. When instead of the system optimal for amino acid analysis a shallower elution gradient was employed some 40 - 50 major and minor components were detected in the 0.5 pH range.

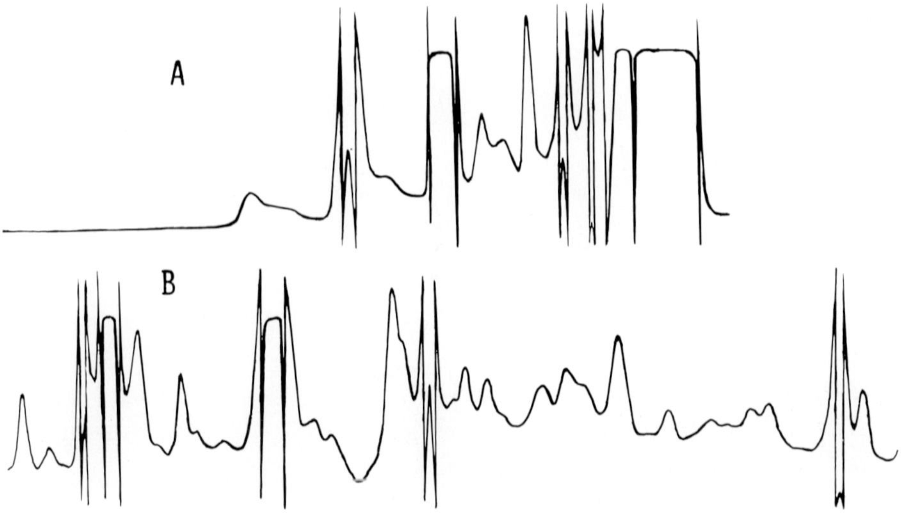

Fig. 4. Ion exchange chromatography of pH 8.5 - 9.0 carrier ampholytes isolated by preparative isoelectric focusing from pH 2 - 11 "Servalyt" ampholytes (see text). A - automated buffer gradient mixing for amino acid analysis, starting at pH 2.58; B - modified buffer gradient starting at pH 4.0. Ninhydrin colour evaluated at 570 nm, linear range expansion for the major peaks.

The complexity of the chromatographic patterns was drastically reduced when instead of the 0.5 pH cuts, carrier ampholytes of a defined pH value were analyzed (Fig. 5). These defined pH value ampholytes were isolated by preparative isoelectric focusing in experiments in which two pH unit carrier ampholytes at a 5 - 10 % concentration were focused on a 40 cm separation distance. After determination of the pH gradient samples with defined pH values, differing by only 0.02 - 0.04 pH, were obtained as 2-3 mm broad strips from the gel layer. Only samples spaced at least 10 mm were selected and a 5 - 7 mm gel layer between the samples was discarded to avoid contamination. Some of the defined pH value ampholytes gave a single peak in chromatography comprising up to 98 % of the eluted ninhydrin-reacting material.

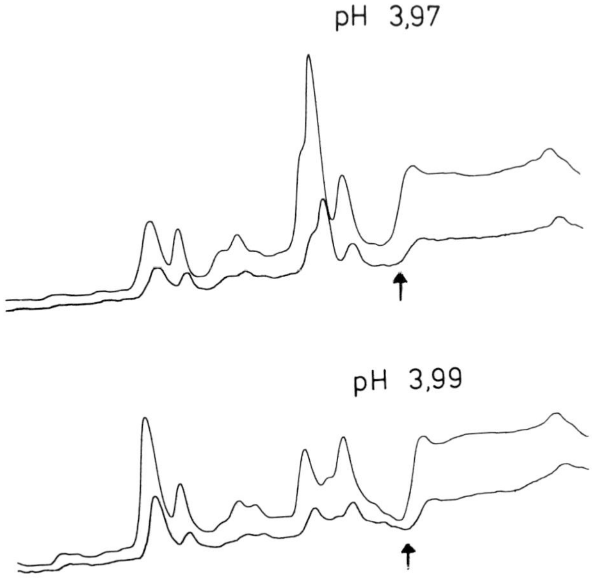

Fig. 5. Ion exchange chromatography of carrier ampholytes with defined pH values, isolated by preparative isoelectric focusing from pH 3 - 5 "Servalyt" ampholytes (see text). Buffer gradient for amino acid analysis. Upper curves - ninhydrin colour at 570 nm, lower curves at 440 nm, ↑ NH_3-plateau.

Other defined pH value ampholytes gave elution profiles consisting of only a few peaks. The excellent resolution obtainable by the system is illustrated by the finding that distinct quantitative changes in the elution profile were detected in ampholytes differing by only 0.02 pH.

Thin-layer gel chromatography: A molecular weight in the range of several hundreds daltons has been claimed for commercial carrier ampholytes but increasing evidence accumulates indicating that an undefined portion of carrier ampholytes with a molecular weight up to 20000 daltons may be present in these preparations (13). High molecular weight carrier ampholytes are undesirable because in preparative work they connot be separated from the focused proteins; in analytical work they could result in artefactual patterns owing to an irreversible interaction with some protein dyes (12).

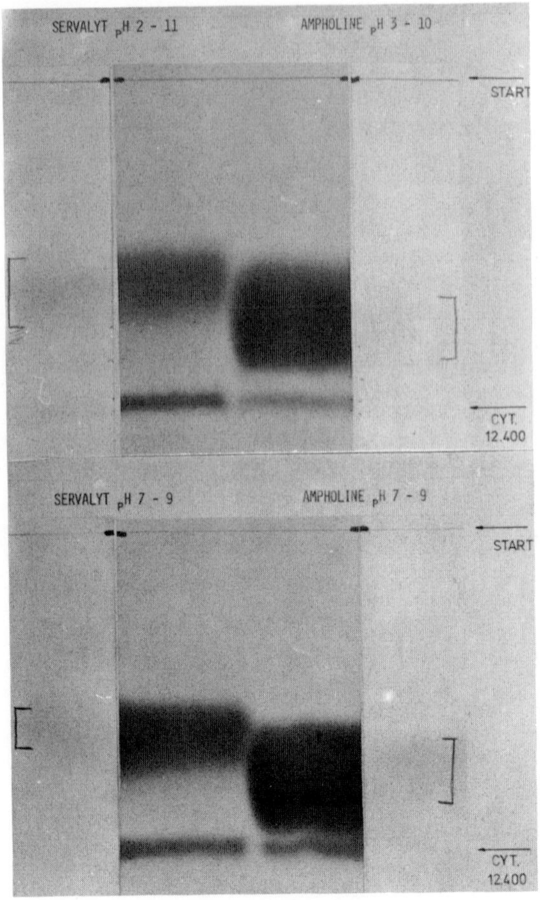

Fig. 6. Thin-layer gel-chromatography on Bio-Gel P-4of 5 % "Servalyt" and"Ampholine" carrier ampholytes with 1 % added cytochrome c. In the middle staining with Amido Black 10B, at both edges visualization with lactose. Maximum visible and fluorescent colour with lactose indicated by the vertical bars.

Thin-layer gel chromatography on Bio-Gel P-4 or P-6 provides a very simple approach to study the molecular weight distribution in commercial and laboratory-made carrier ampholytes. Staining with Amido Black 10 B has proved to be most suitable for detecting high molecular weight ampholytes in thin-layer gel chromatography because the dye-ampholyte complex does not dissociate easily. Staining with Amido Black depends strongly on the chemistry of the carrier ampholytes, and on an equal weight basis "Servalyt" ampholytes were found to take up much less of the dye than "Ampholine". Cytochrome c (mol.weight 12400 daltons) was employed as an internal marker in each sample to facilitate comparison and to detect viscosity effects. It can be seen from Figure 6 that in "Servalyt" ampholytes the major portion has a lower molecular weight than the corresponding "Ampholine" preparations. For "Servalyt" carrier ampholytes Amido Black 10B staining virtually parallels lactose visualization, while for "Ampholine", in the extension of the material stained with Amido Black 10B and visualized with lactose, an additional heavy staining is observed corresponding to a material of higher molecular weight. Differences in the molecular weight distribution are particularly evident when alkaline narrow pH range carrier ampholytes are compared. Column gel chromatography of wide range ampholytes has revealed that in the leading limb of the main, partly asymmetrical peak an pH increase can be detected confirming that alkaline carrier ampholytes have a higher molecular weight. Experiments with different markers in the range of 300 to >3000 daltons suggest that the molecular weight of the major portion of commercial carrier ampholytes exceeds notably the value of several hundreds daltons (B.J. Radola, unpublished results).

The following conclusions may be derived from the experiments described in this report: By the combined approach employing different visualization reactions and ion exchange chromatography it has been possible to demonstrate the correctness of previous estimates assuming several hundreds of ampholyte species in commercial wide range carrier ampholytes. Visualization reactions revealed some 150 major and minor ampholyte species in the pH 3-5 range. The results of ion exchange chromatography suggest that in the neutral and alkaline pH range some 50-70 species are present per pH unit. The ion exchange chromatographic analysis of carrier ampholytes can be considerably improved by using modified conditions for elution instead of buffer gradients devised for optimun resolution of amino acids. Thin-layer gel chromatography provides a rapid and conveniant method for studying the size distribution of synthetic carrier ampholytes. The new visualization reactions could together with the chromatographic methods contribute to a better standardization of commercially available carrier ampholytes and may also prove useful in following the synthesis of carrier ampholytes by new methods (17,18).

REFERENCES

1. Rilbe, H.: Trends in instrumental and methodological developments of isoelectric focusing. In: "Protides of the Biological Fluids" (H. Peeters, Ed.) 17, 369-382 (1970)

2. Vesterberg, O.: Physicochemical properties of the carrier ampholytes and some biochemical applications. Ann. N.Y.Sci. 10, 55-76 (1973)

3. Righetti, P.G., Pagani, M., Gianazza, E.: Characterization of synthetic carrier ampholytes for isoelectric focusing. J. Chromatog. 109, 341-356 (1975)

4. Rilbe, H.: Rapid isoelectric focusing in density gradient columns. Ann. N.Y.Sci. 209, 80-93 (1973)

5. Jonsson, M., Pettersson,S., Rilbe, H.: Scanning in small density gradient columns. I. Use of a standard spectrophotometer cuvette for focusing. Chemical modification of proteins by migrating reactive ions. Anal. Biochem. 51, 557-576 (1973)

6. Fries, E.: Visualization of protein zones in preparative electrophoresis and carrier ampholyte zones on isoelectric focusing in gel slabs by two light refraction methods. Anal. Biochem. 70, 124-135 (1976)

7. Frater, R.: Behavior of Ampholines during isoelectric focusing. Anal. Biochem. 38, 536-538 (1970)

8. Felgenhauer, K., Pak, S.J.: Detection of Ampholine patterns. Ann. N.Y.Sci. 209, 147-153 (1973)

9. Radola, B.J.: Isoelectric focusing in layers of granulated gels. I. Thin-layer isoelectric focusing. Biochim. Biophys. Acta 295, 412-429 (1973)

10. Brown, R.K., Lull, J.M., Lowenkron, S., Bagshaw, J.C., Vinogradov, S.N.: Analysis of carrier ampholytes by ion exchange chromatography. Anal. Biochem. 71, 325-332 (1976)

11. Brown, R.K., Caspers, M.L., Vinogradov, S.N.: Carrier ampholytes distribution. This volume.

12. Otavsky, W.I., Drysdale, J.W.: Recent staining artifacts with LKB "Ampholines" on gel isoelectric focusing. Anal. Biochem. 65, 533-536 (1976)

13. Baumann, G., Chrambach, A.: Lack of interaction between hormonal proteins and synthetic carrier ampholytes (Ampholine). Anal. Biochem. 64, 530-536 (1975)

14. Radola, B.J.: Isoelectric focusing in layers of granulated gels. II. Preparative isoelectric focusing. Biochim. Biophys. Acta 386, 181-195 (1975)

15. Tschesche, H., Frank, C., Ebert, H.: Submicro-determination of amino acids by a single column analysis in a 4-hour run using automated buffer gradient mixing. J. Chromatog. 85, 35-43 (1973)

16. Radola, B.J.: Thin-layer gel filtration of proteins. I. Method. J. Chromatog. 38, 61-77 (1968)

17. Grubhofer, N., Borja, C.: Synthesis of carrier ampholytes for isoelectric focusing containing sulfonic and phosphonic acid groups covering a wide pH range. This volume.

18. Just, W.W., Werner, G.: Continuous isoelectric focusing: Further studies on human red blood cells and on the separation of rat liver light mitochondrial fraction. This volume.

Synthesis of Carrier Ampholytes for Isoelectric Focusing Containing Sulfonic and Phosphonic Acid Groups Covering a Wide pH Range

N. Grubhofer and C. Borja

SUMMARY

Carrier ampholytes for isoelectric focusing with improved separation characteristics and wider pH-range have been obtained by reacting a newly synthesized oligo-amine with propane sultone and chloromethyl phosphonic acid with subsequent electrolysis of the reaction products in multiple diaphragm chambers. In order to obtain ampholytes focusing at very alkaline pH, guanidino groups have been attached to part of compounds.

INTRODUCTION

Information on synthetic carrier ampholytes was first published in August, 1964 through a Swedish patent application filed by LKB-Produkter AB, Bromma, Sweden, Vesterberg as inventor (1), and other patent applications followed in 1965 (2, 3). The principal claim of the applicants was "polyprotic aminocarboxylic acids suitable as carrier ampholytes having at least four nitrogen atoms and at least one carboxylic group in the molecule". In 1969, three years after the commercial introduction of carrier ampholytes by LKB, Vesterberg (4) presented more data on the synthesis and electrophoretic separation of carrier ampholytes. This paper mentioned that conventional pentaethylene hexamine which was used as a starting material was not really an ideal polyamine. The claims of the Swedish patent and the methods described by Vinogradov (5) and Righetti (6) were basically the same as these reported by Vesterberg (4).

The primary purpose of the work described here was to prepare carrier ampholytes which would not be in violation of the existing patent rights and which would have the advantage of a wider pH-range.

In 1970, Grubhofer and Pogacar made the first experiments to synthesize polyaminophosphonic and polyaminosulfonic acids and in July, 1971 a patent (7) was filed claiming aminosulfonic acids as suitable carrier ampholytes. Another patent was filed by Grubhofer (8) on polyaminophosphonic carrier ampholytes and mixed derivatives with sulfonic and carboxylic groups in 1972.

Pogacar and Jarecki (9) presented some remarks on "inorganic acidic ampholytes", their term for polyaminophosphonic and polyaminosulfonic acids suitable as carrier ampholytes. Unfortunately, the authors seem to have forgotten to mention the name of the actual originator of this project.

The list of properties characterizing good carrier ampholytes are:

1. The product should have a large number of individual compounds in the pH-range 2-11.

2. A range of narrow pH-cuts should be available for spreading the fractionation at desired intervals.

3. The compounds should have low optical density in the visible region and at 280 nm and also if possible at 260 nm in order to facilitate evaluation in the density gradient column.

4. The ampholytes should possess low molecular weights (preferably below 1000) and should not stain with various dyes used in polyacrylamide and thin layer isoelectric focusing.

5. The materials should develop a linear, regular pH-gradient and should yield clear and sharp separation of protein mixtures.

For attaining the purposes cited above the amine (pentaethylene hexamine) used as an example in the patent literature (1, 2, 3, 4) and also in other publications (5, 6) is not very suitable. The first task therefore consisted in synthesizing an amine which has more isomers and better optical properties, but still possesses the favourable molecular weight of pentaethylene hexamine.

Such an amine has been synthesized by condensing ethylenimine with propylene diamine under conditions based on the work of Könnecke and Heise (10), that do not lead to high molecular products. Purification was done by distillation under high vacuum, thus eliminating compounds with molecular weights higher than 400. The reaction is illustrated as follows:

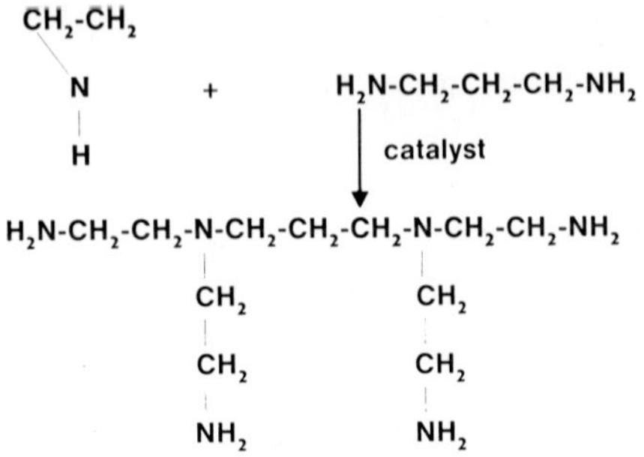

Alkylation of such amines with compounds that introduce sulfonic and phosphonic acid groups is relatively easy and is achieved with propane sultone and chloromethyl phosphonic acid yielding products with the following formula as a representative example:

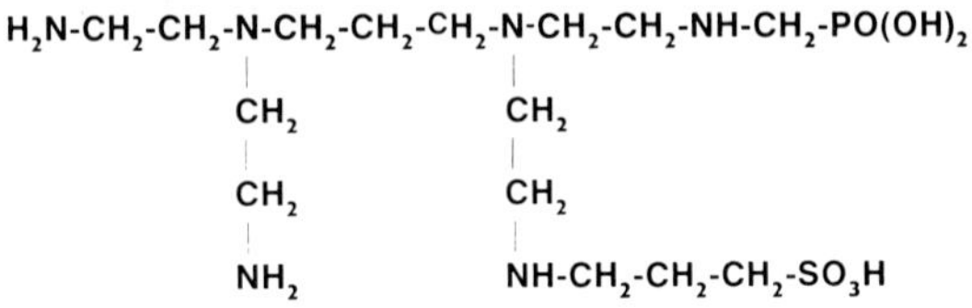

The products obtained are fairly similar to the ones produced by using acrylic or metacrylic acids and other similar unsaturated carboxylic acids as described in the patent literature (1).

The reaction products with the various alkylating agents introducing acidic groups cannot be used as such but have to be submitted to electrolysis in a fairly dilute solution. In the present work electrolysis chambers have been used which are divided by diaphragms made of ceramic material. It is not sufficient to prepare only one reaction mixture, but a number of mixtures must be used ranging from 0.8 acid equivalents per mole nitrogen down to 0.1 acid equivalents per mole nitrogen. Very basic components can be prepared by reacting the amine with 0.1 mole propane sultone, followed by 0.2 moles O-methyl isourea. Selection of the proper fractions and blending them under constant testing make preparation of larger quantities of carrier ampholytes a highly empirical procedure. The thin-layer method described by Radola (12) was particularly helpful in routine testing and screening of ampholyte products.

MATERIALS AND METHODS

Materials:

Ethylenimine was obtained from DOW Chemical Co., Midland, Michigan, or Chemirad Corp., Port Washington, New York. For some time already, this material was not available due to the restrictions that were imposed on account of its carcinogenic properties. The material was synthesized therefore from amino ethane sulfonic acid according to a method described in the literature (11). Pentaethylene hexamine has been obtained from Union Carbide Corp., but it is another chemical that is now practically unavailable. Propane sultone (the next candidate for scarcity due to anticancer restrictions) has been obtained from Deutsche Shell AG, chloromethyl phosphonic acid from Stauffer Chemical Co. The production of this material however has been discontinued recently. Other ways to introduce phosphonic groups are under investigation, but it may even become necessary to omit phosphonic groups at all, although they are quite useful, especially in the alkaline range.
All materials have been used without further purification.

Methods:

The amine has been prepared by heating a mixture of a 20 % (V/V) solution of propylene diamine and 4 % (V/V) of 32 % hydrochloric acid in water to 80°C and introducing ethylenimine by means of a bellows pump into the mixture which is contained in a flask equipped with stirrer and thermometer. Preliminary experiments were performed to determine the conditions of the reaction by following the specific viscosity and terminating the reaction (addition of caustic soda) when its viscosity corresponded to that of pentaethylene hexamine. The water was removed by means of a rotary evaporator, and the viscous mixture was distilled on a thin film fractionation apparatus (Schott u. Gen., Mainz). The amine mixture thus obtained had a boiling range between 90 and 240°C at 0.1 torr, about 15 % had higher boiling points and about 10 % had lower boiling points. It had an optical density of 0.10 at 260 nm, 0.04 at 280 nm and 0.01 at 340 nm. The molecular weight determined osmometrically was between 250 and 300. The NH_2-equivalent weight as determined with trinitrobenzene sulfonic acid was ca. 70.

Reaction with propane sultone.

A typical reaction procedure to introduce 0.5 equivalents sulfonic group per atom nitrogen: Into a 20 liter 4-necked flask equipped with stirrer, reflux condenser and thermometer, 3 kg amine and 12 liter water were mixed. The mixture was stirred and heated to 40°, then 4 kg propane sultone was pumped into the stirred contents of the flask by means of a Masterflex tubing pump carrying silicon tubing. The speed of addition is adjusted in such a way that the temperature of the mixture did not exceed 65°. After complete addition of the propane sultone the temperature was increased to 75° and kept at this temperature for one hour. The flask contained 17,6 liters with about 7.6 kg of synthesized ampholytes.

Reaction with chloromethyl phosphonic acid.

A typical reaction for introducing 0.25 equivalents phosphonic groups per atom nitrogen: 2 kg amine and 500 ml water were placed in a polyenthylene bottle equipped with a stirrer and cooled with tap water. 800 g chloromethyl phosphonic acid was added slowly for about 3 hours, keeping the reaction below boiling point. After the reaction, the polyethylene bottle was loosely stoppered and put into an oven at 70°C for 72 hours. The hot viscous reaction product was poured into 20 liters of hot deionized water.

Electrolysis of a 2 % solution was done in chambers made of polyvinyl chloride (inner dimensions: 67 cm long, 40 cm wide, 30 cm deep), with 20 diaphragms (DIAPOR: Schuhmacher'sche Fabrik, Bietigheim), 30 x 40 cm, with a distance of 3 cm from each other. The electrodes used were platinum coated tantalum grids (DEGUSSA, Pforzheim). Each of the 21 cells was fitted with a bent glass tube, through which cold water (10°C) was circulated. The normal schedule for electrolysis was as follows:

time	voltage
15 hours	200
4 hours	700

At the end of the run the pH of the content of each cell was determined. Test analyses were prepared by means of the Radola-method (12).

ELECTROLYSIS CHAMBER

PART OF ELECTROLYSIS PLANT

For the broad range ampholytes, the contents of the cells which have pH in the range of 3-8.5 were pooled and were fortified with various pH 2-4 and pH 9-11 fractions obtained by re-electrolysis.

In order to obtain narrow pH-cuts the contents of the cells obtained from the first electrolysis with pH-range of 2-3.9, 4-5.9 up to 10-11.9 were collected and pooled. The two pH-unit fractions were re-electrolyzed and from these, one pH-unit fractions were obtained ranging from 2-11. This single pH-fractions were concentrated to 40 % and used for blending the desired narrow range carrier ampholytes. All the ampholytes after concentration were hot treated with charcoal and filtered while hot through Servapor A 40 filter discs with 0.5 micron pores (Serva-Technik, Heidelberg) in order to remove fine particles of charcoal, bacteria and spores.

RESULTS AND DISCUSSIONS

The figures 1-4 show how a pH-gradient (—•—•—•—) forms in the electrolysis chambers and how the contents of the individual cells perform in the Radola thin layer system using a test mixture of 9 proteins ranging from pI 3.5 (amylo glucosidase) to pI 10.5 (cytochrome C).

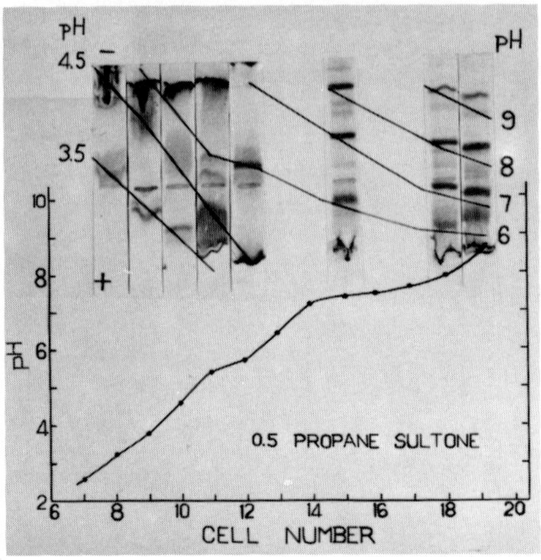

FIGURE 1

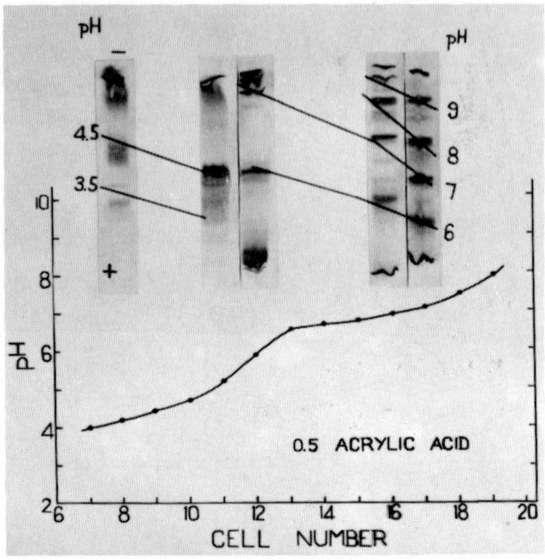

FIGURE 2

Figures 1 and 2 show the result obtained with propane sultone as compared with acrylic acid.

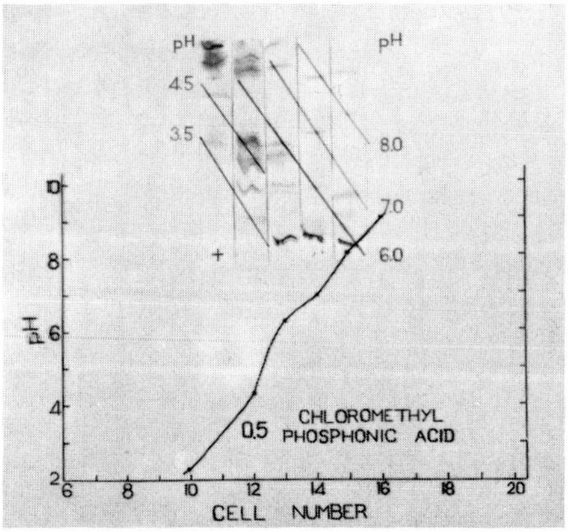

FIGURE 3

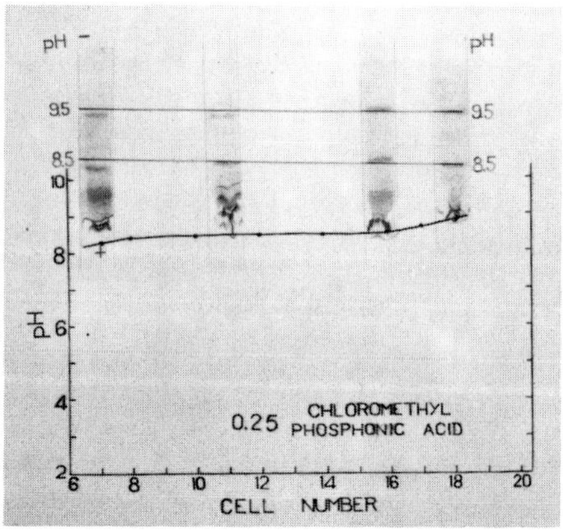

FIGURE 4

Figures 3 and 4 show the results obtained with 0.5 and 0.25 equivalents phosphonic acid per atom nitrogen. It can be seen that higher degrees of substitution with phosphonic acid give only unsatisfactory yields of useful carrier ampholytes. The lower degrees of substitution give a very good yield of narrow-range ampholytes. With 0.35 chloromethyl phosphonic acid the ampholytes between pH 6 and pH 7 can be obtained in good yields.

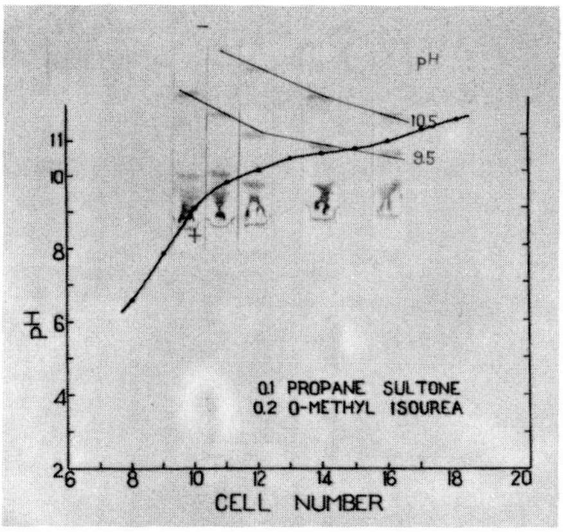

FIGURE 5

Figure 5 is a typical example of how a pH-gradient (—•—•—•—) is formed from the electrolysis of a very alkaline mixture (0.1 equivalents propane sultone per mole nitrogen and 0.2 mole O-methyl isourea per mole nitrogen of the amine). The separation of cytochrome C (pI = 10.5) from ribonuclease (pI = 9.5) is clearly indicated. The ampholytes obtained in the experiments as shown in figure 5 were used to prepare the alkaline part of a broad range mixture or to prepare small pH-cuts.

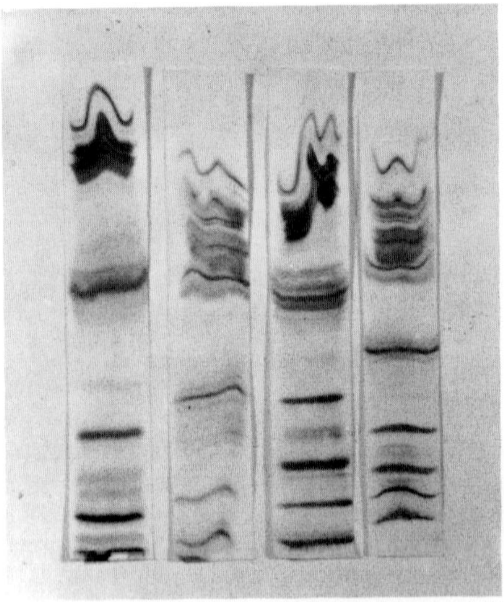

a b c d

FIGURE 6

a = pH 3-8.5

b = pH 3-8.5 and pH 2-4

c = pH 3-8.5 and pH 9-11

d = pH 3-8.5, pH 2-4 and pH 9-11

Figure 6 shows isoelectric focusing of a protein test mixture with pH 3-8.5 mixture (a) and how the mixture is improved by adding pH 2-4 (b) and pH 9-11 fractions (c) and both fractions (d). All ampholyte mixtures used in the experiments did not contain any carboxylic acid groups.

The technical assistance of Miss S. Scheibe is deeply appreciated.

REFERENCES

(1) Swedish Patent Application 10273; August 8, 1964.

(2) German Patent Application 1493939; August 10, 1965.

(3) US Patent Application 1106818; August 4, 1965.

(4) Vesterberg, O.: Synthesis and Isoelectric Fractionation of Carrier Ampholytes. Acta Chem. Scand. 23, 2653-2666 (1969).

(5) Vinogradov, S. N., Lowenkron, S., Andonian, M. R., Bagshaw, J., Felgenhauer, K. and Pak, S. J.: Synthetic Ampholytes for the Isoelectric Focusing of Proteins. Biochem. Biophys. Res. Commun. 54, 501-506 (1973).

(6) Righetti, P. G., Pagani, M. and Gianazzo, E.: Characterization of Synthetic Carrier Ampholytes for Isoelectric Focusing. J. Chromatog. 109, 341-356 (1975).

(7) Grubhofer, N. and Pogacar, P.: Offenlegungsschrift 2137617; July 28, 1971.

(8) Grubhofer, N.: Offenlegungsschrift 2230743; June 23, 1972.

(9) Pogacar, P. and Jarecki, R. (1972) in Electrophoresis and Isoelectric Focusing in Polyacrylamide Gel (R. C. Allen and H. R. Maurer, eds.) Walter de Gruyter, Berlin, pp. 153-158.

(10) Könnecke, H.-G. and Heise, M.: Über die Polymerisation des Äthylenimins. J. Prakt. Chem. 9, 232-236 (1959).

(11) Allen, C. F. H., Spangler, F. W. and Webster, E. R.: Organic Synthesis, John Wiley & Sons, Inc., New York/London 1963, rev. ed. of annual volumes 30-39, Vol. 4, pp. 433-435.

(12) Radola, B. J.: Isoelectric Focusing in Layer of Granulated Gels. I. Thin Layer Isoelectric Focusing of Proteins. Bio chem. Biophys. Acta 295, 412-428 (1973).

3. Analytical Methodology

Staining of Proteins in Polyacrylamide Gels

O. Vesterberg and L. Hansén

SUMMARY

Four procedures are described for staining of protein bands after isoelectric focusing in polyacrylamide gels. In one procedure which is based on an earlier publication by Vesterberg (2) Coomassie Brilliant Blue R 250 is used, whereas in the others the G 250 form is used in an aqueous solution of perchloric acid according to modifications of the method described by Reisner et al. (1). The detection limit for protein seems to be lower by these procedures than with earlier published methods. Thus by using isoelectric focusing in flat beds of polyacrylamide and the new staining procedures it is possible to detect less than 0.2 µg of many proteins. One of the procedures is especially rapid, allowing detection of proteins without destaining within an hour after starting fixation of proteins. Procedures for preservation of the protein patterns and determination of proteins by scanning in a densitometer are also described.

INTRODUCTION

Many methods have been published for staining of proteins after electrophoretic separation procedures in polyacrylamide gels. Each of them has its draw-backs. For example, some are very time consuming, while others have too low sensitivity for amny purposes. Moreover, after isoelectric focusing removal of carrier ampholytes by tedious leaching procedures is often required. During efforts to improve the art we found that it was not only possible to reduce the time required for staining and destaining but also to improve the sensitivity. Initially, a modification of the method recently described by Reisner et al. (1), using Coomassie Blue G 250, was tried.

PROCEDURES AND RESULTS

Isoelectric focusing of proteins was performed as described earlier by Vesterberg (3) in flat beds of polyacrylamide gel with the dimensions 110 x 250 x 2 mm using LKB Multiphor apparatus. Acrylamide and N, N′ - methylene bisacrylamide were purchased from British Drug House (Poole, England), and were used unrecrystallized. Coomassie Brilliant Blue, R 250 and G 250, were obtained from ICI (Manchester, England). Ovalbumin and β-lactoglobulin were obtained from Sigma (St. Louis, USA), and serum albumin from Kabi (Stockholm, Sweden). β_2-microglobulin was prepared from urine by isoelectric focusing as to be described in a forthcoming paper. Hemoglobin was derived from red cell hemolysates. Solutions of the mentioned proteins in TRIS acetate buffer 0.002 M, pH 7.5, were designated as "standard proteins". Dilutions of the standard proteins were made in the buffer mentioned. The protein samples were applied on the gel by soaking into pieces of porous cotton paper. (Paratex II/80 Lohman, Fahr/Rhein.) The solutions used for fixing and staining are presented in the Tables 1-2. Procedures A-D are presented here for the first time, while Procedure E is exactly the one described earlier by Vesterberg (2, 3). Fixing and staining were performed at 60°C, each for 30 minutes.

Table 1

Solutions for fixing of proteins in polyacrylamide gels

Note that no fixing is performed prior to staining in procedure A and E.

The ingredients per 1 000 ml are shown

Ingredients	Procedures				
	A	B	C	D	E
Trichloroacetic acid (g)	-	147	147	147	-
Sulphosalicyclic acid (g)	-	44	44	44	-
Water (ml)	-	910	910	910	-

Table 2

Staining solutions (1000 ml)

Ingredients	Procedures				
	A	B	C	D	E
Coomassie G 250 (g)	-	0.4	0.4	0.4	-
" R "	0.9	-	-	-	0.9
Trichloroacetic acid (g)	32	-	-	-	32
Sulphosalicyclic acid (g)	107	-	-	-	107
urea (g)	-	-	-	39	-
methanol (ml)	268	-	-	-	268
water (ml)	664	970	970	932	664
perchloric acid (70 % w/v; ml)	-	30	30	29	-

It is preferable to dissolve Coomassie Blue first in methanol in Procedures A and E, and in about 800 ml of water in Procedures B-D before mixing with the other ingredients. In Procedure D urea is also added to the water. The perchloric acid is slowly added with vigorous mechanical stirring immediatly before the use of the staining solution. Volumes are made up to 1 000 ml with water. With Procedures B and C it is necessary to stir the solutions for one hour and to filter the solution through a small pore filter paper. In this respect the staining solution for Procedure D is more simple to prepare because it does not use extra steps. To prevent fading of the stained protein bands in Procedure B it is recommended that the gels are stored in a solution made up of 1 % staining solution in water.

Destaining was not necessary with Procedures B. With the other procedures destaining was done with the solutions given in Table 3. Fresh destaining solution was exchanged for the destaining wash during working time of the day at least once every hour.

Table 3

Destaining solutions (1000 ml)

Ingredients	Procedures				
	A	B	C	D	E
acetic acid (99 %, ml)	50	-	50	50	80
ethanol (99 %, ml)	70	-	70	70	255
ethyl acetate (ml)	100	-	100	100	-
water (ml)	780	-	780	780	665

The total times required for each method are shown in Table 4. Destaining was quicker if made with slow agitation in a rocking bath (Heto, Danmark). It should be noted that Procedure B is very rapid and that the result can be read already 1 hour after starting fixation although some background stain is still left. This residual stain may be removed by continuing with Procedure C. The fixation step of Procedure B can be omitted at the expense of sensitivity.

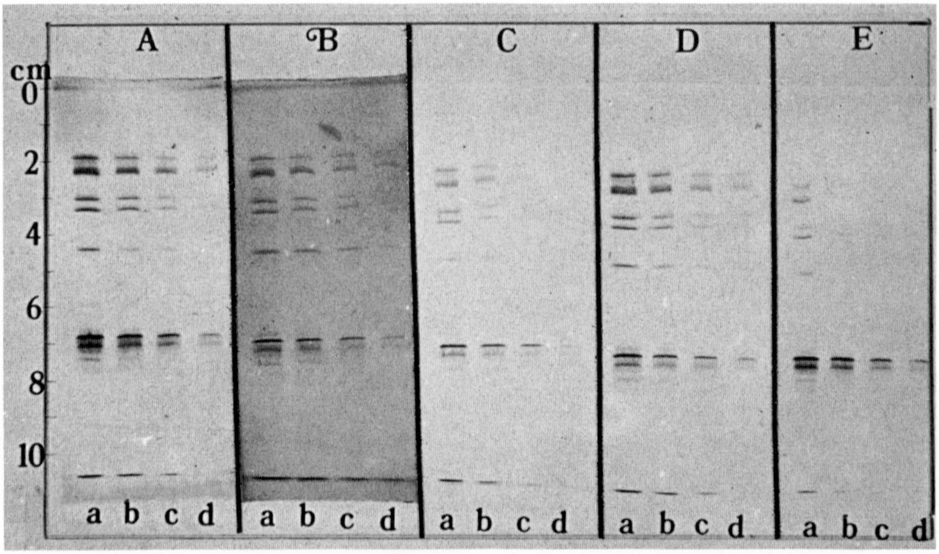

Fig 1.

Stained protein zones in gels with Procedures A-D of this report and also of Procedure E which is identical to the method described by Vesterberg (2, 3). Each gel is indicated by a reference capital letter. A cm scale is shown to the left with 0 at the anode. Of each protein solution 25 µl was soaked into pieces of porous cotton paper. From left to right, the following proteins (with the amounts in µg) were applied:

Protein	a	b	c	d
ovalbumin	4.0	2.0	1.0	0.5
serumalbumin (double band)	4.0	2.0	1.0	0.5
β-lactoglobulin (2 bands)	4.0	2.0	1.0	0.5
β_2-microglobulin	3.0	1.5	0.8	0.4
hemoglobin (2 major bands and 1 minor)	16	8	4	2
cytochrome c	8.0	4.0	2.0	1.0

The proteins are listed in the order they appear from top to bottom.

Table 4

Times in hours for the various steps in the Procedures A-E

Procedure	Fixing	Staining	Destaining	Totaltime x)
A	-	0.5	6 - 48	6.5 - 48
B	0.5	0.5	-	1
C	0.5	0.5	6 - 23	7 - 24
D	0.5	0.5	4 - 22	5 - 23
E	-	0.5	24 - 48	24 - 48

Typical results of stained zones are shown in Fig 1. Procedure D was the most sensitive. The increased sensitivity probably results from the presence of urea which by unfolding the proteins increases the number of binding sites for the stain. It was possible to detect less than 0.2 µg of β_2-microglobulin.

After soaking the gels for thirty minutes in a solution of polyethylene glycol (5 % w/v, MW 20,000) and glycerol (5 % v/v) in destaining solution, the gels could be placed on a glass plate, covered with a dialysis membrane and dried. In the dried form they can be stored for years without noticeable change in quality.

It is also possible to quantitate proteins by densitometry after staining by the described methods. The absorption maxima are 570 nm and 600 nm for the R 250 and G 250 forms of Coomassie Brilliant Blue, respectively. Typical densitograms are shown in Fig 2.

---

x) A time range is given because the time required depends on many factors, e.g. frequency of exchange of destaining solution. Furthermore, the proteins can often be seen quite well when the background is not completely destained, i.e. the shorter time of the interval. The longer time of the interval refers to a more complete destaining of the background.

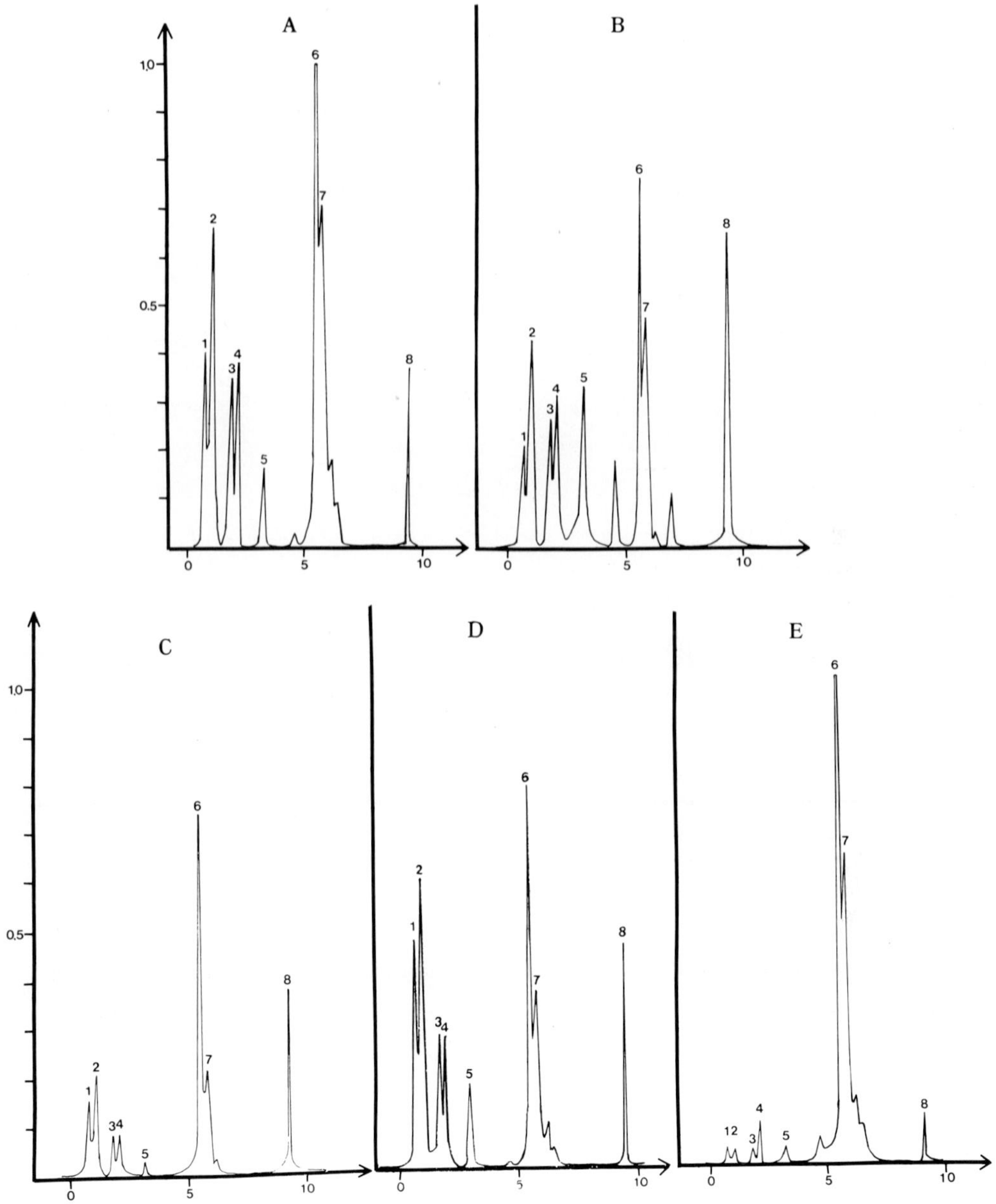

Fig. 2. Densitometric recordings of stained protein bands after staining by the Procedures A-E. Scanning was performed on the protein sample shown at 'b' in Fig 1. with a Zeiss Chromatogram Scanner using a slit 2 x 0.02 mm.

Standard curves for ovalbumin derived from densitogram data of the different staining procedures are shown in Fig 3. The linear range is from 0.4 to 4 µg. This may be compared with earlier publications citing difficulties experienced by others (4).

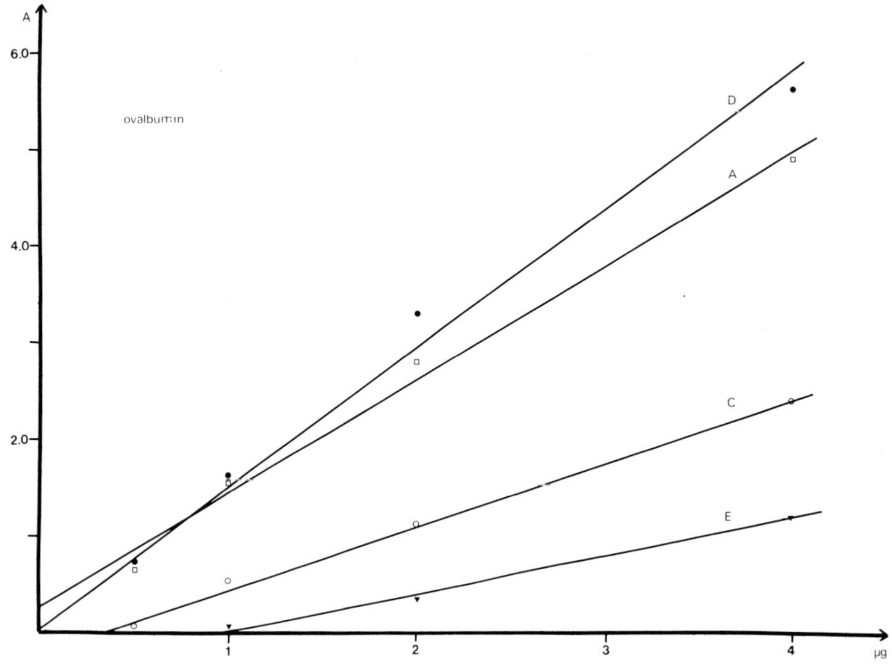

Fig 3. Standard curves for various amounts of ovalbumin. The values on the ordinate are peak height times zone width.

DISCUSSION

For several reasons it is not easy to compare the staining methods presented in this report with other published methods. Other reports have often indicated several alternative procedures without exactly specifying strict conditions (1, 5). Many investigators seem to have devoted little effort to optimization. Furthermore, different investigators seem to be satisfied by different degrees of destaining of the background, perhaps because it is difficult to arrive at exactly the same degree of destaining. This partly explains why it is difficult to compare times needed for destaining in various published procedures.

Table 5

Examples of total time for fixing, staining and destaining of proteins after separation by various procedures in polyacrylamide gels

Procedure according to	Separation Technique	Stain	Hours[x)]
Gorovsky et al. (6)	Electrophoresis	Fast Green	48-72
Fishbein (4)	"	Coomassie R 250	36
Bertolini et al. (5)	"	-"-	24
Vesterberg (2)	Isoelectric focusing	-"-	24-48
This report Procedure D	-"-	Coomassie G 250	5-23
-"- " B	-"-	-"-	1

Table 5 gives an idea about the times used. The longer destaining is continued, the higher is the chance that protein zones will also lose gradually their stain. Indeed, it has recently been shown by Bertolini et al. (5) that very significant amounts of Coomassie Brilliant Blue R 250 stain were lost from protein zones even before the background was destained. The losses were different for different destaining procedures. Such factors provide partial explanation of the difficulties in comparing the sensitivity between different procedures.

Other problems arise from the fact that different investigators have used gels of different geometry, i.e., rods and flat-beds. When comparing sensitivities given in different publications this must be considered.

The determination of the detection limit for a given method is also complicated by the fact that the protein concentration is not constant

x) See note of Table 4.

in different parts of a protein zone. Fortunately the above mentioned difficulties are largely obviated when comparison of sensitivity is made for protein zones focused under identical conditions, e.g., in a flat bed of polyacrylamide gel as performed in this report. Advantages of using a sensitive procedure are presented in Table 6. We hope that the rapid and/or the very sensitive procedures described in this study for detection of proteins in polyacrylamide gels after isoelectric focusing and also various forms of electrophoresis could be of great value for many biochemists.

Table 6

Advantages with a sensitive protein staining technique

1. The volume of protein solution applied to the gel can be kept low which gives the following benefits:
   - 1.1 Many methods are possible for application
   - 1.2 Small pieces of absorbent material can be used, whereby the risk of adsorption is minimized
   - 1.3 The salt concentration can be relatively high
   - 1.4 Decreased need for concentration, which can cause losses of a selective and/or variable nature
2. When the total amount of protein applied is low there is less tendency for precipitation of certain substances, e.g. proteins and nucleic acids.

REFERENCES

1. Reisner, A.H., Nemes, P., Bucholtz, C.: The use of Coomassie Brilliant Blue G 250 perchloric acid solution for staining in electrophoresis and isoelectric focusing on polyacrylamide gels. Anal. Biochem. 64, 509-516 (1975).

2. Vesterberg, O.: Isoelectric focusing of proteins in polyacrylamide gels. Biochim. Biophys. Acta 257, 11-19 (1972).

3. Vesterberg, O.: Isoelectric focusing of proteins in thin layers of polyacrylamide gel. Science Tools 20, 22-29 (1973).

4. Fishbein, W. N.: Quantitative densitometry of 1-50 µg protein in acrylamide gel slabs with Coomassie Blue. Anal. Biochem. 46, 388-401 (1972).

5. Bertolini, M.J., Tankersley, L.D., Schroeder, D.D.: Staining and destaining polyacrylamide gels: A comparison of Coomassie Blue and Fast Green protein dyes. Anal. Biochem. 71, 6-13 (1976).

6. Gorovsky, M.A., Carlson, K., Rosenbaum, J.L.: Simple method for quantitive densitometry of polyacrylamide gels using Fast Green. Anal. Biochem. 35, 359-370 (1970).

Isoelectric Focusing of Complex Protein Mixtures in the Nanogram Range and Enzyme Kinetics of Dehydrogenases Following IEF

G. Bispink and V. Neuhoff

SUMMARY

Micro amounts of LDH-isoenzymes and complex protein mixtures (e.g., water soluble proteins of brain, heart and skeletal muscle, kidney, liver and serum) can be fractionated by IEF in 2, 5, 10 or 20 µl capillaries. IEF can be adapted to any problem of separation by mixing servalytes of suitable pH ranges. A few micrograms of fresh tissue are sufficient for several separations of soluble proteins. Individual proteins not sufficiently fractionated for evaluation by one dimensional IEF can be eluted from the gel and refractionated in a micro gradient gel. By combining IEF in capillaries, with subsequent incubation of the microgels in a suitable tetrazolium assay and the continuous registration of the time-dependent formation of the formazan band enzyme kinetics for isolated isoenzymes in a single microgel can be analysed. The IEF of complex protein mixtures as well as the enzyme kinetics are highly reproducible and can be carried out in approximately 30-45 min.

INTRODUCTION

Isoelectric focusing is a well known method for the separation of proteins. Usually 5% polyacrylamide gels with a diameter of 5 mm and a length of 65 mm are used (1-4) and for a single separation in this macroscale 30 - 300 µg of protein are necessary (3). In 1972 Grossbach (5) first separated bovine serum albumin and ß-lactoglobulin by IEF in capillaries with a length of 65 mm and an inner diameter of 50, 100 and 300 µm. In 1973 Gainer (6) described a micromethod for isoelectric focusing of six proteins and has shown that during IEF in capillaries a pH gradient is formed as in macro IEF-gels.

A procedure for micro-isoelectric focusing of lactatedehydrogenase isoenzymes from biological material in 5 µl capillaries has been described by Quentin and Neuhoff in 1972 (7). That enzyme kinetics can also be analysed in microgels using a suitable tetrazolium assay was shown by Cremer et al. 1972 (8). Combining and modifying both methods, and using a computer directed scanning microscope for measurement and evaluation, the sensitivity and reproducibility for the determination of enzyme kinetics of LDH isoenzymes, separated by IEF in capillaries can be increased significantly. A further variation of this method also allows separation of complex protein mixtures in the nanogram range.

MATERIAL and METHODS

Acrylamide was obtained from Merck-Schuchard, N,N-methylene-bisacrylamide, N,N,N',N',-tetramethyl-ethylene diamine (TEMED) and ammonium peroxide disulfate from Fluka. As carrier ampholytes Servalyt (Serva) and Ampholine (LKB) were used. High purity of acrylamide and bisacrylamide is required for a good separation quality. Therefore, acrylamide and bisacrylamide were recrystallized as described by Loening (9). Gels were prepared in commercially available glass capillaries (Brandt, Wertheim) with constant volumes of either 5 or 10 µl. The capillaries were routinely precleaned according to the method of Rüchel et al. (10).

The different gel compositions for IEF of isoenzymes or complex protein mixtures are listed in Table 1. Preparation of the gels and handling of the capillaries were as described previously (11,12). For separation, the diluted sample containing 25% glycerine and 0,4% Servalyt was applied on top of the

Table 1: Gel composition for IEF of isoenzymes or complex protein mixtures

	protein mixture	LDH iso-enzymes
4g acrylamide +0,18g bisacrylamide ad 10 ml	0,30ml	-
4g acrylamide +0,10g bisacrylamide ad 10 ml	-	0,35ml
TEMED 0,06 ml/10ml H_2O	0,30ml	0,30ml
Servalyt 40%	0,25ml	-
Servalyt 40% or Ampholine 40%	-	0,20ml
ammonium peroxide disulfate 15mg/10ml H_2O	0,40ml	0,60ml
glycerine 99%	-	0,30ml
H_2O	1,25ml	0,25ml
	2,50ml	2,00ml

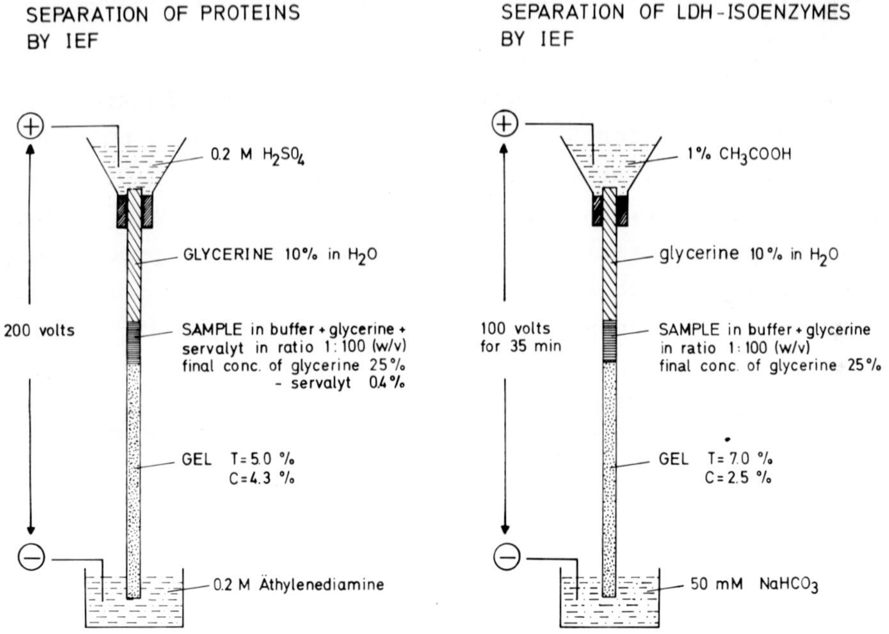

Fig. 1: Scheme of separation systems for IEF of complex protein mixtures and LDH isoenzymes.

gel and overlayered with water or 10% glycerine in water (compare Fig. 1).

Separation of water soluble proteins was performed with 0.2 M H_2SO_4 and 0.2 M ethylene diamine as ionic carrier, while 1% acetic acid and 50 mM sodium bicarbonate was used for separation of LDH isoenzymes (see Fig. 1). For complex protein mixtures a constant voltage of 200 V for approximately 10 min was used. The current flow during focusing decreased from approx. 200 μAmp. at the beginning to 20 - 30 μAmp when electrophoresis was terminated. For parallel runs of up to ten gels a power supply allowing current flow control for every single gel was used (12). Therefore the use of marker dyes, e.g., xylene cyanole FF, naphtolgrün-B, naphtolgelb-S, ponceau-S or a marker protein, e.g., cytochrome C, was not necessary.

After fractionation, the gels were immediately transferred to ice cold 10% TCA for 5 min using water pressure to push out the gels of the capillaries (12) and then stained with 0,05% coomassie brillant blue (Biomol) and 0,1% Cu II sulfate in acetic acid, ethanol, water (10:25:65) using a modified procedure of Righetti and Drysdale (13). The gels were stained for 20 min at 37°C in the coomassie solution and thereafter differentiated in acetic acid, ethanol, water (10:10:80) at room temperature.

For standardisation a protein mixture from Serva containing 7 crystalline proteins was used. Water soluble tissue proteins were prepared from different organ tissues of rat and guinea pig and human oncocytoma of thyroid gland (11).

For isoelectric focusing of LDH isoenzymes a const. voltage of 100 V was applied for 35 min. The sample must be suitable diluted for measuring the activity of the isolated isoenzymes. The addition of Servalyt or Ampholine to the sample was ommitted. However an overlayering of the sample with 10%

glycerine in H_2O was absolutely necessary for a separation without artifacts. After fractionation the gel was fixed in a special incubation chamber (14) which was filled with a tetrazolium mixture according to the method of Quentin and Neuhoff (7). The gel was then focused in the light beam (540 nm) of a computer directed ZEISS scanning microscope and measurements with 50 μm steps were carried out at defined time intervalls (14). The linear increasing peak areas of the final formazan peak at a gel locus with enzymatic activity was calculated and the slope of the resulting regression line is the quantitative measure of the enzymatic activity which can be calibrated in common enzyme units.

RESULTS and DISCUSSION

Fig. 2 shows the fractionation of 0,052 μg and 0,078 μg of the protein test mixture in 10 μl capillaries. In Fig. 2a a pH gradient of 2-11 was used and in 2b a mixture of different pH ranges which has the effect of spreading the protein peaks according to the variance of pH. The amount of total protein fractionated in Fig. 2b is almost the maximum for a 10 μl gel as can be recognized by the broadness of several peaks.

The fractionation of water soluble proteins from different organs is shown in Fig. 3. Protein amounts corresponding to 12,5 μg of wet weight tissue are fractionated in gels a and b where 10 μl capillaries are used, and 6 μg in gel c where 5 μl capillaries were employed. The effect of variations in the pH gradient is again demonstrated by Fig. 3 which shows the separation of proteins in a gel containing 2% Servalyt pH 2 - 11 + 2% pH 4 - 6 (Fig. 3a, gel 1), whilst all other gels had a pH range from 2 - 11.

In Fig. 4 the refractionation of a single protein peak obtained after IEF of the water soluble proteins from a human oncocytoma of the thyroid gland is shown. In gel a the same

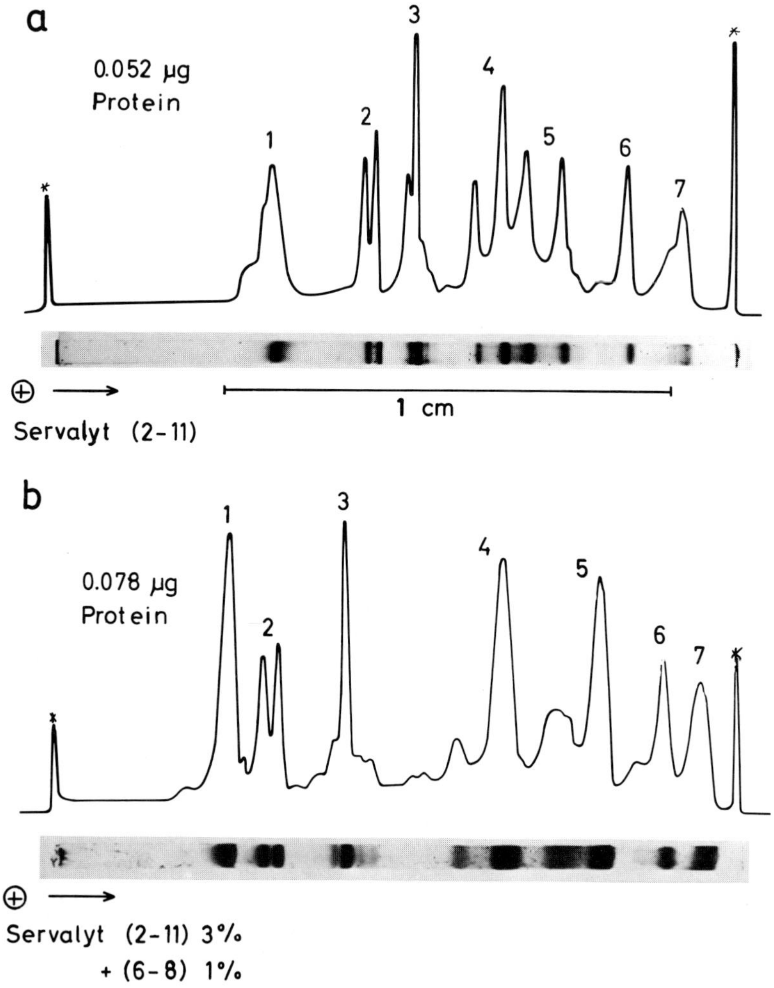

Fig. 2: Separation of a protein test mixture (Serva) in 10 µl capillaries on 5% acrylamide gels containing 4% Servalyt of different pH ranges. 1: bovine serum albumin, 2: ß-lactoglobulin, 3: conalbumin, 4: myoglobin from horse, 5: myoglobin from whale, 6: ribonuclease, 7: cytochrome C.

*: represents the bottom (cathodic side) and the top (anodic side) of the gels.

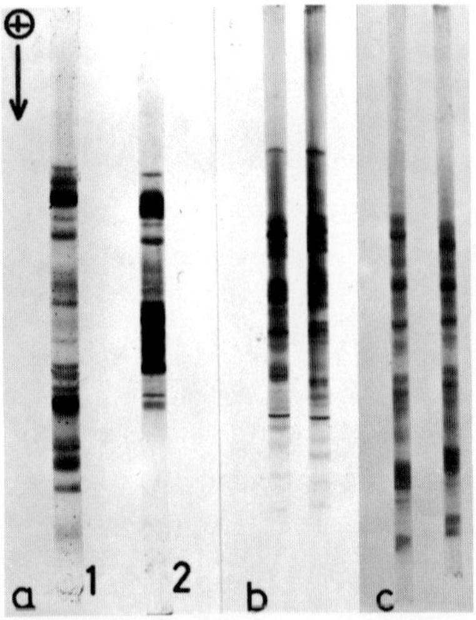

Fig. 3: Separation of water soluble proteins in 10 µl (a and b) and 5 µl gels (c). In a the extract corresponds to 12,5 µg wet weight of rat heart, in b 12,5 µg of rat cerebellum, and in c 6 µg of skeletal muscle from guinea pig. The pH range in the 5% acrylamide gels was 2% Servalyt pH 2-11 + 2% of pH 4-6 in gel 1, whilst in all other gels the pH gradient was 2-11.

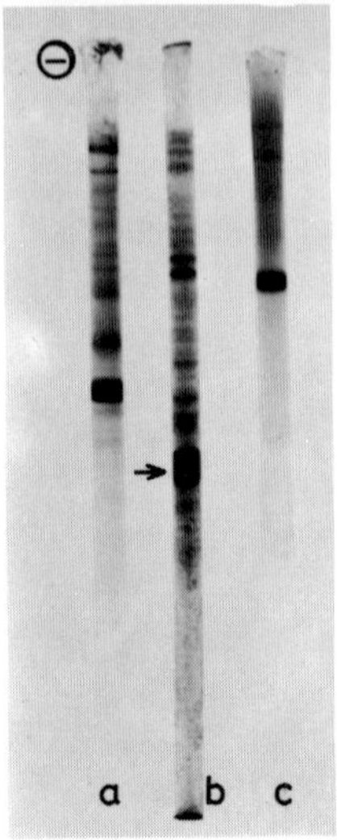

Fig. 4: Fractionation of water soluble proteins of a human oncocytoma of the thyroid gland in a 1-33% micro gradient gel (10 µl, gel a) and by IEF (gel b, pH 2-11). The peak marked with an arrow was after staining with bromphenol blue cut out of the IEF gel under a stereomicroscope and after electrophoretic elution direct refractionated in a 1-33% micro gradient gel (c). Observe that after refractionation several additional protein peaks of high molecular weight are detectable.

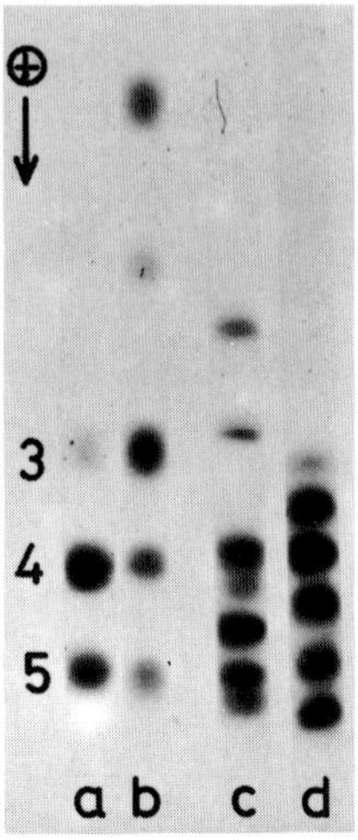

Fig. 5: Fractionation of LDH isoenzymes from the pineal gland (gel a) and the cerebrum (b - d) of the rat. IEF in gel a and b was performed under optimal conditions for isoenzymes, and in gel c and d under conditions normally used for the fractionation of complex protein mixtures. Additionally, the acrylamide used for gels c and d was not recrystallized. For further details see text. Observe the splitting of the five isoenzymes in gel b to produce additional peaks in gel c and d.

amount of extract (12,5 µg wet weight tissue) was fractionated in a 1 - 33% microgradient gel (10,12) as in the IEF gel b. The peak marked with the arrow in the IEF gel (b) was electrophoretically eluted according to Neuhoff and Schill (15) and directly refractionated in a 1 - 33% microgradient gel (c) with the effect that further protein peaks are detectable.

The fractionation of LDH isoenzymes in 10 µl IEF gels of pineal gland (a) and cerebrum (b) of the rat is shown in Fig. 5. The amount of extract fractionated in each gel corresponds to 12,5 µg wet weight tissue. The isoenzyme peaks are demonstrated after 15 min of incubation in the tetrazolium assay. In pineal gland only the LDH isoenzymes 3, 4 and 5 are detectable but in the cerebrum all five are present although in different amounts. The gels c and d in Fig. 5 show the effect of IEF on the same extract from the cerebrum of the rat as in b, but not performed under the optimal conditions described in the material and method section. For both of these gels acrylamide was used which was not recrystallized and instead of overlayering the samples with 10% glycerine they were overlayered with dist. water and brought in contact with the electrode solution. Gel c was left for 10 min in this situation before 100 V were applied for 30 min. 200 V were applied to gel d for 10 min directly after filling the capillary with sample and water. For gel a and b 1% acidic acid and for c and d 0,2 M H_2SO_4 was used at the anode. At the cathodic side of gel a and b 50 mM bicarbonate was used and for gel c and d 0,2 M ethylene diamine solution. In other words the sample in gel a and b was fractionated under optimal conditions for LDH isoenzymes and in gel c and d under suitable conditions for separation of complex protein mixtures. After IEF, gel c and d were incubated for 30 min in the tetrazolium mixture. Under these separation conditions new and not identifiable enzymatic active protein peaks are visible which are most likely artifacts produced by incorrect separation

Fig. 6:

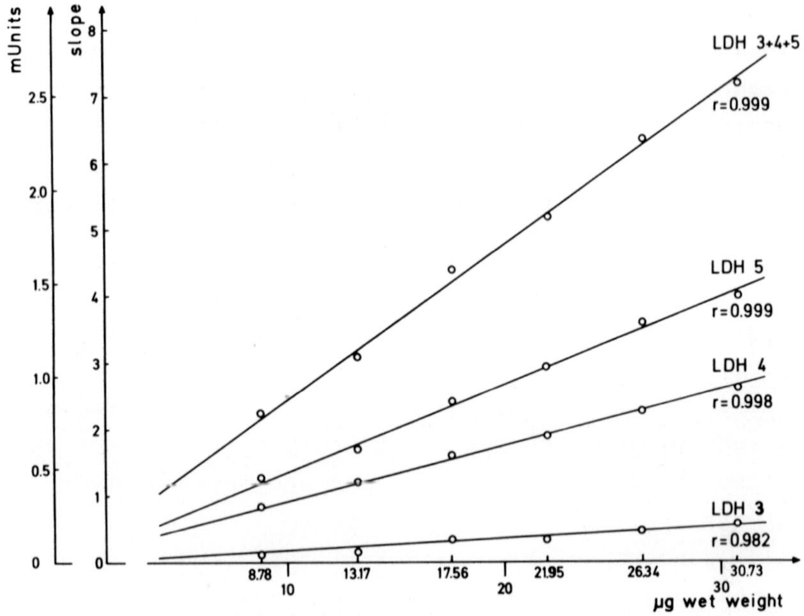

Enzyme kinetics of LDH isoenzymes from the pineal gland of rat measured with a computer directed ZEISS scanning microscope during incubation in the tetrazolium assay mixture. Each point of the regression lines for LDH 3-5 represents the slope of a regression line obtained after incubation in the tetrazolium mixture of IEF gels in which the given amount of wet weight tissue was fractionated.

Fig. 7:

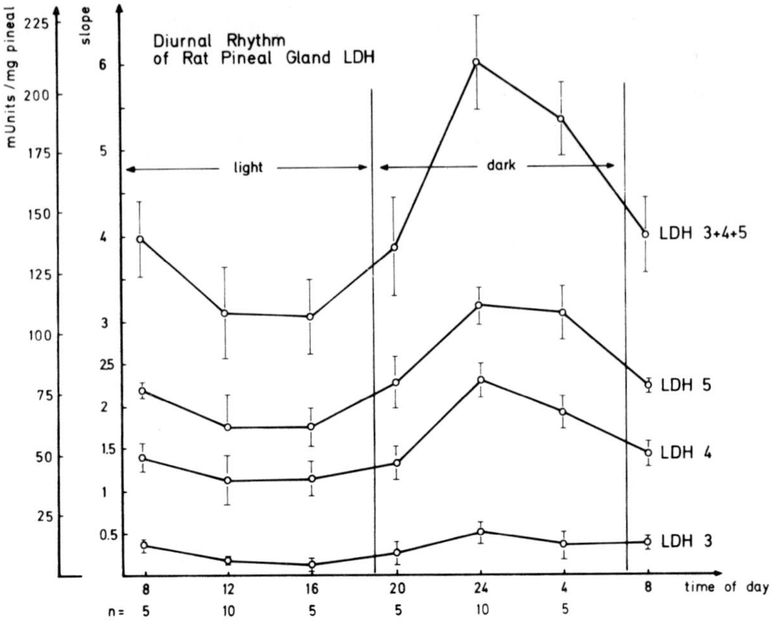

Diurnal rhythm of the activity of LDH isoenzymes 3-5 from the pineal gland of the rat.

conditions. This observation however may be useful for producing further split products or different subunit combinations of the isoenzymes. For further analysis of these effects a combination of IEF with isolation of single fractions and then refractionation in microgradient gels may be useful since molecular weight determinations in microgradient gels are easily carried out (10,12).

Fig. 6 demonstrates that enzyme kinetics can be reproducibly performed in a single gel for all isoenzymes fractionated. There is a strong correlation between the micrograms of wet weight tissue used for fractionation of the isoenzymes and the slope which is a measure of the enzymatic activity. An example of the applicability of this method to a biological problem (14) concerning the behavior of the 3 LDH isoenzymes from pineal gland of the rat analysed during the day and night cycle is shown in Fig. 7. There are remarkable and significant differences in the LDH activities during day and night.

The examples demonstrated here clearly show that with IEF in micro scale a reproducible fractionation of minute amounts of complex protein mixtures is possible. The resolution power of these micro IEF gels is at least as good as in macro gels. Performance of enzyme kinetics demonstrated here for LDH isoenzymes is only possible in micro gels since their small diameters cause no diffusion problems during incubation in a suitable assay mixture. Furthermore by creating suitable pH gradients, fractionation in micro IEF gels can easily be adapted to any problem of separation.

The concentration of 5% acrylamide with a crosslinking of 4,3% bisacrylamide has proven to be optimal since in such gels no sieving effect occurs for proteins up to a molecular weight of 500 000. If, for example, ferritin is applied to the anodic as well as the cathodic side of an IEF gel, it migrates from both sides to its isoelectric point to form one single band. If the concentration of acrylamide is higher than 7,5%, proteins with molecular weights higher than 140 000, e.g. LDH isoenzymes, are retarded in their mobility due to sieving effects. To avoid diffusion of single protein bands during staining or incubation in enzyme assay mixtures it is recommended that gel concentrations just below sieving conditions are used.

Micro-IEF can not only be combined with refractionation of isolated protein peaks in micro gradient gels (11), with their high resolution power (10), but also with a second fractionation in slab gels (16). Even if large amounts of material are available for analysis, micromethods are highly recommended since they are easily carried out and reproducible results are obtained in a very short time.

REFERENCES

1. Haglund, H.: Isoelectric Focusing in pH Gradients - A Technique for Fractionation and Characterization of Ampholytes. Methods of Biochem. Anal. 19, 1-104 (1971)

2. Righetti, P.G. & Drysdale, J.W.: Small-Scale Fractionation of Proteins and Nucleic Acids by Isoelectric Focusing in Polyacrylamide Gels. Ann.N.Y. Acad. Sci. 209, 163-186 (1973)

3. Maurer, H.R.: Disc Electrophoresis and Related Techniques of Polyacrylamide Gel Electrophoresis (Fischbach, K, Editor) W. de Gruyter, Berlin (1971)

4. Vesterberg, O.: Isoelectric Focusing of Proteins in Polyacrylamide Gels. Biochim. Biophys. Acta 257, 11-19 (1972)

5. Grossbach, U.: Microelectrofocusing of Proteins in Capillary Gels. Biochem. Biophys. Research Commun. 49, 667-672 (1972)

6. Gainer, H.: Isoelectric Focusing of Proteins at the 10^{-10} to 10^{-9}-g level. Anal. Biochem. 51, 646-650 (1973)

7. Quentin, C.D. & Neuhoff, V.: Micro-Isoelectric Focusing for the Detection of LDH Isoenzymes in different Brain Regions of Rabbit. Intern. J. Neuroscience, 4, 17-24 (1972)

8. Cremer, Th., Dames, W., Neuhoff, V.: Micro-disc Electrophoresis and Quantitative Assay of Glucose-6-phosphate Dehydrogenase at the Cellular Level. Hoppe-Seyler's Z. physiol. Chem. 353, 1317-1329 (1972)

9. Loening, U.E.: The Fractionation of High-Molecular-Weight Ribonucleic Acid by Polyacrylamide-Gel Electrophoresis Biochem. J. 102, 251-257 (1967)

10. Rüchel, R., Mesecke, S., Wolfrum, D.J., Neuhoff, V.: Mikroelektrophorese an kontinuierlichen Polyacrylamid-Gradientengelen, I: Herstellung und Eigenschaften von Gelgradienten in Kapillaren, ihre Anwendung zur Proteinfraktionierung und Molekulargewichtsbestimmung. Hoppe-Seyler's Z. physiol.Chem. 354, 1351-1368 (1973)

11. Bispink, G. & Neuhoff, V.: Isoelektrische Fokussierung in Mikrogelen zur Fraktionierung komplexer Proteingemische im Nanogrammbereich. Hoppe-Seyler's. Z. physiol. Chem. 357, 991-997 (1976)

12. Neuhoff, V.: Micromethods in Molecular Biology, Biochemistry and Biophysics (Neuhoff, V., Editor) Bd.14, Springer-Verlag, Berlin, Heidelberg, New York (1973)

13. Righetti, P.G. & Drysdale, J.W.: Isoelectric Focusing in Gels. J. Chromatogr. 98, 271-321 (1974)

14. Bispink, G. & Neuhoff, V.: in preparation.

15. Neuhoff, V. & Schill, W.-B.: Kombinierte Mikro-Disk-Elektrophorese und Mikro-Immunpräzipitation von Proteinen. Hoppe-Seyler's Z. physiol. Chem. 349, 795-800 (1968)

16. Rüchel, R.: in preparation.

Soft Laser Scanning Densitometer Compatible with the High Resolution Obtained by Electrofocusing

R.A. Zeineh

ABSTRACT

Conventional scanning densitometers are designed for clinical quantitation of serum proteins that are separated by cellulose acetate electrophoresis. Their resolution is not compatible with that of electrofocusing. Modifications of existing spectrophotometers were of limited success.

The new scanner utilizes a thin line laser beam. The laser tracing of closely spaced black lines on an illustration plate demonstrated high resolution and reliable determination of zone spreading during defocusing. For comparison, the laser and conventional scanners were used to trace the electrophoregrams of purified catalase and of red blood cells hemolysate.

INTRODUCTION

Most of the present scanners are designed for the clinic in order to quantitate serum proteins after electrophoresis on cellulose acetate. The resolution of such scanners is not compatible with that of disc electrophoresis (1,2) or the recent technique of electrofocusing (3,4). Spectrophotometers and other instruments have been modified by researchers and by the manufacturers to improve resolution and to permit faithful quantitation of electrophoregrams. Success was limited. First, most of the modifications resulted in a limited application. Tubes and narrow strips only could be scanned. Slabs, autoradiographs and the two dimentional plates need to be scanned also. Second, conventional scanners utilize the white light source and the slit system which provides the incident beam with adjustable width. Closely stacked bands of stained proteins with visible good separation on the electrophoregram reveals overlaping or complete fusion upon densitometric tracing with conventional scanners. Thus, high resolution obtained by electrofocusing may be lost upon quantitative scanning.

Among the number of instruments and modifications that have been described for quantitive analysis of electrophoregrams are the projector-type enlargement (5), photographic enlargement, the microscope system detection, photon counting and the production of parallel incident beam. Each modification was associated with a drawback that prohibited it's adoption.

Futhermore, high resolution provides accuracy in the determination of zone spreading of protein bands. Zone spreading which occurs during the defocussing stage (absence of electric field) is related to diffusion coeffiecient which is in turn related to Molecular Weight of the protein (6,7).

The present soft laser scanning densitometer provides high resolution suitable for quantitative analysis of the electrophoregram and for the determination of zone spreading of the protein bands.

MATERIALS AND METHODS

Carrier ampholyte pH 2-9, PAG plates pH 2-11 and Multiphor Electrofocusing unit model 2117-010 were obtained from LKB, Bromma, Sweden. Servalytes pH 2-11 & 7-9 were obtained from Serva Technik, Heidelberg, Germany. Acrylamide (#5521), N,N^1 methylene bisacrylamide (#8383), Tris 3 amino-2 hydroxymethyl-1,3 propanediol (# 4833) N,N,N^{1}N^1-tetramethyle-thylenediamine (#8178,TEMED) riboflavin (#11151), and glycine-ammonia free (#445) were obtained from Eastman, Rochester, New York. The soft laser scanning densitometer, model SL-504 with DC-500 computing integrator and two dimentional scanner, model SL-503-2D, with a T.V. screen display and plotter, were obtained from Biomed Instruments, Chicago, Illinois. The Clifford Scanning densitometer was obtained from Corning, Natick, Massachusets.

Five groups of solid parallel black lines at different spacings were photographically reduced to serve as an illustration plate. This plate simulates protein bands on electrophoregram. The plate was scanned with varying beam widths in order to demonstrate resolution and zone spread determination. The peak width on the graphical recording, was measured at 61% of maximium height. The result is then multiplied by the scan ratio 1:7 to give the line width on the plate. The results of line width determination were compared to those obtained by direct measurement using the microscope with 10 x 10 magnification and a reference scale of 0.01 mm divisions.

In order to determine the value of the scanner from an application point of view, electrophoregrams of catalase and hemoglobin were scanned. Electrofocusing of purified calf catalase on thin plate was kindly supplied by LKB. Human red blood cells were washed 3 times with isotonic sucrose, and then lysed in distilled water. The cell walls were centrifuged out. Aliquots of 1,2,5,10,15 and 20 ul of the hemolysate, 15 mg/ml were applied to PAG plates pH 2-11 and to 7% polyacrylamide vertical slabs, 2mm thick and the pH ranges of 2-9, 2-11 and 7-9. After 1 hour of electrofocusing, the slabs were stained with coomassie brilliant blue. The electrophoregrams were scanned by the scanner model SL-504 using it's built-in white light source and by the Clifford Scanning Densitometer. Resolution and zone spread determination were compared to those of the laser scanning at 633 nm wavelength.

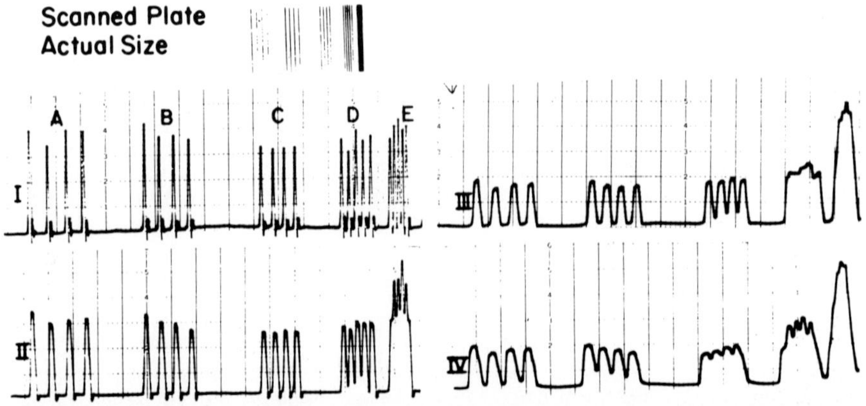

Fig. 1. Actual size picture of the illustration plate (top) and it's Densitometric tracing, (bottom) by the soft laser scanner Model SL-504. The plate has five groups of lines. From left to right, each group, A,B, C, D & E, has 4,4,4,5 and 5 lines and the interspacing in between the lines of each group are 900, 700, 500,300, and 120 u respectively. Laser beam widths of 50,300,400 and 500 u produced graphical tracings I, II, III, IV, respectively. The five lines in group E were misrepresented by four peaks and two shoulders, tracing II. As the beam width increased, resolution decreased, tracings III & IV.

RESULTS AND DISCUSSION

The Effect of Beam Width on Scanner Resolution. The graphical tracing of the illustration plate with 3mm long and 50, 300, & 500 u wide laser beam is seen in Fig. 1. Tracing with 10,50 or 100 u beam revealed good resolution for all the groups A-E. Upon optimal alignment, a better resolution was observed with the narrow beam of 10 u width. Resolution decreased upon increasing the beam width. Upon scanning with a 300 u beam, the five lines at the right side of the plate appeared as 4 peaks with two shoulders, group E, tracing II. This false resolution is due to the fact that the beam width 300 u is equal to the total sum of two adjacent lines plus the interspacing; 85, 85 & 120 u. Using a 400 u beam, groups E and D were not resolved, tracing III. Group D has a plateau shape because the beam width equals the total sum of the line width 85 u, plus the interspacing; 300 u. The tracing with a 500 u beam, revealed a significant loss of resolution in groups, C, D & E, tracing IV. Most conventional scanners utilizing the white light source and the slit system use a beam width of 200-400 u (8) and their resolution is equivalent to tracings II and III, Fig. 1. Smaller slits either produce interference or does not permit a detectible amount of light.

Eventhough the mechanical response of the recording pen in the laser scanner, model SL-504 is fast, 0.03 seconds, full deflection, and the photoelectric response is 0.002 m sec, the carriage speed of 42 cm/min was rather too fast for resolving spacing of 100 u magnitude or less.

The resolution obtained by the optical system and retained by the photoelectric response is lost by the mechanical drag of the recording pen. Slowing down the carriage speed from 42 to 7 cm/min in SL-504 laser scanner improved resolution. The use of digital data aquisition by an on-line computor, Model SL-503-2D eliminates the problem of mechanical drag. However, the major advantage of the on-line computor is extended performance. Data could be stored for later analyses and manipulations. Flexibility, convenience, and accuracy are provided. For example, any segment of interest on the graphical tracing could be electronically expanded in order to demonstrate a finely resolved pattern, to explore subtle differences and to allow accurate measurement of zone spreading or spacing between adjacent bands, Fig. 2.

Zone Spread Determination from Graphical Tracing. According to the microscopic measurement, the average width of the 22 uniform lines on the illustration plate was 85 ± 6 u. The average peak width in graphical tracings I, II, III and IV produced by using 50, 300, 400 and 500 u beams were 0.66, 0.95, 1.45 and 2.11 mm respectively. The respective calculated values for line width were 95,134, 207 and 301 u with standard deviations of 23, 28, 36 and 57 respectively. Line width determination, similar to zone spread evaluation was more accurate when a narrow light beam is used for tracing. The line width value of 95 u of the tracing produced by 50 u beams compares well with 85 u microscopic determination. Line width determinations from tracings produced with 10, 50 and 100 u laser beams in scanner Model SL-504 revealed values of insignificant differences, 102,95 and 96 u respectively. Lack of significant difference could be attribute to unattained alignment of lines on the plate relative to the incident beam. Similar determination on another illustration plate was made by the scanner Model SL-503-2D. The line width calculated from expanded segments of the graphical tracing produced by 10, 50 and 100 u beams averaged 91, 97 and 108 u respectively. The microscopic value of average line width was 87 u.

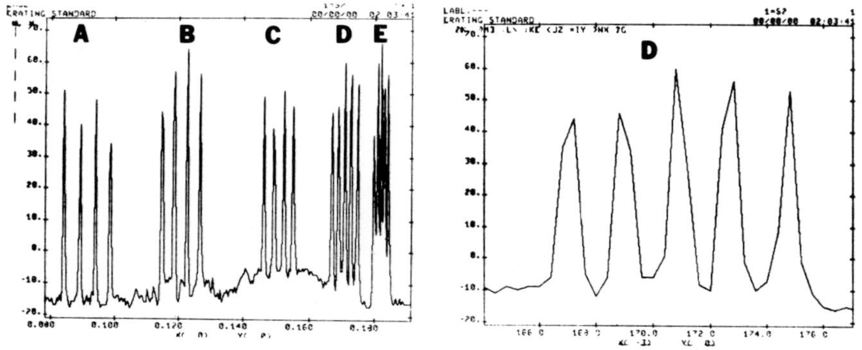

Fig. 2. Densitometric tracing of another illustration plate with scanner SL-503-2D using 2 mm long and 100 u wide beam. The pattern appearing on the TV screen and later recorded by the plotter is seen, left. A segment of interest, group D was expanded electronically, right. This selective expansion permits accurate determination of zone spreading and demonstrates details and fine resolution.

Scanning of Catalase and Hemoglobin Electrophoregrams. The densitometric tracings of catalase electrophoregram revealed that the narrow laser beams retain the resolution obtained by electrofocusing and permits a valid determination of zone spreading. Optimal resolution was obtained with 3 mm long and 50 u wide laser beam. Spot and narrower beams improved resolution but the graphic tracing was not smooth. Scanning with white light systems at various slit widths did not faithfully retain the fine resolution of catalase. Parallel results were observed upon the densitometric tracings of hemoglobin electrophoregrams. The area under each peak was proportional to the protein content of the corresponding band unless the maximal O.D. of that stained band exceeds 1.8. Since the scanner response is linear up to 3.2 O.D. then the non-linear recording of the hemoglobin electrophoregram may be associated with staining procdure.

REFERENCES

1. Ornstein, I. Disc electrophoresis. Ann. N.Y. Acad. Sci. 121, 321-349 (1964)

2. Davis, B.J. Disc electrophoresis:method and application to humans serum proteins. Ann. N.Y. Acad. Sci. 121, 404-427 (1964).

3. Svenson, H. Isoelectric fractionation. analysis amd characterization of ampholytes in natural pH gradients. Acta Chem. Scand. 15, 325-341 (1961)

4. Taber, H. W. and Sherman, F. Spectrophotometric analyzers for disc electrophoresis: studies on yeast cytochrome C. Ann. N.Y. Acad. Sci. 121, 600-615 (1964).

5. Verbanov, V. S., Spaaskii, G. A. and Leontev, V. K. Modification of MF-4 microphotometer for densitometry of polyacrylamide gel electrophoregrams. Biomed. Eng. 7(3), 188-189 (1974).

6. Catsimpoolas, N. Transient state isoelectric focusing experimental determination of apparent diffusion coefficients in polyacrylamide gels. Anal. Biochem. 54, 79-87 (1973).

7. Catsimpoolas, N. Transient state isoelectric focusing digital measurement of zone position, zone area, segmental pH gradient and isoelectric point as a function of time. Anal. Biochem. 54, 66-78 (1973).

8. Zeineh, R. A., Nijm, W. P. and Al-Azzawi, F. Soft laser scanning densitometer for quantitation of tube isoelectric focusing. Amer. Lab. 7(2), 51-58 (1975).

Immunocore Electrofocusing: A Separation and Detection Technique Amenable to Scanning Densitometry

R.A. Zeineh

ABSTRACT

Electrofocusing of serum proteins was performed on a hollow cylinder of polyacrylamide gel. The core of the cylinder was then filled with agar containing antisera. The precipitin bands formed in the agar core were investigated with soft laser scanning densitometry. The technique offers improvements in resolution and sensitivity.

INTRODUCTION

The resolution obtained by electrofocusing is superior to that of other electrophoretic methods (1). Staining is widely used for detection and scanning densitometry for quantitation. However, loss of resolution is encountered during the defocusing stage prior to fixation. Adjacent bands may fuse together or may overlap by diffusion. The rocket type technique of immunological detection is not attenuated by diffusion, is more sensitive than staining and amplifies the resolution obtained by electrofocusing. Immunocore electrofocusing revealed similar type of improvement (2,3). Furthermore, the precipitin bands formed in the agar core could be analyzed by densitometric tracing.

In this study, the densitometric tracing of precipitin bands of immunocore electrofocusing (ICEF) is evaluated.

MATERIALS AND METHODS

Acrylamide, N,N^1-methylene bis acrylamide, Tris(3 amino-2-(hydroxymethyl) -1, 3 propanediol), N,N,N^1,N^1-tetramethylethylenediamine (TEMED), riboflavin and glycine-ammonia free were obtained from Eastman, Rochester, N.Y. Agar was obtained from Difco, Detroit, Michigan, Ampholine pH 3-10 from LKB, Bromma, Sweden, Servalyte pH 2-9 from Serva Technik, Heidelberg, Germany, Serum albumin and transferrin and their specific antisera from Boehringer, Mannheim,California. The purified albumin and transferrin were further purified and the antisera was adsorbed. The immunocore electrofocusing cell and the soft laser scanning densitometer model SL-504 were obtained from Biomed Instruments, Chicago, Illinois.

A 14 x 0.6 (ID) cm glass tube open at both ends is capped at the lower end and 15 x 0.35 (D) cm plastic rod is centered in the tube. The protein sample was mixed with 7% polyacrylamide containing ampholine pH 3-10 and then pipetted in the tube around the plastic rod and electrofocused.

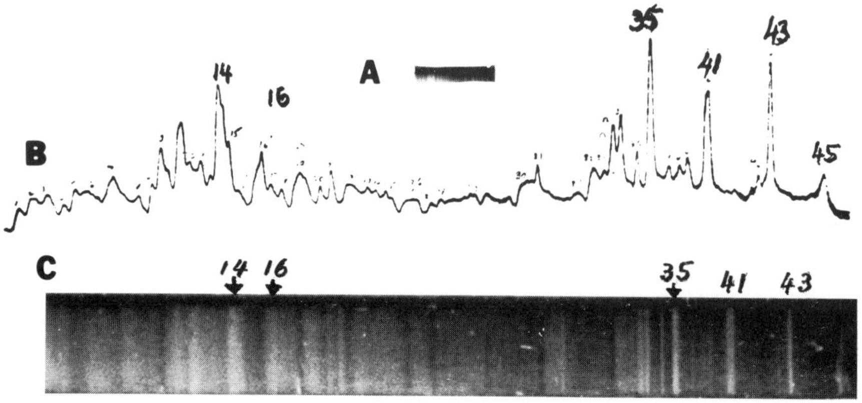

Fig. 1. ICEF of human serum against polyvalent antisera. The core segment (A) of 2.5 cm length contains 45 sharp precipitin bands as seen in laser tracing (B) and in 10x magnification (C). Spherical aberration and field depth have restricted camera focusing to the right side of (A).

The rod was withdrawn and the core was filled with 1 gr% agar in pH 8.1 buffer containing antisera. Precipitin bands that developed in the agar core were analyzed by scanning. Varying amount of serum or antigen were used in order to determine the linear range of valid quantitation. A 15x 0.4x0.15 cm plastic piece was centered in the tube with one side touching the inner wall and was used instead of the round rod.

RESULTS AND DISCUSSION

ICEF of human serum against polyvalent antisera revealed 146 discrete precipitin bands. The densitometric tracing made by the soft laser scanner revealed compatible resolution, Fig. 1. The linear range for quantitating individual bands of serum orosomucoid was 1-18 ug. A linear range 0.1-1.5 ug was achieved by diluting the antisera. Antigen excess results diffuse wide band that gradually splits into two tha diffusely migrate sideways to zones of equivalence. The use of a rectangular plastic piece instead of the round rod revealed precipitin spikes of varying heights and densities. The ICEF retains and amplifies the resolution of electrofocusing but not as efficient as electrofocusing followed by rocketting.

FEFERENCES

1. Dale, G. and Latner, A. L. Isoelectric focusing on plyacrylamide gel. Lancet 1, 874-848 (1968).

2. Zeineh, R. A., Barrett, B., Niemirowski, L. and Fiorella, B. J. Turnover rate of orosomucoid in the dog with sterile abscess. Am. J. Physiol. 222, 1226-1232 (1972).

3. Zeineh, R. A., Mbawa, E. H, Pillay, V. K. G.,Fiorella, B. J. and Dunea, G. Immunocore electrophoresis of urinary albumins. J. Lab. Clin. Med. 82, 326-333 (1973).

Rapid, Convenient and Economical Procedures for the Determination of the Isoelectric Spectra of Proteins

D.H. Leaback

ABSTRACT

A new column for the density-gradient-stabilized isoelectric focusing is described which permits the rapid determination of the isoelectric spectra of proteins for which sensitive and specific assay procedures are available. The design of the column and the materials employed permit the separation of the protein species in a few hours using a few millilitres of density gradient. The technique is rapid and very economical in terms of the amounts of protein sample and carrier ampholytes required. The use of the column has been illustrated by the separation of the A and B variants of a mammalian β-N-acetyl-D-hexosaminidase.

INTRODUCTION

Recent years have been notable for the development of isoelectric focusing procedures for rapid, convenient and relatively economical separations of proteins in thin layers of polyacrylamide gel. Such procedures can give indications of the isoelectric spectra of protein mixtures but are not well suited to the quantitative determination of the isoelectric spectrum of an assayable protein of particular interest.

In 1970, Grant and Leaback [1] described how a commercially-available column designed for electrophoresis could be employed for the determination of the isoelectric spectrum of readily-assayable proteins after separation by density-gradient-stabilized isoelectric focusing. A simpler version of this column was described by Fawcett [2] in 1975. In the present communication a further modification of this apparatus is described together with its use in rapid, convenient and economical procedures for the measurement of the isoelectric spectra of proteins that can be assayed readily in small samples from the contents of the density-gradient-stabilized columns.

EXPERIMENTAL

Materials

The enzyme sample used was the crude concentrated extract prepared as described by Robinson and Leaback [3]. The carrier ampholytes employed were purchased from Messrs. L.K.B. Instruments and other reagents were of AR Grade.

Description of the New Apparatus for Isoelectric Focusing.

A schematic diagram of the new column for isoelectric focusing is shown in Fig. 1. The apparatus comprises a central thin-walled, borosilicate glass column CC' (1cm int. diam x 40cm long) surrounded by a perspex coolant jacket (J) through which water at about 5°C circulates after passing through the small glass coil (G) immersed in the dense, electrolyte solution in the lower electrode compartment (L). Coolant water enters at I and leaves the system at O. A potential difference is applied to the contents of the column by means of the terminals T which are connected to the platinum wire electrodes E contained in the upper (U) and lower (L) electrode compartments. The voltage is applied to the lower end of the density gradient through three 4mm diam. holes (H) in the glass column CC' over which a porous polyethylene sleeve (S) has been fitted snugly and into which a semipermeable membrane of polyacrylamide gel has been impregnated. (v.i.) A moulded silicon-rubber cap F is fitted to the column at lower end of the column before and during electrolysis.

SOLUTIONS FOR COLUMN

The following solutions proved satisfactory for the particular applications described here and should serve as a guide for other isoelectric focusing separations.

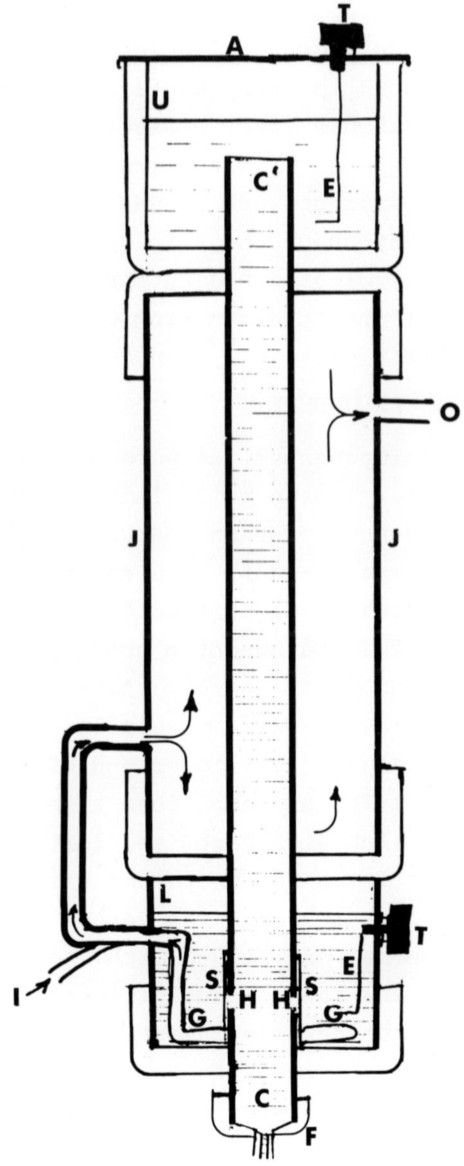

Fig. 1. Schematic diagram of the new column for density gradient stabilized isoelectric focusing. Central glass column (CC') with 4mm diam. holes (H) surrounded by porous polythene sleeve (S) impregnated with polyacrylamide gel. Coolant inlet (I) into glass coil (G) to coolant jacket (J) with outlet (O). Terminals (T) connected to electrodes (E) immersed in electrolytes contained in upper (U) and lower (L) electrode vessels. Rubber cap (F) and plastic plate (A).

Anode (upper) electrode solution. 150ml of 1% (w/v) aqueous orthopha acid. Cathode (lower) electrode and dense support solution. NaOH 1% (w/v) in dense sucrose solution. Dense sucrose solution. Sucrose (84g) dissolved in water (126ml). Density gradient solutions. For the gradient (approximately 9ml) employed in the present work it was convenient to make up 5ml of the dense component (containing 200µl of 40% pH3-10 carrier ampholytes in 5ml of dense sucrose solution) and 5ml of the less dense component (containing 0.25g sucrose and 50µl of 40% w/v carrier ampholyte solution).

ISOELECTRIC FOCUSING PROCEDURE

The first task is to impregnate the porous polythene sleeve S with a suitable polyacrylamide gel. This is done by immersing the porous sleeve (on the lower end of the glass column but less the electrode compartment L) in a small plastic vial containing a fresh solution of acrylamide (1.8g), N,N'-methylene bisacrylamide (0.2g) and ammonium persulphate (20mg) in water (20ml) containing 0.1ml Temed. After 30min. the vial is removed and excess gel removed carefully from the membrane and from inside the column. The lower end of the column is then immersed in water for an hour, then stoppered at C and the central column filled with water and left to stand overnight. If the water level in the column has not fallen by more than about 1cm overnight the gel membrane is satisfactory and the column is drained and fitted with the lower electrode compartment assembly and the coolant circulated at 5°. After fitting the lower electrode compartment, the central column is filled with dense lower electrolyte solution to within about 10cm from top by means of a peristaltic pump (pumping at about 2ml/min) connected to the cap (F). The density gradient is then layered carefully on the dense support solution within the column (adding the 50µl sample at an appropriate point). The gradient is lowered by about 10cm by pumping support solution from the lower end of the column and the upper electrode compartment is added together with electrolyte and the electrode assembly (A). Electrolysis is carried out at a constant 1000V (initially at about 2.5W) and equilibrium was achieved in 3-4hr. whereupon the power is disconnected followed by removal of the upper electrolyte solution and electrode assembly. A cap similar to that shown (F) at the lower end is then fitted to the top of the column C' and the contents of the column collected by upward displacement using a peristaltic pump working at 2ml/min; 100μl fractions are collected and pH measurements carried out immediately.

Other procedures. β-N-acetyl-D-hexosaminidase (EC.3.2.1.30) assays were carried out using 5μl samples and 4-methylumbelliferyl 2-acetamido-2-deoxy-β-D-glucopyranoside (1mM) as substrate but otherwise as described earlier [4]. Isoelectric focusing gels were prepared, run and stained prior to scanning with a Joyce-Loebl densitometer Mark IIIa using the red filter supplied by the manufacturers.

RESULTS

Fig. 2 shows the results of enzyme and pH measurements on the fractionated contents of the new electrophoresis column after isoelectric focusing of the epididymal enzyme. The results show a preponderance of the basic (B) variant of the enzyme and compare favourably with results on similar preparations separated previously using an LKB 8101 isoelectric focusing column-which used over ten times the quantity of carrier ampholytes and which took about six times longer to achieve isoelectric focusing equilibrium [3,6]. As pointed out by the author [7] the focusing of large, basic proteins is particularly difficult, so that the 'B' form of the pig epididymal enzyme (with a molecular weight of about 130,000 daltons and pI of about 8.2)[3,5] is probably a good test of the capabilities of the new column.

In fig. 3 is shown the results of the densitometric scan of a stained gel of the proteins present in the same crude epididymal enzyme preparation-but after transposition of the absorp-

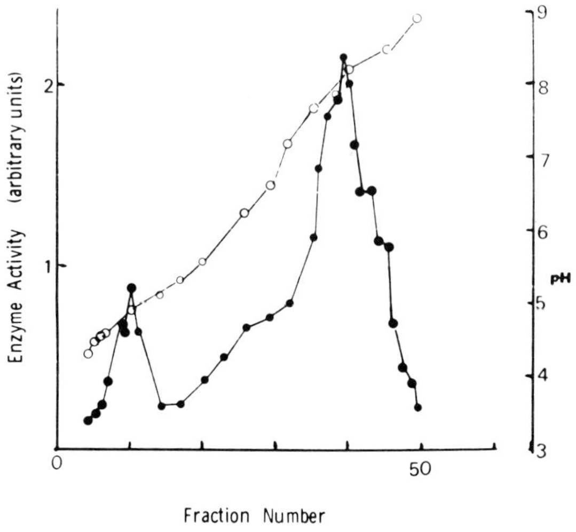

Fig. 2. pH (O—O) and β-N-acetyl-D-glucosaminidase activity (●—●) in fractions (100µl) eluted from the new density-gradient stabilized column after isoelectric focusing for 4hr. at 5° using a potential difference of 1000V.

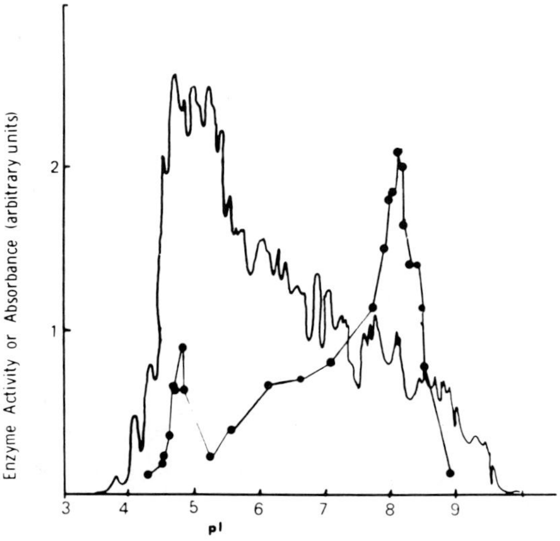

Fig. 3. Isoelectric spectra of crude pig epididymal β-N-acetyl -D-glucosaminidase activity (●—●) and of the stainable proteins present in the enzyme preparation (—) determined respectively by isoelectric focusing in the new density gradient stabilized column and in a thin-layer polyacrylamide gel and displayed here on a linear pI scale.

tion contour onto a linear pH abscissa. The results shown in fig. 2 have been transposed onto the same linear pH abscissa, so that fig. 3 readily permits the comparison of the isoelectric spectrum of the particular enzyme with that of all the (stainable) proteins present in the crude preparation.

DISCUSSION

It is now fully a decade since Vesterberg and Svensson [8] introduced the jacketed glass column which, in its commercial form (principally the LKB 8101 column) has been used extensively for analytical isoelectric focusing, or for relatively small preparations of proteins after separation according to the isoelectric focusing principle. Such columns are expensive to fabricate; difficult to repair; inflexible in operation (e.g. can only be emptied by downward displacement); present a complex internal contour from which it is difficult to obtain non-turbulent elution of the separated proteins; and are not economical in the protein samples required nor in the (expensive) carrier ampholytes necessary to generate an adequate pH gradient. Density gradient stabilized isoelectric focusing is not well suited to the determination of the proteins in crude solutions (due to the interference of carrier ampholytes with most general methods for detecting proteins), but this is less important now that thin-layer gel techniques permit rapid isoelectric focusing of proteins and can give a semi-quantitative assessment of the isoelectric spectrum of all the stainable proteins in the sample (see fig. 3). Although there are some interesting developments in preparative isoelectric focusing [2], such approaches are severely limited by the expense of the carrier ampholytes, and the need to remove these materials from the separated proteins. In the opinion of the author the most important role that isoelectric focusing has to play in the preparations of proteins on any substantial scale, is in the guidance that the method can give in monitoring the progress of the purification and in the choice of conditions for ion-exchange chromatography [3].

For this purpose an isoelectric spectra of the proteins involved are invaluable for guidance to a strategy for ion-exchange chromatography. An economical, density-stabilized analytical column for isoelectric focusing would seem to be the best way of arriving at such information.

Grant and Leaback [1] described procedures whereby a commecial version of a column (originally designed by Brake, Allington and Langille [9]) could be adapted for analytical isoelectric focusing. While this column had a number of attractions (upward displacement, simple internal contours and an internal capacity of up to about 13ml), it was expensive; it committed a U.V. monitor exclusively to this use; it had two fragile and difficult-to-replace cellulose acetate membranes, and had very large capacity electrode compartments; it was fabricated with a considerable number of (potential leakable) '0'-ring seals and certain of the materials employed in the construction (e.g. Teflon central tube) did not readily transfer heat. After discussion with the Manufacturers, certain (but by no means all) of these features were modified and another version (Isco model 212) was produced which was designated as a column for isoelectric focusing. The essentials of the use of these Isco columns have been described more recently by Leaback and Wrigley [10].

The new column described here is not difficult to construct; it has one, readily-repairable membrane; it has simple internal contours; the material of the central column was chosen to facilitate heat transfer; the design of the column permits (if required) the removal of the column contents by either upward of downward displacement; as with the Isco columns [1, 10], it is possible to add a U.V. monitor and an extension tube to the top of the column (at C') and therefore offers the option of monitoring the progress of the focusing before finally deciding to remove the gradient from the column.

ACKNOWLEDGEMENTS

The author is indebted to Mr. Steven Creme for able assistance and to Dr. G.H. Beaven (M.R.C., Mill Hill) for the use of the scanning densitometer.

REFERENCES

[1] Grant, G.M. and Leaback, D.H., The Use of the Isco Density Gradient Electrophoresis Columns for Isoelectric Focusing. Shandon Instrument Application No. 31, 1-10 (1970).

[2] Fawcett, J.S. Some recent developments in preparative Isoelectric Focusing in "Isoelectric Focusing", Arbuthnott, J.P. and Beeley, J.A. (eds.) Butterworths, London, 23-43 (1975).

[3] Robinson, H.K. and Leaback, D.H., A New Systematic approach to the isolation of proteins illustrated by the Purification of a mammalian β-N-acetyl-D-hexosaminidase . Biochem. J. 143, 143-148 (1974).

[4] Leaback, D.H. and Walker, P.G., The Fluorimetric Assay of β-N-acetyl-D-hexosaminidase . Biochem. J. 78, 151-156 (1961).

[5] Leaback, D.H. and Robinson, H.K., A new procedure for the two-dimensional Display of the molecular-size-electric charge characteristics of native proteins in crude mixtures . FEBS. Lett. 40, 192-195 (1974).

[6] Leaback, D.H. and Robinson, H.K., Ampholyte Displacement Chromatography - A new technique for the separation of proteins illustrated by the resolution of β-N-acetyl-D-hexosaminidase isoenzymes unresolvable by isoelectric focusing or conventional Ion-exchange chromatography . Biochem. Biophys, Res. Commun. 67, 248-254 (1975).

[7] Leaback, D.H. Extreme pH ranges in "Progress in Isoelectric Focusing and Isotachophoresis", P.G. Righetti (ed.), North Holland Publ. Co., Amsterdam, p 376, (1975).

[8] Vesterberg, O., and Svensson, H., Isoelectric Fractionation analysis and characterization of ampholytes in natural pH gradients . Acta. Chem. Scand. 20, 820-834 (1966).

[9] Brake, M.K., Allington, R.W., and Langille, F.A., Mobility measurements by photometric analysis of zone electrophoresis in a sucrose gradient column . Anal. Biochem. 25, 30-39 (1968).

[10] Leaback, D.H. and Wrigley, C.W., Isoelectric Focusing of Proteins in "Chromatographic and Electrophoretic Techniques", 4th ed. Vol. 2, I. Smith (ed) Heineman, London, 272-320 (1976).

A Simple Method of Choosing Optimum pH-Conditions for Electrophoresis

Å. Rosengren, B. Bjellqvist and V. Gasparic

The present work describes the possibility of performing electrophoresis perpendicular to an Ampholine® pH-gradient, focused in a flat bed of polyacrylamide gel. In a focused pH-gradient all Ampholine® molecules are close to their isoelectric points, and thus their electrophoretic mobilities are very low. This implies also that, if an electric field is applied perpendicularly to the pH-gradient, an appreciable time will elapse during which the pH-values in the gel will remain practically unchanged. During this time, electrophoresis can be run perpendicularly to the pH-gradient at approximately constant and known pH-values. This can be useful in a number of ways.

1. Samples can be applied all along the focused pH-gradient. This gives, in one single experiment, the mobility variation of sample components in the pH-interval created by the focused Ampholine®. This type of information is of great help in choosing optimum conditions for disc-electrophoresis, isotachophoresis and conventional electrophoresis.

2. The mobility variation curves correspond approximately to titration curves, and thus can be used to give the same type of information as titration curves. A minimum amount of sample is needed, and titration curves are obtained for all protein components in the sample. Denaturation and rearrangement reactions can be spotted as diffusiveness and/or discontinuities in the mobility variation curves.

3. The point where the mobility variation curve crosses the application line is the isoelectric point for the corresponding component in the sample. Comparison of band pattern, obtained by isoelectric focusing and by the method described, makes it possible to recognize if any extra bands have been created in isoelectric focusing, owing to conditions unsuitable for the sample.

4. In a focused Ampholine® gel, electrophoresis can be performed at any desired discrete pH-value, followed by isoelectric focusing in the same gel.

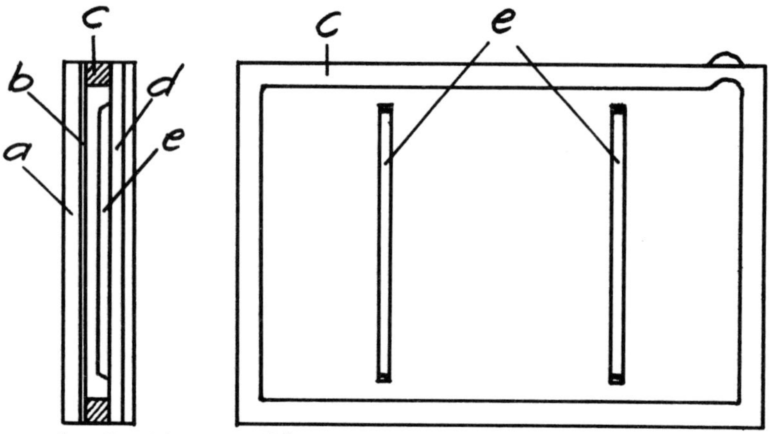

Figure 1, Casting mould
a) glass plates 260x125x2mm, b) polyester foil, c) rubber gasket 2mm thick, d) Plexiglas® milled off to leave two bars, e) 1.7 height, 1.2mm width and 106mm length. The bars are situated 70mm from the short sides of the plate.

Materials and methods

The experiments were performed on LKB 2117 Multiphor, connected to a constant power supply, LKB-Biochrom 2103 Power Supply. The behaviour of focused pH-gradients, when an electric field was applied perpendicularly to the gradient, was examined on Ampholine® PAG-plates pH 3.5 to 9.5 (LKB-Aminkemi 1804-101). The electrophoretic experiments were performed on cast gels, the mould for which is shown in figure 1.

The composition of the gels was T=4% and C=3%, and the final Ampholine® concentration was 2.4%. The Ampholine® mixture used contained 70% Ampholine® pH 3.5-10 (LKB 1809-101), 10% Ampholine® pH 2.5-4 (LKB 1809-106), 10% Ampholine® pH 7-9 (LKB 1809-136) and 10% Ampholine® pH 9-11 (LKB 1809-146).

In all steps of the experiments, the same electrode solutions and the same type of electrode strips were used. 1 M H_3PO_4 was used as anodic solution and 1 M NaOH as cathodic solution. The paper used as strips was Grycksbo 1220, Stora Kopparberg, Sweden.

When the aim of the experiment was to generate mobility variation curves, the sample was applied all along the slot. In order to perform electrophoresis at any desired pH-value between pH 3.5 and 10, the slot was filled in on both sides of the desired pH-value after the initial focusing step, using small pieces of polyacrylamide gel.

Experimental

The behaviour of focused pH-gradients was examined when an electric field was applied perpendicularly to the gradient. Focusing was performed on the commercial PAG-plate in accordance with the instructions for isoelectric focusing. Focusing was interrupted as soon as an approximately linear pH-gradient had developed, when the voltage had reached 1150V and the current 25 mA. Immediately after focusing, the gel plates were cut with a scalpel along the electrode strips. The strips and the gel underneath were removed from the cooling plate. A 20 mm wide gel strip, covering the whole pH-interval, was cut out, transferred to another cooling plate and used for pH-measurements. The gel was cut down to a length of 110 mm and turned 90° on the cooling plate. New electrode strips were placed on the gel, and an electric field was applied perpendicularly to the pH-gradient created in the isoelectric focusing.

Owing to the high conductivity in the acidic range, the applied field strength had to be restricted to 30 V/cm.

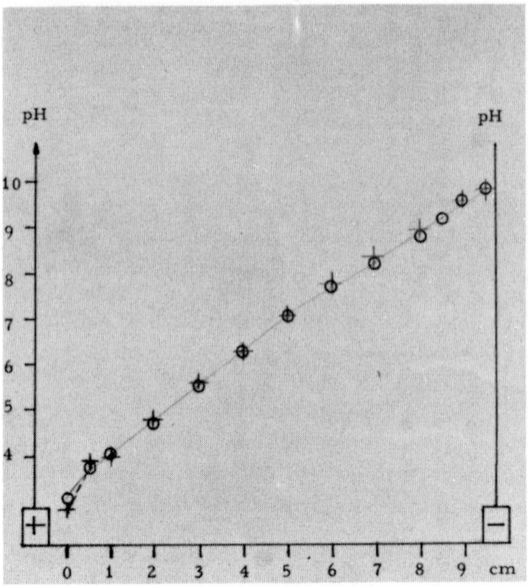

Figure 2
The pH-gradient half-way between the elctrodes before (o) and after (+) an electric field of 30V/cm has been applied perpendicularly to the gradient for 60 minutes.

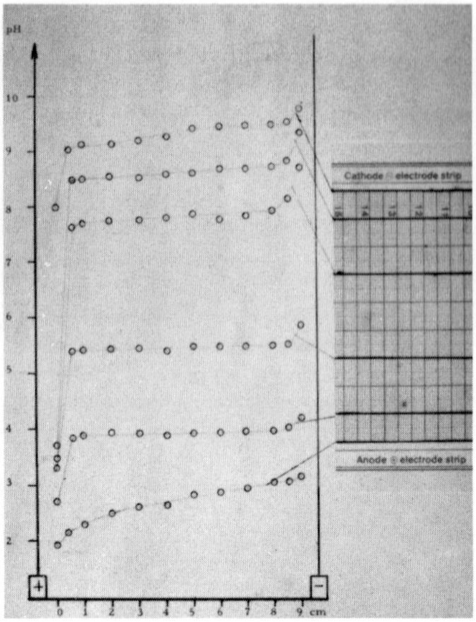

Figure 3
pH-variations perpendicularly to the prefocused pH-gradient when an electric field of 30V/cm has been applied for 60 minutes. On the right positions during focusing.

Figure 2 shows the pH-gradient after the initial focusing as well as the gradient measured half way between the electrodes after the electric field has been applied perpendicularly to the gradient for 60 minutes. A small change of pH-value is observed only in the region of the gel which has been very close to the anode during focusing.

Figure 3 shows the pH-variations along the expected iso-pH-lines in the same experiment. In the region of pH 4 to 9, the behaviour of Ampholine® is close to ideal. Only in the region of the gel which has been close to the anode during focusing does the pH-variation become appreciable.

When 2 mm gels with cast slots were focused, it was found important to suck out excess liquid from the electrode strips. Otherwise electrode solutions leak out into the slots and this causes severe disturbances of the pH-gradient. The 2 mm gels were focused to a final voltage of 600V and a final current of 25 mA. After the focusing, the gels were cut 3 mm from the cathode strip and 6-8 mm from the anode strip.

To prepare the gel for electrophoresis, one alternative is to cut the gel in two halves, which are then turned 90° on the cooling plate, in order to run two electrophoretic experiments in parallel. Another possibility is to apply the electrode strips on the short sides of the focused gel and run the electrophoretic experiments in series. Electrophoresis was performed at 60V/cm for 10 minutes. To ensure the uniformity of sample application all along the slot, it is convenient to add Bromphenol Blue to the sample before application.

Figures 4, 5 and 6 show mobility variation curves of proteins generated in these types of experiments.

When crossed electrophoresis-isoelectric focusing was performed, the desired pH-value for electrophoresis was located after initial isoelectric focusing. After electrophoresis, the electrode strips and the gel underneath were cut off. New, wider electrode strips were used in order to allow the electric field to be applied in the same direction as in the initial focusing step. Figure 7 shows the resolution obtained with human serum as sample.

Discussion

The results have confirmed that no major changes in the pH-gradient take place when electrophoresis is performed perpendicularly to focused Ampholine® gradients.

This type of experiment gives information which could be used in order to choose a suitable pH for different types

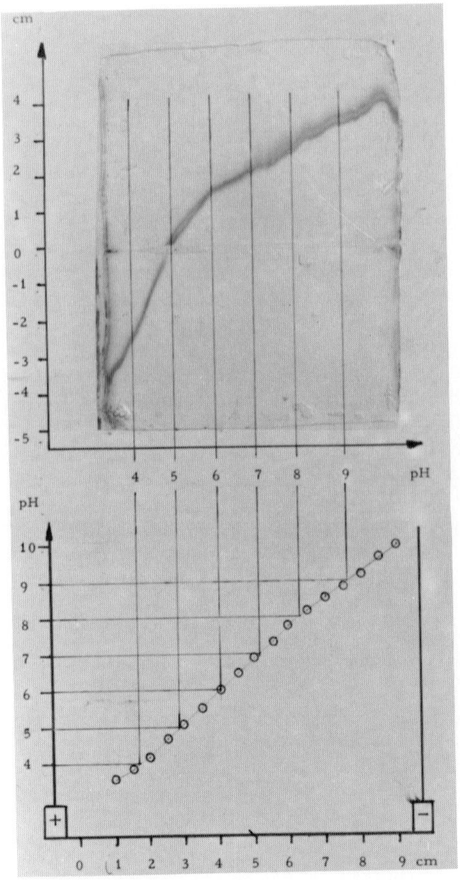

Figure 4
Electrophoresis of β-lactoglobulin perpendicularly to a focused Ampholine® pH-gradient. Electrophoresis: 10 minutes, 60 V/cm.

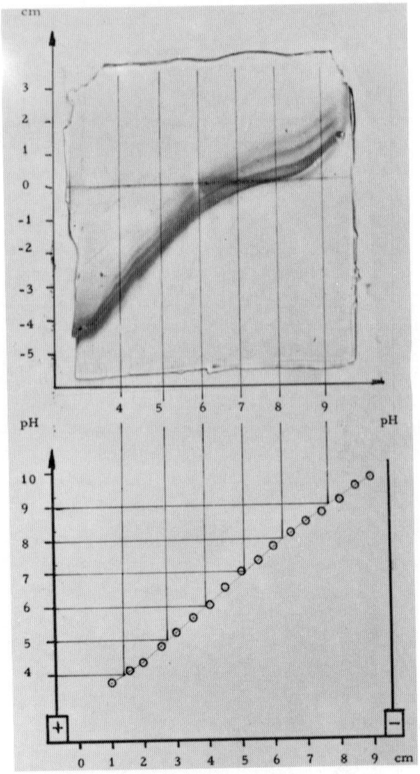

Figure 5
Electrophoresis of sperm whale myoglobin perpendicularly to a focused Ampholine® pH-gradient. Electrophoresis:10 minutes, 60 V/cm.

of electrophoresis, and also to select spacers for isotachophoresis as well as lower and upper stacking limits in disc-electrophoresis.

Without discussing the relevance of the bands obtained in the myoglobin experiment (figure 5), it can easily be seen that the best electrophoretic separation is obtained around pH 8.5 and a lot of resolution is lost at pH 9.

pH-dependent reactions, which might confuse comparison between separations obtained at different pH-values, can be spotted as discontinuities in the discrete mobility variation curves separated by diffuse zones.

Inspection of the β-lactoglobulin mobility variation curves (figure 4) shows that isoelectric focusing should reveal two isomers of the protein, in accordance with obtained results (1). Above pH 6, these two bands coincide. At still higher pH-values (between 7 and 8.2) the protein appears as one diffuse band, which divides into two discrete bands at pH-values above approximately 8.5. The two bands which can be observed in electrophoresis around pH 9 are most likely not identical with the isomers observed in isoelectric focusing. To our knowledge, this is the only simple method to correctly correlate results from electrophoretic experiments run at different pH-values, as well as results obtained by electrophoresis with those obtained by isoelectric focusing.

The point where a mobility variation curve crosses the application line is the isoelectric point of a component in the sample. As a relatively high field strength is recommended in the electrophoretic step, the time needed to generate the mobility variation curves is approximately 10 minutes only, which allows pI determination also for fairly unstable proteins.

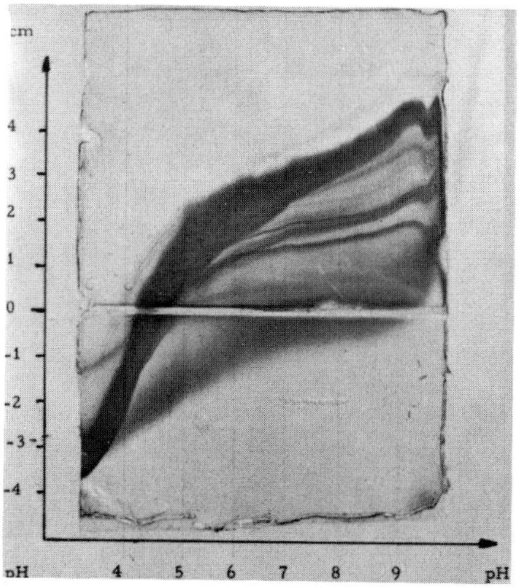

Figure 6
Electrophoresis of human serum perpendicularly to a focused Ampholine® pH-gradient. Electrophoresis: 10 minutes, 60 V/cm.

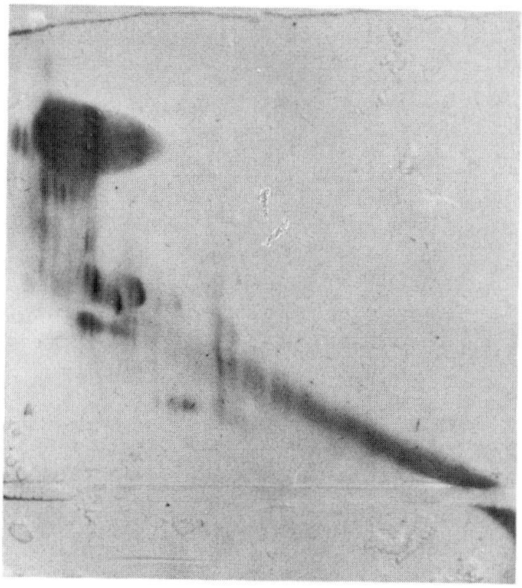

Figure 7
Crossed electrophoresis-isoelectric focusing
Sample: 2 μl human serum
Application pH: 8.90-9.10
Electrophoresis: 10 minutes, 60V/cm
Final focusing: 90 minutes

For proteins with low buffering capacity close to their isoelectric points, especially if they are focusing in the alkaline region of the pH-gradient where drift of the gradient is often observed, the method described offers an alternative way of determining isoelectric points, without any interference by drift phenomena.

The mobility variation curves contain the same type of information as titration curves (2). The high buffering capacity of β-lactoglobulin (figure 4) and low buffering capacity of sperm whale myoglobin (figure 5) at their respective isoelectric points is evident from the slopes of the curves around the application line (3). The curves should also provide information about the nature and number of ionized groups in the molecule. When β-lactoglobulin is run perpendicularly to a pH-gradient (figure 4), the resulting curves indicate that the difference between the two forms of β-lactoglobulin detected by isoelectric focusing is that one imidazolyl group is missing in the acidic variant. At about pH 6, the imidazolyl group is converted to its unionized form and this causes the curves to coincide.

The advantages of crossed electrophoresis-isoelectric focusing in the same gel are obvious. Electrophoresis can be run at the optimum pH-value and is followed by isoelectric focusing, which is a concentrating method. This should mean that resolution and sensitivity are optimum for the combination of these two techniques. The method is also convenient as there is no need to transfer the sample from one gel to another.

The present work covers some but not all possiblities created by using ampholytes at their isoelectric points as electrophoretic buffers. For example, both electrophoresis and isoelectric focusing are often used as criteria of the purity of proteins purified by other methods. The generation of the mobility variation curves should also reveal possible impurities as well as artefacts introduced by the method used.

References

1. *Rilbe, H.; Petterson, S.: Preparative Isoelectric Focusing in Short Density Columns with Vertical Cooling, In: Arbuthnott, J.P., Beeley, J.A. (ed.), Isoelectric Focusing, London: Butterworth & Co., (1975) p. 44-57.*

2. *Kenchington, A.W.: Analytical Information from Titration Curves. In: Alexander, P., Block, R.J. (ed.), Analytical Methods of Protein Chemistry, Vol. 2, Pergamon Press (1960), p. 353-388.*

3. *Fredriksson, S.: On the Temperature Dependence of Isoelectric Points of Proteins with Special References to Isoelectric Focusing. In: Proc. Int. Symp. Electrofocusing and Isotachophoresis, 1976, Hamburg Berlin: Walter de Gruyter (1976).*

pH Determinations in Isoelectric Focusing with an Iridium Electrode

E. Gianazza, P.G. Righetti, S. Bordi and G. Papeschi

SUMMARY

A new method for pH determinations in gels, using an Ir/IrO_2 electrode,is described. This electrode shows a linear relationship potential/pH and a constant temperature coefficient between 0 and 40°C. Its advantages are: fast response times, even at 4°C; high reproducibility and readings on a very small surface area (500 μ diameter).

INTRODUCTION

The pI of a protein determined by isoelectric focusing (IEF) also represents its isoionic point in the absence of complex-forming ions (1). The isoionic point is a measure of the intrinsic acidity of a pure protein,as it is defined as that pH which does not change on addition of a small amount of pure protein (2). It should be remembered that pI values are temperature-dependent and usually decrease with temperature (3). The difference in pI for the same protein, measured at 25 and 4°C, could be as high as 0.5 pH units, the higher value being obtained at the lower temperature. This difference is usually more pronounced in alkaline regions, and when a protein has a pI value close to the pK of some of its functional groups. Ideally, pH measurements should be made at the same temperature used during the IEF separation, since in IEF the temperature coefficient,dpI/dT, refers to the corresponding carrier ampholytes rather than to the protein contained in a given fraction. In fact, at the usual Ampholine concentration (1%), the buffering capacity of carrier ampholytes in the isoelectric state will normally be sufficient to permit them to dictate the pH even in the presence of as much as 1% protein (4,5). This means that the pH value assigned to the concentration maximum of a focused protein at a temperature different from the IEF temperature will refer to the corresponding ampholyte fraction rather than to the protein (6). Furthermore, different protolytic groups are known to display widely different degrees of temperature dependence in their dissociation constants (as a consequence of large differences in their standard heats of ionization) (7). Thus, once a protein has been focused at 4°C, a pI measurement at 20 or at 25°C may not represent the true pI of the protein or the pI of the Ampholine surrounding it.

In preparative IEF in liquid support media, automated pH readings *via* a flow cell have been described by Jonsson *et al.* (8), Secchi (9) and Strongin *et al.* (10). In microcolumns, Fredriksson (11) has used a microglass electrode shaped as a horizontal capillary.

When performing IEF in solid support media, it is convenient to be able to measure directly the pH gradient along the gel length.One of the first electrodes developed for pH measurements on gel surfaces is the antimony micro-electrode, in conjunction with a calomel reference electrode, reported by Beeley *et al.* (12). One important property of this electrode is its rapid equilibration time (less the 10 sec), even at low temperatures, which makes it very attractive for use at 4°C. However, one disadvantage is its low reproducibility, given by Beeley *et al.* (12) as $<$0.25 pH units, in comparison with the resolution of 0.02 pH units afforded by IEF. Alternatively, on gel slabs, the pH can be measured directly on the gel surface with a flat membrane electrode, such as LOT Type 403-30-M8 from Ingold. However, this electrode has a high surface area (membrane diameter, 8 mm).

Instead of performing pH measurements, it is possible to use a calibrated mixture of pH markers (proteins or dyes), covering the pH range of interest. Studies on the use of pH markers have been reported by Conway-Jacobs and Lewin (13), Nakhleh *et al.* (14), Bours (15) and Radola (16). These markers have been tabulated by Righetti and Caravaggio (17).

We report here a new approach to pH measurements, an iridium needle electrode. The advantages of this electrode are: 1) fast response times, even at 4°C; 2) high reproducibility (within 5 hundreths of a pH unit); 3) readings on an extremely small surface area (500 μ diameter). The physicochemical properties of this electrode have been reported by Papeschi *et al.* (18).

RESULTS AND DISCUSSION

Fig.1 shows how to calibrate the Ir/IrO_2 electrode. To this purpose a combined glass electrode is dipped in a thermostated vessel and connected to a pH meter. The potential of the iridium electrode is read on a high input impedance voltmeter against a calomel electrode. If needed, especially for alkaline pH readings, the vessel can be kept under N_2. Fig.2 shows

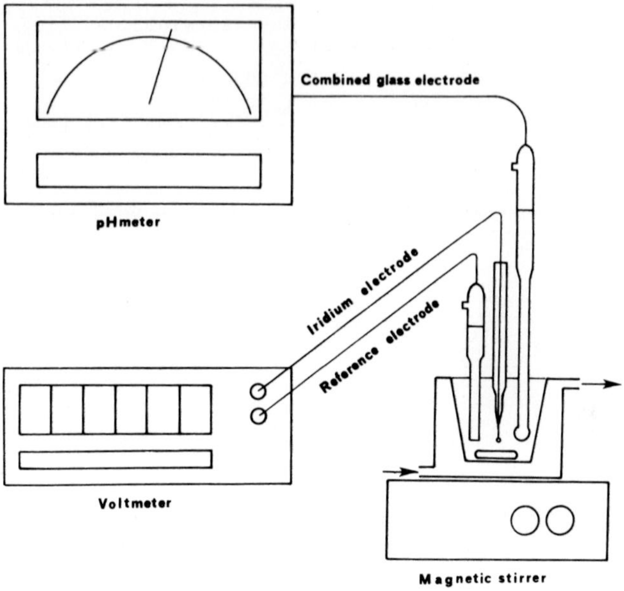

Fig. 1. Experimental assembly for calibration of the iridium electrode against a combined glass electrode. The arrows indicate flow of coolant from a thermostat.

how to perform pH readings with the Ir/IrO_2 electrode on a gel slab. To this purpose, we have built a bridge with a ruler upon which the electrode can slide. We usually perform readings at 5 mm intervals. Each site of pH reading is marked with a 1-2 mm hole bored in the gel with the aid of a gel puncher connected to a water suction line.

Fig.3 shows the calibration of the Ir/IrO_2 electrode at 25 and 4°C against standard buffers and against Ampholine solutions. There is a linear rela-

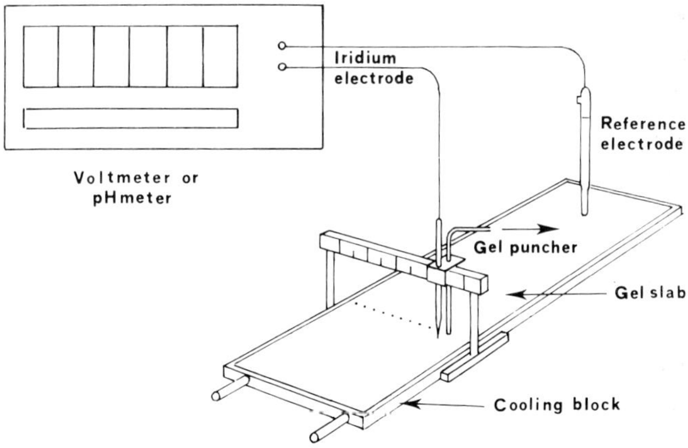

Fig.2. Assembly for pH readings on IEF gel slabs. A bridge with a ruler allows sliding of the iridium electrode at known intervals on the gel surface. Gel discs (1-2 mm diameter) are punched in the slab with the aid of a puncher connected to a vacuum line, at the site of each pH measurement. The gel is kept on the cooling block, at the same IEF temperature. The reference electrode was pre-equilibrated at the temperature of the thermostatic unit.

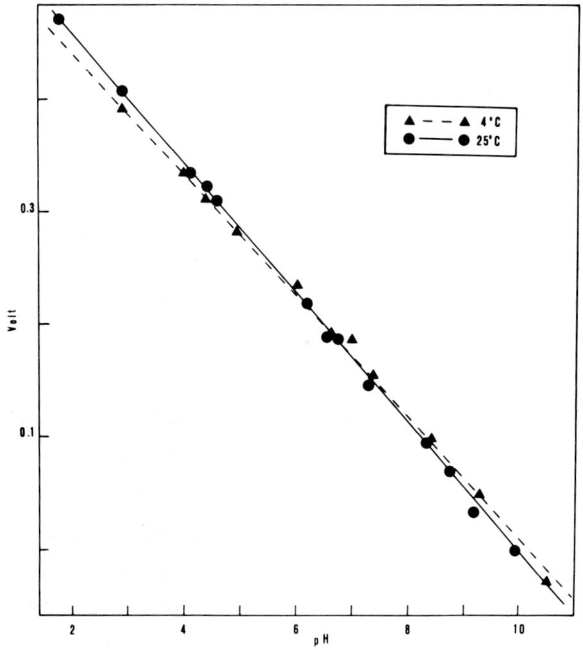

Fig.3. Calibration of the iridium electrode at 25 and 4°C against standard buffers and carrier ampholytes. The LKB Ampholine ranges were 1% solutions of: pH 2.5-4;pH 3.5-5;pH 4-6;pH 5-7;pH 6-8;pH 7-9;pH 8.5-10;pH 9-11. The standard buffers(Radiometer) were:pH 1.68;pH 4.01;pH 6.50 and pH 9.18.

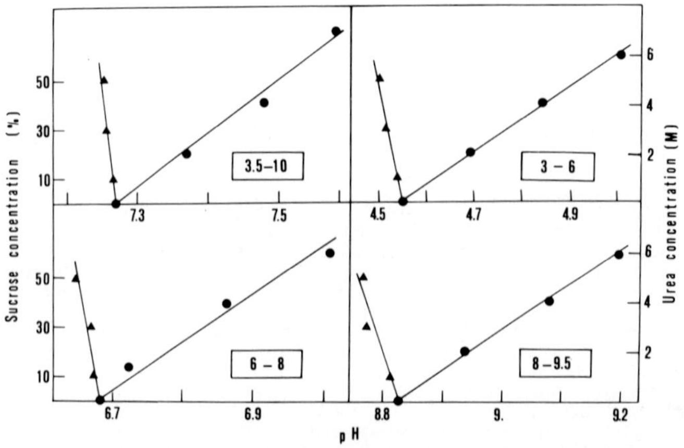

Fig.4. Calibration of the iridium electrode against increasing concentrations of sucrose and urea. Four different Ampholine ranges (pH 3.5-10;pH 3-6;pH 6-8;pH 8-9.5) were made, respectively, either 2,4 or 6 M in urea or 10, 30 or 50% in sucrose. In all experiments Ampholine concentration was 1%. The pH was read at 25°C.

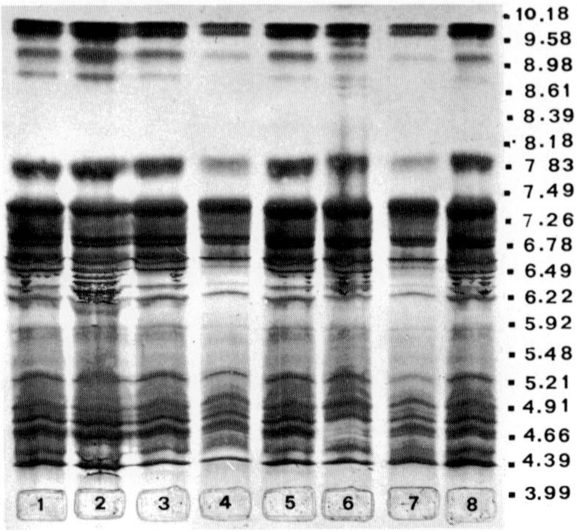

Fig.5. Focusing of rat pancreatic juice after stimulation with caerulein (samples 1-5) or with secretin (samples 6-8). Focusing in a 5% acrylamide gel slab,2% Ampholine pH 3.5-10, at 4°C for 3 hours at 13 W. Sample load: approximately 100 µg.Staining: Coomassie Brilliant Blue R-250. The center of each black dot represents the actual site of pH measurement with the iridium electrode. The cathode is uppermost. (Unpublished experiments with V.Scortecci and C.Poma).

tionship between the potential of the iridium electrode and the pH measured with a glass electrode in the same solution. The few Ampholine points which do not fall on the line deviate by less than 5 hundreths of a pH unit.

Fig.4 shows the calibration of the Ir/IrO_2 electrode against increasing concentrations of sucrose and urea, which are the two most common additives in IEF; in four different Ampholine solutions. The pH increase in urea ranges, for the various Ampholine pH ranges used, from 5 to 7.5 hundreths of a pH unit/unit of urea molarity. This is in agreement with the findings of Ui (19). However, contrary to what is generally believed, we find a small pH decrement with increasing sucrose concentrations. In going from 0 to 50% sucrose, the pH of 1% Ampholine solution decreases by about 0.02 to 0.05 pH units.

Fig.5 shows the focusing of rat pancreatic juice, after stimulation with caerulein and secretin, in a pH range 3.5-10. The black dots represent the actual site of pH measurement with the Ir/IrO_2 needle. By reading the pH *in situ*, we find that the alkaline pH ranges usually give higher pH values than the ones obtained by cutting and eluting gel slices. This is due to: 1) measurements taken at 4°C instead of room temperature; 2) prevention of CO_2 absorption from the atmosphere.

ACKNOWLEDGEMENTS

E.Gianazza is a fellow of the Foundation Anna Villa Rusconi (Varese, Italy). We thank Dr. M. Luzzana for help and advice.

REFERENCES

1. Vesterberg,O.: Isoelectric focusing of proteins. *Svensk. Kem. Tidskr.80*, 213-225 (1968).

2. Rilbe,H.: Historical and theoretical aspects of isoelectric focusing. *Ann. N. Y. Acad. Sci. 209*, 11-22 (1973).

3. Vesterberg, O., Svensson, H.: Isoelectric fractionation, analysis and characterization of ampholytes in natural pH gradients. IV. Further studies on the resolving power in connection with separation of myoglobins. *Acta Chem. Scand. 20*, 820-834 (1966).

4. Vesterberg,O.: Physico chemical properties of the carrier ampholytes and some biochemical applications. *Ann. N. Y. Acad. Sci. 209*, 23-33 (1973).

5. Davies, H.: Some physical and chemical properties of the Ampholine chemicals. *Prot. Biol. Fluids 17*, 389-396 (1970).

6. Fredriksson, S.: Isoelectric focusing in small density-gradient columns. *Ph.D. Thesis*, Chalmers Institute of Technology, University of Gothenburg, Gothenburg, Sweden (1975).

7. King, E. J.: *Acid-Base Equilibria*, Pergamon Press, Oxford (1965).

8. Jonsson, M., Petterson, E., Rilbe, H.: Isoelectric fractionation, analysis and characterization of ampholytes in natural pH gradients.VIII. Continuous recording of pH and light absorbance of column effluent after isoelectric focusing. *Acta Chem. Scand. 23*, 1553-1559 (1969).

9. Secchi, C.: A new method for pH determination in isoelectric focusing experiments. *Anal. Biochem.51*, 448-455 (1973).

10. Strongin, A. J. A., Balduev, A. P., Levin, E. D.: A simple device for pH measurement in flowing liquids. *Sci. Tools 20*, 34-35 (1973).

11. Fredriksson, S.: Scanning isoelectric focusing in small density-gradient columns.II. Microfractionation of column contents. Evaluation of pH course. Isoelectric points of β-lactoglobulins A and B. *Anal. Biochem. 50*, 575-585 (1972).

12. Beeley, J. A., Stevenson, S.M. and Beeley, J.G.: Polyacrylamide gel isoelectric focusing of proteins: determination of isoelectric points using an antimony electrode. *Biochim. Biophys. Acta 285*, 293-300 (1972).

13. Conway-Jacobs, A., Lewin, L. M.: Isoelectric focusing in acrylamide gels: use of amphoteric dyes as internal markers for determination of isoelectric points. *Anal. Biochem. 43*, 394-400 (1971).

14. Nakhleh, E. T., Samra, S.A., Awdeh, Z.L.: Isoelectric focusing of phenantroline iron complexes and their possible use as pH markers. *Anal. Biochem. 49*, 218-224 (1972).

15. Bours, J.: The use of protein markers in thin layer isoelectric focusing. *Sci. Tools 20*, 29-34 (1973).

16. Radola, B. J.: Isoelectric focusing in layers of granulated gels.I. Thin-layer isoelectric focusing of proteins. *Biochim. Biophys. Acta 295*, 412-428 (1973).

17. Righetti,P. G., Caravaggio, T.: Isoelectric points and molecular weights of proteins:a table. *J. Chromatogr. 127*, 1-28 (1976).

18. Papeschi,G., Bordi, S., Beni, C., Ventura, L.: Use of an Ir electrode for direct measurement of pI of proteins after isoelectric focusing in polyacrylamide gels. *Biochim. Biophys. Acta,* in press (1976).

19. Ui, N.: Isoelectric points and conformation of proteins.I. Effect of urea on the behaviour of some proteins in isoelectric focusing. *Biochim. Biophys. Acta 229*, 567-581 (1971).

Anomalous Behaviour of Horseradish Peroxidase in Isoelectric Focusing

H. Delincée and B.J. Radola

ABSTRACT

Focusing in gel media, namely thin-layers of granulated gels (Sephadex and Bio-Gel), thin slabs and rods of continuously polymerized polyacrylamide gels, revealed that horseradish peroxidase did not reach equilibrium under conditions which gave constant pI values for a number of pH marker proteins. After prolonged gel focusing equilibrium conditions were approached, and the isoelectric point of the predominant peroxidase isoenzyme converged to a value of 8.4. By density gradient focusing and free electrofocusing, a pI value of 9.1 was determined; this value being confirmed by the estimation of the isoionic point after desalting on ion-exchangers. The discrepancy in isoelectric points determined by different focusing techniques could be traced to the interference by carbon dioxide.

INTRODUCTION

Horseradish peroxidase has been fractionated by several techniques of focusing, namely gel focusing and density gradient focusing (1-4). The focusing patterns were quite different and even more surprising was that the isoelectric points determined by the density gradient technique disagreed with those obtained by the gel method. This was true particularly for the predominant component which had an isoelectric point of about 9.1 in density gradient focusing and a value of 7.1 in granulated gels when measured under standardized conditions (1,4). This discrepancy in isoelectric points is far beyond the error limits for pI determination in isoelectric focusing. Thin-layer refocusing of the column fractions showed that the predominant component obtained by density gradient focusing corresponded to the main isoenzyme in the thin-layer pattern. Thus, the isoelectric point seemed to depend on the system employed for its determination, and an attempt was made to elucidate the reason for the anomalous behaviour of horseradish peroxidase.

MATERIAL AND METHODS

Horseradish peroxidase with an absorbance ratio 403/280 nm of 0.6 was purchased from Boehringer (Mannheim, FRG). Fractionation into individual isoenzymes was carried out as described previously (4). Lactoperoxidase with 412/280 nm greater than 0.6 came from Calbiochem (Lucerne, Switzerland).

Thin-layer isoelectric focusing in granulated gels, detection of peroxidase activity, determination of isoelectric points at 25 $^{\circ}$C were carried out as described previously (1,4-7). Density gradient focusing was performed in a 110 ml LKB column (8). Free electrofocusing was accomplished in a coil of a 100 cm long polyethylene tubing (9), without addition of sucrose. Focusing in continuously polymerized polyacrylamide gels (5% T, 3% C) with 2% Ampholine was carried out either in rods (Ø 4-5 mm, 65 mm long) in a disc electrophoresis apparatus or in thin slabs (120 x 200 x 1-2 mm) in a "Double-Chamber" (Desaga, Heidelberg, FRG). For pI determination, the pH of the visible brown peroxidase zones were measured directly on the gel surface, in the case of rods after slicing and disintegrating the gel.

The estimation of isoionic points was performed by the procedure of Dintzis (10), modified according to de Bruin and van Os (11). A small column (100 x 2.5 mm) was filled with a mixed bed of equilibrated Amberlite IR-120 and IRA-400, and protein solution (>1%) was pumped continuously through the column. A thermostated glass electrode (filled with paste to reduce ion release) in the circuit measured the pH of the desalted solution.

RESULTS AND DISCUSSION

Table I summarizes the results of density gradient focusing under different conditions; the isoelectric points did not change on prolonged focusing. When the polarity of the column was reversed or narrow pH range ampholytes were employed, which both resulted in a different final position in the density gradient, the pI values of the main isoenzyme of horseradish peroxidase were not affected. By contrast, the isoelectric points of the secondary component depended on the polarity of the column and the ampholyte range, a finding difficult to interprete.

In addition to sucrose, a number of other substances were employed for stabilization in the column: ethyleneglycol, polyvinylpyrrolidone (MW 10 000), ficoll and also dextran (MW 150 000). The soluble dextran was employed to obtain in the column a stabilization medium with a similar chemical structure as on the Sephadex thin-layer plate. Irrespective of the stabilization medium used, the predominant component focused at an

Table I.

ISOELECTRIC POINTS OF HORSERADISH PEROXIDASE DETERMINED BY DENSITY GRADIENT FOCUSING

Focusing in the 110-ml LKB column
50 -100 mg of horseradish peroxidase were applied in the middle of the column

1% Ampholine pH range	Top electrode	Density gradient	Focusing time and voltage	Isoelectric points (25°) Predominant component	Secondary component	Minor component
3 - 10	Anode	Sucrose	48 h 300 V	9.1	7.3	5.8
"	"	"	64 h 300 V	9.2	7.2	5.8
"	Cathode	"	68 h 300 V	9.3	7.3	5.9
7 - 10	Anode	Sucrose	24 h 300 V 44 h 600 V	9.2	8.3	~5.9
"	Cathode	"	"	9.1	7.6	~5.7
8 - 10	Anode	Sucrose	72 h 600 V	9.3	-	-
"	"	"	144 h 600 V	9.2	in N_2 atm.	
3 - 10	Anode	Ethyleneglycol	68 h 300 V	9.1	7.2	5.8
"	Cathode	"	"	9.0	7.2	6.0
7 - 10	"	"	24 h 300 V 44 h 600 V	9.1	7.6	~5.8
"	"	PVP	24 h 300 V 120h 800 V	~9.1	-	-
"	"	Ficoll	"	~9.2	-	-
"	"	Dextran 150	"	~9.3	-	-
"	"	Sephadex G - 200 Fine	"	~9.3	-	-
"	"	Sephadex G - 200 Fine Sucrose	"	~9.4	-	-

isoelectric point of about 9. On refocusing in thin-layers, this component was in each case shown to be identical with the main isoenzyme in the thin-layer pattern with a pI value of 7.1. To rule out an effect of specific adsorption or orientation of the peroxidase to the solid matrix in thin-layer focusing, the LKB column was filled with Sephadex, and peroxidase focused without and with a superimposed sucrose gradient. Although elution was very tiresome, the isoelectric point of the main component was again close to 9.

The estimation of alkaline pI values (>8.5) needs caution because of the interference by atmospheric carbon dioxide. When the density gradient columns were eluted and the pH was measured in a glove box in a nitrogen atmosphere, the pH values in the region up to about 9 agreed with those determined in air, if the pH measurements were done immediately after elution. If the eluted fractions were allowed to stand in air, decreasing pI values were obtained, e.g., for the main component of horseradish peroxidase the pI value of 9.1 determined immediately after elution dropped to 8.8 after 16 h. This influence of carbon dioxide might be one of the reasons for the varying pI values described for the main component of horseradish peroxidase (2-4).

Free electrofocusing in a coil of polyethylene tubing was performed in order to eliminate any possible influence of the stabilizing medium. Experiments with pH marker proteins gave pI values similar to those in density gradient focusing (6). Free electrofocusing of horseradish peroxidase yielded for the main component a pI value of 9.1, which did not change when the tubing was filled with Sephadex.

When the isoionic point of peroxidase was estimated with an independent method, namely the procedure of Dintzis (10), by which the protein is completely desalted, the isolated predominant isoenzyme yielded an isoionic point of 9.1; for the secondary component a value of 6.5 was found. Thus, the isoionic point for the main component confirmed the pI value obtained by density gradient focusing, whereas the isoionic point of 6.5 for the secondary component disagreed with the column values, but was identical with the thin-layer value.

Thin-layer isoelectric focusing under conditions of drastically changed focusing time and voltage revealed that the predominant and the more alkaline isoenzymes did not reach equilibrium under conditions which gave constant pI values for the pH marker proteins (Table II). The upper part of Table II shows our previously published results, which we considered to represent equilibrium values on the basis of the following criteria: a) similar pI values were obtained at a 20 and 40 cm separation distance, b) different focusing times, and c) different carrier ampholyte pH ranges did not influence the results (1). However, longer focusing time and increased voltage resulted in a steady increase of the isoelectric points of the predominant and the more alkaline isoenzymes. The pH drift towards the cathode impeded the pH measurements and after prolonged focusing the most alkaline isoenzymes were lost. Extensive focusing yielded a pI value of about 8.4 for the predominant isoenzyme. Essentially, the same value was obtained in experiments involving variation of carrier ampholyte pH range (pH 3-10, pH 6-8, pH 7-10), different concentration of carrier ampholyte (1%, 1.5%, 2%), addition of sucrose (10%, 20%, 30%), different types of carrier ampholyte (Ampholine or Servalyt), and variation of the gel support (Sephadex G-75, G-100, G-200 and the polyacrylamide Bio-Gel P-60).

These results suggested that even on prolonged focusing horseradish peroxidase might not achieve equilibrium, but the cathodic pH drift hindered further measurements. Occasionally, it is claimed that equilibrium in focusing is reached, when the protein zones of samples applied at different positions in the pH gradient join. When at a 20 cm separation distance the predominant isoenzyme was applied 5 cm from the anode and cathode, the zones merged after about 20 h of focusing (4 h at 200 V followed by 16 h at 800 V). Under these conditions a pI value of 8.4 was measured, which did not change on further focusing.

Table II.

APPARENT ISOELECTRIC POINTS OF HORSERADISH PEROXIDASE DETERMINED BY THIN-LAYER ISOELECTRIC FOCUSING IN SEPHADEX

Ampholine pH range	Separation distance	Focusing time and voltage	Horseradish peroxidase isoenzyme pattern										
1% 3.5-10 " "	20 cm 40 cm 40 cm	7 h 300 V 18 h 400 V 24 h 400 V	8.4 8.4 8.4	8.2 8.25 8.2	8.0 8.0 8.05	7.6 7.65 7.6	7.1 7.15 7.1	6.5 6.55 6.5	6.0 6.0 5.95	5.3 5.2 5.3	5.0 5.0 5.0	4.0 4.1 4.0	3.6 3.7 3.8
1% 5-9	20 cm	1 h 300 V 5 h 600 V	8.4	8.2	8.0	7.65	7.2	6.45	6.0	5.3	4.9	-	-
1% 3.5-10	20 cm	4 h 200 V 3 h 800 V	8.8	8.6	8.4	7.9	7.6	6.5	6.0	5.3	5.0	4.0	3.7
"	"	4 h 200V 4 h 800V	9.0				7.9	6.5					
"	"	4 h 200V 16 h 800V	migrated out				8.3	6.6					
"	"	4 h 200V 20 h 800 V	-				8.4	6.4					
"	"	8 h 200V 16 h 800V	-				8.3	6.5					
"	"	18 h 200 V 8 h 800 V	-				8.4	6.5					
1% 7-10	20 cm	4h 200 V 3h 800 V	9.1				8.0	~6.3					
"	"	4h 400 V 7h 1000 V	9.2				8.2	~6.4					
"	"	4h 400 V 16h 1000 V	-				8.4	~6.6					
"	"	16h 400 V 8h 1000 V	-				8.3	~6.5					
2% 7-10	"	16h 400 8h 1000	-				8.3	~6.5					

The isoelectric points in Table II were determined at 25°C (with the exception of the pI's at 20°C reported in the upper part of the table). Focusing was usually performed on 20x10 cm plates in a 0.06 cm layer of Sephadex G-75 Superfine. The peroxidase solution was applied as a 50 mm long streak next to a 18 mm long zone of pH marker proteins. After focusing, peroxidase activity was detected with a substrate-impregnated 20x2 cm guide strip, and the pI was measured at 25°C either directly on the plate next to the visualized peroxidase zones, or small streaks (20x2 mm) were removed with a spatula, and liquified with 20-50 µl of double distilled CO_2-free water. A Ø 2 mm flat membrane microelectrode (type LoT 273-M2, Ingold, Frankfurt, FRG) was used. For peroxidase average values from at least three runs are given. In all these experiments the isoelectric points of the pH marker proteins (6) were practically identical, e.g., for the main components of cytochrome c 9.25±0.10, for sperm whale myoglobin 8.20±0.12 and for horse myoglobin 7.35±0.08 was found (at 25°).

A few experiments with focusing in thin slabs and rods of continuously polymerized polyacrylamide gels revealed a similar behaviour of horseradish peroxidase. Equilibrium was not achieved under conditions giving constant pI values for the pH marker proteins, and on prolonged focusing the pI of the main peroxidase isoenzyme reached values up to about 8.4.

Particularly for alkaline proteins discrepancies between isoelectric points obtained by the density gradient technique and isoelectric points determined in gel focusing have been described, lower values being consistently noted for the gel technique, e.g., for cytochrome c (6, 12-14) and ribonuclease (6). In addition to the experiments with horseradish peroxidase we have found that lactoperoxidase exhibits an anomalous behaviour in gel focusing, too. The isoelectric points determined by thin-layer isoelectric focusing in Sephadex were by one pH unit lower than the values of about 9.5 measured after density gradient focusing (15). For this enzyme we also noted an increase (about 0.2 pH units) of its pI values on prolonged focusing. It was recognized, however, that the relative equilibrium position of the horseradish and lactoperoxidase to that of ribonuclease and cytochrome c corresponds in the thin-layer pattern closely to that in density gradient focusing. Experiments in which thin-layer isoelectric focusing and the subsequent pH measurements were carried out in a glove box in nitrogen atmosphere revealed that the deviation of isoelectric points of alkaline proteins to lower values in the thin-layer technique is mainly due to the interference by carbon dioxide. In the CO_2-free atmosphere the following isoelectric points were found: for the main component of horseradish peroxidase 9.1, for ribonuclease 9.3 and for cytochrome c 10.1 (all values at 25°C). These values are in excellent agreement with the isoelectric points obtained by the density gradient technique.

These experiments and also observations of other workers (16, 17) indicate that the determination of isoelectric points in focusing should be approached with caution. Our results demonstrate that it is important for a particular protein to reach equilibrium, which can differ considerably from equilibrium for pH marker proteins. The long focusing time required by horseradish peroxidase to reach equilibrium may be ascribed to its low buffering capacity in the region about its isoelectric point (18). Presumably, lactoperoxidase also has a low electrophoretic mobility near its pI value. The influence of CO_2 on isoelectric focusing and pH measurements in the alkaline region, especially in thin-layer focusing, was not given sufficient consideration in the past. The deviation of the pI values of a number of alkaline proteins cited above as well as the fact that the pH gradients reported for gel electrofocusing rarely reach the nominal pH values at the cathode, can be attributed to the interference by CO_2. If for some reason thin-layer isoelectric focusing in a CO_2-free atmosphere is not practicable, the pI values of unknown proteins may be estimated by extrapolation from the apparent equilibrium positions of alkaline marker proteins.

REFERENCES

1. Delincée, H., Radola, B.J.: Thin-layer isoelectric focusing on Sephadex layers of horseradish peroxidase. Biochim. Biophys. Acta 200, 404-407 (1970).
2. Paul, K.-G., Stigbrand, T.: Four isoperoxidases from horse radish root. Acta Chem. Scand. 24, 3607-3617 (1970).
3. Welinder, K.G., Smillie, L.B., Schonbaum, G.R.: Amino acid sequence studies of horseradish peroxidase. I. Tryptic peptides. Can. J. Biochem. 50, 44-62 (1972).
4. Delincée, H., Radola, B.J.: Fractionation of horseradish peroxidase by preparative isoelectric focusing, gel chromatography and ion-exchange chromatography. Eur. J. Biochem. 52, 321-330 (1975).
5. Delincée, H., Radola, B.J.: Detection of peroxidase by the print technique in thin-layer isoelectric focusing. Anal. Biochem. 48, 536-545 (1972).
6. Radola, B.J.: Isoelectric focusing in layers of granulated gels. I. Thin-layer isoelectric focusing of proteins. Biochim. Biophys. Acta 295, 412-428 (1973).
7. Radola, B.J.: Isoelectric focusing in layers of granulated gels. II. Preparative isoelectric focusing. Biochim. Biophys. Acta 386, 181-195 (1975).
8. Haglund, H.: Isoelectric focusing in pH gradients - A technique for fractionation and characterization of ampholytes. Methods Biochem. Anal. 19, 1-104 (1971).
9. Macko, V., Stegemann, H.: Free electrofocusing in a coil of polyethylene tubing. Anal. Biochem. 37, 186-190 (1970).
10. Dintzis, H.W.: Ph.D. thesis, Harvard University 1952. Described in Nozaki, Y., Tanford, C.: Examination of titration behaviour. Methods Enzymol. 11, 715-734 (1967).
11. de Bruin, S.H., van Os, G.A.J.: Charge fluctuations in isoionic protein solutions. Rec. Trav. Chim. 88, 17-29 (1969).
12. Righetti, P.G., Drysdale, J.W.: Small-scale fractionation of proteins and nucleic acids by isoelectric focusing in polyacrylamide gels. Ann. N.Y. Acad. Sci. 209, 163-186 (1973).
13. Bobb, D.: The use of gel isoelectrofocusing in monitoring chemical modifications of proteins. Ann. N.Y. Acad. Sci. 209, 225-236 (1973).

14. Moody, A.J., Frandsen, E.K., Sundby, F.: Fractionation of gut glucagon-like activities by isoelectric focusing in polyacrylamide gel. In Righetti, P.G. (ed.): Progress in isoelectric focusing and isotachophoresis, 179-192 (1975).

15. Carlström, A. Vesterberg, O.: Isoelectric focusing and separation of the subcomponents of lactoperoxidase.
Acta Chem. Scand. 21, 271-278 (1967).

16. Wadström, T., Smyth, C.J.: Zymogram methods applied to thin-layer isoelectric focusing in polyacrylamide gel.
Science Tools 20, 17-21 (1973).

17. Robinson, H.K.: Comparison of different techniques for isoelectric focusing on polyacrylamide gel slabs using bacterial asparaginases.
Anal. Biochem. 49, 353-366 (1972).

18. Fredriksson, S., Rilbe, H.: These proceedings.

Does Lipoprotein Lipase Bind Ampholytes?

G. Bengtsson and T. Olivecrona

ABSTRACT

Lipoprotein lipase (LPL, purified from bovine milk) formed one or several peaks on isoelectric focusing in sucrose-stabilized pH-gradients. The position of these peaks depended on how the experiment was carried out, suggesting that the enzyme could bind acidic ampholytes to form complexes with isoelectric points intermediary between that of the enzyme and that of the ampholytes which it had bound. In contrast, denatured LPL focused in 5 M urea only around pH 9. Ampholytes decreased the binding of LPL to heparin-Sepharose and had effects on the enzyme activity similar to those that heparin has, suggesting that the binding of ampholytes to the enzyme may be related to its previously known ability to bind certain other polyanions (e.g. heparin) tightly.

The enzyme lipoprotein lipase (LPL) catalyzes the hydrolysis of triglycerides in very low density lipoproteins and in chylomicrons. This reaction takes place at the capillary endothelium in certain extrahepatic tissues, and this is probably the only way in which these tissues can utilize lipids from the circulating triglyceride-rich plasma lipoproteins (review 1). High activity of this enzyme can be found in such lipid-metabolizing tissues as heart and sceletal muscle, lactating mammary gland and in adipose tissue. Characteristically, the activity of the enzyme is enhanced severalfold by a protein cofactor which is a component of certain plasma lipoproteins (2). Another characteristic feature of this enzyme is that it is released from the capillary endothelium into the circulating blood when heparin is injected. The mechanism of this release is still unknown, but it has been shown that the enzyme binds to heparin and to some related sulphated polysaccharides (review 3). This binding can be utilized for purification of LPL from tissue extracts or from post-heparin plasma by affinity chromatography on immobilized heparin. The present experiments were carried out with enzyme purified from bovine milk. This is a

convenient starting material from which LPL can be purified to near homogeneity (4). It is a glycoprotein which in aqueous solution probably is a noncovalent dimer of two identical 55000 dalton chains.
On isoelectric focusing in a sucrose stabilized pH gradient with Ampholine[R] pH 3.5-10, the enzyme activity was recovered in good yield in peaks at pH 4.5 and 5.5 (Fig 1). In a similar experiment with LPL purified from pig adipose tissue Bensadoun et al. (5) found a single peak at pH 4.0. In our experiment a small amount of inactive enzyme protein was also found around pH 9. For analytical purposes we wished to develop a method for isoelectric focusing of LPL in polyacrylamide gel slabs. Previous experiments with gel electrophoresis had shown that it is usually necessary to include urea or a detergent when LPL is run. Therefore, gel isoelectric focusing experiments were carried out in 5 M urea (with Ampholine pH 3.5-10). Much to our surprise, the slabs showed faint bands around pH 9, but no detectable bands at pH 4-6. Therefore also a column isoelectric focusing experiment was run in 5 M urea. Under these conditions, the enzyme

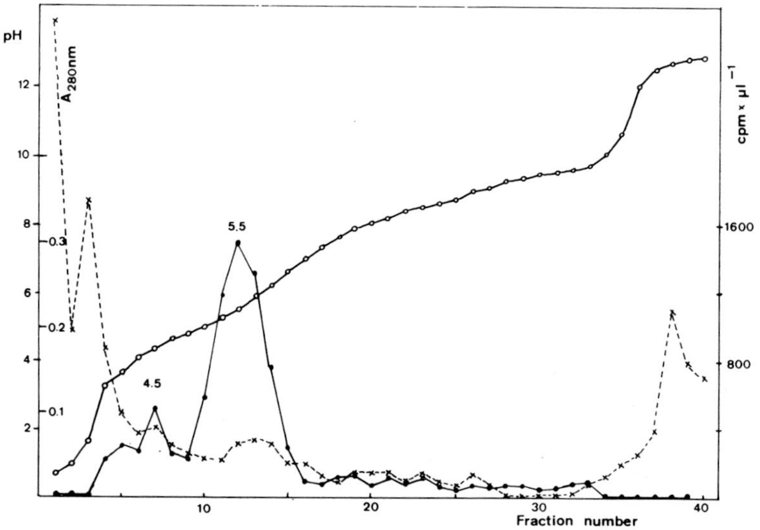

Fig 1. Isoelectric focusing of LPL mixed with pH 3.5-10 ampholytes. 1 ml of purified LPL (0.2 mg protein) in 0.005 M Na-veronal buffer pH 7.4 1.2 M NaCl was mixed into a sucrose gradient (110 ml) containing 0.9% (w/v) Ampholine pH 3.5-10. Focusing was for 30 h. At the end of the run the voltage was 800 V and the current 2.5 mamp. pH (○—○), absorbancy at 280 nm (×--×) and LPL activity (●—●) (6) were determined on fractions. They were also analysed by SDS gel electrophoresis.

looses all activity and can only be detected by (e.g.) SDS gel electrophoresis of the fractions. As in the gel experiments, enzyme protein was found only around pH 9. These experiments thus showed that in 5M urea (denatured) LPL focused around pH 9, whereas the (native) enzyme focused at a much lower pH.

There are several possible explanations for the anomalous behaviour of LPL on isoelectric focusing. One is that the native enzyme bound acidic ampholytes. If so, the pH at which it focuses would depend on the nature of the ampholytes present. To test this, LPL was mixed with pH 6-11 ampholytes and focused in a column (Fig 2). In this experiment, enzyme activity was recovered in good yield in two main peaks at pH 7.5 and 8.5, i.e. at a much higher pH than before. Recall that in the first experiment with pH 3.5-10 ampholytes (Fig 1), no peaks of enzyme activity were found at pH 7.5-8.5, but only at 4.5-5.5. The only difference between the two experiments was the nature of the ampholytes.

In further experiments the enzyme was applied at different levels of preformed pH-gradients in columns. When it was applied in the alkaline part of the gradient, enzyme protein but no activity was found around

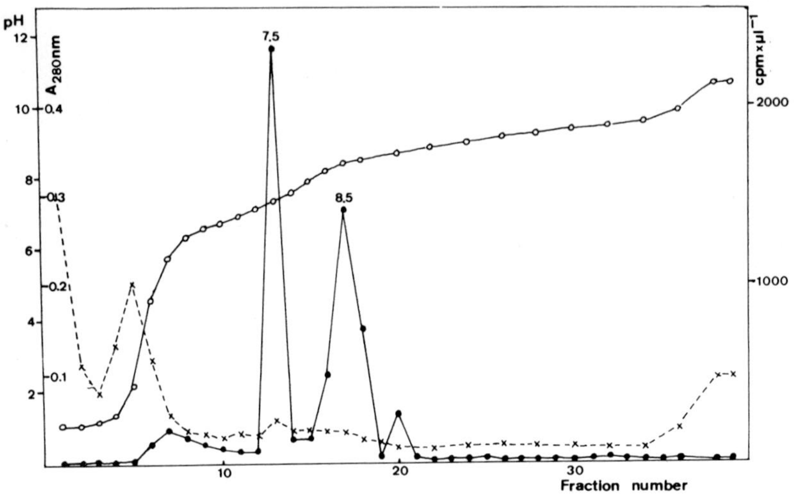

Fig 2. Isoelectric focusing of LPL mixed with pH 6-11 ampholytes. Conditions were as in fig 1 but the ampholytes used were a mixture of Ampholine pH 7-9, 8-10, and 9-11 and a cut between pH 6-7.5 from a focusing of Ampholine pH 3.5-10.

pH 9, i.e. at the same level where the enzyme focused in the presence of 5M urea. Previous experiments have shown that the enzyme is rapidly inactivated at pH above 8.5 (7). It is likely that when it was applied in the alkaline part of the gradient it soon lost activity and that the protein found around pH 9 was denatured enzyme. In contrast, when the enzyme was applied in the acidic part of the gradient (Fig 3), the enzyme activity was recovered in good yield in three peaks at pH 6.0, 7.2, and 9.5. This suggests that the enzyme bound ampholytes at or above the level where it was applied and that the complexes so formed migrated to higher pH where they were isoelectric.

All experiments were carried out with focusing times as long as or longer than usually recommended for routine systems. No time studies were made to see if the enzyme remained at the pH where it initially focused or if it moved towards higher pH. The latter behaviour may be expected and could account for the multiple peaks of enzyme activity found in some of the experiments.

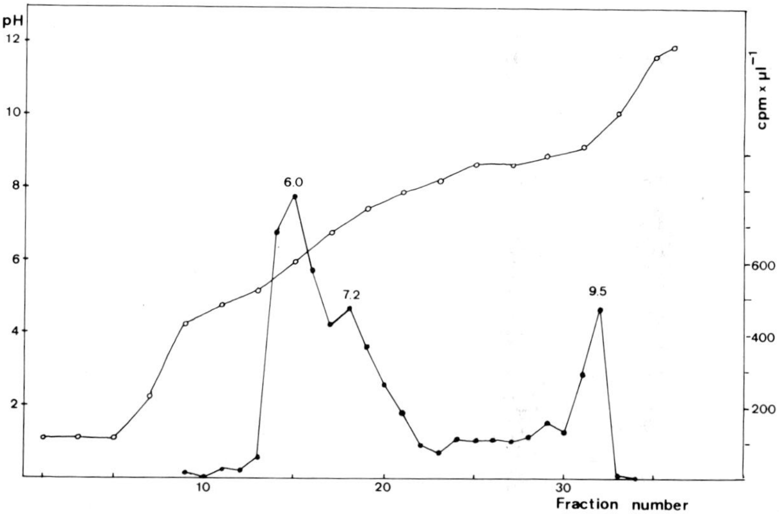

Fig 3. Isoelectric focusing of LPL applied to the acidic part of a preformed pH-gradient.
0.9% (w/v) Ampholine pH 3.5-10 in a sucrose gradient was prefocused for 36 h. Then a thin polyethylene catheter was placed with the tip in the acidic part of the gradient (estimated level - pH 5). About 5 ml of the gradient was removed, mixed with 1 ml LPL at the same sucrose concentration, and then reinjected at the same level of the column. Focusing was then continued for 18 h.

Our tentative interpretation of these results was that LPL binds certain ampholytes rather tightly. This interpretation would be strengthened if evidence for such binding could also be obtained in other systems. It was previously known that LPL binds heparin and some related acidic polysaccharides (3), and it appeared that ampholytes might bind to the "heparin-binding site" on the enzyme. This possibility was tested with three different approaches.

(1) When ampholytes were added to a system containing LPL bound to heparin-Sepharose, enzyme was almost completely displaced to the liquid phase (table 1), suggesting that the ampholytes could successfully complete with heparin for binding sites on the enzyme and/or with the enzyme for binding sites on heparin.

Table 1. Effect of pH 3.5-10 ampholytes on the binding of LPL to heparin-Sepharose.

	Addition to liquid phase (mg/ml) Ampholine				Heparin	
	0	0.1	1.1	6.1	.05	0.5
Heparin-Sepharose	26	16	120	939	689	896
No gel	894	966	1020	1125		

To 100 µl of heparin-Sepharose in a total volume of 1 ml 10 mM Tris-HCl, pH 8.4 containing 0.5% (w/v) sodium deoxycholate and 1% (w/v) bovine serum albumin was added about 15 µg purified LPL. The tubes were gently shaken for 2 h at 4° and were then briefly centrifuged and a small aliquot of the liquid phase taken for determination of LPL activity. Successive additions of 0.1, 1, and 5 mg Ampholine pH 3.5-10 adjusted to pH 8.4 (or of heparin) were then made. Between each addition the sample was slowly shaken at 4° for 20 min. Samples of the liquid phase were removed after a brief centrifugation immediately before the next addition of Ampholine (or heparin) was made.
This system differs from that used in many previous studies of the binding of LPL to various polysaccharides (3) in that the ionic strength was much less. In the presence of 0.3-0.5 M NaCl as has been used previously the ampholytes had little effect. Since LPL precipitates out of solution at low salt concentrations it was necessary to include deoxycholate in the present system to keep the enzyme in solution.
The figures are LPL activity in the liquid phase expressed in counts per minute obtained in an assay using radioactive triolein as the substrate (6).

(2) Addition of heparin often stimulates the activity of LPL, especially in crude systems (3). Likewise, addition of ampholytes stimulated the enzyme activity (table 2). In this assay the enzyme activity was strongly inhibited by low concentrations of pyrophosphate (8). The basis for this inhibition is not known but heparin efficiently relieved it. Likewise, ampholytes also relieved the pyrophosphate inhibition. Although we do not have enough information on the kinetic properties of this enzyme to interpret these effects in any detail, the fact that ampholytes had similar effects to heparin support the view that they bind to the enzyme at its "heparin-binding site".

Table 2. Effects of pH 3.5-10 ampholytes on the activity of LPL.

Heparin 0.1 mg/ml	Ampholine 1 mg/ml	Pyrophosphate 5 mM	LPL activity CPM
0	0	0	2591
+	0	0	3948
0	+	0	3153
0	0	+	450
+	0	+	3124
0	+	+	2729

The assay system was essentially the same as that described by Hernell, Egelrud and Olivecrona (6), but the concentration of NaCl was lowered to 0.012 M. At the NaCl concentration (0.102 M) of the original method the effects of the ampholytes were much less.

(3) LPL is not normally found in the circulating blood, but acts at the capillary endothelium. However, injection of heparin causes release of the enzyme into the blood. To test if the ampholytes could also release lipase into the blood, 200 mg Ampholine pH 3.5-10 was injected intravenously to a rat. A small but significant lipase activity appeared in the blood, but the amount found was much less than after heparin injection.

Taken together these studies give considerable but indirect evidence for binding of ampholytes to lipoprotein lipase. Apparently the binding is tight enough to cause formation of complexes that appear as peaks on isoelectric focusing. Formation of such complexes is thought not to occur with most other proteins but may be related to the ability of LPL to bind certain other polyanions (e.g. heparin) tightly.

ACKNOWLEDGEMENTS

This work was supported by a grant from the Swedish Medical Research Council (13X-00727).

REFERENCES

1. Scow, R.O., Blanchette-Mackie, E.J. and Smith, L.: Role of capillary endothelium in the clearance of chylomicrons from blood: Model for lipid transport by lateral diffusion in cell membranes. Circulation Res. In press (1976).
2. Havel, R.J., Shore, V.G., Shore, B., and Bier, D.M.: Role of specific glycopeptides of human serum lipoproteins in the activation of lipoprotein lipase. Circulation Res. 27:595-600 (1970).
3. Olivecrona, T., Bengtsson, G., Marklund, S.E., Lindahl, U., and Höök, M.: Heparin-Lipoprotein lipase interactions. Fed. Proc. In press (1976).
4. Egelrud, T., and Olivecrona, T.: The purification of a lipoprotein lipase from bovine skim milk. J. Biol. Chem. 247:6212-6217 (1972).
5. Bensadoun, A., Ehnholm, C., Steinberg, D., and Brown, W.V.: Purification and characterization of lipoprotein lipase from pig adipose tissue. J. Biol. Chem. 249:2220-2227 (1974).
6. Hernell, O., Egelrud, T., and Olivecrona, T.: Serum-stimulated lipases (lipoprotein lipases). Immunological crossreaction between the bovine and the human enzymes. Biochim. Biophys. Acta 381:233-241 (1975).
7. Castberg, H.B., Egelrud, T., Solberg, P., and Olivecrona, T.: Lipases in bovine milk and the relationship between the lipoprotein lipase and tributyrate hydrolyzing activities in cream and skim milk. J. Dairy Res. 42:255-266 (1975).
8. Korn, E.D., and Quigley, T.W.: Lipoprotein lipase of chicken adipose tissue. J. Biol. Chem. 226:833-839 (1957).

4. Biochemical Applications

Zein: Macromolecular Properties, Biosynthesis and Genetic Regulation

P.G. Righetti, E. Gianazza, F. Salamini, E. Galante, A. Viotti and C. Soave

INTRODUCTION

In cereals, most of the proteins are accumulated in the endosperm cells, which represent approximately 90% of the seed weight. It has been customary to divide endosperm proteins into four classes, based on their solubilities: albumins, water-soluble; globulins, salt-soluble; prolamines, alcohol-soluble and glutelins, alkali-soluble (1). Zein, the maize prolamine, accounts for 60-70% of the total endosperm protein. It is extracted by 70% ethanol plus 2-mercapto ethanol. Total zein is very rich in Glu (≃20%), Leu (≃18%), Pro (≃11%), Ala (≃11%), and extremely deficient in lysine and tryptophan (1).

Ideally, for the attainment of a maize mutant with a high nutritional value, it would be desirable to alter genetically the amino acid composition of zeins. This has not been possible so far, however Mertz *et al.* (2) and Nelson *et al.* (3) have found two mutants, *opaque2* (*o2*) and *floury2* (*fl2*) with an almost double Lys content in the endosperm. Similar high Lys mutants have been isolated in the case of barley (4) and sorghum (5). However, these mutants do not act by improving the quality of zein, but simply by repressing its synthesis. This has led in turn to a lower protein yield per seed and to a higher susceptibility of the *o2* mutant toward plant pathogens.

Little is known on the actual action mechanism of these genes mostly because the molecular properties and the genetic organization of zein are

still largely unknown. We have undertaken a program aimed at:
1) investigating the physico-chemical properties of zein molecules;
2) understanding the mechanism of zein synthesis and its genetic regulation;
3) studying the mechanism of packaging and unpackaging of zein granules.

METHODS

Zein extraction, isoelectric focusing, SDS-electrophoresis and *nucleic acid analysis* were done as previously described (6-11).

RESULTS AND DISCUSSION

Formation of protein bodies

Zeins are synthesized by membrane-bound polysomes and discharged within cisternae, where they form granules (protein bodies) (12). A close inspection of Fig. 1, in fact, reveals that: 1) a mature zein granule (Fig. 1, lower center) presents a homogeneous protein distribution within a fully-extended single membrane; 2) near-to-maturation granules present a homogeneous central core surrounded by smaller grains in the periphery. The granules do not fill completely the cisternae; 3) nascent protein bodies are represented by a floppy cisterna partially filled with small zein granules; 4) polysomes are seen bound to all membranes. It thus appears that, once

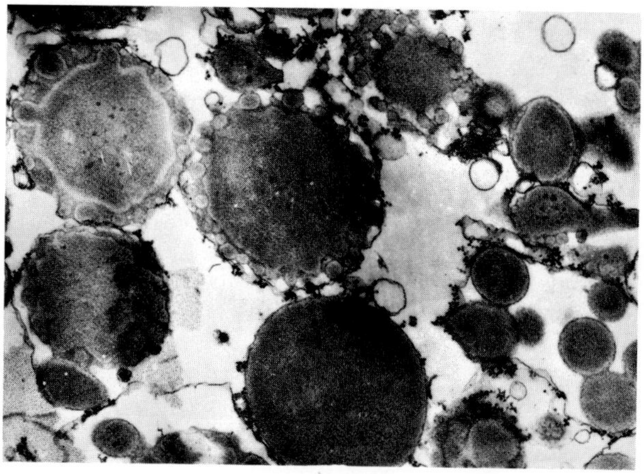

Fig. 1 Electron micrograph of protein bodies isolated in a sucrose density gradient (x 70,000).

the vacuoles are made and ribosomes bound to it, zein granule formation proceeds *via* subsequent coalescence steps from small granules.

Size distribution

When total zein extracted from maize endosperm is analyzed by SDS-gel electrophoresis, it is usually resolved into three major polypeptide chains, having the following M.W.'s (18,13): 23,000 (Z23), 21,000 (Z21) and 13,500 (Z13.5) daltons. By densitometry of the Coomassie Blue-stained bands, the amount of Z23 + Z21 accounts generally for 80% of the total (Fig. 2A). In addition to these chains, there are also some minor subunits, having M.W.'s: 22,000 (Z22), 20,000 (Z20) and 9,600 (Z9.6), appearing in some of the inbred lines studied (see Fig. 2B). The amino acid composition of Z23 + Z21, Z13.5 and Z9.6 fractions is reported in Table I. From the data in this table we can draw the following conclusions:
1) judging from the paucity of basic amino acids and from the relative abundance of Glu, Pro, and Ala, Z13.5 and Z9.6 can indeed be classified as zein chains; 2) Z13.5 and Z9.6 chains have marked decrements in the levels of Leu, Ile and Phe, which could reduce their hydrophobicity in comparison with the heavier chains; 3) Z13.5 and Z9.6 chains are very rich in sulfur-containing amino acids, as compared to Z23 and Z21 chains; 4) Z23 and Z21 chains determine the typical amino acid composition of zeins. Furthermore their amino acid compositions are very similar to each other (14).

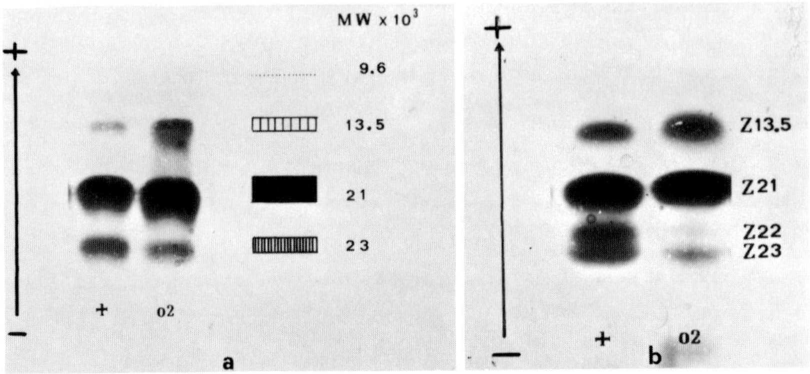

Fig. 2 SDS-PAGE pattern of zeins from normal and *opaque*-2 maize. (a) : inbred line T220 ; (b) : inbred line NC230.

Table I Amino acid composition of zein chains from SDS-electrophoresis

Amino acid	Z23+Z21 chains +	Z23+Z21 chains *o2*	Z13.5 chains +	Z13.5 chains *o2*	Z9.6 chains +
Lys	0.5	1.1	0.3	0.7	0.4
His	3.2	1.3	1.2	1.6	1.2
Arg	1.9	1.5	2.4	2.1	2.0
Asp	5.6	4.6	2.5	1.7	2.2
Thr	3.2	2.9	3.9	3.4	4.1
Ser	7.1	6.8	6.0	6.3	5.8
Glu	17.3	18.5	17.9	20.5	22.9
Pro	11.3	11.8	11.9	11.9	11.6
Gly	4.8	3.8	9.0	8.9	8.6
Cys	traces	traces	traces	traces	1.4
Ala	11.0	18.3	10.4	10.2	10.7
Val	3.8	3.8	3.4	4.1	3.6
Met	traces	0.3	5.3	4.9	3.9
Ile	2.5	3.4	1.3	1.2	1.1
Leu	19.4	18.0	11.6	9.1	9.6
Tyr	3.4	2.2	4.7	3.9	3.4
Phe	4.3	4.8	2.8	3.2	2.5

Samples were from W64A normal (+) and *o2* inbred lines. Values in µM%. Each entry is the average of duplicate runs.

Charge distribution

By IEF analysis zeins have been fractionated into several components which focus in the pH range 6 to 9. After an extensive analysis of various maize lines, at least 28 positions appear to be occupied by specific zein bands (13) (Fig. 3).

The frequency of the 28 IEF components among 36 inbred lines has been determined (13). Bands No. 7, 13, 14, 16, 17, 20, 23 and 26 are present practically in the IEF pattern of every inbred. The other bands, instead, are not always present and their frequency is highly variable. These additional components contribute to characterize the IEF pattern of each inbred studied.

Furthermore we have tried to correlate the charge distribution observed in IEF with the size heterogeneity observed by SDS-electrophoresis (8). To this purpose, Z23 + Z21 and Z13.5 chains, obtained by SDS-gel filtration,

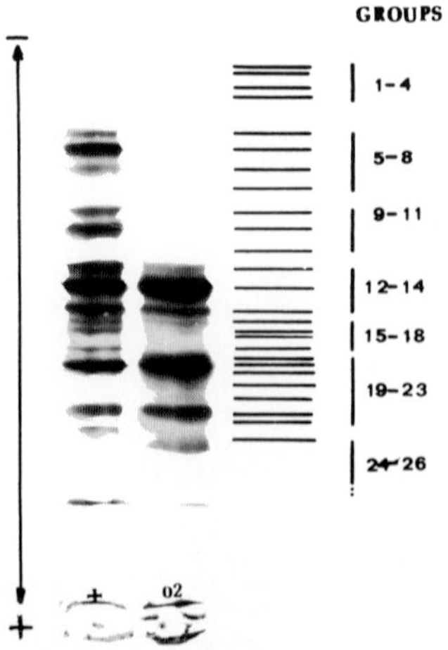

Fig. 3 IEF pattern of the normal and *opaque*-2 versions of the inbred line L1047. The 28 positions observed after analysis of 36 inbreds are reported on the right side of the figure. The bands were divided into 7 groups i.e. into clusters of bands exhibiting similar pI's.

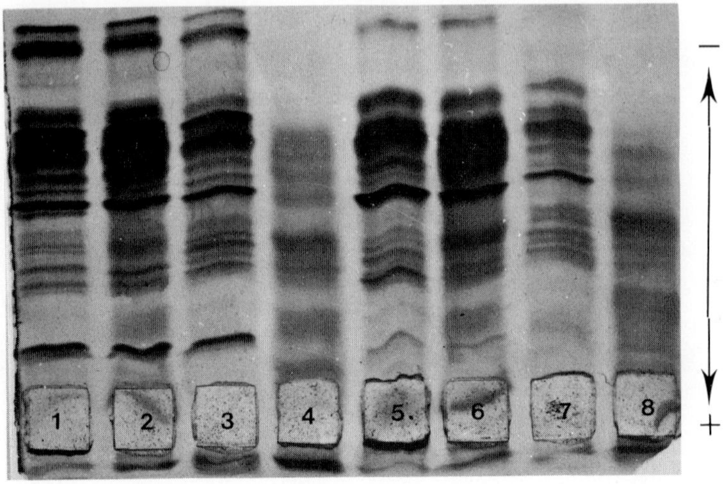

Fig. 4 IEF pattern of zeins from W64A inbred line in both normal and *opaque*-2 versions. Sample 1 refers to total zein extracted from meal, sample 2 to zein from protein bodies, sample 3 to Z23+Z21 chains, sample 4 to Z13.5 subunits. Samples 5-8 are the same from the *opaque*-2 mutant.

have been rerun in IEF. As seen in Fig. 4, while the higher M.W. components are practically indistinguishable from total zein, Z13.5 chains lack the more alkaline bands. The same happens also with the *o2* mutant.

Source of charge heterogeneity

Zeins have been fractionated by preparative IEF in a sorbitol density gradient (15). Much of the same heterogeneity observed in analytical IEF could be reproduced in the preparative column. At least height protein bands, precipitated at their pI, could be distinguished in the column. This means that the heterogeneity observed by gel IEF is not due to the sieving properties of the gel matrix nor to Ampholine-protein interaction, since in the preparative run the Ampholine/protein ratio has been greatly altered as compared to analytical runs.

Amino acid analysis of the total zein and of fractions A (the most acidic, pI ca 5) through H (the most basic, pI ca 8.5) from the preparative IEF column has shown a progressive change in some residue composition. Since in this case zeins are constituted essentially of chains of 23,000 and 21,000 daltons, which correspond to an average of about 190 amino acids per molecule, we have calculated a variation in at least 18 residues in going from peak A to H (see table II). In particular, there are positive changes in Ala (+4 residues), Ser (+4 residues) and negative changes in Val (-4 residues) and Tyr (-6 residues). This suggests that the zein heterogeneity demonstrated by IEF can be due to spot mutations in some genes responsible for zein synthesis. However these changes, which mostly affect neutral or hydrophobic amino acids, are insufficient to explain the broad spectrum of pI's observed in IEF.

In search for other possible causes of zein charge heterogeneity we have stained focused gels for glycoproteins with the periodic acid-Schiff stain (16). No sugar moiety could be found in any of the zein bands, not even in the most acidic components. However, when the gel was stained for lipoproteins with Sudan Black (17), a rather acidic protein, having pI $\simeq$ 3, and representing approximately 3-5% of the total zein population, was found to stain quite intensively. This lipoprotein contains a carotenoid covalently bound to the polypeptide backbone and was found to be a component of the membrane which envelops the zein protein bodies in the endosperm.

Table II: *Amino acid composition of zein fractions from preparative IEF*

Amino acid	Total zein	Relative amounts (μM%) Fraction A	B	C	D	E	F	H
Lys	0.3	0.2	0.8	0.3	0.2	0.2	0.3	0.2
His	1.2	1.1	1.1	1.2	0.8	1.2	1.0	0.9
Arg	1.3	1.6	1.4	1.4	1.2	0.9	1.0	1.2
Asp	4.7	5.1	4.9	4.9	4.7	4.8	4.7	4.7
Thr	3.0	3.1	3.0	3.1	2.4	2.3	2.4	2.4
Ser	5.2	4.4	4.9	4.9	5.3	5.3	6.6	6.6
Glu	21.9	19.0	21.0	20.9	22.4	21.4	22.3	21.2
Pro	9.8	8.9	10.5	8.2	9.4	10.3	9.7	9.7
Gly	2.9	2.9	2.3	2.3	2.2	3.4	2.4	2.1
Ala	11.7	11.5	12.5	12.5	12.6	13.4	13.5	13.7
Cys	traces	traces	traces	traces	traces	traces	traces	traces
Val	4.2	5.8	5.0	4.9	3.7	4.2	3.6	3.6
Met	0.6	1.3	traces	0.5	0.4	0.3	traces	traces
Ile	3.9	4.2	4.1	3.8	3.8	3.4	4.9	4.1
Leu	17.5	18.4	19.1	18.4	20.6	19.1	19.7	20.1
Tyr	4.1	6.2	5.4	5.4	3.8	2.7	3.1	3.1
Phe	6.4	7.6	6.8	7.5	6.2	7.1	6.4	6.3

Data from H56 inbred line; fraction A through H refer to the most acidic until the most basic zein components. Each entry is the average of duplicate runs.

Deamidation could, of course, be one of the major causes of zein heterogeneity as observed by IEF. To check for that, we have tried chemical deamidation of zeins *in vitro* (9). However, zeins were found to be quite resistent to chemical treatment and complete deamidation could only be achieved when zein was incubated at 50°C, at pH 9.5, for 22 days (see Fig. 5). We had previously calculated that since zein (based on an average M.W. of 22,000 daltons) contains approximately 52 glutamic and aspartic acid residues as acidic groups and 3 arginine, 2 histidine and less than one lysine group as basic residues per chain, it should have a theoretical pI of 3.2. This is in fact the pI obtained upon complete deamidation (see sample 11 and 12 in Fig. 5). On the other hand, since most native zeins, as isolated from maize endosperm, focus in the pH range 7-8, this means that at least 90% of the glutamic and aspartic acids must be present as glutamine and asparagine residues. Our results thus suggest that *in vitro* deamidation is a quite unlikely event. If any deamidation occurs, it must be a cellular event, occurring in the cell cytoplasm perhaps during zein synthesis or during subsequent packaging into granules.

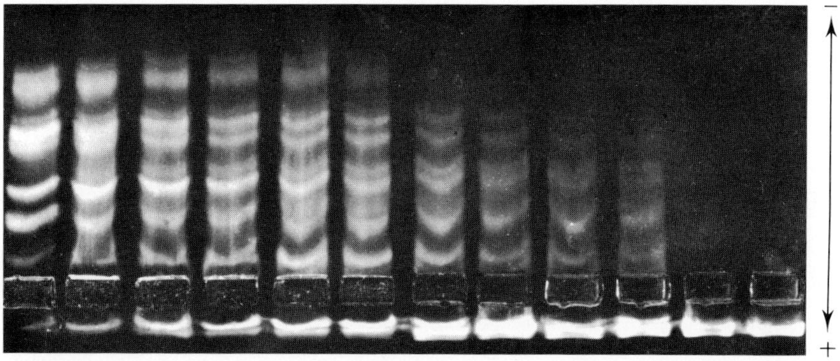

Fig. 5 IEF pattern of zein treated at 55°C in 10 mM borate buffer pH 9.5. From left to right the incubation time changes from 0 to 22 days.

Biosynthesis

Fig.6 shows the kinetics of accumulation of zein in normal and *o2* endosperms from an inbred and a hybrid line of maize. In both cases zein synthesis begins around the 14th day and then proceeds linearly until the 35th day. However, in the *o2* samples the rate of synthesis is dramatically reduced.

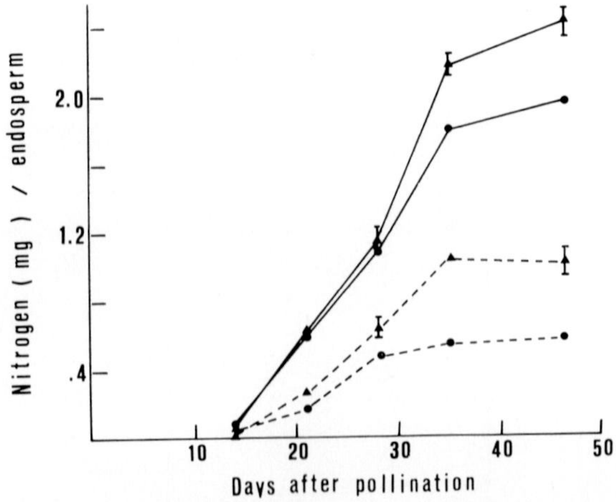

Fig. 6 Kinetics of accumulation of zein in normal (continous line) and *opaque*-2 endosperm (dotted line) from an hibrid (▲) and an inbred line (●).

These accumulation kinetics are well correlated with the profiles of total and protein body-bound polysomes which synthesize zein, isolated from normal and *o2* maizes. The evidence for the specificity of zein synthesis by the bound polysomes comes from *in vitro* experiments: radioactive lysine and proline were incorporated in the newly synthesized proteins in the ratio expected for the known composition of zeins. As shown in Fig. 7 the two po-

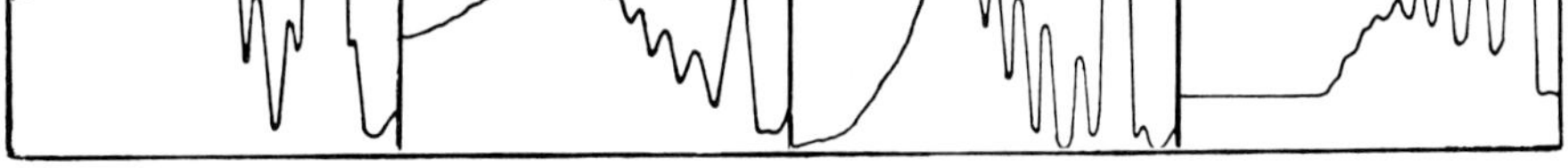

Fig. 7 Protein body-bound polysome profiles from normal and *opaque*-2 samples at 21 and 35 days after pollination. Normal: samples a-c; *o2*: samples b-d.

Table III: *Average reduction effect of the o2 allele on the synthesis of the IEF zein components*

Genotypes		mg of zein / g d.m.							
			IEF components						
		Total	1-4	5-8	9-11	12-14	15-18	19-23	24-26
Normal	Mean	60.92	0.69	8.38	4.38	14.67	11.64	17.06	3.66
	Range	32.5-84.1	0-5.7	1.1-21.5	0-11.8	2.9-28.9	3.6-20.3	2.2-30.7	0-11.1
Opaque-2	Mean	32.39	0.12	2.26	1.60	5.44	6.62	12.62	3.48
	Range	8.2-57.4	0-1.4	0-8.2	0-5.2	0-19.0	1.9-19.9	3.7-25.1	0-7.3
o2 vs normal. (% of reduction)		46.83	82.60	73.03	63.47	62.91	43.12	25.98	4.91

lysome populations, in normal maize, are separated into several components with a major class around esa-octamer (11). In *o2* maize, these polysome profiles at day 21 are qualitatively unchanged as compared to normal maize. However, after 35 days, when zein accumulation in *o2* maize has practically stopped, the polysome profiles seem to have been extensively disaggregated, with a predominance of the lighter classes.

We have also tried to extract the zein mRNA from total cell and protein body-bound polysomes. By formamide gel electrophoresis, mRNA from total cell appears to be quite heterodisperse in size, while mRNA from bound polysomes appears to be more homogeneous, with a main band having a M.W. of approximately 3.5 - 4 x 10^5 daltons. This corresponds to a total length of approximately 1200 nucleotides, which would code for a chain of about 400 amino acids. This is just about twice the size of a zein chain, as determined by SDS-electrophoresis. It would thus appear that zein mRNA is dicistronic. Alternatively, this could be a monocistronic mRNA with a long poly-A sequence and/or repetitive sequences at one or both ends. We can not exclude however a M.W. overextimation of zein mRNA due to macromolecular aggregation during electrophoresis as shown in the case of ovalbumin mRNA (18).

Genetics

Endosperm cells are triploid since they have two copies of the maternal and one copy of the paternal chromosomes. We have done reciprocal crosses between inbred maize lines, choosen on the basis of marked differences in their zein IEF patterns. In Fig. 8 the patterns of the parental lines and of the two reciprocal progenies are compared. Clearly, the F_1 patterns are additive, being constituted by all the bands present in the parental lines. The amount of each component correlates with the gene doses present in each crossing experiment. This suggests that the IEF bands correspond to a system of structural genes acting in an additive way.

Various genetic factors (*opaque-2*, *opaque-7*, *floury-2*, etc.), segregating with Mendelian ratios controll the endospermic levels of zeins (2, 3, 19, 20). The effects of the *o2* allele on the synthesis of the IEF components in 36 inbred lines are reported in Table III.

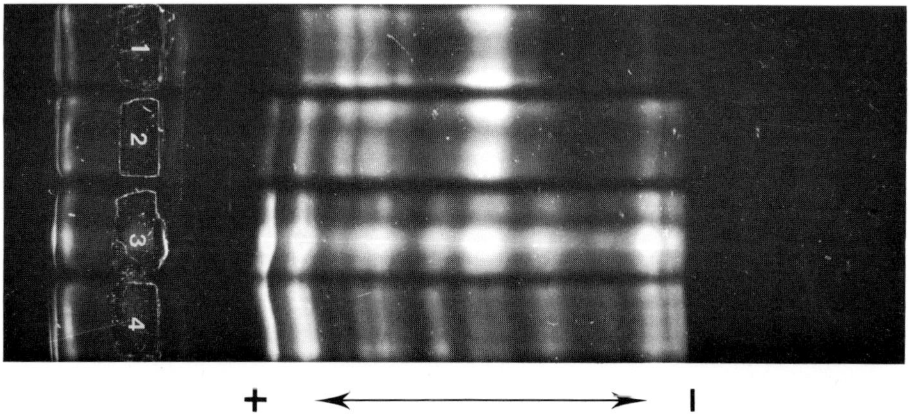

Fig. 8 IEF pattern of zein from parental lines and their two reciprocal F_1 progenies. 1): Ea2121; 2): Ea2121 x Lo38; 3): Lo38 x Ea2121; 4): Lo38.

Both in the normal and *o2* versions, zein components belonging to groups 12-14, 15-18 and 19-23 predominate. A strong variability exists among inbreds in the amount of zein present in each group. The *o2* allele negatively affects all the IEF bands but the extent of reduction shows a sharp trend from the most alkaline (group 1-4, reduced by 82.6%) to the most acidic components (group 24-26, reduced by only 4.9%).

Finally, we have compared the effects of *o2* with other major genes, such as *o7* and *fl2*, which also depress zein accumulation in maize endosperm. By analysis of their IEF patterns, it was evident that, in the case of the three inbreds investigated, *o2* reduces to a higher extent the most alkaline bands, while *o7* and *fl2* have a uniform effect on all the IEF components.

REFERENCES

1. Wall,J.S.: Cereal proteins, in *Proteins and their reactions*, Symposium on Foods (Schultz,H.W. and Anglemier,A.F.,eds.) pp.315-341, AVI Publi. Co., West Port, Conn. (1964).

2. Mertz,E.T., Bates,L.C. and Nelson,O.E.: Mutant gene that changes protein composition and increases lysine content of maize endosperm. *Science* 145, 279-280 (1964).

3. Nelson,O.E., Mertz,E.T. and Bates,L.S.: Second gene affecting the amino acid pattern of maize endosperm proteins. *Science* 150, 1469 (1965).

4. Doll,H., Køie,B. and Eggum,B.O.: Induced high lysine mutants in barley. *Radiat. Bot.* 14, 73-80 (1974).

5. Singh,R. and Axtell,J.D.: High lysine mutant gene (hl) that improves protein quality and biological value of grain sorghum. *Crop Sci.* 13, 535-539 (1974).

6. Soave,C., Pioli,F., Viotti,A., Salamini,F. and Righetti,P.G.: Synthesis and heterogeneity of endosperm proteins in normal and *opaque-2* maize. *Maydica* 20, 83-94 (1975).

7. Righetti,P.G. and Righetti,A.B.B.: High voltage analytical and preparative isoelectric focusing, in *Isoelectric focusing* (Arbuthnott,J.P. and Beeley,J.A., eds.) pp. 114-131, Butterworths, London (1975).

8. Gianazza,E., Righetti,P.G., Pioli,F., Galante,E. and Soave,C.: Size and charge heterogeneity of zein in normal and *opaque-2* maize endosperm. *Maydica,* in press.

9. Righetti,P.G., Gianazza,E., Viotti,A. and Soave,C.: Heterogeneity of storage proteins in maize. *Biochim. Biophys. Acta,* submitted for publication.

10. Laemli,U.K.: Cleavage of structural proteins during assembly of the head of bacteriophage T4. *Nature* 227, 680-685 (1970).

11. Viotti,A., Sala,E., Alberi,P. and Soave,C.: RNA metabolism and polysome profiles during seed development in normal and *opaque-2* maize endosperms. *Maydica* 20, 111-124 (1975).

12. Khoo,V. and Wolf,M.J.: Origin and development of protein granules in maize endosperm. *Amer.J.Bot.* 57, 1042-1050 (1970).

13. Soave,C., Righetti,P.G., Lorenzoni,C., Gentinetta, E. and Salamini,F.: Expressivity of the *opaque-2* gene at the level of zein molecular components. *Maydica,* in press.

14. Lee,K.H., Jones,R.A., Dalby,A. and Tsai,C.Y.: Genetic regulation of storage protein synthesis in maize endosperm. *Biochem.Genet.,* in press.

15. Gianazza,E., Viglienghi,V., Righetti,P.G., Salamini,F. and Soave,C.: Amino acid composition of zein molecular components. *Phytochemistry,* submitted for publication.

16. Hebert,J.P. and Strobbel,B.: Double staining for proteins and glycoproteins, *LKB Application Note* N° 151 (1974).

17. Prat.J.P., Lamy,J.N. and Weil,J.P.: Coloration des lipoprotéines après électrophorèse en gel de polyacrilamide. *Bull.Soc.Chim.Biol.* 51, 1367 (1969).

18. Savio,L.C.Woo,Rosen,J.M., Liarakos,C.D., Choi,J.C., Busch,H., Means, A.R., O'Malley,B.W. and Robberson,D.L.: Physical and chemical characterization of purified ovalbumin messenger RNA.*J.Biol.Chem.* 250,7027-7039 (1975).

19. McWhirter,K.S.: A floury endosperm, high lysine locus, in chromosome 10. *Maize Genet.Coop.News Letter* 45, 184 (1971).

20. Ma,Y. and Nelson,O.E.: Amino acid composition and storage proteins in two new high-lysine mutants in maize. *Cer.Chem.* 52, 412-419 (1975).

Growth and Isoelectric Patterns of Peroxidase in Tobacco Tissue Cultures Under the Influence of Growth Regulator Systems

W. Rücker and J. Markotai

Summary

Comparative studies on callus growth of tobacco tissue and pertinent enzyme determinations have shown that under the simultaneous influence of three different growth regulators growth and isoelectric peroxidase patterns are in close correlation one with the other. The influence of some combinations of growth regulators, that do not alter growth can nonetheless become visible through quantitative differences in the activity of some zones of the peroxidase isoenzyme systems.

Introduction

Numerous results of experiments on the simultaneous action of two different growth regulators on growth and differentiation in tobacco tissue cultures demonstrate, that the ratio of the compounds is as important as their absolute concentration (1-3).

The question arises as to whether similar results can be obtained with combination of more than two compounds. Our present study therefore deals with the simultaneous effect of three phytohormones: indoleacetic acid, gibberellin A_3 and benzylaminopurine. They are applied in various concentrations and proportions. Callus growth and the isoelectric peroxidase patterns of tobacco tissue cultures in vitro are the object to be studied. Each of these hormones induce and stimulate in tobacco tissue cultures various morphogenetic and physiological processes in a different way and to a different extend (1-7).

Material and Methods

The original tissue for the stock cultures is pith tissue of Nicotiana tabacum var. Wisconsin. The cultures have been maintained in the basal medium of Murashige and Skoog (8) with 3% saccharose and 10^{-6} g/ml 2,4 D. In this medium the tissue achieve a high growth rate but no organ differentiation.

For the experiments the concentration of the three growth substances: IAA, GA_3 and BAP was varied by means of the triangle methods (Fig.1). The concentration for each phytohormone ranged from 10^{-8} g/ml to 10^{-5} g/ml (media 1 - 10) and in a second series of experiments from 10^{-9} to 10^{-6} g/ml (media 1a - 10a).

The callus growth was estimated by measuring the weight of the fresh and the dry tissue following a cultivation period of three weeks.

The peroxidase systems were separated by IEF in layer of granulated gels (9-12). It was performed with the carrier ampholite: Ampholine pH 3,5 - 10 and with Sephadex G 75 Superfine on 20 x 20 cm glass plates. The peroxidase was detected by taking a print with buffered (11) paper impregnated with a methanolic solution containing 1% urea peroxide and 1% o-dianicidine. For the IEF tissue amounts of each sample were used giving equal total staining. The amounts of the tissue to be analyzed corresponded to 0,84 - 2,25 mg dry weight of the tissue.

Results and Discussion

The results of the series of experiments carried out with the two different concentration ranges will be reported on separately.

Abbreviations. IAA, indoleacetic acid; GA_3, gibberellin A_3; BAP, benzylaminopurine; 2,4 D, 2,4-dichlorphenoxyacetic acid; IEF, isoelectric focusing.

Experiments with concentrations of the growth regulators ranging from 10^{-8} to 10^{-5} g/ml (media 1 - 10). The results of callus growth are summarized in Fig. 1. They are best represented by distinguishing two different groups of growth patterns. The combinations of growth substances in the media 1 - 5 and in the medium 10 cause similar growth patterns (group I). They differ markedly from the cultures grown in the media 6 - 9 (group II) concerning the growth rate as well as the percentage of dry matter. The low dry weight and the high ratio of dry matter of group of II indicate that tissue growth is inhibited and the processes of cell expansion are more inhibited than those of cell division. Passing through the media 9,8,7 or 6,7 growth inhibition get more pronounced. These effects seem to be dependent on the dose of the phytohormones as well as on their ratio: with increasing BAP concentration the dry weight becomes lower an the dry matter higher. The BAP effects can be antagonized by lower proportion of GA_3 to BAP than IAA to BAP.

The influence of the ten different media on the peroxidase patterns of the tissues led to the following results (see Fig.2 and 3): the isoelectric peroxidase pattern of tobacco tissue cultures reveal two groups with a total of 6 - 8 zones in the acid range (group A and B) and an additional group with 4 - 6 zones in the alkaline range (group C). Although the groups A and B are closely adjacent, it might be assumed on the basis of some other work (13) and of our results that we are here confronted with two clearly distinguishable isoenzymes (14).

Marked quantitative and qualitative changes of the peroxidase patterns occur in the tissues with inhibited growth rate (group II of growth pattern i.e. cultures grown in the media 6 - 9): in the acid range the influence of increasing BAP concentrations cause quantitative changes of the peroxidase systems (media 6 + 9), the further increase results in qualitative changes of the enzyme patterns, too (media 8 + 7); it can be seen (Fig. 2 + 3) that the activity shifts from group A to group B: the first most acidic zones become weaker

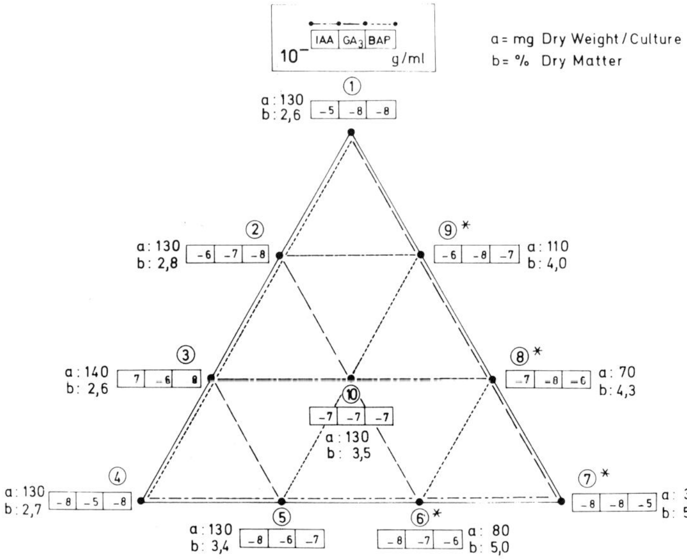

Fig. 1 Triangle with the ten different combinations of the phytohormones, media 1 - 10. The concentrations of IAA, GA_3 and BAP are given as negative exponents. They range from -8 to -5 i.e. 10^{-8} - 10^{-5} g/ml. a and b represent the tissue growth: a show the mean of dry weight and b the dry matter in per cent of fresh weight. * group II; without * group I of growth pattern.

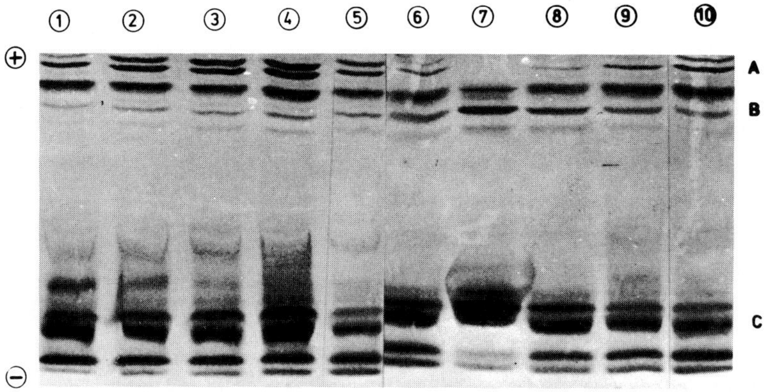

Fig. 2 The peroxidase isoelectric patterns of tobacco tissue cultures grown in the ten different media 1 - 10; in the acid range: group A and group B, in the alkaline range: group C.

until they finally disappear completely. At the same time the color of the zones of group B become more intensive. Particularly one of these zones get more and more prominent. This is striking in medium 7. As mentioned before the low dry weight and high content of dry matter of the tissue grown in this medium indicate that the processes of cell expansion are stopped while cell division still goes on. Therefore, it might be possible that this isoenzyme has an important role in the regulation of endogenous growth hormones, especially of the content of endogenous IAA (15,16).

Within the alkaline area of peroxidase patterns (group C) changes are ascertained only in the tissue with the highest growth inhibition, grown under the influence of the highest BAP concentration (medium 7). But the exact verification of the changes of the alkaline peroxidase isoenzymes seems to be difficult because of the unstable character of these isoenzymes from tobacco tissue cultures (17).

No connection could be noted between the results of growth and the proportion of the phytohormones IAA and GA_3; however, a comparison of the isoenzyme patterns of peroxidase one with the other reveal mutual influence of these two growth substances: in the media with the lowest BAP concentration i.e. the media 1 - 4: the most acidic zones have unequal intensity, following high IAA and low GA_3 concentrations (media 1 + 2). The color intensity, however, is similar with the reverse ratio of these hormones (media 3 + 4).

<u>Experiments with concentrations of growth regulators ranging from 10^{-9} to 10^{-6} g/ml; media 1a - 10a</u>. Also here, on the basis of growth rate as well as of the amount of dry matter, the tissue can be arranged into two groups of growth patterns, similar to group I and II described before (Fig. 1). Considering the isoelectric peroxidase patterns we also observe close correspondence with the growth results. In contrast to the experiments mentioned before the inhibition of growth was less pronounced, due to the application of lower concentrations of the growth hormones. The changes of the peroxidase

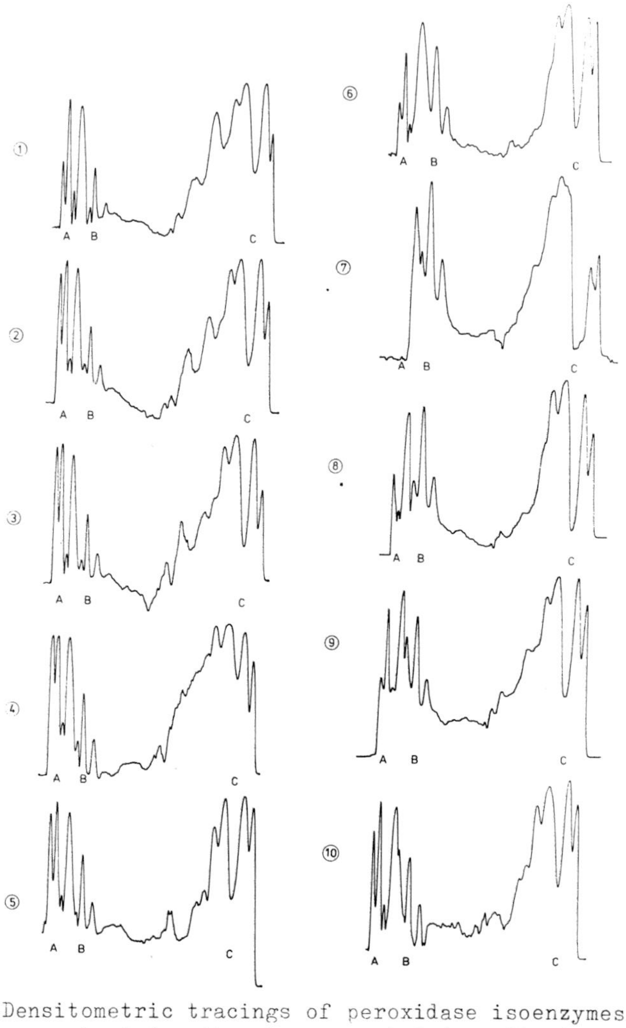

Fig. 3 Densitometric tracings of peroxidase isoenzymes obtained with a Zeiss spectral densitometer operated in reflectance. For the legends 1 - 10 and for A, B and C see Fig. 2.

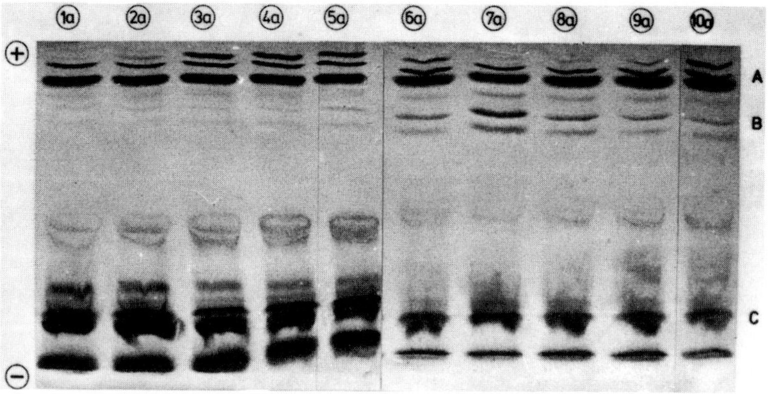

Fig. 4 Isoelectric peroxidase patterns from tobacco tissue cultures, grown in media 1a - 10a: series of experiments with concentrations of growth regulators ranging from 10^{-9} - 10^{-6} g/ml.

patterns (Fig. 4) in the cultures in the group of II, i.e. in the cultures grown in the media 6a - 9a are in correspondence with the growth results less impressive as compared with the previous series of experiments (Fig. 2). Both series of experiments, i.e. those of concentration ranges from 10^{-8} - 10^{-5} g/ml and from 10^{-9} - 10^{-6} g/ml respectively, reveal identical shifts of the activity of the zones from group A to those of group B. This effect is most striking in medium 7 as well as in medium 7a. In both cases one zone of group B is strongly pronounced. Differences in the peroxidase patterns according to the different ratio of IAA to GA_3, almost totally conforming to those found in the media 1 - 4 (Fig. 2) can bė observed in the media 1a - 4a (Fig. 4); they refer the first acid zones in the peroxidase pattern.

These results of the two series of experiments with different concentration ranges of growth regulators show a correspondence of the growth rate as well as of the peroxidase patterns and thus underscore the importance of the quantitative proportions of the several individual growth regulators for their effects in combination.

Acknowledgments

The skillful technical assistance of Mrs. B. Grauwald is appreciated. We thank the 'Fonds zur Förderung der wissenschaftlichen Forschung' of the Austrian Government, and the Austrian National Bank for support of this work, and the Austrian Research Society for Nuclear Energy for providing facilities in the Reactor Centre, Seibersdorf.

References.

1. Skoog, F., Miller, C.O.: Chemical regulation of growth and organ formation in plant tissue cultured in vitro. Symp.Soc.exp.Biol. 11, 118-130 (1957).

2. Schaeffer, G.W., Smith, H.H.: Auxin.Kinetin interaction in tissue culture of Nicotiana species and tumor conditioned hybrids. Plant Physiol. 38, 291-297 (1963).

3. Skoog, F.: Aspects of growth factor interactions in morphogenesis of tobacco tissue cultures. Colloques internationaux C.N.R.S. 193, 115-135 (1971).

4. Patau, K., Das, N.K., Skoog, F.: Induction of DNA synthesis by kinetin and indoleacetic acid in excised tobacco pith tissue. Physiol.Plant. 10, 949-966 (1957).

5. Fox, J.E.: Growth factor requirements and chromosome number in tobacco tissue cultures. Physiol.Plant. 16, 793-803 (1963).

6. Lee, T.T.: Interaction of cytokinin, auxin, and gibberellin on peroxidase isoenzymes in tobacco tissue cultured in vitro. Can. J. Bot. 50, 2471-2477 (1972).

7. Mäder, M., Münch, P., Bopp, M.: Regulation und Bedeutung der Peroxidase-Musteränderungen in sproßdifferenzierenden Kalluskulturen von Nicotiana tabacum L. Planta 123, 257-265 (1975).

8. Murashige, T., Skoog, F.: A revised medium for rapid growth and bio assays with tobacco tissue cultures. Physiol.Plant. 15, 473-497 (1962).

9. Radola, B.J.: Isoelectric focusing in layers of granulated gels. I. Thin-layer isoelectric focusing of proteins. Biochim.Biophys. Acta 295, 412-428 (1973).

10. Radola, B.J.: Analytical and preparative isoelectric focusing in gel-stabilized layers. Annals New York Acad.Sci. 209, 127-143 (1973).

11. Rücker, W., Radola, B.J.: Isoelectric patterns of peroxidase isoenzymes from tobacco tissue cultures. Planta (Berl.) 99, 192-198 (1971).

12. Delincée, H., Radola, B.J.: Detection of peroxidase by the print technique in thin-layer isoelectric focusing. Anal. Biochem. 48, 536-545 (1972).

13. Mäder, M., Bopp, M.: Neue Vorstellungen zum Problem der Isoperoxidasen anhand der Trennung durch Disc-Electrophorese und isoelektrische Fokussierung. Planta (Berl.) 128, 247-253 (1976).

14. Scandalios, J.G.: Isoenzymes in development and differentiation. Ann.Rev.Plant Physiol. 25, 225-258 (1974).

15. Meudt, W.J., Stecher, K.J.: Promotion of peroxidase activity in the cell wall of Nicotiana. Plant Physiol. 50, 157-160 (1972).

16. Schneider, E.A., Wightman, F.: Metabolism of auxin in higher plants. Ann.Rev.Plant Physiol. 25, 487-513 (1974).

17. Rücker, W.: Unpublished results.

A Method for Simultaneous Analysis of Several Genetic Loci in Mice After Electrofocusing

K.R. Narayanan and A.S. Ray

Several techniques are being currently evaluated in an effort to provide an additional method for mutagenicity testing in mammals. These include starch or polyacrylamide gel electrophoresis, isoelectricfocusing, and two-dimensional technique comprising isoelectricfocusing and electrophoresis to detect structural variants of enzymes and proteins, enzymatic assays to detect heritable enzymatic and thermolabile variants, and chromatographic and electrophoretic techniques to detect heritable metabolic disorders in mice (5,6,9,10,11,12,13,14,15,16, 17). The ultimate aim in every case is to provide a system which is able to screen for a maximum number of genetic loci within the shortest possible time with a minimum on animal requirements.

One of the best techniques that is currently available in the mammals for mutagenicity testing and risk estimation is the specific locus method in mice (3,4). The technique essentially consists of treating mice which are homozygous dominant for 7 genetic loci -- non-agouti, brown, chinchilla, pink-eyed, dilute, short-ear, and piebald. The treated animals are mated with untreated animals of a tester strain which are homozygous recessive for these 7 loci. Normally, the F_1's are heterozygous and show the dominant phenotype of the treated parent. However, if a mutation occurs in the treated parent--for example, causing S/S to mutate to s/S, then one would get the recessive phenotype of the untreated tester stock for this locus in the F_1 generation. The advantages of this technique include the simplicity of the experiment, ease of detection of mutants in the F_1 generation, ability to screen large number of animals rapidly, and the ability to conduct further genetic analysis of the mutants. However, because of the small number of genetic loci involved, one needs a large animal population for an experiment. Therefore, several biochemical techniques are being currently evaluated to provide an additional method for mutagenicity testing.

Valcovic and Malling (16, 17) used starch gel electrophoresis to detect mutations in 9 polymorphic enzyme systems between DBA and C57 Black mice. Their preliminary results show that the mutation frequency in the biochemical multiple loci method is very close to the frequency obtained with the specific locus method (8). This shows that the resolution of the electrophoretic procedures is as good as the specific locus method with an added advantage that the mutants can also be analyzed biochemically. However, the experimental procedure for electrophoresis is more complicated than the specific locus method, requires more trained personnel, and needs as much time and animals as the specific locus method. Therefore, we started a program on the evaluation of isoelectricfocusing for mutagenicity testing. Some of our additions to the electrofocusing procedure to facilitate simultaneous analysis of several genetic loci in mice are presented here.

Any biochemical technique that is evaluated for mutagenicity testing should fulfill the following criteria: (1) able to detect structural or functional alteration(s) of gene products with a high resolution, (2) able to screen for atleast 25-30 genetic loci, (3) technique should be fast enough to allow screening of 75-100 animals per day, (4) sample preparation should be simple and fast enough to allow processing 75-100 samples per day, (5) should require small sample volume, and (6) the technique should lend itself to automation, especially at the evaluation level.

Polyacrylamide gel electrofocusing is a powerful analytical tool for separation of macromolecules. One can successfully separate molecules which differ as little as 0.01 pH units in their isoelectric point (2, 7). Furthermore, with flat gel electrofocusing using 1 mm gels, it is a relatively quick method with very good separation and capable of processing relatively large number of samples. Fig. 1 shows a pH gradient between 3.5 - 9.5 that was formed in 90 minutes including 30 minutes of prefocusing without the sample. Further details regarding experimental details are given in the legend. We have also prefocused gels in the morning and used them in the afternoon or prefocused the previous evening and used them the next morning without apparent ill-effects.

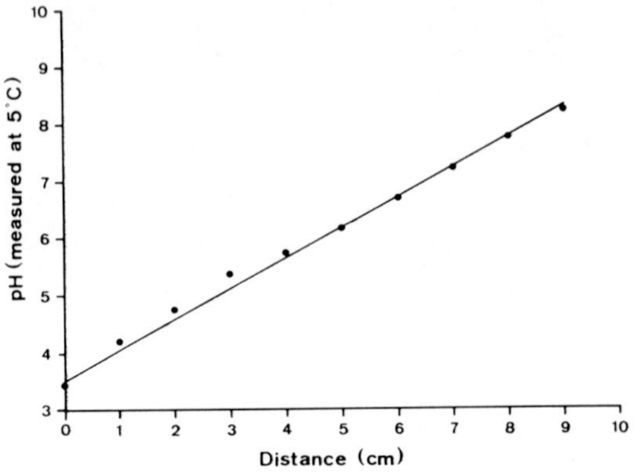

pH Gradient After 90 min. of Isoelectricfocusing.

Fig. 1. pH GRADIENT AFTER ELECTROFOCUSING. LKB PAGplate was prefocused for 30 minutes at 25 W. The cooling plate was thermostated at 2°C with Colora Kryostat WK 5. After prefocusing, samples were applied at the anodal end on filter paper. New electrolyte strips (1M phosphoric acid at anode and 1M sodium hydroxide at the cathode) were applied. Electrofocusing was now done for 60 minutes at 35 W. The LKB power supply 2103 was used. At the end of the run, pH at 1 cm intervals was measured with a surface electrode, Ingold LOT, pH 0-11, 0°C - 70°C, 7.0 Ag (3 Mol/L). Orion model 601 pH meter equipped with automatic temperature compensator was used.

Fig. 2 shows the kind of separation we get with our focusing procedure. These are non-specific tissue esterases from liver of the mouse strain 101. All the animals were approximately 30 days old. It is apparent that we get very sharp and clear separation of the various bands. Further details regarding sample preparation, staining techniques etc. are given in the legend.

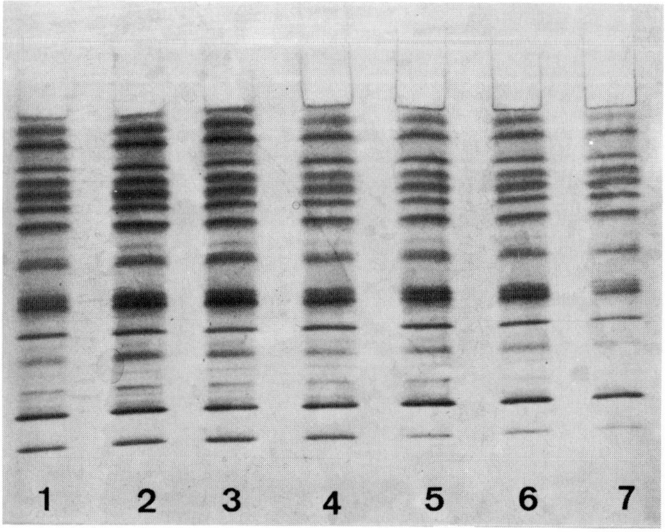

Fig. 2. NON-SPECIFIC TISSUE ESTERASES FROM LIVER OF MOUSE STRAIN 101. Each sample is from a different animal. The organs were immediately homogenized with Ultra-Turrax homogenizer, centrifuged at 100,000 g for 90 min. at 4°C, and stored at -70°C until used. The protein content of the homogenates was measured by Lowry's method. The sample was diluted in such a way that it contained approximately 10 mg/ml protein. The sample volume was approximately 15 microliter. The staining solution for esterases consisted of 96 ml Tris buffer, 0.05 M, pH 6.6 ; 4 ml α-naphthy acetate (dissolved in acetone) ; 150 mg Fast Red TR. The staining solution was freshly prepared and filtered prior to use. The staining was done for 5-8 minutes at room temperature. Other electrofocusing details were the same as in Fig. 1.

It is known that electrofocusing can separate molecules with minor differences which are not resolvable by normal electrophoresis. Fig. 3 shows electrophoretic separation of liver LDH from 5 animals of the mouse strain (101 x C3H) F_1. All the samples look identical except for some minor quantitative differences. Fig. 4 shows liver LDH from the same 5 animals, but separeted with isoelectricfocusing. One can note an additional band on sample 3 with a high isoelectric point of around 9.0. This is a mutant which could not be resolved with electrophoresis. It is apparent from Figs. 1 through 4 that electro-

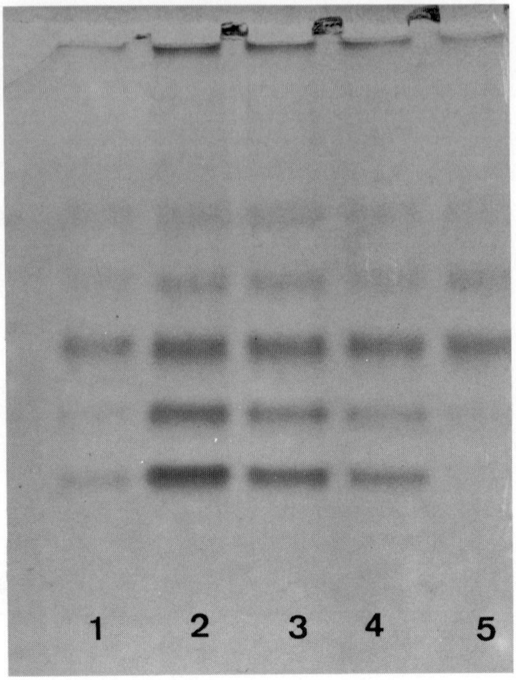

Fig. 3. ELECTROPHORETIC SEPARATION OF LIVER LDH FROM MOUSE STRAIN (101 x C3H) F_1. The sample preparation was the same as in Fig. 2. Electrophoresis was carried out with an Ortec system with 4 1/2, 7 1/2 % gel. The gel buffer consisted of 1.5 M Tris-citrate, pH 9.0. The tank buffer was 0.065 M Tris-borate, pH 9.0. Electrophoresis was carried out for 75 minutes with Ortec model 4100, pulsed constant power supply. The staining solution for LDH consisted of : 1 g lithium lactate ; 40 mg NAD ; 40 mg NBT ; 4 mg PMS, all dissolved individually in 100 ml Tris-glycine buffer, 0.05 M, pH 8.4. The gel was stained at 37°C in the dark for 15-20 minutes.

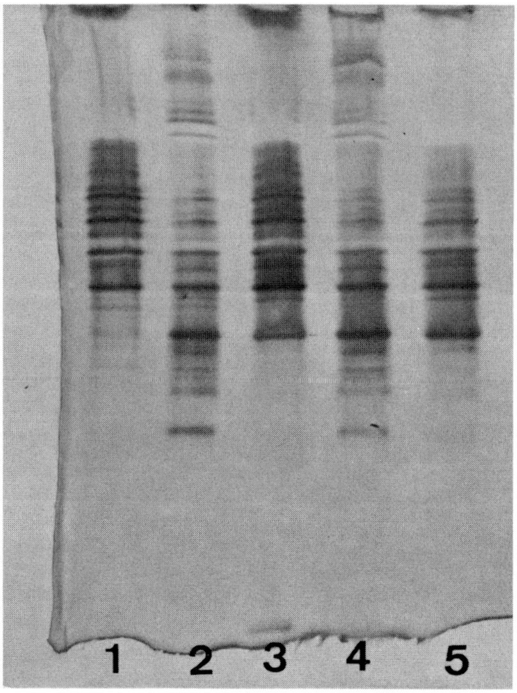

Fig. 4. SEPARATION OF LIVER LDH FROM MOUSE STRAIN (101 x C3H) F_1 WITH ISOELECTRICFOCUSING. The sample preparation and focusing details same as in Figs. 1 and 2. Details of LDH staining were the same as in Fig. 3.

focusing offers a method capable of high resolution. We can easily screen 96 samples per day and optimal separation is obtained with microgram quantities of tissue extracts. Moreover, the gels can be easily analyzed through automated densitometry. Besides the resolution, electrofocusing has another advantage over electrophoresis. For optimal separation in electrophoresis, different enzyme systems need different pH and buffer systems. On the other hand, in electrofocusing, a single pH gradient of 3.5 - 9.5, can separate a wide variety of enzymes and proteins. In other words, this technique fulfills 5 of the 6 criteria that were listed as essential if a technique were to be an addition to the specific locus method. There are, however, some drawbacks with electrofocusing, especially in comparison to some electrophoretic procedures. These include, (1) one gel can be stained for only one enzyme or protein. In comparison, a starch gel can be sliced horizontally into 3 or 4 slices and stained for different enzymes. In other words, the number of genetic loci that can be analyzed simultaneously is limited. (2) there are several enzyme systems which stain poorly in a polyacrylamide base, (3) several enzymes need incorporation of substrates into the gel for staining purposes. In a focusing gel one can add only uncharged substrates, and (4) storage of polyacrylamide gels for future reference and documentation is not easy. Of all the disadvantages, the most critical is the fact that only a limited number of genetic loci can be analyzed simultaneously. However, with our modifications, we have overcome every one of the disadvantages listed above.

Arvidson and Wadström (1), Vesterberg (18), Vesterberg and Eriksson (19), Wadström and Smyth (20) and others have used agar or agarose overlay methods successfully for a variety of staining purposes, especially for detection of enzymes that act on large substrates such as proteins. These reports prompted us to try, instead of agar overlay, agar contact replicas. Actually the method is very similar in principle to the replica plating technique used routinely in bacteriology. Glass plates of the same dimension as the focused gel are coated with 1 - 1.5% agar and allowed to set. At the end of focusing, the electrolyte strips and filter paper pieces used for sample application are removed and the focused gel is gently laid on top of the agar coated plate making sure there are no air bubbles. We also use an atomizer to moisten the agar plates, especially if they have been standing on the bench top too long. The trick here is the copying surface should neither be too dry nor too moist. Moreover, if the acrylamide gel is on a glass plate, it would be better to have the agar copy on a Mylar plastic sheet so that there is perfect contact between the acrylamide surface and the agar surface. The time required to form a good replica would depend on several factors such as the amount of the sample in the focused gel, whether it is the first, second, or the sixth copy etc. Usually, after 1-2 minutes (slightly more after the third copy), the gel can be removed and laid on top of another agar plate to make another copy. Actually, by leaving the focused gel on Multiphore unit, one can make replicas for a longer time without the danger of diffusion in the original gel, because one can refocus for 1-2 minutes between replica printing. Fig. 5 shows a 5th and 6th copy stained for non-specific esterases. One can notice that both copies are identical. These are liver esterases from 101 male mice, 25 days old.

Before the implications of this method are discussed, one should answer the question as to how good is the transfer of the macromolecules from the original to the copy. The transfer is very

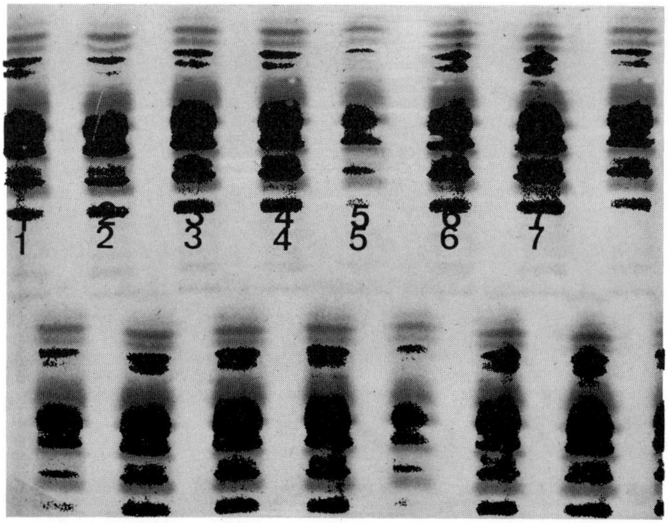

Fig. 5. AGAR CONTACT REPLICA OF A FOCUSED GEL STAINED FOR NON-SPECIFIC ESTERASES. A fifth and sixth copy of a focused gel are photographed together. The copying time was approximately 3 minutes each. The staining techniques were the same as before.

good. Actually, one can find the transfer of even ampholine on copies. LKB Multiphore is very ideal for such contact replicas because the sample is applied on filter papers on the surface of the gel. Even after separation, most of the macromolecules are on the surface of the gel. The number of copies that can be made and the number of enzyme systems that can be detected by contact replicas would depend on several factors. These include (1) the concentration and the amount of sample applied on the gel. One cannot overload the sample to make more copies because one has to strike a balance between the sample amount and optimal separation, (2) the sensitivity of the enzyme system that is being stained for. It would be preferable that the original or the first copy be used for protein staining and later copies for enzymes. Very sensitive enzyme systems such as esterases can be stained even as a sixth copy. It is also conceivable that one uses UV or fluorescent scanning methods. Then, the number of copies that can be made and analyzed will be more. With appropriate incorporation of substrates in the copy, one can analyze a variety of enzymes. With appropriate modifications, even immunelectrofocusing can be simultaneously done. Essentially by such modifications, the range of detection system can be lowered to nanogram ranges. Not only does this modification increase the number of genetic loci that can be analyzed simultaneously in mutagenicity testing but it has potential in clinics etc. From a practical stand point, the agar plates after staining can be left in 7-10% acetic acid for a few hours and dried. This would facilitate storage, densitometry, and documentation. We have recently started a large scale experiment with mice to test whether this technique is suitable for routine mutagenicity testing and/or risk estimation.

ACKNOWLEDGEMENTS

Able technical assistance of Miss Gunthild Borgwedel and Miss Elisabeth Kaspar is gratefully acknowledged.

REFERENCES

1. Arvidson, S., Wadström, T. Detection of proteolytic activity after isoelectric focusing in polyacrylamide gel. Biochim.Biophys. Acta 310, 418-420 (1973).
2. Catsimpoolas, N. Isoelectric focusing and isotachophoresis of proteins. Separation Sci. 8, 71-121 (1973).
3. Ehling, U. H. The multiple loci method. In "Chemical Mutagenesis in Mammals and Man", F. Vogel and G. Röhrborn (Eds), Springer Verlag, Berlin-Heidelberg-New York, 156-161 (1970).
4. Ehling, U. H. Dose-response relationship of specific locus mutations in mice. Arch. Toxicol. 32, 19-25 (1974).
5. Klose, J. Protein mapping by combined isoelectric focusing and electrophoresis of mouse tissues. A novel approach to testing for induced point mutations in mammals. Humangenetik 26, 231-243 (1975).
6. Klose, J. Protein-Mapping-Methode bei Mäusen. Paper presented at 3rd Annual Meeting of the Gesellschaft für Umwelt-Mutationsforschung on July 1-2, 1976 at Neuherberg, Federal Republic of Germany.
7. Latner, A. L. Isoelectric focusing in liquid and gels as applied to clinical chemistry. Adv. clin. Chem. 17, 193-250 (1975).
8. Malling, H.V., Valcovic, L. R. Electrophoretic mobility assay in the mouse. Paper presented at 3rd Annual Meeting of the Gesellschaft für Umwelt-Mutationsforschung on July 1-2, 1976 at Neuherberg, Federal Republic of Germany.
9. Mareschal, J., Valcovic, L.R., Malling, H.V. Personal Communication.
10. Narayanan, K. R. Detection of biochemical variants in mice. GSF-Bericht M 199, 27 (1974).
11. Narayanan, K. R. Fingerprinting as a tool for biochemical characterization of different strains of mice. GSF-Bericht B 626, 44 (1975).
12. Narayanan, K. R. Detection of biochemical mutants in mice. Paper presented at 3rd Annual Meeting of the Gesellschaft

für Umwelt-Mutationsforschung on July 1-2, 1976 at Neuherberg, Federal Republic of Germany.

13. Pretsch, W., Narayanan, K.R. Untersuchung erblicher Stoffwechselstörungen bei der Maus (in press).

14. Siciliano, M.J., Humphrey, R. M. Electrophoretic approach to mutagenesis in mammalian cell lines. Mutation Res. 31, 323 (1975).

15. Trippa, J., Loverre, A., Catamo, A. Thermostability studies for investigating non-electrophoretic polymorphic alleles in Drosophila melanogaster. Nature 260, 42-44 (1976).

16. Valcovic, L.R., Malling, H.V. An approach to measuring germinal mutations in the mouse. Environ. Health Perspect. 6, 201-205 (1973).

17. Valcovic, L.R., Malling, H.V. Mutation detection in the biochemical specific locus system. Mutation Res. 31,338 (1975).

18. Vesterberg, O. Isoelectric focusing of proteins in polyacrylamide gels. Biochim. Biophys. Acta 257, 11-19 (1972).

19. Vesterberg, O., Eriksson, R. Detection of fibrinolytic activity by a zymogram technique after isoelectric focusing. Biochim. Biophys. Acta 285, 393-397 (1972).

20. Wadström, T., Smyth, C.J. Some applications of a zymogram method in combination with isoelectric focusing in polyacrylamide gels. In,"Isoelectric Focusing", Arbuthnott, J. P., Beeley, J. A. (Eds), Butterworths, London, 152-177 (1975).

Phylogenetic Differences in Isoelectric Components of Liver Acid Phosphatase and LDH Between Genera of Two Closely Related Rodent Families (Muridae and Microtidae)

A. Kubicz and L. Wolanska

ABSTRACT

The isoelectric patterns of liver acid phosphatase and LDH within and between genera of two closely related rodent families were compared. Species within one genus showed similar overall patterns, but essential differences appeared between the compared genera. They were expressed as charge differences on the enzyme molecule and as different ability to LDH subunit combination.

INTRODUCTION

Studies of multiplicity of biologically active proteins in respect to phylogeny and evolution are of wide biological interest. Thus observations of variations in protein components may give us some informations on the events occuring at the molecular level. Isoelectric focusing on thin layers of polyacrylamide gel seems to be very useful for tracing variations expressed as charge differences on the protein molecules. The present work is part of our comparative studies concerning postembryonic and phylogenetic differentiation of some enzymes (1, 2, 3). Here, we present the phylogenetic relationship within and between genera belonging to two closely related rodent families, Muridae and Microtidae, on the basis of two enzymes: the acid phosphatase and lactic dehydrogenase. Although comparative studies on LDH isoenzymes in a number of different rodents were already reported (4, 5) they were not concerned with systematic investigations of all species belonging to one genus and compared with another one from a closely related family.

MATERIAL AND METHODS

Investigations were performed on adult, male animals which were all local species except M. pensylvanicus. The Muridae species were live-trapped from wild populations, the Microtidae species were mostly born in the laboratory. The systematic background of the investigated animals is presented in Tabl. 1.

Order	Rodentia	
Family:	Muridae	Microtidae
Genus:	Apodemus	Microtus
Species:	A. agrarius	M. arvalis
	A. microps	M. agrestis
	A. sylvaticus	M. oeconomus
	A. flavicolis	M. pensylvanicus

Tabl. 1. Systematic background of investigated animals.

Liver extracts were made by homogenization with 2 volumes of cold, 0.1 M acetic buffer, pH 5.0 (for acid phosphatase) or with 0.1 M phosphate buffer, pH 7.5 (for LDH). After centrifugation at 12 000 xg for 30 minutes the supernatants were applied to focusing experiments. Isoelectric focusing on thin-layer polyacrylamide gel and determination of pH gradient were carried out after Awdeh et al. (6). The gel composition and preparation were according to the instruction of the LKB Multiphor manufacturer (7), except that the final concentration of Ampholine (LKB Produkter, pH range 3.5 - 9.0) was 2.5 %. 0.1 ml of the extracts were applied on filter paper near the anode. The initial current was 15 mA until the voltage raised to 400 V which was further kept constant. The focusing time was 14 h at 4^{0} C. After termination of the run and after cutting disks of gel for pH determinations, the slabs were equilibrated at 4^{0} C with cold, 0.2 M acetic buffer, pH 5.0 (for acid phosphatase) or with 0.1 M phosphate buffer, pH 7.5 (for LDH).

Staining for acid phosphatase activity: 1-naphthyl phosphate and Fast Blue B (1 mg of each per 1 ml 0.2 M acetate buffer, pH 5.0). Staining for LDH was accomplished by combining the methods of Papadopoulos and Kintzios (8) and Dietz and Lubrano (9) as described in our previous paper (2). The incubation mixture contained: 0.1 M phosphate buffer, pH 7.5 - 15 ml; 1 M Na-lactate - 0.8 ml; INT 1 mg/ml - 2.5 ml; PMS 1 mg/ml - 0.25 ml; NAD 10 mg/ml - 0.8 ml; 0.005 M $MgCl_2$ - 2 ml; 0.1 M NaCl - 2 ml. Incubation at dark, at 37° C.

RESULTS AND DISCUSSION

Acid Phosphatase

Specific staining for acid phosphatase showed a number of multiple enzyme components covering the pH range from about 4.3 to about 2.7. The most prominent bands were focused between pH 6.3 to 6.6 or 7.3 (depending on the

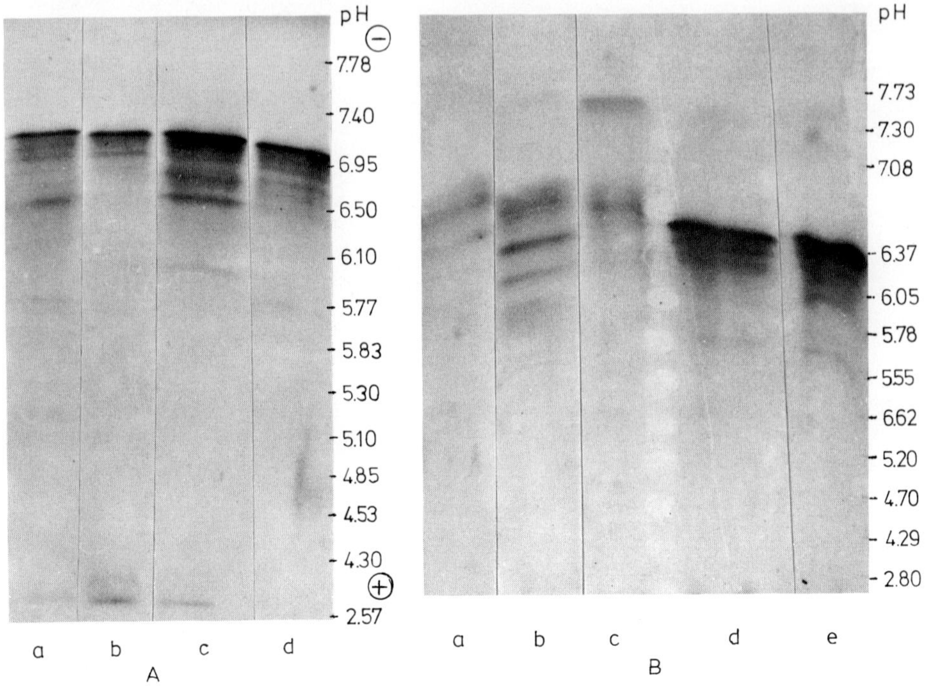

Fig. 1: Acid phosphatase zymograms after isoelectric focusing of liver extracts of Microtus (A) and Apodemus (B) species. Staining with 1-naphthyl phosphate and Fast Blue B.
A: a, M.pensylvanicus; b, M.oeconomus; c, M.arvalis; d, M.agrestis.
B: a, A.agrarius; b, A.flavicolis; c, A.microps; d, M.arvalis; e, M.oeconomus.

genus). Fig. 1 A shows the isoelectric spectra of liver extracts from the Microtus species. Although similar in the overall pattern, characteristic species differences were observed which were mainly concerned with the number and intensity of bands. Even sibling species, M.oeconomus and M.agrestis (Fig. 1 A, b and 1 A, c resp.) can easily be distinguished on basis of their zymograms. These differences are not so essentially expressed by ordinary disc electrophoresis.

Characteristic species variations were also demonstrated in zymogram patterns between members of the genus Apodemus (Fig. 1 B, a, b, c) although

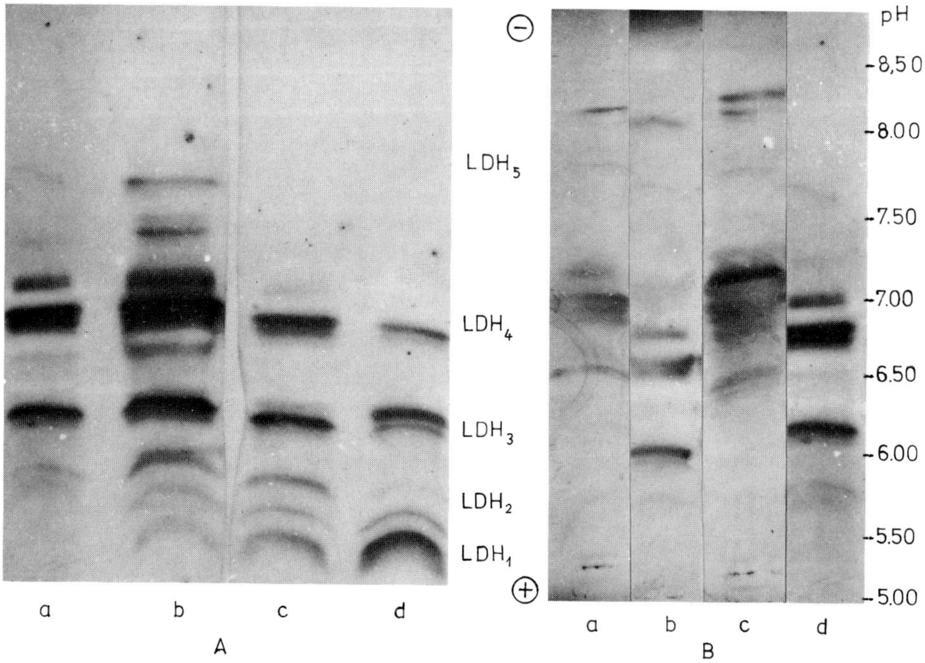

Fig. 2: Isoelectric focusing patterns of LDH from Apodemus species.
A: Organs from A.microps: a, liver; b, muscle; c, kidney; d, heart.
B: Liver extracts from: a, A.sylvaticus; b, A.flavicolis; c, A.agrarius; d, A.microps.

the overall pattern seems to be also similar.

Very essential differences appear when the two genera are compared. In all experiments we obtained higher pH values for the dominating cathodic bands of the examined Muridae species than for the corresponding bands of the

Microtidae species. This is examplified in a parallel run of the compared genera and demonstrated in Fig. 1 B.

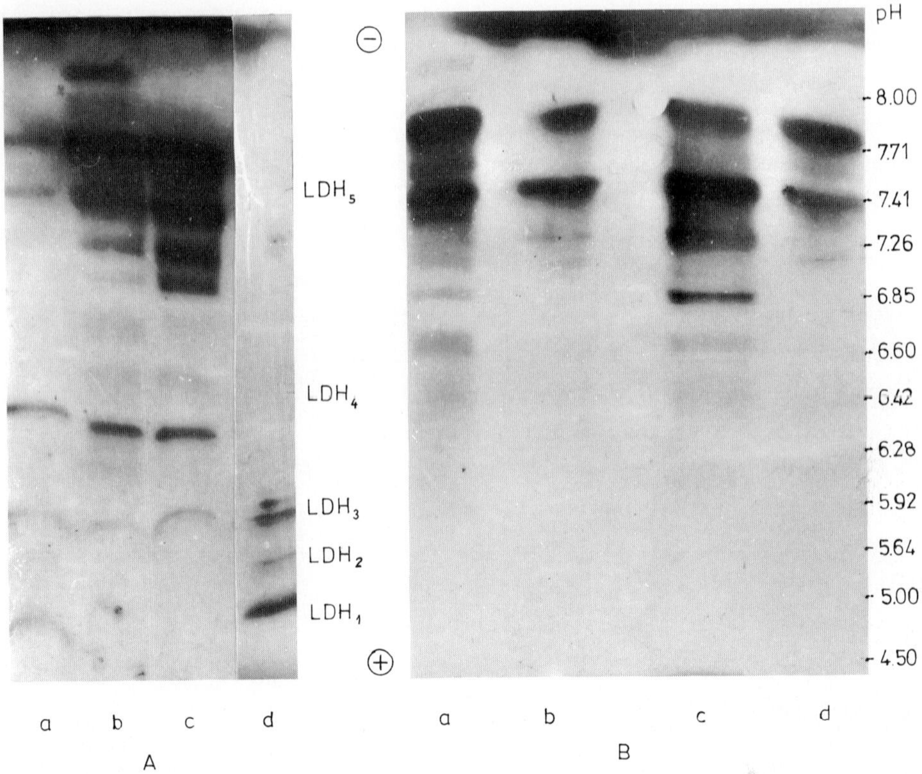

Fig. 3: Isoelectric spectra of LDH from Microtus species.
A: Different organs of Microtus arvalis: a, liver; b, muscle, c, kidney; d, heart.
B: Liver extracts from: a, M.arvalis; b, M.agrestis; c, M.oeconomus; d, M.pensylvanicus.

Lactic dehydrogenase

LDH revealed a very complex pattern after isoelectric focusing. Additional subbands occured in each LDH isoenzyme zone, the most heterogenous were those of LDH-4 and LDH-5, depending on the genus. To facilitate identification of the particular LDH isoenzyme zones, parallel runs of extracts from various organs of one animal were performed (Fig. 2 A, 3 A). Fig. 2 B demonstrates the LDH pattern in livers of the Apodemus genus and Fig. 3 B that of the Microtus genus. Species differences within the genera are evident from the isoelectric patterns; also here, sibling species (A.flavicolis - A.sylvaticus and M.oeconomus - M.agrestis), which are very difficult to distinguish on morphological basis, show quite dissimilar patterns. Species within one genus do not show marked deviations from their overall LDH pattern, but essential differences appear when comparisons are made between the genera. In general, the enzyme activity in Apodemus is focused predominantly in the hybrid LDH-4 zone and only 1 - 2 faint bands are visualized in the LDH-5 region. On the contrary, in the Microtus genus the activity is concentrated in the LDH-5 zone. It seems that there are some restrictions in hybrid formation between the H and M subunits, as the H subunits are produced, although LDH-1 and LDH-2 are hardly visible on the zymograms of Microtus species. The presented data on the behaviour of two enzymes suggest that phylogenetic differentiation on the molecular level is expressed as charge differences on the enzyme molecule and as different ability to subunit combination. We have also shown that on the basis of minor or essential differences of isoelectric components one can indicate relationship or distance on the molecular level. This may be of importance in taxonomic classification and evolutionary studies.

ACKNOWLEDGEMENT

This work has been partly supported by grants No PR-I/16; No PM-II/15 and No 21/1-Z/76.

The authors are indebted to Mrs. Danuta Kowalczyk and Mrs. Zofia Grzybowska for valuable technical assistance.

REFERENCES

1. Kubicz, A., Cisowska, I., Kowalczyk, D.: Studies on some enzyme systems, their tissue specifity and changes in electrophoretic patterns during postembryonic development of the common vole, Microtus arvalis. I. Acid Phosphatase. Zoologica Poloniae 24, 319-328 (1975).

2. Wolańska, L., Kubicz, A.: ditto. II. Lactate dehydrogenase. Zoologica Poloniae 25, 81-90 (1975).

3. Kubicz, A., Bartecka, Z., Wolańska, L.: ditto. III. Non-specific esterases. Zoologica Poloniae 25, 91-97 (1975).

4. Baur, E.W., Pattie, D.L.: Lactate dehydrogenase genes in rodents. Nature 218, 341-343 (1968).

5. Jacobson, K.B., Murphy, J.B., Dunaway, P.B.: Phylogenetic relationships of rodents on the basis of isoenzymes of lactate dehydrogenase. Comp. Biochem. Physiol. 29, 1135-1144 (1969).

6. Awdeh, Z.L., Williamson, A.R., Askonas, B.A.: Isoelectric focusing in polyacrylamide gel and its application to immunoglobulins. Nature 219, 66-70 (1968).

7. LKB 2117 Multiphor, Application Note (1973).

8. Papadopoulos, N.M., Kintzios, J.A.: Quantitative electrophoretic determination of lactate dehydrogenase isoenzymes. Amer. J. Clin. Pathol. 47, 96-102 (1967).

9. Dietz, A.A., Lubrano, T.: Separation and quantitation of lactic dehydrogenase by disc electrophoresis. Anal. Biochem. 20, 246-257 (1967).

Fractionation of Nucleic Acids on Isoelectric Focusing

J.W. Drysdale

This article describes a fractionation of nucleic acids during IEF in acidic pH ranges. Banding patterns of simple oligonucleotides are entirely consistent with a genuine isoelectric fractionation. However, larger nucleic acids such as tRNA and mRNA give rise to multiple components with apparently anomolous pIs. Some of this heterogeneity may represent a partial, but reversible, denaturation in addition to a possible separation on the basis of primary structure.

INTRODUCTION

This article considers an odd, but perhaps a potentially useful fractionation of nucleic acids that occurs during isoelectric focusing (IEF) in acidic pH ranges (1,2). These studies arose from a search for alternative methods for fractionating RNA species. Unlike protein fractionations, most methods for RNA fractionation are presently based on differences in size rather than charge. Such methods often might not allow separation of important species such as messenger RNA of similar size. It therefore seemed on interest to determine whether unique primary structures in nucleic acids would give molecules separable by IEF due to differences in their surface charge. Although this might seem an unlikely proposition in view of the acidic nature of nucleic acids, preliminary experiments with well-defined oligonucleotides gave encouraging results and led us to explore the behaviour of more complex polynucleotides such as transfer RNA, ribosomal RNA, and messenger RNA. Some of these results are reported here. For further details see refs. 1 and 2.

MATERIALS AND METHODS

Oligonucleotides were obtained from Miles Laboratory and Sigma Chemical Co., St. Louis, Missouri. Specific isoaccepting species of tRNA (Phe_2, F-met, Arg) from E. coli were generously supplied by Dr. A. D. Kelmers, Oak Ridge National Laboratories. 4S, 9S, 18S and 28S RNA were prepared by sucrose density gradient centrifugation of phenol extracts of whole rabbit reticulocytes (2). Preparations enriched in messenger RNA (mRNA) for globin chains were prepared by fractionating 9S RNA on oligo d(T) cellulose columns (3).

Gel electrofocusing was performed in small cylinders (10 cm x 0.3 cm) of polyacrylamide gel in apparatus from MRA, Boston, Mass (4). Gels to be fractionated were cast in syringe-type tubes (5). Gels to be scanned were cast in quartz tubes for direct scanning. Controls containing ampholyte but no sample were run in parallel to provide a correction for the absorbancy of ampholytes at 260 nm (1). Fractionations of oligonucleotides and tRNA were achieved in acrylamide gel with N,N'-methylene bisacrylamide cross-linking (T = 5%, C = 4%) and polymerized with persulfate. More porous gels (T = 3.5%, C = 4%) were used with rabbit reticulocyte RNA species to minimize molecular sieving effects. Although most of the RNA banded after 4 hours under a potential difference of 400 v, a total period of 18 hours was used routinely to ensure that large molecules reached their equilibrium positions.

Protein Synthesis Assay: After electrofocusing, gels were extruded into 20 µl aliquots through a #20 "snub nosed" syringe needle (5) and the eluted RNA was used to direct protein synthesis in a mRNA-dependent cell-free system (2). By characterizing the labelled peptides by electrophoresis and ion exchange chromatography (6), it was possible to identify different globin mRNA species.

RESULTS

Simple di- and trinucleotides behaved in a straightforward and predictable manner. ApA, ApU, GpC, CpC, CpCpC and GpCpC all focused as discrete single peaks in the pH range 3-5. A typical experiment showing the banding of GpC, CpC, GpCpC and CpCpC over the background "noise" from "Ampholine" is shown in Fig. 1. The apparent pIs of these nucleotides were entirely consistent with an isoelectric fractionation arising from protonation of the bases to balance the acidic phosphates. For example, the pKa of Gp, Ap and Cp are 2.4, 3.7 and 4.3 resp.(7),

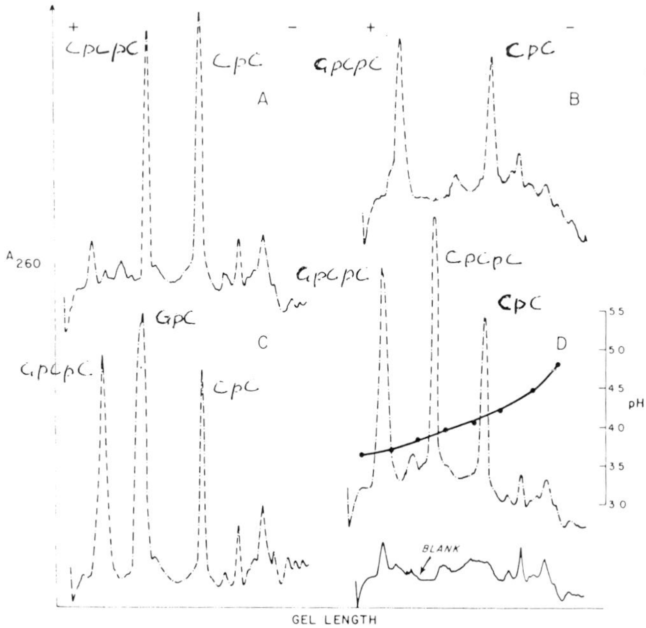

Fig. 1. Separation of di and trinucleotides by gel electrofocusing in the pH 3-5 range.(From Drysdale, J.W. and Righetti, P.(1) by permission of the American Chemical Society.)

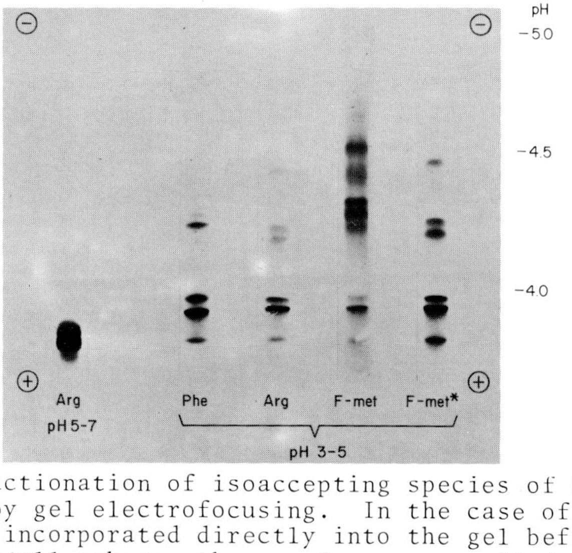

Fig. 2. Fractionation of isoaccepting species of Phe, Arg, and fMet-tRNA by gel electrofocusing. In the case of fMet*, the sample was incorporated directly into the gel before polymerization. In all others, the samples were applied to the top of the gel in a buffering layer of 2% ampholyte.(from Drysdale and Righetti (1) by permission of the American Chemical Soc.)

and the apparent pI of the nucleotides increased in the order ApU, ApA, GpC, CpC. Since U has no protonatable function, the pI of GpU was below the pH range of the ampholytes. The apparent pI values of ApA (3.8) and CpC (4.3) are close to the pKa of the free bases suggesting that their isoelectric forms corresponded to half protonation of both bases. Similarly the apparent pI of CpCpC was close to the value expected for 30% protonation of all three bases.

Experiments with tRNA: Banding patterns of larger polynucleotides and the basis for their behaviour proved considerably more complicated. Initial experiments with highly purified preparations of isoaccepting species of transfer RNA (tRNA) showed that all species formed stable banding patterns in the pH range 3-5. However, all gave considerably more complex patterns than was expected from their known purity (Fig. 2). For example, $tRNA^{fMet}$, which was at least 97% pure as judged by chromatography and amino acid accepting activity, separated into about eight components in the pH range 3.5-4.5. The patterns were reproducible with regard to the number of components and their apparent pIs, but the relative distribution of some components varied in different experiments. Much of this variability arose from the mode of sample application. When samples were premixed throughout the gel, proportionately higher amounts of the more acidic forms were obtained than when samples were applied at the cathode to preformed gels.

In addition to the heterogeneity, one of the most striking aspects of the patterns was the overall similarity in the number and distribution of components given by different tRNA species. For example, $tRNA^{Arg}$ and $tRNA_2^{Phe}$ each resolved into about 5 components. Many components from one species appeared to have counterparts in the other, but their relative distributions were quite different. Only about 8 components were resolved from a mixture of both species (1). Although elements unique to each species could be discerned on close inspection, the positions of several components coincided so closely that they were not readily distinguishable.

The focused components were usually visible as white, opaque regions in the gel, so that it was a relatively simple matter to isolate substantial quantities for further experimentation. When refocused, each of the major components redistributed to generate many of the other components in the original mixture (1). Analysis of the biological activity of the various forms from a single tRNA species showed that their amino acid accepting ability was a function of their pI. The specific activity of the most basic fractions (pI 4.5) was almost 90% of that of the original sample, but the specific activity of other components decreased with decreasing pI so that the material banding at pH 3.5 was only 20% as active as the original (1). The biological activity of all fractions could be restored essentially to control levels by allowing them to renature after annealing at 65°. These experiments suggest that the multiple forms are not degradation products, but different conformational states caused by a partial, but reversible denaturation.

Experiments with globin mRNA: Preliminary experiments with a crude extract of RNA from rabbit reticulocytes showed multiple bands of RNA in the pH range 3-6 (Fig. 3). Many peaks of viable mRNA were found throughout the gel. Most were well separated from the bulk of the cell RNA (ribosomal) and transfer RNA (Fig. 4). Subsequent experiments indicated that most of these mRNA activities coded for intact globin chains, implying that the activities represented intact mRNA molecules.

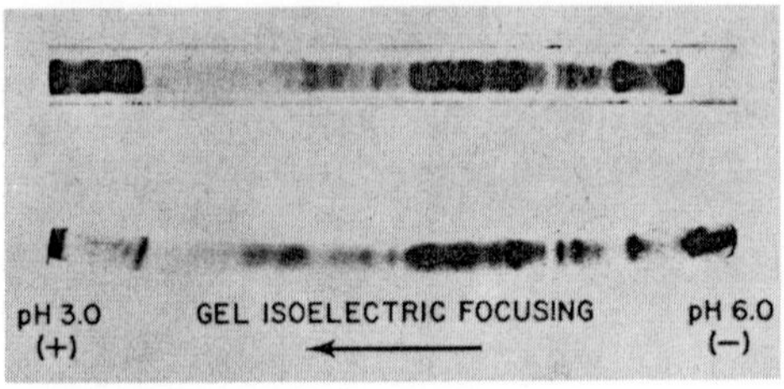

Fig. 3. Banding pattern of reticulocyte RNAs separated by isoelectric focusing in polyacrylamide gels. 1.0 A_{260} units of unfractionated reticulocyte lysate RNA (upper gel) or A.4 A_{260} units of reticulyte crude "9S" (lower gel) were focused for 24 h in 3.6% acrylamide gels over a pH range 3-6 Toulidene Blue stain. (From Drysdale and Shafritz (2). By permission of the American Chemical Society.

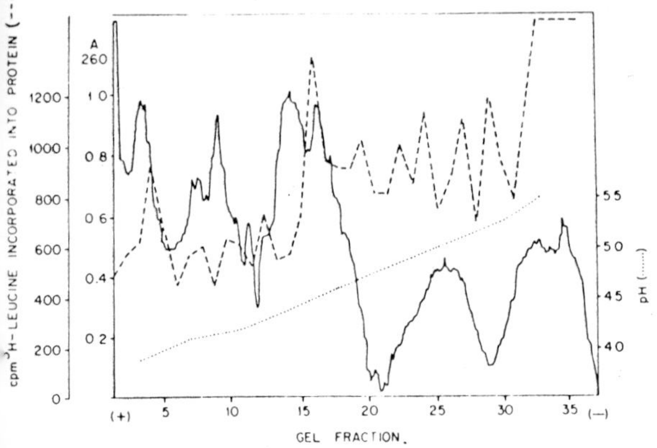

Fig. 4. 0.4 A_{260} units RNA from a reticulocyte lysate were fractionated in replicate gels in the pH 3-6 range. After focusing for 24 h one gel was scanned at 260 nm to identify major RNA peaks. Another gel was sectioned, the RNA extracted and tested for mRNA activity (---) (From Drysdale and Shafritz (2). By permission of the American Chemical Society).

Figure 5 shows a fractionation of mRNA activities from unfractionated RNA from a reticulocyte lysate and from a preparation enriched with α mRNA. By determining the relative distribution of radioactivity incorporated into the α and β chains of globin in a cell-free system programmed by these mRNAs, it was possible to characterize the predominant mRNA species in each peak. Table 1 shows that some of the focused mRNAs were substantially enriched for either the α or the β mRNA. A major peak of α mRNA was found near pH 4.7 while the major peak of β mRNA was near pH 5.2. However, several minor peaks of both mRNA species were scattered throughout the gel. Since some peaks of mRNA activity contained similar ratios of α and β globin chains, the complexity may be greater than was demonstrable by our methods of gel sectioning.

Our experience with tRNA led us to investigate whether some of the complexity might be attributable to partial denaturation. To investigate this possibility, globin mRNA was focused in replicate gels, the gels fractionated and the RNA eluted as before. One set was heated to 65°C for 10 min and then cooled slowly to permit reannealing of RNA molecules. Both sets were then assayed for mRNA activity. The results (Fig. 6) showed little, if any, difference in the mRNA activity profiles. Minor differences between the two patterns are within experimental limits for replicate gels. Certainly, any differences were considerably less than occurs with tRNA where 1-5 fold increases in biological activity of some forms

TABLE 1

α/β RATIOS OF PRODUCTS WITH RETICULOCYTE mRNA FRACTIONS SEPARATED BY ISOELECTRIC FOCUSING

pI	Lysate mRNA		Supernatant mRNA	
	Fraction*	α/β	Fraction*	α/β
3.9		—	A	0.83
4.3	1	0.87		—
4.4		—	B	1.32
4.7		—	C	2.72
4.8	2	1.03	D	2.75
4.9	3	0.69		—
5.0		—	E	1.86
5.1	4	0.71	F	1.36
5.2	5	0.39	G	1.24

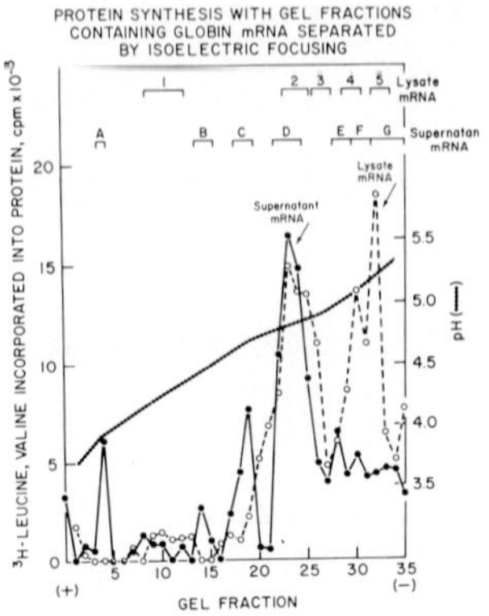

Fig. 5. Distribution of mRNA activities from unfractionated lysate RNA and a preparation enriched in mRNA. (From Drysdale and Shafritz (2) by permission of the American Chemical Society)

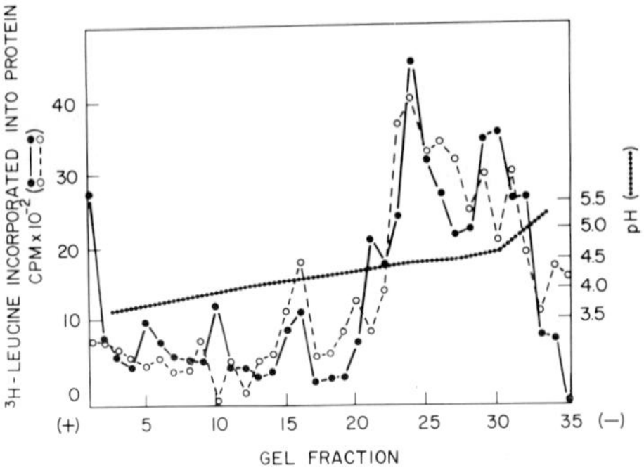

Fig. 6. Effect of annealing on the pattern of messenger activity with focused reticulocyte RNA fractions. (●—●, without annealing, o- - - -o, with annealing. (From Drysdale and Shafritz (2), by permission of the American Chemical Society.)

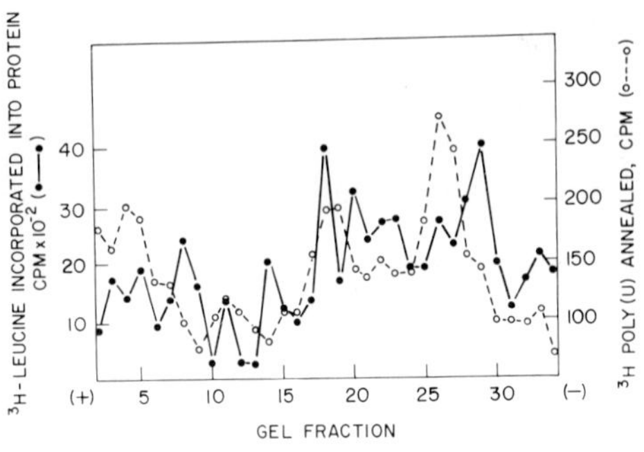

Fig. 7. Distribution of messenger activity versus poly(A) content of isoelectric focused reticulocyte RNA. 2.0 A_{260} units of unfractionated reticulocyte lysate RNA was focused in the pH range 3-6. (From Drysdale and Shafritz (2) by permi sion of the American Chemical Soc.

were observed on reannealing. These results indicate that the multiple forms of mRNA are unlikely to be due to partial denaturation as a result of the extraction or focusing procedures, unless separated fractions renature spontaneously after elution, a possibility which we cannot exclude.

Another potential source of heterogeneity comes from poly(A). It is known that globin mRNA has variable amounts of poly(A) at the 3' end of the molecule. We felt that this variability in poly(A) content together with changes in pKa of A with increasing length of poly(A) (8), might contribute to the heterogeneous globin mRNA focusing profile. While this appeared to be so (Fig. 7), there was no clear correlation of poly(A) content with the apparent pI of the multiple forms of globin mRNA. Several bands of mRNA appeared to contain little, if any, poly(A). It would be of interest to know whether these bands represent discrete α and β mRNAs separated on the basis of their primary structure.

Several factors may, therefore, contribute to the heterogeneity in globin mRNAs on IEF: (1) primary structure differences in the coding and regulatory sequences, (2) variability in poly(A) chain length, (3) denaturation and/or aggregation during RNA extraction or on IEF.

The physico-chemical basis for these and other fractionations of polynucleotides also still eludes us. It is known that nucleic acids can be fractionated by pH precipitation in acidic pH ranges. However, since we obtain similar banding patterns on IEF in granulated gels (Pevikon) which allow focusing of particulate matter (9), non-specific precipitation is unlikely to be the basis for our observations in polyacrylamide gel. Our experiments with tRNA suggest that the banding patterns may arise from complex formation with ampholytes since RNA is known to bind strongly to polyamines (10). Specific interactions with different ampholytes might account for the apparently anomolous pI range of polynucleotides yet still allow unique distributions for different RNA species. On the other hand, Guyen and Chrambach (11) have recently reported that 16S ribosomal RNA from E. coli can be 'focused' by IEF without carrier ampholytes. The apparent pI of this RNA is similar to that obtained in our experiments with the mammalian species. Clearly no all-embracing explanation is yet at hand. However, in an often pragmatic world of science, lack of understanding of basic mechanisms need not preclude useful application of empirical observations.

REFERENCES

1. Drysdale, J.W. and Righetti, P.G.: Biochemistry, 11, 4044-4052 (1972).

2. Shafritz, D.A. and Drysdale, J.W.: Biochemistry 14, 61-68 (1975).

3. Aviv, H. and Leder, P.: Proc. Natl. Acad. Sci. U.S.A. 69, 1408-1412 (1972).

4. Righetti, P.G. and Drysdale, J.W.: Biochim. Biophys. Acta 236, 17-28 (1971).

5. Bagshaw, J.C., Drysdale, J.W. and Malt, R.A.: Ann. N. Y.

Acad. Sci. 209, 363-371 (1973).

6. Dintzis, H. M.: Proc. Natl. Acad. Sci. U.S.A. 47, 247-261 (1961).

7. Jordan, D.O.: Chemistry of Nucleic Acids, Washington, D.C. Butterworth (1960)

8. Voet, D. and Rich, A.: Progr. Nucl. Acid Res. Mol. Biol. 10, 183-265 (1970).

9. Otavsky, W. and Drysdale, J.W.: Anal. Biochem. (in press).

10. Hirschman, S.Q., Lenz, M. and Felsenfeld, G.: Biopolymers 5, 227 (1967).

11. Guyen, N. and Chrambach, A.: Anal. Biochem. 74, 145 (1976).

5. Clinical Applications

Polyacrylamide Gel Isoelectric Focusing of Alpha-1-Proteinase Inhibitors (α_1 Pi) Phenotypes: I. Optimal Condition, Histochemical and Proteinase Probe Studies

R.C. Allen, P.M. Oulla, P. Arnaud and J.S. Baumstark

ABSTRACT

Short separation times achieved by high field strengths were made possible by improved heat transfer using Berrylium oxide support plates have lead to improved resolution of α_1Pi phenotypes. To verify the proteins and glycoproteins resolved in the α_1Pi region as true proteinase inhibitors trypsin probe techniques are described, which can be carried out either by premixing trypsin and serum prior to isoelectric focusing or directly on the gel using crossed isoelectric focusing of trypsin and serum. Using these procedures one glycoprotein in the M-4 region has been found not to be an inhibitor and a second glycoprotein found only in the Z-2 region of serum from patients with deficiency related cirrhosis has been found to be a proteinase inhibitor.

INTRODUCTION

The use of isoelectric focusing on thin polyacrylamide gel slabs for the identification of alpha 1 proteinase inhibitor (α_1Pi) phenotypes has proven to be a superior method to that of the so-called acid starch gel method [1,2,3,4]. The various isoelectric focusing systems described, while all offering high resolving power, have lead to some confusion as to whether all of the glycoproteins resolved in each case are actually proteinase inhibitors and thus true allele products. Allen et al. [1] and Pierce et al. [3] have reported proteins with isoelectric points similar to M allele products in non-M phenotypes and proteins present which do not appear to be antiproteinases respectively. This has lead to confusion in phenotyping of approximately five per cent of routine samples in regard to the origin and function of these proteins. Whether they represent sample denaturation, haptoglobin 1-1 variants, physiopathological changes, or actually represent previously unresolved Pi allele products or new subtypes must be answered.

To properly assess this question both maximal resolution of the α_1Pi allele region is necessary as are direct methods to determine which proteins are true proteinase inhibitors. Two problems are presented in answering this question. First, optimization of this basic separation system is required to assure reproducibility and maximal resolution with minimal separation time for clinical use. Second, direct biochemical techniques are required to substantiate the identity of each protein separated in the α_1Pi region.

Maximal resolution in minimal separation time requires a high field strength to overcome the low conductance of commercially available ampholytes, particularly in the α_1Pi region of isoelectric points between pH 4 and 5 [5]. This requires additional cooling capacity and improved conductance of Joule heating away from the system, greater than that which can be obtained from media support plates made of glass. To this end we have investigated, and employed as part of this study, Berrylium oxide plates to more efficiently dissipate heat using both a circulating refrigerated bath and thremoelectric cooling system.

The second problem has been approached utilizing a specific histochemical stain for glycoproteins and specific sugar moities as well as the development of methodology to indicate which glycoproteins in the α_1Pi region will complex with trypsin by employing a direct proteinase probe in conjunction with histochemical methods previously reported [7]. These were developed as additional methods to the immunological techniques described separately by Arnaud et al. [6].

METHODS AND MATERIALS

Isoelectric Focusing. Isoelectric focusing was performed on a standard LKB multiphore system as previously described [1] with the exception that power settings of 20 watts for the first 20 minutes, 25 watts for 90 minutes and 30 watts for 10 minutes with a cooling plate temperature of 3°C were employed. Gels were cast as previously described [1] with pH 3.5 to 5, 3.5 to 8 ampholine and LKB standard prepared gels for Pi phenotyping were also utilized. The cathode buffer in all cases was pH 5 to 8 ampholine diluted 1:100. Maximum gel surface temperatures as measured by an electrically shielded thermocouple were 20°C. Power was supplied by an LKB 2103 Constant Power Supply.

Both types of gradients were also run using a Berrylium oxide plate with dimensions of 10 cm x 11.2 cm x 3.5 mm thick as a support. Cooling was obtained on both a standard LKB cooling plate connected to a Lauda refrigerated circulating bath or a thermoelectric cooling device employing two CP 2 31-06 "Frigichip"* modules mounted on air cooled heat transfer surfaces

---

*TM Melcor, Inc., Trenton, New Jersey, USA

and each capable of transferring 102 BTU/hr at 14 amps. Surface temperatures of 3°C on the Berrylium oxide support plates as measured by a thermocouple were attained prior to starting focusing. Separations were carried out at 30 watts constant power for 60 minutes with a maximum gel surface temperature during the run of 21°C.

Gradient pH Determination. Following focusing the gel was cut to a length of 9.5 cm. One cm was cut off the right side of the gel and a second vertical cut made 3 cm from the edge. The gel was then cut into 5 mm sections from the cathodal to anodal ends using a marked template. Each section was placed, in order, into screw-cap vials containing 3 ml of boiled distilled water and capped immediately. After overnight diffusion of ampholine out of the gel sections, the pH of each vial was measured to the nearest 0.001 pH unit on a Corning Model 112 Research pH Meter. All vials were covered during measurement with parafilm to prevent the introduction of CO_2.

Crossed Isoelectric Focusing. The pH gradient was partially established for a period of 20 minutes at 20 watts of constant power which provides a pH of 4.68 in the crossing zone at the time of sample application on pH 3.5 to 5 gradients. Filter paper tabs 8 x 20 mm for sample applications were then placed on the gel surface 1 cm anodal to the cathodal wick. An additional 8 x 10 mm tab was placed 5 mm below each sample tab extending over one half of the serum migration path. Sera from standard reference phenotypes were placed in 40 µl amounts on the upper tabs. Twenty microliters of bovine pancreatic trypsin*, at varying molar enzyme to inhibitor ratios, with enzyme activity being determined by the active site titration method of Chase and Shaw [7], was applied to the lower 8 x 10 mm tabs. A lower control tab containing distilled water was used as a control. Electrofocusing was carried out for an additional 45 minutes, the tabs removed, and the gradient sharpened for an additional 10 minutes at 30 watts of power. At this time bromphenol blue dye placed on the cathodal wick buffer on 3.5 to 5 pH gradients and on a filter paper tab a pH 3.5 to 8 gradient was completely focused as a yellow band just cathodal to the anode buffer wick.

Serum Incubation of Serum and Trypsin. Forty microliters of serum were incubated with 10 µl of trypsin in PBS for 20 minutes at 37°C. An E/I excess of 2.36:1 for a standard MM containing 3 mg/ml of antitrypsin was employed for MM serum. For other α_1Pi phenotypes the ratio for the standard MM was utilized which gave an enzyme ranging from 3.54 to 7.08 depending

---

*2x crystalized, Sigma

on the phenotypes tested. For comparison 10 µl of PBS was added to control sera. Isoelectric focusing was carried out directly as in the phenotyping method, i.e. without prior gradient establishment for two hours. Staining of protein and Periodic Acid-Schiff stains for glycoproteins in the α_1Pi allele products was performed as previously described [8].

RESULTS

Comparisons of both the LKB prepared gels for α_1Pi phenotyping and pH 3.5 to 5 gradient gels separated on glass supporting media for two hours at power levels of 20, 25 and 30 watts for 20, 90 and 10 minutes respectively are shown in Figures 1a and 2a. In Figures 1b and 2b separations carried out for 1 hour at 30 watts on 10 cm wide gels are shown as a comparison.

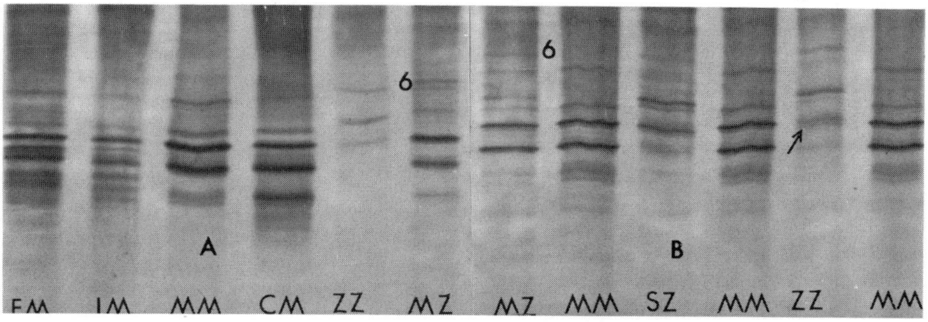

Figure 1. LKB prepared gels showing various phenotypes separations were carried out on glass plates for two hours at a maximum power of 1.25 watts per cm gel width in 1a. In 1b samples were separated for one hour at 3.0 watts per cm. Note the difference in resolution of the Z-6 band between 1a and b and the Z-2 doublet (arrow) in 1b.

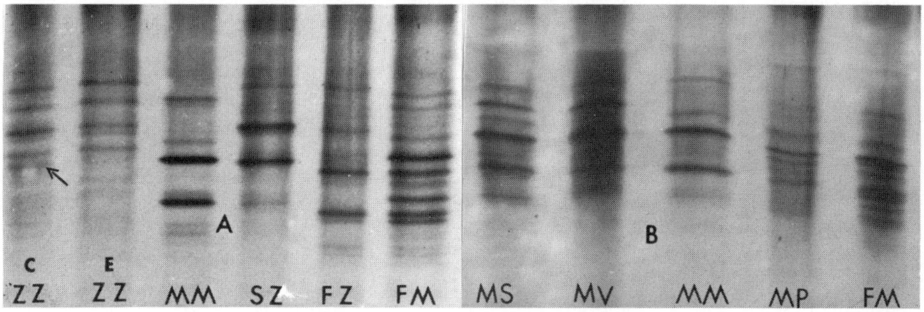

Figure 2. Various phenotypes on a pH 3.5 - 5 gradient. In 2a 1.25 watts per cm power, 2b 3.0 watts per cm. Note difference between the two ZZ c = cirrhosis, E = emphysema in Z2 doublet (arrow).

Since different plate widths were used a comparison of power per cm width of gel indicates that on glass supported media the maximum power of 1.25 watts per cm width was achieved with a gel surface temperature of 20°C at maximum power. On Berrylium oxide* supported media cooled either by a circulating bath or thermoelectrically a maximum power of 3 watts per cm width was achieved with a gel surface temperature of 21°C. Similar power on glass supported media lead to gel drying and curling with resultant ruin of the gel and pattern.

Comparison of the MZ pattern achieved with LKB prepared gels with both methods, Figure 3, indicates that the resolution achieved at the higher power setting markedly improves the separation of the M7-Z4 and M5-Z6 bands, a major problem in identification using this system at low field strengths.

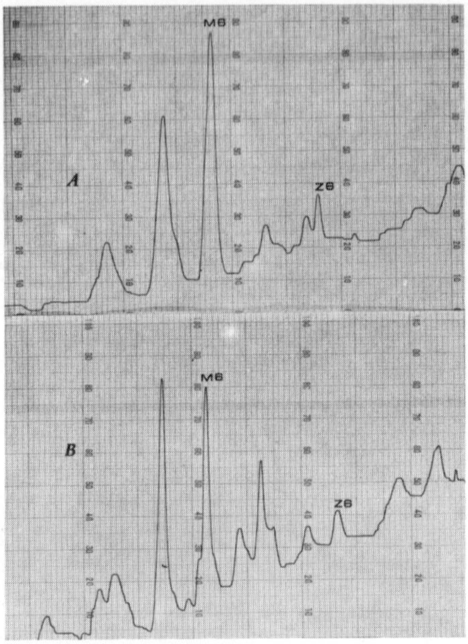

Figure 3. Separation of MZ phenotype at 1.25 watts per cm width maximum power A, and 3.0 watts per cm width in B.

A comparison of pH gradient profiles shown in Figure 4 for both pH 3.5 to 5 gels and LKB prepared gels indicates good linearity over the entire Pi region of isoelectric points with the LKB gradient shifted to the right of a pH 3.5 to 5 gradient but with similar slope.

Specific binding by crossed focusing of trypsin and a standard MM serum on partially prefocused gels is shown in Figure 5 where varying amounts of trypsin excess were employed and the pH of the crossing zone was 4.62.

---

*Berlox K-150, National Berrylia Corp., Haskell, New Jersey, USA

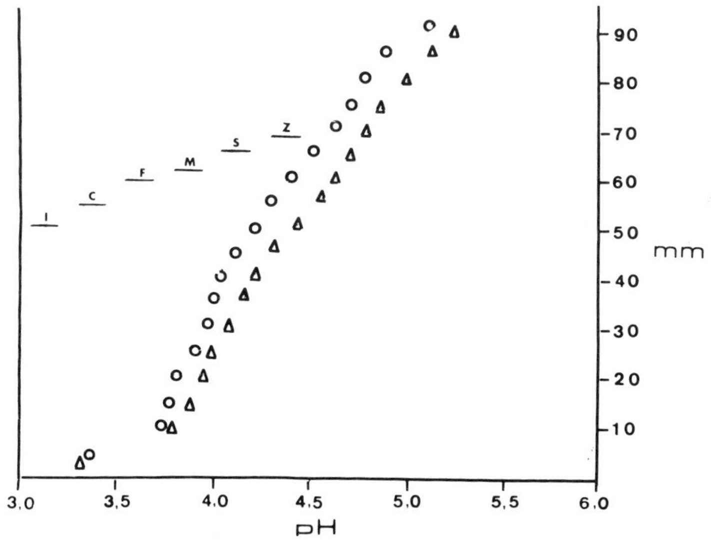

Figure 4. Gradient from LKB pH 3.5 to 5 ampholine open circles and prepared α_1Pi commercial gel triangles. Major bands are shown for various phenotypes on pH 3.5 to 5 gels, shown in relative position on pH gradient.

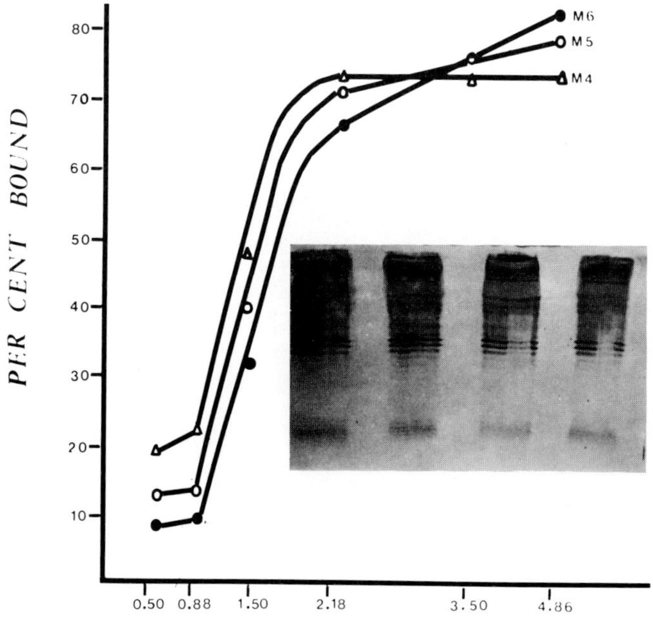

Figure 5. Effect of increasing molar concentrations of trypsin on allele products M4, 5 and 6 density following crossed focusing. The insert shows 0.50, 0.80, 3.50 and 4.86 E/I molar concentrations.

The per cent binding of bands M-4, 5 and 6 was determined by quantitative microdensitometry by comparison of the reacted and control side of the separation.

More effective binding was apparent as evidenced by almost complete disappearance of allele products with enzyme to inhibitor ratios (E/I) of 2.4 when trypsin was incubated with serum at pH 7.2 than when allowed to react by crossed focusing as indicated in Figure 6. Both members of the Z2 doublet found in infantile cirrhosis are completely removed following either crossed or direct binding procedures and indicate that this additional band in the cirrhotics is a true antiproteinase. All other allele products of C, F, I, M, S and Z appear to complex strongly with trypsin although at E/I excesses sufficient to completely remove all other Z allele products some Z6 remains.

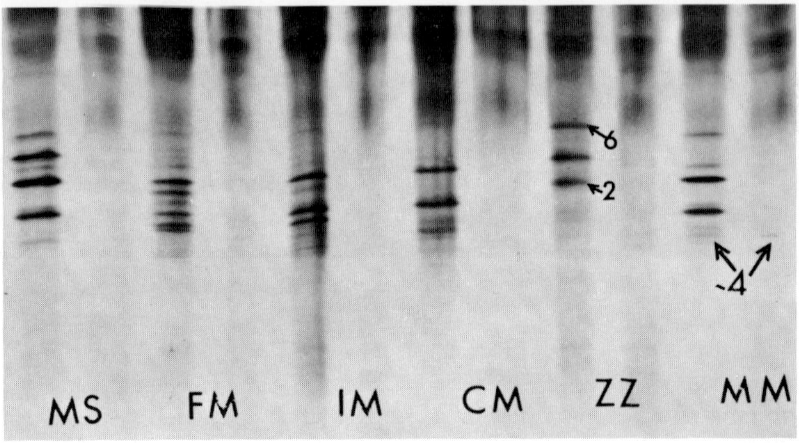

Figure 6. Control sera and sera incubated with trypsin excess showing diminution of allele products. Bands M4 and Z6 are marked by arrows.

The M4 region does not all bind trypsin which is further verified by PAS staining of separations carried out on 3.5 to 8 gradients where a PAS positive band remains and also a new glycoprotein region, presumably bound α_1Pi allele products, appears as shown in Figure 7A. Thus a portion of the M4 region while a glycoprotein does not appear to have antiproteinase activity. In Figure 7B protein is a coomassie blue protein stain on a similar pH 3.5 to 8 gradient showing additional proteins at higher isoelectric points in the path of the migrating trypsin indicating a major complex at similar isoelectric points as in 7A and the effect of pH in the crossing region on binding.

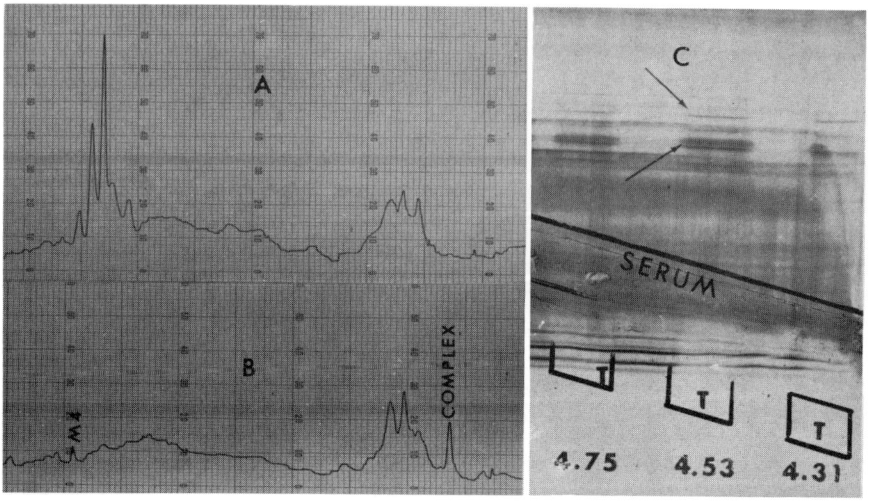

Figure 7. A. Microdensitometry of 20 μl of Pi type MM serum containing 3.1 mg/ml alpha-1 protease inhibitor. B. A similar amount of Pi MM serum cross focused against trypsin in an E/I ratio of 4.86. Separations were carried out on a pH 3.5 to 8 gradient and stained for glycoproteins with PAS. C. Effect of pH in crossing zone on binding showing pH at crossing zone center marked below. Arrows point to new protein zones in the trypsin path corresponding to major complex peak in B, coomassie blue stain.

DISCUSSION

The use of Berrylium oxide support plates for either commercially prepared gels or gels cast thereon offers a support that is some 300 times more efficient as a heat conductor compared to glass and allows higher field strengths to be used with markedly reduced separation time. Under optimal conditions we have been able using thermo-electric cooling to achieve complete focusing under 30 minutes with surface temperatures on the gel of 5-8°C at maximum field strength which occurs when the gradient is established. The problem of anodal and cathodal gradient drift, which is time and field strength dependent, as described by Chambach [9] appears to offer no problem for the system described either to reproducibility of gradient or band location from run to run. However, we have employed ampholines as the cathode buffer rather than materials with high pK values. No effect was found on the isoelectric point of M6 which was 4.520 ± 0.010 on Berrylium oxide plates and 4.502 ± 0.005 at 22°C at lower field strengths on glass supported pH 3.5 to 5 gels. With LKB prepared gels the location of bands M5 and M6 were identical with only the Z6 band being more anodal in the gels subjected to the higher field strengths, Figure 2 which improves phenotyping. Also peak width at one half peak height was significantly better a high field strength thus improving resolution.

Pierce [3] has reported that some sera show additional protein zones in the α_1Pi region thus leading to potential confusion on Pi phenotyping. He has shown that the sample may often be converted to a standard pattern by the addition of cysteine. However, this procedure does not indicate if such proteins are proteinase inhibitors or if overlapping bands are present. With the presently reported methods, for example, the multizone M4 region in both fresh and frozen stored serum consists of three mannose - containing glycoproteins [7]. The most anodal and cathodal glycoproteins bind trypsin whereas the central one does not or only weakly. The latter protein therefore, by presently applied criteria is probably not α_1Pi protein.

The marked increase in the anodal band of doublet Z2 region found in many α_1Pi type ZZ infantile cirrhotics [1, 10] has been found not only to be a glycoprotein containing mannose [7] but also to be a proteinase inhibitor, as evidenced by its binding of trypsin using both the crossing procedure and direct binding and thus is a true α_1Pi allele product characteristic of infantile cirrhosis.

The studies reported here indicate that both procedures, either direct binding of crossed isoelectric focusing or direct binding of trypsin with the alpha-1 isoinhibitors, may serve as additional valuable tools with which to assess more acurately the presence of proteins other than α_1Pi which may appear in the α_1Pi region. Such proteins may represent denaturation, haptoglobin of type of 1-1 [6], new types, subtypes, or possibly result from unrelated pathophysiological changes. Thus any of the latter may be differentiated by the methods described here with each sample serving directly as its own control in the crossing method or by side by side comparisons of trypsin mixed with serum prior to electrophoresis and a serum control plus saline.

ACKNOWLEDGEMENTS

The authors wish to express their gratitude to Dr. Peter Fleischner of the National Berrylia Corporation for supplying the Berlox K-150 support plates used in this study.

REFERENCES

1. Allen, R.C., Harley, R.A. and Talamo, R.C.: A new method for determination of alpha-1-antitrypsin phenotypes using isoelectric focusing on polyacrylamide gel slabs. Am. J. Clin. Path. 62, 732-739 (1974).

2. Arnaud, P., Chapuis-Cellier, C. and Creyssel, R.: Polymorphisme de l'alpha a antitrypsine plamatique (system Pi) mise en evidence par electrofacalization sur gel de polyacrylamide. Comp. Rend. Soc. Biol. 168, 58-64 (1974).

3. Pierce, J.A., Jeppsson, J.O., Laurell, C.B.: Alpha-1 antitrypsin phenotypes determined by isoelectric focusing of cysteine-antitrypsin disulfes in serum. Anal. Biochem. In Press.

4. Keuppers, F.: Determination of α 1 antitrypsin by isoelectric focusing in polyacrylamide gels. J. Lab. Clin. Med. 88, 151-155 (1976).

5. Frater, R.: Behavior of ampholines during isoelectric focusing. Anal. Biochem. 38, 536-538 (1970).

6. Arnaud, P., Chapuis-Cellier, C., Koistinen, J., Allen, R.C., Fudenberg, H.H.: Polyacrylamide gel isoelectric focusing applied to alpha-1 proteinase Inhibitor (A_1Pi) phenotyping - Immunochemical procedures. This Proceeding.

7. Allen, R.C., Spicer, S.S. and Zehr, D.: Concanavalin A-horseradish peroxidase bridge staining of alpha-1 glycoproteins separated by isoelectric focusing on polyacrylamide gel. J. Histochem. Cytochem. In Press.

8. Chase, T. Jr. and Shaw, E.: p-Nitrophenyl -pl-Guanidinobenzoate HCl: A new active site titrant for trypsin. Biophys. Res. Comm. 29, 508-514 (1967).

9. Chrambach, A., Doerr, P., Finlayson, G.R., Miles, L.E.M., Sherins, R., and Robard, D.: Instability of pH gradients formed by isoelectric focusing in polyacrylamide gel. Ann. N.Y. Acad. Sci. 209, 44-64 (1973).

10. Allan, R.C.: Molecular studies of allele products associated with infantile cirrhosis and emphysema. In conference, The role of protease and proteinase inhibitors. Airlie, Virginia, USA (1976).

Polyacrylamide Gel Isoelectric Focusing (PAGIF) Applied to Alpha-1-Proteinase Inhibitor (α_1 Pi) Phenotyping: II. Immunochemical Procedures

P. Arnaud, C. Chapuis-Cellier, G.B. Wilson, J. Koistinen, R.C. Allen and H.H. Fudenberg

ABSTRACT

When direct interpretation of PAGIF results is used for positive identification of Pi types of Alpha 1 - proteinase inhibitor, problems arise in approximately 5 to 10 % of the sera routinely studied. Two complementary immunochemical procedures, crossed immunoelectrophoresis and a new method of print immunofixation, can be applied in such cases and are particularly valuable for confirmation of potential new Pi phenotypes detected by PAGIF

INTRODUCTION

Polyacrylamide gel isoelectric focusing (PAGIF) (1,2,3) represents a major improvement over the techniques of acid-starch gel electrophoresis (4) and agarose electrophoresis at pH 5.15 (5) used previously for studies of the microheterogeneity and polymorphism of Alpha 1 - proteinase inhibitor (A_1Pi, formerly termed "Alpha 1 - antitrypsin")(6). The major advantages of PAGIF are the rapidity of the run, the reproducibility of the results, and the high resolution of the bands that constitute the Pi system (7).After careful comparison with the results obtained from isoelectric focusing and acid-starch gel electrophoresis, we have used electrofocusing exclusively for routine Pi typing over the course of the past 3 years during which we have analyzed over 15,000 different sera. In our hands, PAGIF permits accurate determination of Pi phenotypes for 90 - 95 % of all samples. For the remaining 5 - 10 % , however, PAGIF alone does not provide clear-cut identification. Thus, two additional immunochemical procedures, crossed immunoelectrophoresis and print immunofixation, are necessary.

MATERIALS AND METHODS

Control Sera

Pi phenotypes were analyzed in sera from subjects of population genetic studies or from patients with various clinical disorders. The Pi types were established by comparison with patterns published in the literature

---

Work supported in part by grants from U E R Biologie Humaine, Université Claude Bernard, Lyon, France, from Fondation de l'Industrie Pharmaceutique pour la Recherche, Paris, France, and from the U.S. Public Health Service (NIH Grants HD-09938 and AI-09145). This is publication no. 44 from the Department of Basic and Clinical Immunology and Microbiology, Medical University of South Carolina, U.S.A.

and with control sera (kindly supplied by Dr. M.K. Fagerhol, Blood Bank and Department of Immunohaematology, Ullevaal Hospital, Oslo, Norway). Some rare serum samples not encountered in our studies were obtained from other investigators (B M phenotype (8) from Dr. J.P. Martin, U 78 INSERM, Bois-Guillaume, France ; and M W_2 and M Y_2 phenotypes (9) from Dr.P.J.L.Cook Human Biochemical Genetics Unit, MRC, London, England). Sera were dispensed in aliquots of 20 μl, stored at - 20°C , and thawed only once before use.

Chemicals

Acrylamide and N,N,N',N' - methylene bis acrylamide were obtained from Polyscience, Inc. (Warrington, Pa.). Agarose was produced by L'Industrie Biologique Française (Gennevilliers, France). Specific antisera against A_1Pi were produced either by Behringwerke (GMBH, Mannheim, Germany) or by Meloy (Springfield, Va.). Ampholines* were obtained from L K B (Bromma, Sweden).

Isoelectric Focusing

The basic equipment used was L K B 2117 - 411 Multiphor applied to A_1Pi phenotyping (LKB Application Note 185). Polyacrylamide gels, pH range 4 - 6, were used as previously described (1,3) with some modifications. Occasionally, a pH range of 3.5 - 5 was used for better resolution of the slow Pi types, especially those due to the presence of the P, S, V, W, and Y allele products, or a range of 3.5 - 6 was obtained by mixing Ampholines* 3.5 - 5 and 4 - 6 in equal proportions. Runs were performed with two additional modifications : a constant power unit (LKB 2103 Power Supply) was used for the power source, and a cooling system (Lauda K 2 R, Brinkmann Instruments) was used to provide better refrigeration of the glass-plate supporting the gel. Gels used for analytical purposes were run at 35 W for 1.5 hr (48 samples per gel), and those to be submitted to immunochemical procedures were run at 35 W for 2.5 hr and then at 40 W for 20 min to enhance the distance between the bands (long run technique).

Crossed Immunoelectrophoresis

Procedure. A solution of acrylamide (0.9 %) and agarose (0.7 %)(W / V) in Barbital buffer, 0.1 M, pH 8.8, containing 1mM calcium lactate was heated to boiling, and 21 ml of the solution was poured on a glass-plate(95 x 85 mm). After 15 min at 4°C, the central part of the gel (32 x 85 mm) was cut and removed. After completion of the run, a strip (90 x 4 mm) was cut in the central part of the migration area of PAGIF and placed close to the edge of the acrylamide-agarose gel plate. The central part of the plate and the strip were covered with 7 ml of the same acrylamide-agarose solution to which a specific anti - A_1Pi antiserum had been added. After gelation, the portion including the PAGIF strip was carefully covered with a larger strip of acrylamide-agarose, and the plate (inclusion close to the cathode) was submitted to electrophoresis (70 mA p. plate for 6 hr) on an LKB Multiphor 2117 - 116 apparatus (L K B Application Note 185). Barbital buffer 0.1 M, pH 8.8, was used as the vessel buffer. After washing overnight in saline, the plates were dried, stained with Amido Black, and destained.

Critical Points. (a) The relative amount of antigen and antibody is critical and can produce peaks that are either too high (excess of antigen) or too small (excess of antibody). According to the above procedure, an optimal ratio is obtained when the serum used for the first dimension is diluted to 0.3 mg / ml A_1 Pi , and when 150 μ g of specific antiserum is added to the 7 ml of acrylamide-agarose filling the central part of the gel.

(b) Inclusion of the PAGIF strip under the gel gives more reproducible results than does application of the strip to the surface of the gel as proposed by Soderholm (11). In addition, covering the inclusion with a strip of acrylamide - agarose prevents separation of the gel at the inclusion and suppresses endosmotic flow.
(c) The steps of the procedure must be performed as fast as possible to avoid any diffusion of the already focused bands. Better results are obtain-ed when all the manipulation is performed at 4°C. The PAGIF strip must be handled carefully to avoid fingerprinting.

Immunofixation.

Procedure. After completion of PAGIF runs, gels were kept at 4°C. A strip of cellulose acetate (Sepraphore* III, Gelman Instrument C°) was soaked with specific antiserum and carefully transferred onto the surface of the PAGIF gel. The gel was incubated for 1 hr, and then a glass-plate was put on the gel, covered with a cellulose acetate strip and firmly pressed on it for 30 min.The gel and the strip were then washed for 16 hr in 3 changes of saline (2 liters each time), stained with Coomassie Brilliant Blue (R 250) in methanol - water - acetic acid (5:5:1) and destained in the same methanol water - acetic acid solution.

Critical Points. (a) Patterns obtained on the cellulose strips are much sharper than those from the gel. Thus, the development of immune complexes in the strips is of great importance. The fragility of cellulose acetate is critical. The strip must be manipulated carefully and fingerprinting by direct contacts must be avoided.
(b) For staining of the cellulose acetate strip : one of its end is attach-ed to a metallic clip and the strip is soaked no more than 6 - 8 times in the staining solution and immediately destained. Bending or stretching of the strip must be avoided during these manipulations, and excess staining will cause background that makes the patterns difficult to interpret.

RESULTS AND DISCUSSION

Polyacrylamide Gel Isoelectric Focusing (P A G I F).

Problems of the interpretation of PAGIF results are illustrated in Figure 1:

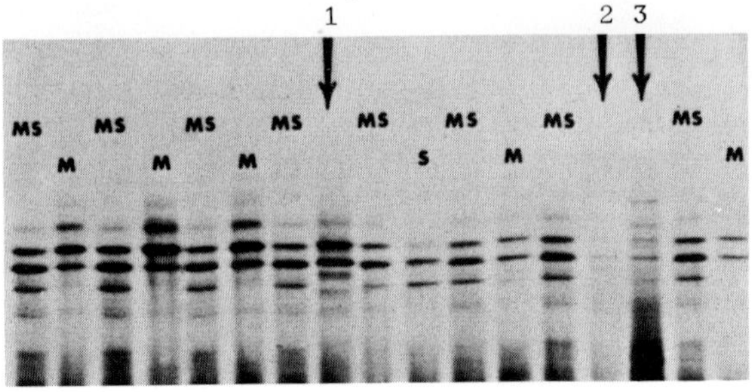

Anode at the top, pH range 4 - 6.

Patterns from a typical PAGIF run for Pi typing.
Questionable patterns are pointed by arrows : in "1", additional bands are present in cathodal part of the Pi bands, in "2", the weakness of the patt-ern suggests an alteration of the serum ; in "3", the fast migrating patt-ern superimposed on the M bands indicates the possibility of a new pheno-type.

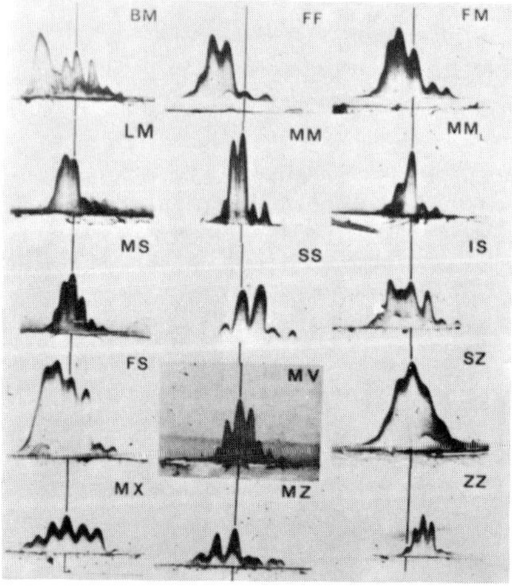

Figure 2.
CIEP patterns of various phenotypes.
1st dimension, anode at the left ;
2nd dimension, anode at the top.
Dark line indicates position of the M 6 band of the M phenotype.

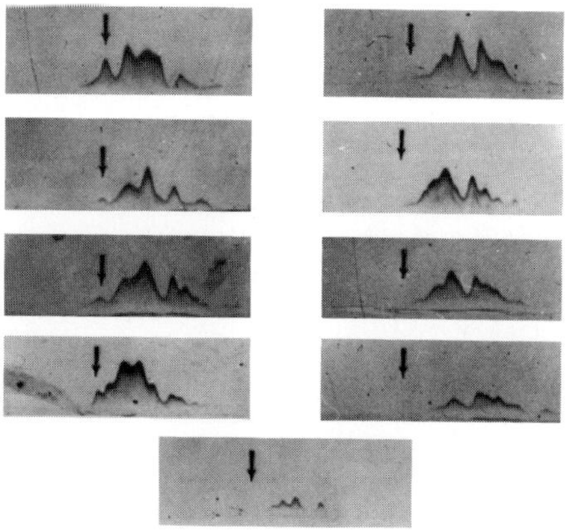

Figure 3. CIEP of altered sera at different steps : position of M 6 band is indicated by an arrow. Density and area of the cathodal pattern eliminate possible confusion with slow phenotypes.
Electrode positions as in Fig. 2

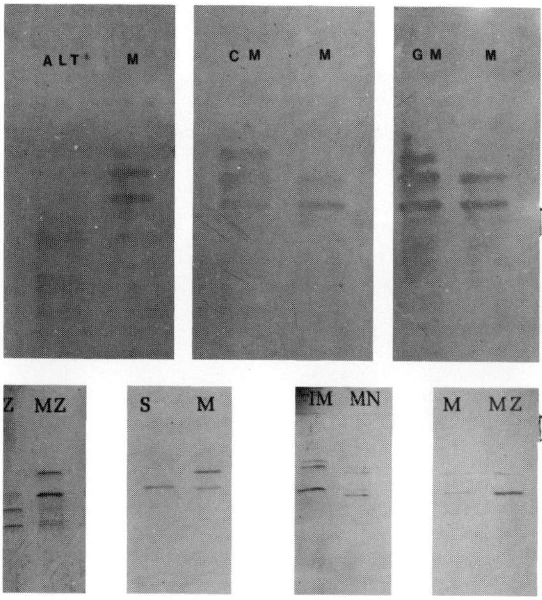

Figure 4.
Results of IF applied to PAGIF of different phenotypes

results obtained in the gel

results obtained from cellulose acetate fixation.

Anode at the top.

Such problems have three major causes.
(a) *Contaminating Proteins*. As reported by Fagerhol (4), haptoglobin of the 1 - 1 type is the major cause of interference for acid-starch gels in the pH zone. Addition of hemoglobin to the serum prior to PAGIF produces hemo-globin-haptoglobin complexes that migrate more cathodally and thus no longer interfere with the Pi zones. However, bands due to other acidic prot-eins in the same zone can sometimes be seen, and we therefore prefer to use CIEP or IF to distinguish additional bands from those really belonging to the A_1 Pi system.- (b) *Deteriorated Serum*. As previously demonstrated (12), A1 Pi is a heat-sensitive protein. Repeated freezing and thawing or prolonged storage at room temperature can modify the Pi serum patterns and shift them to a more basic isoelectric point, producing patterns that can be confused with slow-moving phenotypes.- (c) *Bacterial Contamination*. Alterations of A_1 Pi resulting from bacterial contamination of serum sampl-es can produce a more cathodal migration (13), presumably due to desialyl-ation of the protein by bacterial neuraminidases. From our experience, bacterial contamination is a major source of difficulty in interpreting Pi phenotypes, and in every case a serum alteration was proved immunochem-ically the sample was found to be infected, usually by streptococcus. Immunochemical techniques produce a typical pattern in such cases and the problem can be detected easily.

Crossed Immunoelectrophoresis (C I E P)

Usual Phenotypes. The patterns obtained with CIEP for different phenotypes are illustrated by Fig. 2. The patterns can easily be differentiated with this technique, and thus the phenotypes can be identified accurately. In addition, the results of CIEP and microdensitometry (2) correspond closely.

Altered Sera. Patterns obtained from altered M sera are shown on Fig. 3. The first step is a disappearance of the normal pattern replaced by a more cathodal one. In the second step, the new pattern migrates in a still more cathodal fashion and, finally, the A_1 Pi bands tend to disappear. These patterns can always be identified and distinguished from those due to slow phenotypes, especially Z, but the initial phenotype of the sample cannot be ascertained retrospectively.

Rare Patterns. These can also be identified by CIEP. Several unusual patt-erns (B M, L M, I S, M X) are shown on Fig. 2.

Immunofixation (I F)

Typical results obtained by IF are shown in Fig. 4. Again, the distinction between different patterns is clear and accurate. Preliminary results indic-ate that microdensitometry can also be applied to the specific bands reveal-ed by IF, permitting quantitation of the distribution of the bands in each phenotype.

Comparison of C I E P and I F

CIEP from PAGIF is directly comparable with CIEP from acid-starch gel electr-ophoresis, facilitating comparison of the results obtained with those in literature. Patterns from CIEP can also be superimposed on the graphs obtained from microdensitometric analysis of PAGIF. The technique is in-expensive, requiring less than 100 μl of commercial antiserum per run. However, in the pH range used in these studies, not all the bands are compl-

-etely resolved in the first dimension, leading to some fused peaks. In add--ition CIEP is a delicate technique, and rather long (6 hr after 1st dimens--ion). Overall, however, the IF print technique is simpler and faster than CIEP and does not require special apparatus. It allows direct comparison of the patterns obtained with those on the stained gels and provides a better resolution of the bands. For these reasons we strongly recommend IF as the easiest way to check doubtful phenotypes. It can give the solution in most cases so that CIEP usually is not necessary, except to confirm a new patt--ern in conjonction with IF.

New Patterns.

Two kinds of new patterns can be discussed : those corresponding to "variants" of a described phenotype, which differ from it by an abnormal distribution inside the allele product or present only one additional band proved to be Λ_1 Pi, and those corresponding to a new allele product, characterized by two major bands and some other minor bands with isoelectric points differ--ent from those of other phenotypes. An example of a variant pattern is shown on Fig. 5. There is a relative decrease of the M 4 band and an increase of the M 6 band. This pattern, although similar to the one usually seen in cord-blood, in this case was found in a healthy adult and was shown to be hereditarily transmitted. Two examples of new allele products are also shown in Fig. 5. The first one (b) corresponds to a fast heterozygous variant, with two bands faster than those of the D allele product but slower than those of the B allele product. It could correspond to the C M phenotype, whose existence has been previously suggested (14). The second one (c) corresponds to a slow variant but does not resemble any slow pattern report--ed elsewhere. It is probably a heterozygous phenotype as indicated by the number of peaks obtained with CIEP.

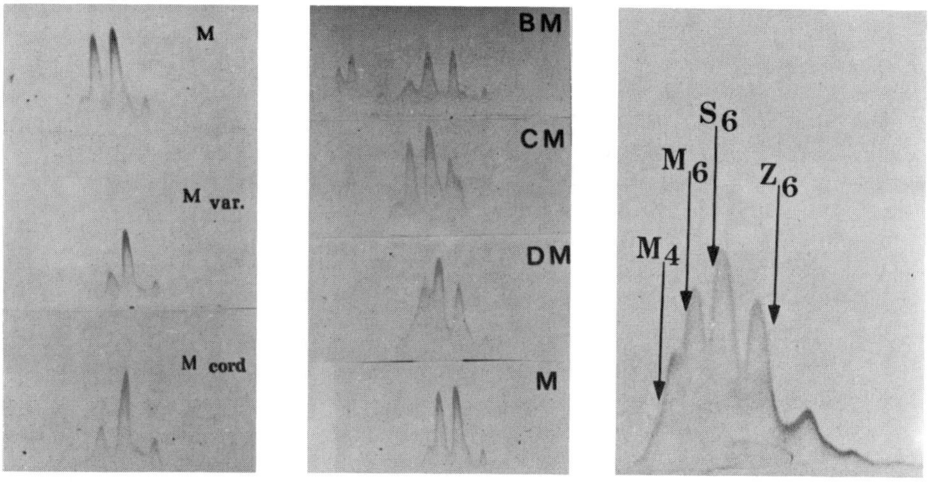

(a.) (b.) (c.)

Figure 5. CIEP patterns of (a.) a variant phenotype characterized by a decr--ease in M 4 and an increase in M 6 .(b.) : a fast heterozygous variant prob--ably corresponding to the C M phenotype.
(c.) a slow variant not yet described.

These results illustrate the difficulty of classifying a new phenotype. To date, 26 allele products and only 46 phenotypes have been reported (7), of which 5 are M variants. An example of a new cathodal M variant, we presently call Mc, is shown on Fig. 6. Strict methodology must be used to confirm the existence of a new allele product, including family studies to ascertain the mode of transmission of the proposed allele and comparison with the results of PAGIF for registered phenotypes with narrow migration patterns followed by CIEP and/or IF, to prove the different migration pattern of the suggested new phenotype.

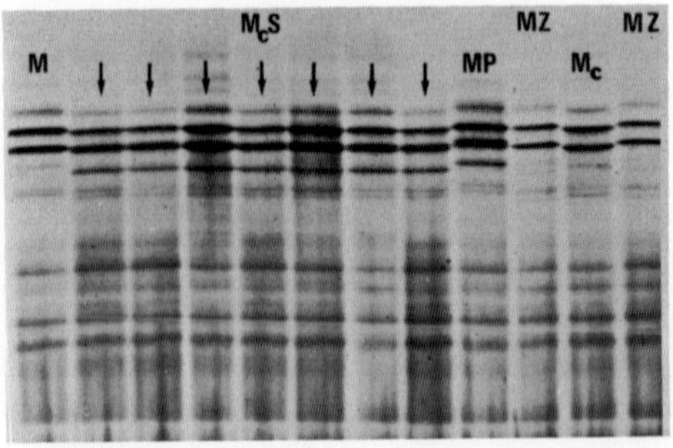

Figure 6.

Patterns of a new M variant, Mc, in the homozygous state and in combination with S allele product (Mc S).

Notice that cathodal translation of bands 7 and 8 can raise a confusion with Z bands.

CONCLUSION

The importance of the A1 Pi system, in addition to its being a powerful genetic marker, resides in the association of its variants with many disease states. Thus, accurate determination of Pi types is essential. In doubtful cases, CIEP and IF can be of great help (in addition to protein-proteinase binding methods) and must serve as routine techniques to complement Pi determination by PAGIF.

REFERENCES

(1) Arnaud,P., Chapuis-Cellier,C., Creyssel,R. : Polymorphisme de l'Alpha 1 antitrypsine plasmatique (système Pi). Mise en évidence par électrofocalisation sur gel de polyacrylamide. C.R.Soc.Biol. 168, 58-64 (1974)

(2) Allen,R.C., Harley,R.A., Talamo,R.C. : A new method for determination of Alpha 1 antitrypsin phenotypes using isoelectric focusing on poly-acrylamide gel slabs. Amer.J.Clin.Pathol. 62, 732-739 (1974)

(3) Arnaud,P., Chapuis-Cellier,C., Creyssel,R. : The Pi system : its study by means of thin-layer gel electrofocusing in polyacrylamide gels. In :"Protides of Biological Fluids" (H.Peeters,Edit.) Pergamon Press, London, pp.515-520 (1975)

(4) Fagerhol,M.K. : The Pi system : genetic variants of serum Alpha 1 anti-trypsin. Ser.Haematol. 1, 153-161 (1968)

... / ..

(5) Laurell,C.B., Persson,U. : Analysis of plasma Alpha 1 antitrypsin variants and their microheterogeneity. Biochim.Biophys. Acta 310, 500-507 (1973)

(6) Pannel,R., Johnson,D., Travis,J. : Isolation and properties of human plasma Alpha 1 proteinase inhibitor. Biochemistry 13, 5439-5445 (1974)

(7) Chapuis-Cellier C. : M.D. Thesis, Université Claude Bernard Lyon I , France (1975)

(8) Martin,J.P., Vandeville,D., Ropartz,C. : Pi B, a new allele of Alpha 1 antitrypsin genetic variants. Biomedicine 19, 395-398 (1973)

(9) Cook,P.J.L. : The Pi system, a family study in England and Scotland. Some genetic problems of the Pi system.
In : "Alpha 1 antitrypsin and Pi system", INSERM Colloquium no 40, pp. 67-70 (1975)

(10) Allen,R.C., Oulla,P.M., Arnaud,P. : In this Symposium.

(11) Soderholm,J., Smyth,C., Waldstrom,T. : A simple and reproducible method for crossed immunoelectrofocusing. Scand.J.Immunol. 4 (Suppl.2) 107-113 (1975)

(12) Lieberman,J. : Heat lability of Alpha 1 antitrypsin variants. Chest 64, 579-584 (1973)

(13) Lieberman,J. : Interference with Alpha 1 antitrypsin studies in stored serum by presumed bacterial proteases. J.Lab.Clin.Med. 86, 7-16 (1975)

(14) Fagerhol,M.K. : Pi typing techniques.
In : "Protides of Biological Fluids" (H.Peeters,Ed.) Pergamon Press, London, pp. 34-37 (1975)

High Resolution Phenotyping of α_1-Antitrypsins by Thin-Layer Polyacrylamide Electrofocusing

J.O. Jeppsson

SUMMARY

The most common phenotypes of α_1-antitrypsin (α_1-protease inhibitor) are segregated by electrofocusing into at least five fractions. Electrofocusing is an extremely helpful tool for differentiating those phenotypes which are correlated to disease. A pH 4-5 gradient permits clear patterns with very few interfering non-antitrypsin bands. The technique can be combined with immunofixation or other functional tests for identification of the α_1-antitrypsin bands.

INTRODUCTION

α_1-Antitrypsin (α_1-AT) is the major protease inhibitor (Pi) in plasma. The normal M-allele is by far the most common. A deficient state with phenotype Pi Z is associated with pulmonary emphysema. The frequency for Z-allele varies in different parts of the world being 0.013 in Americans and 0.026 in Swedish newborns (1). The present extensive primary classification system based on starch gel electrophoresis and crossed immunoelectrophoresis has shown more than twentyfive different variants(2).

Several groups have recently reported isoelectric focusing techniques (3-5) for phenotype determination. Commercially available ampholytes in pH-range 4-6 or 3.5-5 have been used. The results vary considerably largely depending on the acrylamide quality and separation conditions.

We have found that resolution can be increased by using narrow pH gradient since the pI of all antitrypsin variants extend from around 4.4 to 4.7. The problem with interference of other proteins in the Z-region can be reduced.

This paper describes a simple procedure for examination of α_1-antitrypsin variants in clinical work. The data confirm the results from starch gel electrophoresis, but the technique has such resolution that several subtypes of Pi M are easily identified.

MATERIALS AND METHODS

Sample preparation. EDTA-plasma or serum freshly drawn and frozen at -70°C were used. Some samples were reduced with cysteine-HCl. Final cysteine concentration was 15 µmol/ml plasma (6).

Experimental procedure. Acrylamide gels were cast in a mold as earlier described (6). Ampholines (pH range 4-6) were obtained from LKB, Bromma, Sweden. Special ampholines with pH ranges 4.0-4.5 and 4.5-5.0 were a gift from Aminkemi, Bromma, Sweden. Prepared thin-layer gel plates, Ampholine® PAG plates for α_1-antitrypsin were kindly supplied by LKB, Bromma, Sweden. The procedure followed the manufacturer's instruction for LKB 2117 Multiphore apparatus cooled to 10°C with a cryostat. 1.0 M H_3PO_4 was used as key solution at the anode and 0.4 % Ampholines pH 5-8 at the cathode. Sample papers (5x10 mm) containing either 25 µl or 40 µl for Pi Z were placed parallel to the cathodal edge of the gel. A distance of 10 mm was maintained between the cathodal strip and the nearest sample paper. Subsequent papers were separated by only 1 mm to allow optimal comparison. As many as 24 samples may be run on one plate. The LKB 2103 power supply was set for 500 V, 30 W, 50 mA for thirty minutes as prerun

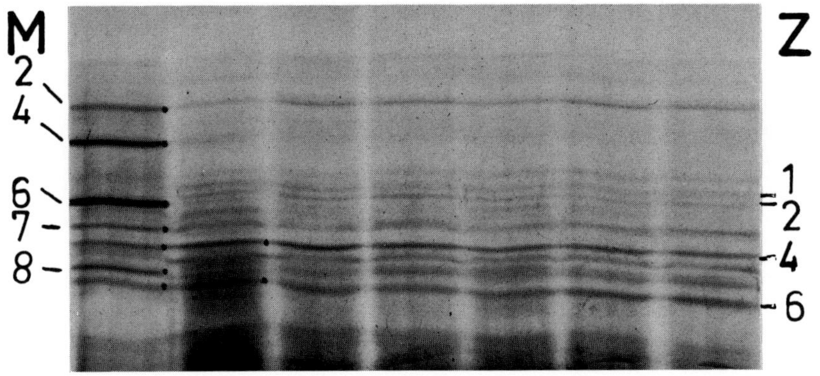

Fig. 1. Patterns from isoelectric focusing of Pi MZ and different Pi Z:s samples. The bands between Z_4 and Z_6 are non-antitrypsin fractions. Z_1-Z_2 bands are always clearly visible anodal to M_6 position.

and 1400 V, 30 W, 50 mA for three hours after application. On completion of electrofocusing pH was sometimes measured at 5 mm intervals with an Ingold (6) surface electrode at 10°C. The gel was then soaked in 3.5 % (w/v) perchloric Acid (PCA) for about 2 h and staining was performed in 0.04 % Coomassie Brilliant Blue G-250 in 3.5 % (w/v) PCA solution over night (7). The α_1-antitrypsin bands are visible within 30 min but increase in intensity as duration of staining is increased.

Alternately, the interference bands were excluded and antitrypsin fractions safely identified by immunofixation (8). A plastic frame was placed on the gel over the area for the expected location of AAT-fractions and filled with antisera. After one hour of incubation at room temperature the gel was then soaked in saline over night and stained as described

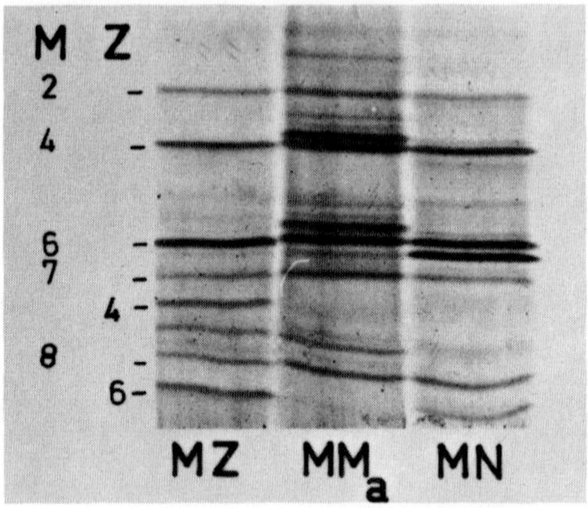

Fig. 2. Representative patterns of MZ, MM_{anodal} and MN (18).

above. A recent method described by Ritchie et al. (9) using a cellulose acetate membrane soaked in antisera is more economical and gives excellent results.

RESULTS

The basic pattern of antitrypsin polymorfism closely resembled results with other methods. A maximum of five major fractions of homozygote Pi M were seen. The pH 4-5 gradient gives a clear picture covering all known variants with pI:s extending from 4.4 to 4.7. The resolution in this system can be adjusted to less than 0.005 pH-unit. This allows clear separation between the M_7 and Z_4 as well as the M_8 and Z_6 bands in a Pi MZ heterozygote. With our gradient the distance between M_7 and Z_4 is approximately 3 mm and between M_2 and M_8 27 mm. A constant non-antitrypsin fraction is located between the M_7 and M_8 fractions. Fig. 1 illustrates a Pi MZ-phenotype and a series of Pi Z patients. All studied Pi Z have besides the normal Z_4 and Z_6 fraction a double band just in front of the location for the M_6 fraction. We have previously stated that the S_4 band is located anodal to the M_6 band clearly visible in the Pi MS heterozygote (6). More experience reveals fast subtypes of Pi M where M_6 has the same pI as S_4. This occurs in MM_{anodal} in fig. 2b. Two types of cathodal subtypes of Pi M are illustrated in fig. 3b and c and one fast in fig. 2. No effort has been made to make frequency studies yet but the subtypes are found in several child-parent studies. All these have been analyzed with and without cysteine reduction to exclude formation of dimers or other mixed disulphides. The Pi MN sample in fig. 2 illustrates another principle for microheterogeneity. In all cases studied so far the change in pI can be followed in all the microheterogenous fractions. The MN-phenotype seems to have a single 4-band a double 6-band a single 7-band and double 8-band suggesting a difference in sialic acid bonding instead of a suggested amino acid substitution which would change pI of all fractions within each phenotype.

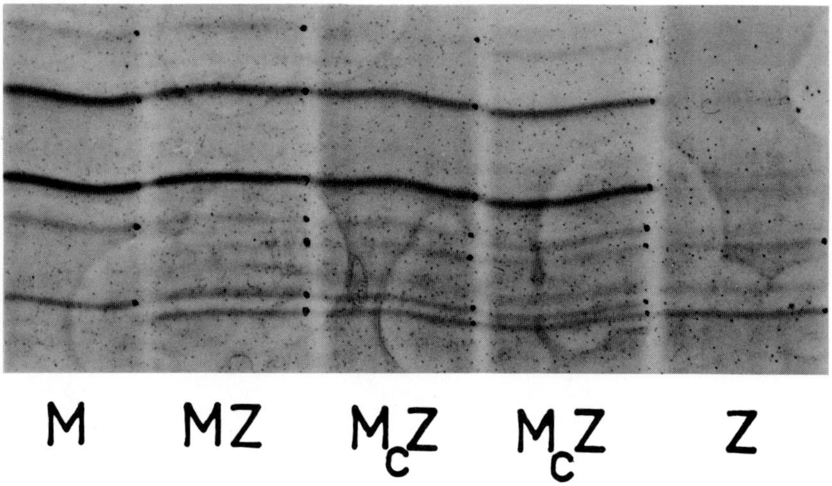

Fig. 3. Normal Pi M plasma compared with "normal" Pi MZ and two cathodal subtypes of M in Pi MZ heterozygotes. A Pi Z sample is included to the right for comparison.

DISCUSSION

Electrofocusing is a simpler and less time consuming technique than starch gel electrophoresis combined with crossed immunoelectrophoresis for the separation and identification of the various phenotypes of α_1AT. Experience has shown that acrylamide quality as well as ampholyte uniformity are extremely important for the quality of results obtained by the thin-layer electrofocusing technique. The pH 4-5 gradient gives better separation of the microheterogeneous fractions in the Z_4 and Z_6 regions as well as less background interference due to other proteins than do wider pH ranges.

Some variable interfering bands invitably occur due to the polymorphism of many plasma proteins. Increased resolution leads to discovery of new phenotypes which must of course be tested according to recommendations of the First International Pi system Workshop at Brügge, Belgium, 1974. Apparently new phenotypes should be subjected to electrofocusing both before and after reduction of the plasma with cysteine since α_1-AT can occur as mixed disulfides giving confusing patterns (6).

We have earlier relied on three functional properties of antitrypsin to identify the microheterogenous fractions of antitrypsin. The inhibition of trypsin tested directly on the gel and reaction with specific antibodies in crossed immunoelectrophoresis are helpful tools (6). The third property, a reactive thiol group, is extremely valuable for enrichment of antitrypsin before electrofocusing and for identification (6). Our recent experience with immunofixation after electrofocusing has shown it to be one of the more simple and rapid methods.

The subtypes of Pi M presented in this work continue earlier observation (10). The presented subtypes are not uncommon and are presumably identical with M_{Lamb} and $M_{Baldwin}$ observed by Johnson (2), Pi N by Constans et al. (11), and M_1 and M_2 by Eriksson and Frantz (12) using various methods. The Z_1-Z_2 banding close to the position of M_6 band are constant in a series of at least sixty Pi Z patients. They have been observed to a varying extent by Allen et al. (5) and Lieberman et al. (13).

Interest has recently been focused on the background of the microheterogeneity of the M-phenotype and secondly the structural differences between the M, S and Z-protein. Recent studies by neuraminidase desialidization show that heterogeneity within each phenotype is mainly caused by differences in sialic acid bonding (14). On the other hand differences between phenotypes are caused by amino acid substitutions resulting in different pI:s. Fingerprint and amino acid analysis of a smaller CNBr-fragment reveal an amino acid substitution of glutamic acid in M-protein to lysine in Z-protein (15). This agrees with results of Yoshida et coworkers (16). The corresponding substitution in the S-protein is glutamic acid to valine (17).

ACKNOWLEDGEMENTS

Plasma Pi MN-sample was a generous gift from Dr. Diane Wilson-Cox, Toronto, Canada. This investigation was supported by grant from the Swedish Medical Research Council (Project No. B76-13X-00581-12B).

REFERENCES

1. *Laurell, C.B., Sveger, T.: Mass Screening of Newborn Swedish Infants for α_1-antitrypsin Deficiency. Amer. J. Human Genetics 27, 213 (1975).*

2. *Johnson, M.: Genetic Typing of α_1-antitrypsin by Immunofication Electrophoresis. Identification of Subtypes of Pi M. J. Lab. Clin Med. 87, 152 (1976).*

3. *Lebas, M.J., Hayem, A., Martin, J.P.: Étude des variants génétiques de l'α_1-antitrypsine en immunoélectrofocalisation bidimensionelle. C. R. Acad. Sci. Paris 278, 2359 (1974).*

4. *Arnaud, P., Chapius-Cellier, C., Creyssel, R.: The Pi System: Its study by means of Thin-layer-gel Electrofocusing in Polyacrylamide Gel. Prot. Biol. Fluids 22,515 (1974).*

5. *Allen, R.C., Harley, R.A., Talamo, R.C.: A New Method for Determination of Alpha-1-antitrypson Phenotypes Using Isoelectric Focusing on Polyacrylamide Gel Slabs. Amer. J. Clin. Pathol. 62, 732 (1974).*

6. *Pierce, J.A., Jeppsson, J.O., Laurell, C.B.: α_1-Antitrypsin Phenotypes Determined by Isoelectric Focusing of the Cysteine-Antitrypsin Mixed Disulfide in Serum. Anal. Biochem. 74, in press (1976).*

7. *Reisner, A.H., Nemes, P., Bucholtz, C.: The Use of Coomassie Brillant Blue G 250 Perchloric Acid Solution for Staining in Electrophoresis. Anal. Biochem. 64, 509 (1975).*

8. *Alper, C.A., Johnson, A.M.: Immunofixation Electrophoresis: A Technique for the Study of Protein Polymorphism. Vox Sang 17, 445 (1969).*

9. *Ritchie, R.F., Smith, R.: Immunofixation. I. General Principles and Application to Agarose Gel Electrophoresis. Clin. Chem. 22, 497 (1976).*

10. *Pierce, J.A., Jeppsson, J.O.: Unpublished Results.*

11. *Constans, J., Vian, M.: Une nouvelle mutation Pi^N au locus Pi dans*

12. *Frantz, R.R., Eriksson, A.W.: $Alpha_1$-antitrypsin: Common Subtypes of Pi M. Hum. Heredity, in press (1976).*

13. *Lieberman, J., Gaidulis, L., Klotz, S.D.: A New Deficient Variant of α_1-antitrypsin (M_{Duarte}). Amer. Rev. Resp. Dis. 113, 31 (1976).*

14. *Jeppsson, J.O., Laurell, C.B.: Properties of α_1-antitrypsin of Types Pi M and Pi Z. Eur. J. Biochem., in press (1976).*

15. *Jeppsson, J.Ol: Amino Acid Substitution Glu $\rightarrow$ Lys in α_1-antitrypsin Pi Z. FEBS Lett 65, 195 (1976).*

16. *Yoshida, A., Lieberman, J., Gaidulis, L., Ewing, C.: Molecular Abnormality of Human $Alpha_1$-antitrypsin Variant (Pi-ZZ) Associated with Plasma Activity Deficiency. Proc. Nat. Acad. Sci. 73, 1324 (1976).*

17. *Owen, M.C., Carrell, R.W.: Alpha-1-antitrypsin: Molecular Abnormality of S Variant. Brit. Med. J. 1, 130 (1976).*

18. *Wilson-Cox, D., Celhoffer, L.: Inherited Variants of α_1-antitrypsin: A New Allele Pi^N. Can. J. Genet. Cytol. 16, 297 (1974).*

Broad-Beta Disease (Hyperlipoproteinaemia Type III): Genetics, Gene Frequency and Diagnosis without Ultracentrifugation

G. Utermann, M. Hess and K.H. Vogelberg

Summary

1. A study by I.E.F. on the protein composition of plasma lipoprotein fractions initially from 15 patients with Broad-Beta Disease and a large control group demonstrated a deficiency of a specific apoprotein (Apo E-III) in the patients lipoproteins system. This protein of mol.wt~39.000 is a normal component of VLDL*.

2. Based on the Apo E marker, a test for diagnosis of Broad-Beta Disease without ultracentrifugation has been developed. The test combines precipitation of VLDL, lipid extraction and I.E.F. of urea-soluble apo-VLDL in a simple and rapid procedure. Application of this test for screening among hyperlipidaemic patients and blood donors revealed the following: a) Patients with Broad-Beta Disease were selectively recognized from the hyperlipidaemic group. b) Five individuals with Apo E-III deficiency detected among 500 blood donors had a dyslipoproteinaemia characteristic for Broad-Beta Disease but no hyperlipidaemia. c) In addition to the "normal" and "deficiency" Apo E patterns two further phenotypes were delineated by I.E.F. and scanning densito-

*Abbreviations:

VLDL:	Very Low Density Lipoproteins (d < 1.006 g/ml)
LDL-1:	Low Density Lipoproteins-1 (d = 1.006-1.019 g/ml)
LDL-2:	Low Density Lipoproteins-2 (d = 1.019-1.063 g/ml)
HDL:	High Density Lipoproteins (d = 1.063-1.21 g/ml)
I.E.F.	Isoelectric Focusing
SDS-PAGE:	Polyacrylamide gel electrophoresis in presence of sodium dodecyl sulphate

metry: the "intermediary" pattern and the +E-IV variant.
3. Family studies indicate that Broad-Beta Disease is transmitted by an autosomal recessive mode of inheritance and that heterozygotes show the "intermediary" Apo E focusing pattern. The data suggest that either probands with Broad-Beta Disease only occasionally decompensate to develop hyperlipidaemia or that the effect of one or more additional gene(s) is required for expression of the more severe forms. Vertical transmission of Broad-Beta Disease occuring occasionally in pedigrees is explained as pseudodominance due to a high frequency of ~0.07 for the main gene for Broad-Beta Disease.

Present concepts of the metabolic role of Apo E allow the conclusion that a defect in the regulation of structure of this protein underlies the familial dyslipoproteinaemia.

Introduction

Broad-Beta Disease is a familial disorder of lipid metabolism in which cholesterol and triglycerides both are elevated in the patients plasma. Clinically the disorder is associated with xanthomatosis and premature arteriosclerosis (1,2). Biochemically it is characterized by the occurrence of atypical cholesterol-rich lipoproteins in the S_f 12-100 range (VLDL and LDL-1) (3). These are considered to be intermediary products or "remnants" in the catabolic conversion of VLDL to LDL (4,5,6). The molecular details in this pathway are not yet completely understood. The initial event is the clearance of triglycerides from VLDL particles by the action of lipoprotein lipase and the concomitant loss of some of the VLDL protein components (Apo C) (7,8). Little is known about the further processing of the cholesterol-rich remnant particles generated by these steps. Two pathways are discussed: Conversion to LDL and uptake and degredation by the liver. The conversion to LDL may require contact with or uptake by liver cells. Broad-Beta Disease might be a tool to study the molecular events involved in these steps since "remnant"

Table I: CLINICAL AND BIOCHEMICAL FEATURES OF PATIENTS WITH BROAD - BETA DISEASE

Patient	Age/Sex	Lipid Concentrations Cholesterol mg/100ml	Triglycerides mg/100ml	Electrophoretic Phenotype	$VLDL_{Ch/TG}$ (Normal < 0.42)	Ch_{VLDL}/TG_{Total} (Normal < 0.3)	Relevant Clinical Findings
1 A.B.	50/F	720	752	III	0.64	0.47	-
2 R.T.	24/M	457	428	III	1.4	0.52	Generalized Xanthomas
3 E.T.	42/M	479	467	III	0.64	0.53	Myocardial Infarction
4 E.M.	68/F	612	843	III	0.66	0.37	Myocardial Infarction, Peripheral Vascular Disease
5 I.J.	71/F	430	709	III	0.66	0.43	Diabetes mellitus
6 G.K.	64/M	380	496	III	0.62	0.48	subclinical Diabetes mellitus, Hyperuricaemia
7 E.R.	39/M	410	865	III	0.53	-	Xanthomas
8 M.B.	59/M	613	366	III	0.84	0.66	Myocardial Infarction, Peripheral Vascular Disease, Xanthomas, subclinical Diabetes
9 S.J.	60/F	630	816	III	0.77	0.66	Myocardial Infarction, Peripheral Vascular Disease, Xanthomas, subclinical Diabetes

like material is accumulated in the plasma of patients with this disorder.

Most of the accumulated lipoproteins occur in the VLDL fraction (d < 1.006 g/ml) upon preparative ultracentrifugation. Accordingly the VLDL of these patients has features different from normal VLDL. It is relatively rich in cholesterol and contains particles that have a slower electrophoretic mobility in paper or agarose gel electrophoresis (Beta-VLDL). Based on these observations three laboratory definitions of the disease have been proposed, namely: 1. presence of Beta-VLDL (1,2), 2. a ratio < 0.42 of cholesterol over triglyceride in in the VLDL fraction (9) and 3. a ratio < 0.3 of cholesterol in VLDL over total plasma triglyceride (10,11). All of these use quantitative criteria instead of a qualitative marker and none is generally accepted. Individuals classified as having

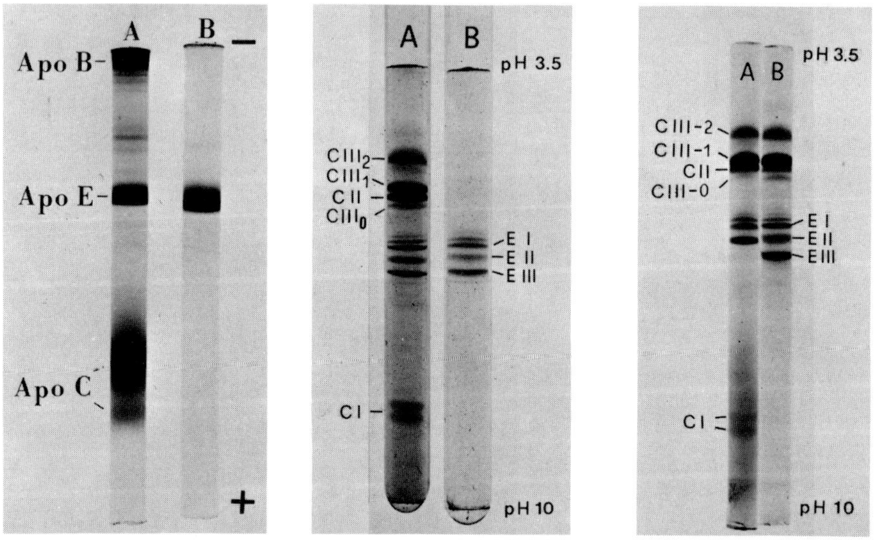

Fig. 1 (left): SDS-PAGE in 10% gels of apo-VLDL (A) and purified Apo E (B). (middle): Analytical I.E.F. in polyacrylamide gels at pH 3.5-10 of urea soluble apo-VLDL (A) and Apo E (B). (right): Analytical I.E.F. in 7,5% polyacrylamide gels of apo-VLDL from patient H.Sch. with Broad-Beta Disease (A) and a control (B). Staining: coomassie brilliant blue

Broad-Beta Disease by one definition are not necessarily indicated by the others (10). Use of the preparative ultracentrifuge is required in any case for evaluation of the disease. As a consequence there is much uncertainty on several aspects and especially on the mode of inheritance of Broad-Beta Disease. Because of these difficulties we started our study with a well characterized group of initial patients who had Broad-Beta Disease according to all contemporary definitions (12). Characteristic features of nine probands are shown in table I. In the course of our study of the underlying defect(s) in Broad-Beta Disease it turned out that I.E.F. could help to resolve some of the clinical, genetic and biochemical problems posed by this challenging disease.

Results and Discussion

I. Biochemical Studies of the Defect in Broad-Beta Disease.

The protein moiety of VLDL (apo-VLDL) is separated into three main fractions by SDS-PAGE. The high mol.wt Apoprotein B, the low mol.wt C-group apoproteins and the arginine-rich Apoprotein E (Fig.1). The latter is a glycoprotein of mol.wt ~39.000 containing about 13% carbohydrate (Utermann, unpublished). It is homogeneous in SDS-PAGE, alkaline disc-electrophoresis and according to immunological criteria (13). In I.E.F., however, Apo E splits into three main bands preliminary designated E-I, E-II, and E-III (Fig.1). The chemical differences between these components have not yet been elucidated.
Isoelectric focusing of apo-VLDL from the patients with Broad Beta Disease showed a deficiency of one of the Apo E "isoproteins" namely Apo E-III and a relative increase in the Apo E-II band (14) (Fig.1). The quantitative composition of urea soluble apo-VLDL was evaluated in five of the probands

Table II: Protein Composition of Urea Soluble Apo-VLDL from Patients with Broad Beta Disease

Patient	urea soluble apoproteins (weight %)		Apo E (% of densitometric area)		
	Apo C	Apo E	E-I	E-II	E-III
H.Sch.	60	40	42	58	-
E.R.	27	73	41	59	-
E.M.	44	56	42	58	-
Ch.J.	41	59	51	49	-
S.J.	45	55	42	58	-
Mean ±SD	43±12	57±12	43.6±4.4	56.4±4.4	-
Control*	82	18	38	23	39

*VLDL isolated from a pool of 12 normolipidemic donors

by scanning densitometry under standardized conditions using pure apoproteins for establishing the calibration curves (14). Table II summarizes the results (data from Ref.14). Total Apo E predominates in the probands apo-VLDL but the Apo E-III component is deficient.

To learn whether Apo E-III deficiency was present under different metabolic conditions and could be used as a stable qualitative marker for Broad-Beta Disease, patients were studied with and without the regime of lipid lowering drugs and the composition of their VLDL was changed by diatary manipulations. The Apo E focusing pattern remained stable under different experimental conditions (12). Of practical

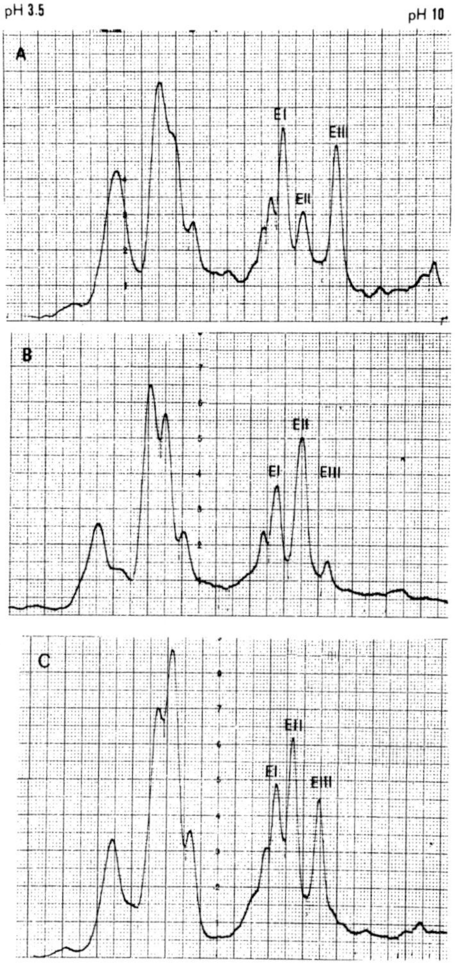

Fig. 2: Densitometric scans of coomassie brilliant blue stained focusing gels. Individual apo-VLDL preparations showing the most common Apo E pattern (A), a deficiency pattern (B) and an "intermediary" pattern (C), were focused in ampholine of pH 3.5-10 on 7.5% polyacrylamide gels.

importance is the fact that Apo E-III deficiency may be demonstrated in a non-fasting state. VLDL isolated from patients four hours after an ad libitum meal showed an increase in the proportion of triglyceride-rich VLDL of normal prebeta mobility. This newly secreted material already showed the defective Apo E. This strongly suggests, that Apo E-III deficiency is a primary event in the disease and not due to secondary modifications as a possible consequence of a long circulation time of partially degraded VLDL particles in the patients plasma.

Most of the individual control VLDL preparations examined during our study showed a quite characteristic distribution of the three Apo E bands of roughly 2:1:2 (E-I:E-II:E-III). However occasionally individuals occured with a ratio of E-II:E-III in the order of ≦ 1:1. The pattern in these individuals is quantitatively intermediary between normal and the E-III deficiency pattern and can be achieved in vitro by mixing both types of apo-VLDL. We will refer to this as the "intermediary" pattern (Fig.2).

II. Family Studies

Four genetic hypothesis have been advanced in the past on the genetic mode of transmission of Broad-Beta Disease:
1. autosomal recessive (2), 2. polygenic (15), 3. mixed heterozygosity for two autosomal genes (2.16) and 4. autosomal dominant (16,17). We now for the first time had the possibility of studying the genetics of Broad-Beta Disease using a qualitative marker at the protein level. Family studies revealed that with one exception none of the children or parents of the probands had Apo E-III deficiency or had Broad-Beta Disease according to classical criteria. Instead all showed the "intermediary" focusing pattern and some had other forms of hyperlipidaemia (Fig.3).

The family data are clearly in favour of an autosomal recessive mode of inheritance of Broad-Beta Disease where the

Fig. 3

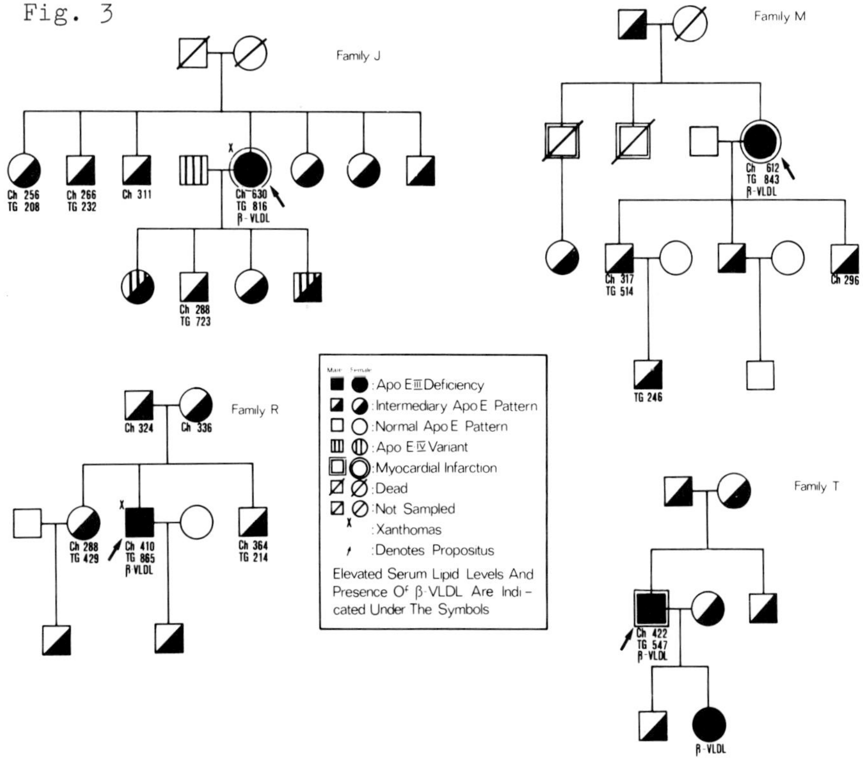

"intermediary" pattern represents the heterozygeous for the disease (18). The exception is shown in Family T.. Here the probands daughter also had Apo E-III deficiency, but the probands parents had not. The occurrence of vertical transmission of Broad-Beta Disease in some families has also been described in the literature in studies using classical methods for definition of the disease (2,15,16,17).
Subsequent studies of the frequency of the different Apo E phenotypes helped to explain this observation. The heterozygeous "intermediary" pattern is frequent, occuring in about 12% of blood donors. Thus occurrence of vertical transmission in the T-family most easily is explained as pseudodominance. This phonomenon possibly has contributed to the previous confusion on the genetics of Broad-Beta Disease.

III. Screening and Delineation of Normolipidaemic Broad-Beta Disease.

For further genetic studies, screening purposes and clinical routine diagnosis a simple and rapid method for the detection of Apo E focusing patterns without prior ultracentrifuation seemed highly desirable. Therefore a precipitation method originally described by Burstein and Scholnick (19) has been adapted to isolate VLDL from a small volume of serum. The total procedure includes heparin-magnesium precipitation of VLDL from only 1 ml of serum, chloroform-methanol extraction and subsequent I.E.F. of urea soluble apo-VLDL (20). The Apo E patterns obtained are identical to those from ultracentrifugally isolated VLDL (Fig.4). Screening among hyperlipidaemic patients showed that those with Broad-Beta Disease could be selectively recognized. In addition to those known to have the disease some more were detected and confirmed by classical methods, where the diagnosis had not been known previously.

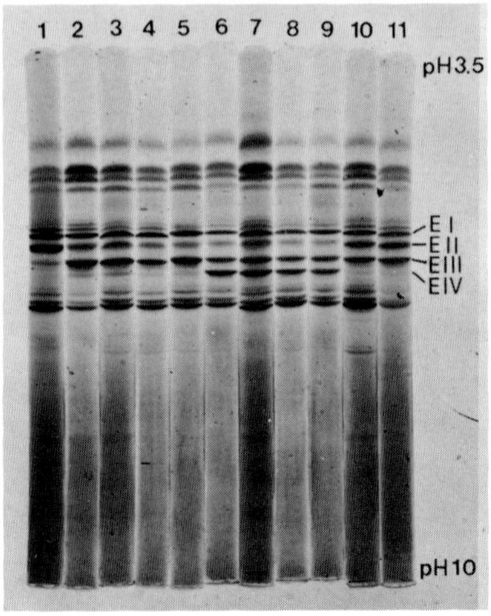

Fig. 4: Analytical I.E.F. in polyacrylamide gels of 11 individual apo-VLDL preparations demonstrating the different Apo E phenotypes: Deficiency pattern (1), normal pattern (2,3,4, 5), intermediary pattern (10,11) and +E-IV variant (6,7,8,9). VLDL was prepared by the heparin-magnesium precipitation method.

To arrive at reliable data on the frequency of Apo E-III deficiency the Apo E patterns were evaluated in 500 blood donors. Results were quite unexpected: Five individuals (1%) had Apo E-III deficiency and ~12% the "intermediary" pattern. Another Apo E phenotype occuring in about 27% of individuals is characterized by the presence of an additional kathodic Apo E band (+E-IV Variant; Fig.4). Preliminary results from family studies indicate, that this marker is transmitted as an autosomal dominant and is genetically independent from the other Apo E variants (see also Fig.3).

Biochemical studies on the probands with Apo E-III deficiency showed that they represented a distinct group with abnormalities in their lipoprotein spectrum characteristic for Broad-Beta Disease. These were: 1. presence of Beta-VLDL in the ultracentrifugal d $<$ 1.006 g/ml fraction, 2. increase in the relative and absolute amount of cholesterol in VLDL, 3. markedly reduced LDL band in polyacrylamide gel electrophoresis and 4. abnormal quantitative distribution of lipoproteins demonstrated by density gradient centrifugation: reduced levels of LDL-2 and elevation of VLDL and LDL-1 levels (20).

However none of these probands had hyperlipidaemia. Rather the total cholesterol was abnormally low in three patients under age 30 ($<$ 125 mg/100 ml) and there is additional evidence that young individuals with Apo E-III deficiency represent a novel genetic form of hypolipidaemia (Utermann, unpublished). Thus by our screening method we were able to detect a group of individuals with a unique lipoprotein profile showing a primary dyslipoproteinaemia but not necessarily hyperlipoproteinaemia.

According to these data the gene frequency for Broad-Beta Disease defined as a dyslipoproteinaemia (including hyper-, normo- and hypolipidaemic subjects) would be about 0.07. This is an order of magnitude higher than previous estimates in the literature (11,16).

There are mainly two possibilities to explain our genetic data. Either hyperlipidaemia occurs only occasionally in probands with Apo E-III deficiency representing a decompensated state as a consequence of unfavourable exogeneous factors, or one or more additional mutant gene(s) like those for hypertriglyceridaemia or combined hyperlipidaemia are operating in those patients with Apo E-III deficiency that have developed gross hyperlipidaemia and xanthomatosis.

Comparative genetic and prevalence studies that are in progress in our laboratory possibly will clarify this situation. One of the most important aspects is, whether the normolipidaemic probands are also at risk for premature development of arteriosclerosis.

References

(1) Fredrickson,D.S., Levy,R.J., Lees,R.S.: Fat transport in lipoproteins - an integrated approach to mechanisms and disorders. New Engl. J.Med. 276, 32, 94, 148, 215 and 273 (1967)

(2) Fredrickson,D.S., Levy,R.J.: Familial Hyperlipoproteinemia. In: The Metabolic Basis of Inherited Disease: ed. Stanburry,J.B., Wyngaarden,J.B., Fredrickson,D.S., Mc Graw-Hill, Inc. N.Y. p.545-614 (1972)

(3) Fredrickson,D.S., Levy,R.J., Lindgreen,F.T.: A comparison of heritable abnormal lipoprotein patterns as defined by two different techniques. J.Clin.Invest. 47, 2446 (1968)

(4) Bilheimer,D.W., Eisenberg,S., Levy,R.J.: Abnormal metabolism of very low density lipoproteins in type III. Circulation 44: II-56 (1971)

(5) Havel,R.J., Kane,J.P.: Primary Dysbetalipoproteinaemia: Predominance of a specific apoprotein species in triglyceride-rich lipoproteins. Proc.Nat. Acad.Sci. USA 70, 2015-2019 (1973)

(6) Hazzard,W.R., Biermann,E.L.: The spectrum of electrophoretic mobility of very low density lipoproteins: role of slower migrating species in endogenous hypertriglyceridemia (type IV hyperlipoproteinaemia) and broad-ß disease (type III). J.Lab.Clin.Med. 86, 239-252 (1975)

(7) Eisenberg,S., Bilheimer,D.W., Levy,R.J., Lindgreen,F.T.: On the metabolic conversion of human plasma very low density lipoproteins to low density lipoprotein. Biochim.Biophys. Acta 326, 361-377 (1973)

(8) Eisenberg,S., Rachmilewitz,D.: Interaction of rat plasma very low density lipoprotein with lipoprotein lipase-rich (postheparin) plasma. J.Lipid Res. 16, 341-351 (1975)

(9) Hazzard,W.R., Porte,D., Bierman,E.L.: Abnormal lipid composition of very low density lipoproteins in diagnosis of broad-beta disease (type III hyperlipoproteinemia). Metabolism 21, 1009-1021 (1972)

(10) Fredrickson,D.S., Morganroth,J., Levy,R.J.: Type III hyperlipoproteinemia: An analysis of two contemporary definitions. Ann.Int.Med. 82, 150-157 (1975)

(11) Mishkel,M.A., Nazir,D.J., Crowther,S.: A longitudinal assessment of lipid ratios in the diagnosis of type III hyperlipoproteinaemia. Clin.Chim.Acta 58, 121-136 (1975)

(12) Utermann,G., Canzler,H., Schoenborn,W., Vogelberg,K.H.: Biochemical studies on the metabolic defect in Broad-Beta disease (hyperlipoproteinemia type III). in press (1976)

(13) Utermann,G.: Isolation and partial characterisation of an arginine-rich apolipoprotein from human plasma very low density lipoproteins: Apolipoprotein E. Hoppe Seyler's Zschr.Physiol.Chem. 356, 1113-1121 (1975)

(14) Utermann,G., Jaeschke,M., Menzel,J.: Familial hyperlipoproteinemia type III: Deficiency of a specific apolipoprotein (Apo E-III) in the very low density lipoproteins Fed.Europ.Biochem.Soc. (FEBS)-Letters 56, 352-355 (1975)

(15) Moser,H., Slack,J., Borrie,P.: Type III hyperlipoproteinemia: a genetic study with an account of the risks of coronary death in first degree relatives (abstract 125). III. International Symposium on Atherosclerosis, West Berlin, Germany, October 24-28 (1973)

(16) Hazzard,W.R., O'Donnell,T.F., Lee,Y.L.: Broad-ß disease (type III hyperlipoproteinemia) in a large kindred. Evidence for a monogenic mechanism. Ann.Int.Med. 82, 141-149 (1975)

(17) Morganroth,G., Levy,R.J., Fredrickson,D.S.: The biochemical, clinical, and genetic features of type III hyperlipoproteinemia. Ann.Int.Med. 82, 158-174 (1975)

(18) Utermann,G., Jaeschke,M., Hees,M., Canzler,H., Vogelberg,K.H.: Broad-Beta Disease (Hyperlipoproteinemia Type III): Evidence for an autosomal recessive main gene effect. in press (1976)

(19) Burstein,M., Scholnick,H.R.: Lipoprotein - Polyanion - Metal Interactions. Advanc.Lipid Res. 11, 67-108 (1973)

(20) Utermann,G., Hees,M., Mühlfellner,G.: A protein marker for Broad-Beta Disease (Hyperlipoproteinaemia Type III): Diagnosis without ultracentrifugation. in press (1976)

Screening for Abnormal Hemoglobins in the Newborn: A Highly Economic Procedure Using Isoelectric Focusing

K. Altland

ABSTRACT

Using electrofocusing as separating principle and multiple syringes as well as multiple cuvets for the multiple handling of blood samples a highly economic technique has been developed to screen for abnormal hemoglobins in the newborn. The method provides the possibility to use dried blood samples as they arrive at the PKU screening laboratories at costs of ca. five cents for material and an average speed of one minute of work per sample analysis including sample collection and documentation of the result.

INTRODUCTION

In recent years screening for abnormal hemoglobins has attracted increasing interest when Valenti in 1973 (1) presented a fetoscope for the withdrawal of fetal blood in the 16th week of gestation and when Kan and coworkers in 1972 (2) developed the methods for the biochemical identification of the fetal hemoglobin phenotype from these samples (3).

Because most hemoglobinopathies and especially the most frequent sickle cell anemia resist to therapy and are poor candidates for secondary prevention the possibility of selective abortion in the cases with bad prognosis may be considered as a real advance in the medical activities against this group of inherited diseases in man.

Reduction of the social and medical load of sickle cell anemia, that is at present reduction of its incidence in newborns by selective abortion in early pregnancy, is effective only when the families at risk are detected before the first baby with a disease is born. This can only be realized when screening programs for detecting high risk families of gene carriers are applied to the total population at risk. Secondary prevention, that is to provide special care to the genetically affected individual in order to prevent or prolong as far as possible the time of onset of the clinically relevant symptoms, again makes necessary the early screening for the candidates.

These screenings are of the large scale type. Many thousands of individuals have to be tested and it is quite obvious that in such a situation especially economic considerations decide whether such screening programs eventually become reality.

For sickle cell anemia screening automated solubility tests have been developed some years ago (4). Up to 120 samples can be tested per hour at low costs for reagents but the machine is expensive and does not detect abnormal hemoglobins other than Hb S.

In 1973 Heredero and coworkers (5) described a screening electrophoretic system in polyacrylamide gels by which hemoglobins can be screened electrophoretically at a speed of 180 samples per hour at costs for consumed material of about US $ 2,5 / 1000 samples. This technique appears to be superior to the automated solubility tests with regard to speed, consumption of material and work and especially to the degree of information since most abnormal hemoglobins with altered charge are ascertained.

However, both test systems have one serious drawback: The solubility test and the screening electrophoresis in PAA gel do not recognize unequivocally gene carriers for Hb S when samples are contaminated in a high proportion with Hb F, and exactly this is the situation in the newborn blood. The blood of newborns, however, is of special interest for several reasons:

1. Secondary prevention of disease if available has the best chance when the affected individual is detected early. This should be at birth.

2. In societies with developed medical care any individual will see the doctor or a hospital at least one time in his life, that is at birth: an excellent occasion to give a drop of blood.
3. Convincing a population outisde the hospital to give a drop of blood is a hard thing and may provoke serious unexpected problems of different type.
4. Blood from all newborns in many countries is already collected within another screening program that is for phenylketonuria (PKU).

It was considered that an attempt for using the blood samples collected for PKU screening also for hemoglobin screening would be highly desirable. Blood for PKU screening usually is soaked and dried on filter paper and sent by normal mail to one of the few testlaboratories in the country. This technique is simple and safe for a screening with regard to aminoacids but problematic for proteins because of denaturation. Tailing is the consequence in separation techniques and electrophoresis in acid citrate-agar is unsatisfactory for this reason.

Electrofocusing has been expected to give better results because of the higher resolu tion power and a higher contrast within non-uniform material. On the other hand thinking on electrofocusing in a large scale screening at the first glance seems to be unrealistic for economical reasons because microtechniques are highly time consuming and other techniques in the usual analytical range suffer from the high prices of the available ampholytes. However, it was tried to optimize the experimental conditions with regard to low costs for both work and material. The result was the following method:

MATERIALS AND METHODS

Areas of ca. 6 mm in diameter were punched out of the PKU test papers and put into a microtiterplate. With a multiple syringe (Fig. 1) as it was developed for the screening electrophoresis system by Heredero and coworkers 50 ul of 0.02 M KCN were applied to the plate for elution of hemoglobins from the papers on a shaking device.

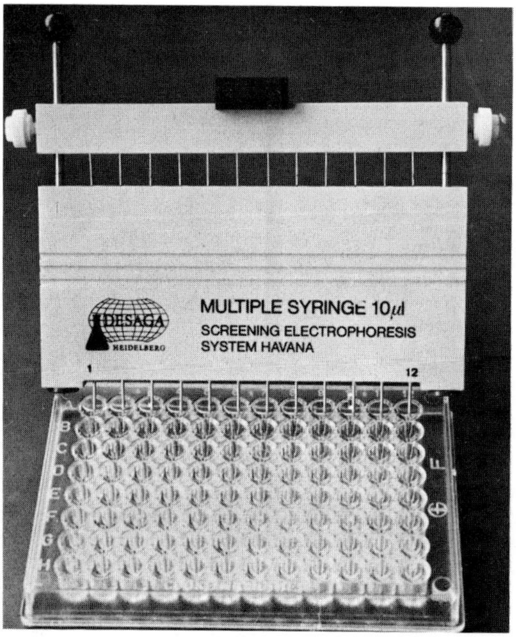

Fig. 1. The microtiterplate and a multiple syringe variable for volumes up to 10 ul used for the multiple transfer of samples from cuvet to gel. A similar multiple syringe for volumes up to 100 ul was used for extraction of blood from PKU testpapers.

Gels for electrofocusing 250 x 100 x 0.8 mm were prepared between a glass and a perspex plate separated by an appropriate perspex frame from deionized acrylamide monomer solutions fro T=5% and C=3% with 1% Servalyte 6-8, 1% Servalyte 7-9, TEMED and NH_4-PS as catalyst.

The ready gel adhering to the glass plate is transferred to the focusing unit Multiphor from LKB or Double Chamber from DESAGA set at 18°C. Two anodic paper wicks Whatman No. 3 were soaked in 1% Servalyte 6-8 and put along both long ends of the gel while another cathodic wick soaked with Servalyte 7-9 was applied along the middle of the gel. Flat band platinum electrodes from DESAGA were used for connecting the wicks with the power supply. Two silicone rubber bands each with 48 circular holes are put on the gel at 1 cm distance parallel to both anodic paper wicks. Electrodes are coverd with a glass plate and prefocusing is started at 30 mA const. current with voltage limitation at 500 V. After 20 min. 10 ul of the eluate from the 96 samples of one microtiterplate were applied to the perforations in the silicone rubber band by means of the multiple syringe seen in Fig. 1. The run is continued for another 90 min. and fotodocumented with a polaroid camera.

RESULTS AND DISCUSSION

The result is seen in Fig. 2. A more detailed view is seen in Fig. 3.

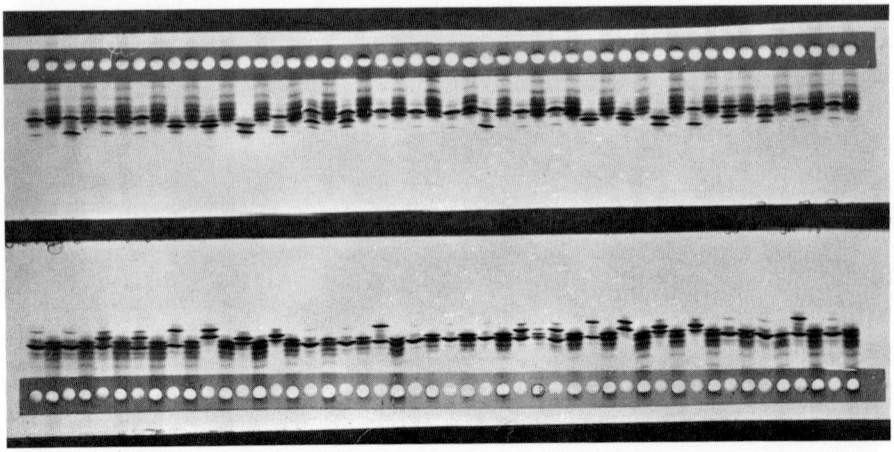

Fig. 2. Reference hemoglobins and hemoglobins extracted from PKU test papers separated by the method described. For identification see Fig. 3.

It comes out that under these conditions a clear identification of the clinically most relevant hemoglobins S, C, E, F, and A is possible.

The material consumed for 1000 analyses amounts to less than US $ 50 mainly consisting of ampholytes. The working time is mainly consumed for preparing the gels, handling the focusing unites, documentation of results and above all for pipeting. Time for pipeting, however, is strongly reduced by the use of the DESAGA multiple syringes (from DESAGA, Heidelberg, FRG) so that under well organized conditions one technician could be able to perform up to 1000 analyses within a day. If there is included another technician's day for collecting blood samples from a PKU screening laboratory, documentation and helping the first in the case of troubles one may arrive at 16 hours of work and US $ 50 in material for 1000 analyses. In other words one technician plus appropriate equipment added to a PKU laboratory with a daily input of some 400 - 500 samples could be able to present the additional information on hemoglobins using this method.

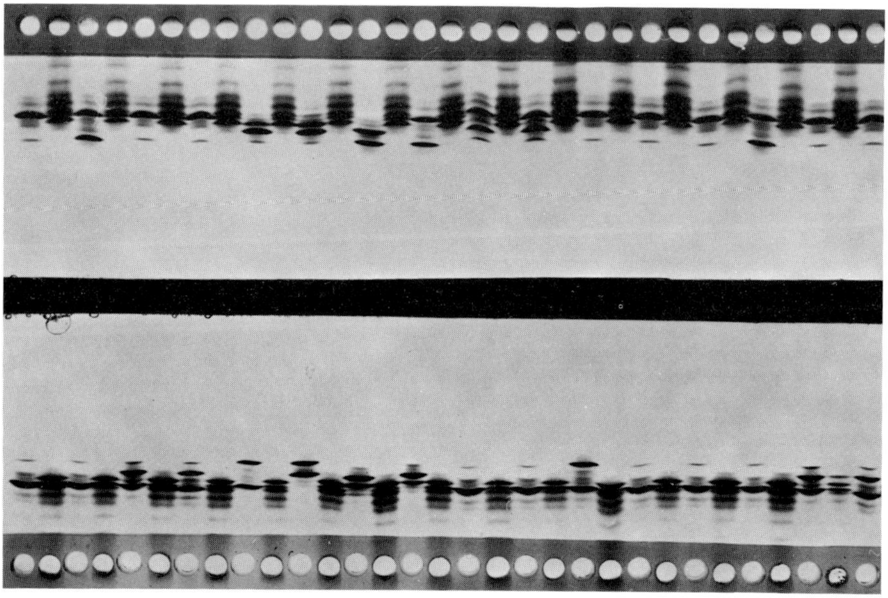

Fig. 3. Detail from Fig. 2. Lower row from left to right: 2,4,6, ... : newborn blood extracted from PKU test papers containing mainly Hb F_{II}, Hb A_{II}, and Hb F_{I}. 1,3,5, ... : reference hemoglobins from full blood specimens never dried. 1,3: from normal adults. 5: phenotype AS with 10% Hb F. 7: phenotype XS with minor Hb F. 9: phenotype AC. 11,13: phenotype S with ca. 13% Hb F. The minor band at the position of Hb C in all reference samples corresponds with Hb A_2. The extra zones in the samples extracted from PKU test papers are found in all samples - washed red cells and full blood, from newborns and adults - after being dried on PKU test papers and tested by the procedure described.

Relative to methods already described this method is considered as highly economic and useful. The principle is not only restricted to hemoglobins but can be applied also in other fields as mutation monitoring at the protein level (6), population genetics, fractionation controles in the preparative isolation of macromolecules, etc..

ACKNOWLEDGEMENT

The reference samples used for the development of the procedure were provided by Dr. B. Colombo and Dr. L. Heredero, University of La Habana. The use of PKU test papers in a large scale screening for human protein variants originates from a suggestion of Dr. F. Vogel, Inst. of Human Genetics, University of Heidelberg.

REFERENCES

1) Valenti, C.: Antenatal detection of hemoglobinopathies. Am. J. Obstet. Gynaecol. 115, 851 (1972).

2) Kan, Y.W., Dozy, A.M., Alter, B.P., Frigoletto, F.D., Nathan, D.G.: Detection of the sickle gene in the human fetus: Potential for intrauterine diagnosis of sickle cell anemia. New England J. Medicine 287, 1 (1972).

3) Alter, B.P., Kan, Y.W., Frigoletto, F.D., Nathan, D.G.: The antenatal diagnosis of the hemoglobinopathies. Clinics in Haematology 3, 509 (1974).

4) Canning, D.M., Crane, R.S., Huntsman, R.G., Yawson, G.I.: An automated screening technique for the detection of sickle-cell haemoglobin. J. Clin. Path. 25, 330 (1972).

5) Heredero, L., Granda, H., Suarez Aguilar, J.A., Altland, K.: An economic high speed electrophoretic screening system for hemoglobin S and other proteins. Humangenetik 21, 167 (1974).

6) Altland, K., Vogel, F. (in preparation).

The Crystallins of the Aging Lens from Five Species Studies by Various Methods of Thin-Layer Isoelectric Focusing *

J. Bours

SUMMARY

Isoelectric focusing in thin layers of polyacrylamide gel was carried out to analyze eye lens crystallins in young and old lenses in order to compare the effects of aging in different species. In the guinea pig, rat, man, ox and dog, the percentage of α-crystallin remains about the same with increasing age, the total β-crystallin percentage increases and the percentage of γ-crystallin and its components decreases. The increase in the total β-crystallin percentage in all five species is due to an increase in the high molecular weight β-crystallins , since the low molecular weight β-crystallins either remain constant or decrease slightly.

A comparison is made between the ready-made Ampholine PAGplate and the lab-made ordinary polyacrylamide gel plates. The time necessary for electrofocusing 10 or 20 samples is 90 minutes. A new method for precipitation of the proteins with trichloroacetic acid as the only agent and for staining with Coomassie Brilliant Blue was adapted for routine use.

Compared with the usual procedures an improvement is achieved by a reduction in separation time and a simpler method of staining.

INTRODUCTION

Since the method of isoelectric focusing in thin layers of polyacrylamide gel was introduced, numerous articles have been published showing results of generally good quality.

A prerequisite to obtaining good results is to perform the experiments at 4°C, and to establish a pH-gradient between the electrodes, which produces exact parallel refractive bands of clusters of ampholytes [1]. When an electric field with a low potential difference is applied between the electrodes for a long time period, these parallel bands remain broad. When higher voltage is used, the number of parallel bands of isofocused ampholytes increases progressively as the voltage is raised.

In this way the resolving power of the method is increased considerably, and hence using a high-voltage method, better results can be produced showing perfect straight bands of single isofocused components. When samples are placed on the gel, the number of components is also likewise increased along with the increase of focused ampholyte bands.

The purpose of this study is to show that good results can be obtained simply and rapidly with lab-made polyacrylamide gels.

---

* Research supported by the 'Deutsche Forschungsgemeinschaft' (No. 249/8: Altersveränderungen von Linsen)

MATERIALS

Preparation of Crystallins from Lens Tissues

Calf and adult bovine eyes were transported from the slaughter house under ice, and the lenses were removed immediately and dissected free from the surrounding tissues. The guinea pig, rat and dog (beagle) lenses were prepared from enucleated eyes immediately after killing the animals. The human lenses were obtained from the surgery department.

Before enucleation, the eyes of guinea pig, rat, man and dog (beagle) were examined under a slit lamp, and considered to be normal and transparent. The age of the lenses obtained was taken from controlled birth dates (rat, beagle-dog and man) and also determined from the fresh weights of whole lenses according to Hockwin (1971) as referred to in [7].

The lenses obtained were homogenized (Ultra-Turrax) in distilled water at 4°C and centrifuged at 38,000 x g for 1 hr at 4°C and the supernatants were lyophilized. The rats (Wistar-AF) were purchased at the Central Institute for Laboratory Breeding, Hannover, F.R.G.

METHODS

Preparation of the Polyacrylamide Gel

A 5% polyacrylamide gel plate was made, with a crosslinking percentage of 2.6 [2]. Stock solutions for preparing the gel were as follows:

(A) Catalyst solution: 1 ml N,N,N',N'-tetramethyl-1,2-diaminomethane (BDH No. 30385), distilled water to 150 ml;

(B) Acrylamide solution: 25 g acrylamide (BDH No. 27045), 0.68 N,N'-methylenebisacrylamide (BDH No. 29227), distilled water to 100 ml;

(C) Photoactivation solution: 2 mg riboflavin (Merck No. 7609), distilled water to 100 ml.

When kept at 4°C, the solutions A and C can be stored for a long time. The solution B should be made fresh each month.

The gel solution was prepared by mixing: 3.9 ml solution A; 8.0 ml solution B;10.5 ml solution C; 12.6 ml distilled water and 1.0 ml Ampholine carrier ampholytes covering the pH range from 3.5 to 9 and consisting of: 0.2 ml pH 3.5-5 (LKB No. 1809-111); 0.4 ml pH 5-7 (No. 1809-121) and 0.4 ml pH 7-9 (No. 1809-136).

A gel mould was prepared with two glass plates (12.5 x 26.5 cm, 2 mm thick) clamped together in a vertical position and separated by an india rubber gasket 1.5 mm thick to contain the gel solution. Using a syringe, the gel mixture of 37 ml was carefully poured between the glass plates. Any air bubbles should be avoided. The solution was polymerized by photoactivation for 2 hr at room temperature, using fluorescent daylight TL tubes placed vertically about 5 cm from the gel. After polymerization, the clamps and the gasket were removed and one of the two glass plates was lifted off with a spatula, allowing air bubbles to form between this plate and the gel while lifting. The polymerized gel adheres firmly to the other glass plate.

Application of the Samples and the Electrode Strips

The crystallin samples were dissolved in distilled water to make a 2%

solution. Sample application pieces of 5 x 10 mm (Whatman 3MM paper) were dipped in the sample solution, each taking up about 12-14µℓ, and applied on the gel, 5 mm apart and 2-3 mm from the anode. Twenty samples can be applied if the short side of the sample application pieces is placed parallel to the electrode strip (Fig 1). Twelve samples can be applied if the long side of the sample application pieces is placed parallel to the electrode strip (Fig 3, 4).
The electrofocusing strips (LKB No. 2117-106) are first cut at a length of 25.5 cm. One is soaked with the anode electrode solution of 1 M phosphoric acid and applied on the gel plate which was already in the Multiphor electrophoresis apparatus (LKB No. 2117) with the cooling plate thermostatted at 4°C and wetted with distilled water. The second strip is soaked with the cathode electrode solution of 1 M NaOH and placed on the polyacrylamide gel at the cathode position.

Isoelectric Focusing Procedure, Comparison of Three Methods:

1. The low-voltage method:
   A polyacrylamide gel plate was made as described above, and after application of the samples, isoelectric focusing was performed with a current of 11 mA (decreasing to 0.8 mA) for 18 hrs at a voltage of 100 V (increasing to 120 V). This method is essentially similar to the one described by Bours [2, 3]. Fixation, staining and destaining was carried out similarly [2, 3, 4].
2. The high-voltage method:
   A polyacrylamide gel plate was made, and after application of the samples, isoelectric focusing was performed for 1 ½ hrs with a constant current of 20 mA, and with a voltage which was increased every 10 minutes from 220V, 360V, 540V, 860V up to a maximum of 1.2 kV at 40 minutes. While maintaining a constant voltage of 1.2 kV, the current decreased after 90 minutes to 10.5 mA.
3. The Ampholine-PAGplate-method:
   The Ampholine polyacrylamide gel plate, commercially available at LKB (No. 1804-101) was handled according to instructions supplied by LKB. Fixation, staining and destaining were performed exactly according to these instructions.

Determination of the pH Gradient

Immediately after the completion of the run, the pH gradient was determined at 4°C with the aid of a combined micro-surface electrode, 6 mm in diameter (Ingold, A.G., Frankfurt, Zürich, type LoT 403-30-M8). The pH gradient can also be conveniently determined by the application of protein markers, as described by Bours [2] and Radola [5].

Fixing and Staining Solutions

Fixing Solution I :	375 ml 33% Trichloroacetic acid solution + 300 ml methanol + 450 ml distilled water
Fixing Solution II:	250 ml 33% Trichloroacetic acid solution + 330 ml methanol + 600 ml distilled water
Staining Solution :	Dissolve 1.15 g Coomassie Brilliant Blue R-250 in 1000 ml of a solution containing: 625 ml ethanol + 200 ml acetic acid + 1675 ml distilled water

Destaining Solution: 625 ml methanol + 200 ml acetic acid + 1675 distilled water

Removal of the Ampholine Carrier Ampholytes; Staining and Destaining of the Gel

The gel slab as used with the high-voltage method was placed immediately in the fixing solution I for 0.5 hr, and then in the fixing solution II for another 0.5 hr. The gel was stained for 1 hr at 60°C in the staining solution. The gel was destained with several changes of the destaining solution until the background was clear; usually overnight was sufficient duration.

Preservation of the Gel

After destaining, the gel was immersed for 10 minutes in 10% acetic acid, to remove excess methanol. Then the gel was preserved by simply covering it with a layer of 7% gelatin and 5% glycerol [3]. The gel was dried for 1-2 days. The gel preserved in this way can be kept for a long time as a hard layer on the glass plate.

RESULTS AND DISCUSSION

Comparison between the Way of Sample Application on the Lab-made and the Ready-made Polyacrylamide gel

A clear difference is shown between the young and the old lens. In the 86-day-old rat lens [Fig 1(b,c,e,f,h)] the γ-crystallins are quite prominent, while also the β-crystallins of higher isoelectric points at pI 6.5-7.0 and of low molecular weight are prominent in this stage. In the 723-day-old rat lens [Fig 1(a,d,g,j)] the concentrations of nearly all γ-crystallin components ($\gamma_1, \gamma_2, \gamma_4 - \gamma_7$) and of the β-crystallin components No. 8-12 are reduced sharply, due to the aging of the lens. The concentration of the β-crystallins of lower isoelectric points is actually increased in the older lens. The percentage of α-crystallins is slightly reduced or may remain constant.

When the sample application pieces are laid on the polyacrylamide gel with the short side (5 mm) parallel to the electrode strip (Fig 1), then 20-24 samples may be analyzed on one plate. When the sample application pieces are prematurely removed from the gel after 10 minutes as is recommended by the manufacturers of the Ampholine PAGplate, an essential part of the protein spectrum, the proteins of higher molecular weights like the α-crystallins are missing entirely from the final pattern. The low molecular weight crystallins form sharp focused zones, and the pre- α-crystallin is discernable very well (Fig 1a,b) while not in the other specimen [Fig 1(c-j)].

In Fig 1(c,d,f,g) the sample papers are dipped in the lens crystallin solution before application. In this way approximately 6-7 μℓ 2% solution [Fig 1(c,d,f,g)] corresponding to 120-140 μg, or 12-14 μℓ 2% solution [Fig 1(a,b)] corresponding to 240-280 μg of protein is applied. In Fig 1(e,h,j), 10 μℓ 2% solution corresponding to 200 μg protein is applied on the gel. There are only small differences between the lab-made polyacrylamide gel [Fig 1(f-j)] and the gel plate which is commercially available [Fig 1(a-e)]. The resolving power is similar but not identical, but the background of the Ampholine PAGplate is difficult to destain and requires more time [Fig 1]. The asterisk shows the position of the gap in the ampholine concentration, which is larger in the ready-made plate than in the lab-made plate (Fig 1) [1].

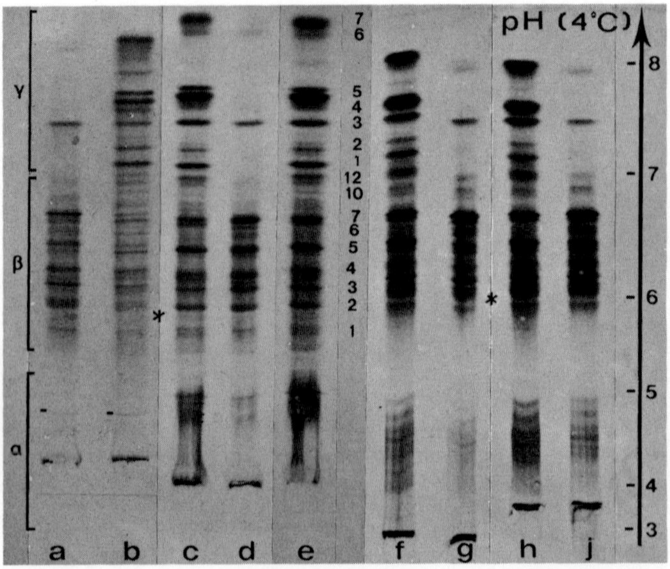

Fig. 1. Thin-layer isoelectric focusing patterns of the crystallins of the aging rat lens: b,c,e,f,h = 86-day-old rat lens; a,d,g,j = 723-day-old rat lens. Various methods are compared: 1. Ampholine PAGplate, 1 mm thick (a-e); 2. polyacrylamide gel plate, 1.5 mm thick (f-j). a,b = sample application pieces dipped in the sample solution and removed after 10 minutes; c,d and f,g = sample application pieces dipped in the sample solution and not removed; e,h,j = amount of sample is 200 μg of protein. The scale shows the pH values along the gel, measured at 4°C. The numbers β_1-β_{12} and γ_1-γ_7 indicate single rat lens crystallin components isofocused. The asterisk shows a gap in the concentration of the ampholine carrier ampholytes. α = α-crystallin; β = β-crystallins; γ = γ-crystallins; the bar in a and b shows the position of pre- α-crystallin.

Comparison between the three isofocusing methods

The resolving power of the first method is upgraded with the application of higher voltage for a shorter time, as shown in Fig 1 and 3. There is an increase in the sharpness of the focused bands. The PAGplate shows a disadvantage compared to the lab-made plate because the precipitation of the α-crystallin occurs most probably due to the higher milliamperage applied over the thinner gel (Fig 3).

Regarding the crystallin composition of the young and old lens, the differences observed with the rat (Fig 1) could also be seen in the other species like guinea pig, man, ox and dog (Fig 2,3,4). In all species there is a progressive decrease with age of the concentration of the γ-crystallins and its components, and a progressive increase of the β-crystallins of higher molecular weight (β_I), while the α-crystallin concentrations are fairly constant (Table 1). The concentrations of the β-crystallin components of higher isoelectric point and lower molecular weight (β_{II},β_{III},β_{IV}) are fairly constant in guinea pig, ox and dog (Table 1). Though only two points in the age curve of these species are shown, these processes develop continuously with age as was reported for the bovine and the dog lens [6].

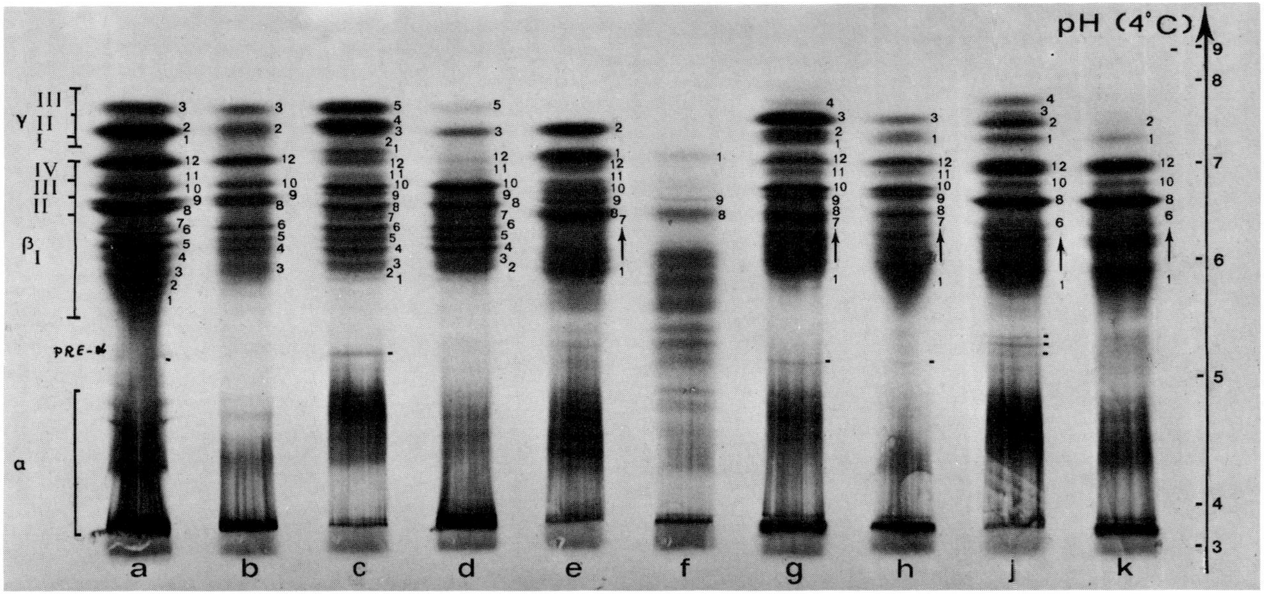

Fig. 2. Thin-layer isoelectric focusing patterns of the crystallins of the aging lens of five species, by the low-voltage method with a 1.5 mm thick polyacrylamide gel plate. The samples a-k amount to 300 μg of protein. The scale shows the pH values along the gel, measured at 4°C. The roman numerals I-IV, and I-III denote the groups of ß- and γ-crystallin components. The numbers $ß_1$-$ß_{12}$ and $γ_1$-$γ_9$ indicate the single crystallin components isofocused. a = 0.5-year-old guinea pig; b = 2.5-year-old guinea pig; c = 86-day-old rat; d = 723-day-old rat; e = 5.5-year-old man; f = 50-year-old man; g = 0.16-year-old calf; h = 17.5-year-old ox; j = 0.77-year-old dog (beagle); k = 9.8-year-old dog (beagle). α = α-crystallin; pre-α and bars = pre- α-crystallin; ß = ß-crystallins; γ = γ-crystallins.

Table 1. Patterns of crystallin composition* of the aging lens of five species as performed by three methods of thin-layer isoelectric focusing

Fig.	Method	Species	Age	α-	pre-α-	β_I	β_{II}	β_{III}	β_{IV}	β_{I-IV}	β_{tot}	γ_I	γ_{II}	γ_{III}	γ_{tot}
				CRYSTALLINS											
2a	LV	Guinea	0.5 Y	40	1	26	8	6	8	22	48	0	8	3	11
3a	APP	pig		23	-	29	14	10	11	35	64	0	11	2	13
4a	HV			13	-	33	15	9	11	35	68	5	12	2	19
2b	LV		2.5 Y	41	-	25	11	6	8	25	50	0	7	2	9
3b	APP			22	-	37	15	8	8	31	68	0	6	4	10
4b	HV			15	-	42	13	8	9	30	72	3	8	2	13
2c	LV	Rat	86 D	35	1	19	9	9	4	22	41	4	13	7	23
3c	APP			27	-	19	8	7	7	22	41	7	17	8	32
4c	HV			20	-	25	11	8	7	26	51	7	13	8	28
2d	LV		723 D	45	-	28	13	9	2	24	52	0	2	1	3
3d	APP			28	-	36	16	10	4	30	66	0	3	3	6
4d	HV			22	-	34	20	10	6	36	70	0	5	3	8
2e	LV	Man	5.5 Y	39	-	28	9	10	2	21	49	7	5	0	12
3e	APP			19	-	39	12	8	4	24	63	8	10	0	18
4e	HV			20	-	33	18	5	6	29	62	8	10	0	18
2f	LV		50 Y	12	-	72	8	3	2	13	85	3	0	0	3
3f	APP			19	-	69	6	2	0	8	77	4	0	0	4
4f	HV			20	-	62	13	0	0	13	75	5	0	0	5
2g	LV	Ox	0.16Y	37	1	27	5	8	7	20	47	7	8	0	15
3g	APP			23	-	37	8	3	5	16	53	11	11	2	24
4g	HV			18	-	33	10	10	7	27	60	8	11	3	22
2h	LV		17.5 Y	34	1	37	6	11	7	24	61	3	1	0	4
3h	APP			25	-	49	10	3	5	18	67	4	3	1	8
4h	HV			21	-	45	10	10	7	27	72	4	3	0	7
2j	LV	Dog	0.77Y	41	2	29	8	3	8	19	48	3	5	1	9
3j	APP	(beagle)		27	-	36	11	2	12	25	61	4	5	3	12
4j	HV			21	-	31	8	7	11	26	57	9	6	6	21
2k	LV		9.8 Y	36	2	39	10	3	8	21	60	2	0	0	2
3k	APP			26	-	48	13	2	8	22	70	3	1	0	4
4k	HV			22	-	37	16	2	11	29	66	4	5	3	12

* The relative concentration is expressed as a percentage of the total, obtained from densitometric tracings of the isofocused lens crystallin components

- = not detected

LV = polyacrylamide gel, run at low voltage during 18 hours

HV = polyacrylamide gel, run at high voltage during 90 minutes

APP = Ampholine PAGplate (LKB Nr. 1804 - 101)

Y = years

D = days

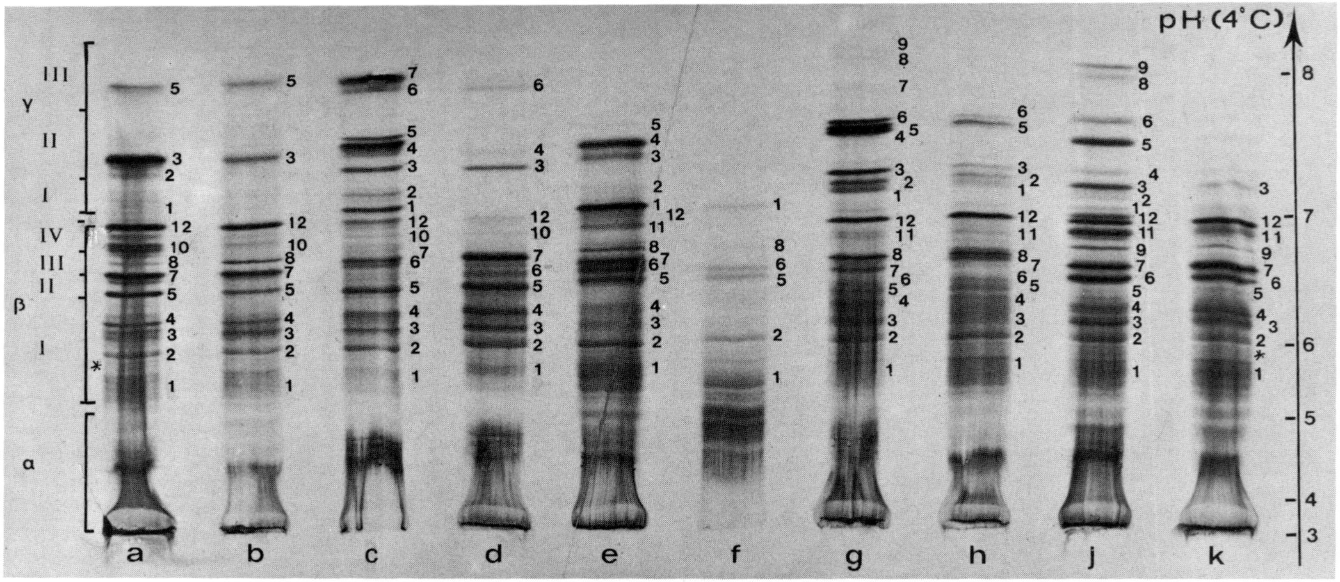

Fig. 3. Thin-layer isoelectric focusing patterns of the crystallins of the aging lens of five species, by the high-voltage method with an Ampholine PAGplate. The samples a-k amount to about 280 μg of protein. The scale shows the pH values along the gel, measured at 4°C. The roman numerals I-IV, and I-III denote the groups of ß- and γ-crystallin components. The numbers β_1-β_{12} and γ_1-γ_9 indicate the single crystallin components isofocused. The asterisk shows a gap in the concentration of the ampholine carrier ampholytes. a = 0.5-year-old guinea pig; b = 2.5-year-old guinea pig; c = 86-day-old rat; d = 723-day-old rat; e = 5.5-year-old man; f = 50-year-old man; g = 0.16-year-old calf; h = 17.5-year-old ox; j = 0.77-year-old dog (beagle); k = 9.8-year-old dog (beagle); α = α-crystallin; ß = ß-crystallins; γ = γ-crystallins.

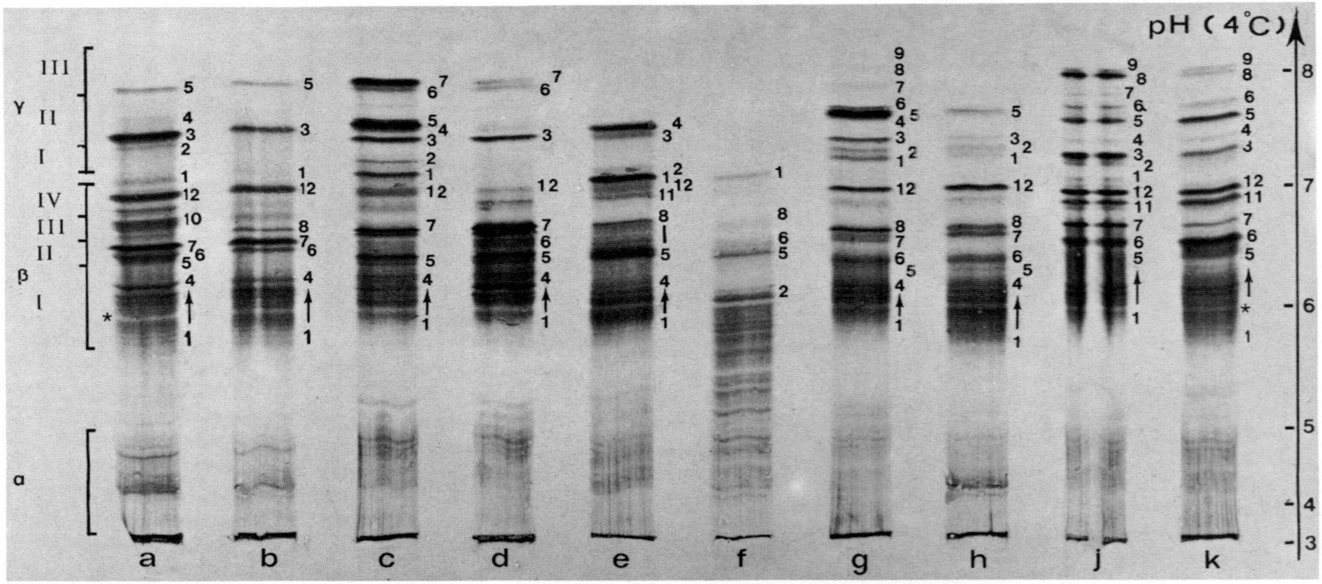

Fig. 4. Thin-layer isoelectric focusing patterns of the crystallins of the aging lens of five species, by the high-voltage method with a 1.5 mm thick polyacrylamide gel plate. The samples a-k amount to about 280 μg of protein. The scale shows the pH values along the gel, measured at 4°C. The roman numerals I-IV, and I-III denote the groups of β- and γ-crystallin components. The numbers β_1-β_{12} and γ_1-γ_9 indicate the single crystallin components isofocused. The asterisk shows a gap in the concentration of the ampholine carrier ampholytes. a = 0.5-year-old guinea pig; b = 2.5-year-old guinea pig; c = 86-day-old rat; d = 723-day-old rat; e = 5.5-year-old man; f = 50-year-old man; g = 0.16-year-old calf; h = 17.5-year-old ox; j = 0.77-year-old dog (beagle); k = 9.8-year-old dog (beagle). α = α-crystallin; β = β-crystallins; γ = γ-crystallins.

Contrary to the other species mentioned, there is a decrease of the low molecular weight β-crystallins in the lenses of rat and man (β_{IV} in Table 1). The β-crystallin component β_{12} is always present in the species guinea pig, rabbit, ox and dog, but almost disappeared in rat and man (Fig 2,3,4; Table 1). Besides the species specificity which was already known for mammals [7] there is a definite difference between rat and man on one side, and guinea pig, ox and dog on the other side.

CONCLUSIONS

1. The broad sample application shows better results (Fig 1, Fig 3, Fig 4);
2. The lab-made polyacrylamide gel combined with the application of higher voltage for a shorter time shows a higher resolving power (Fig 4);
3. The methods of fixation and subsequent deampholinization, staining of protein bands and destaining of the gel is quicker, the gel is less fragile, does not shrink and sticks to the glass plate;
4. The lab-made gel shows equal or even better results, compared to the ready-made PAGplate;
5. Compared to the usual procedures, an improvement is achieved by a reduction of separation time and a simpler way of staining.

ACKNOWLEDGEMENTS

I take pleasure in expressing my thanks to Dr W. Schnitzlein and Dr Astrid Fischer, Department of Toxicology, KNOLL, A.G. Chemische Fabriken, Ludwigshafen for the supply of beagle-dog lenses, and to Dr F. Schlieter of the Eye Clinic, University of Bonn, and Dr H. Kilp of the Eye Clinic, University of Cologne for the human lenses put at my disposal.

I also wish to express my thanks to Prof Dr rer.nat. O. Hockwin and to Dr P. Talbot (The Wellcome Research Laboratories, Beckenham, Kent, England) for reviewing the manuscript. Further I thank Mrs Elke Oellers and Miss Barbara Polenz for the photographic illustrations.

REFERENCES

1. Righetti, P-G.: Preparative isoelectric focusing. Separ. & Purif. Meth. 4, 23-72 (1975).
2. Bours, J.: The use of protein markers in thin-layer isoelectric focusing . Science Tools 20, 29-34 (1973).
3. Bours, J.: Isoelectric focusing of lens crystallins in thin-layer polyacrylamide gels. J. Chromatogr. 60, 225-233 (1971).
4. Bours, J.: Isoelectric focusing of lens crystallins. In: Progress in isoelectric focusing and isotachophoresis (Ed. P.G. Righetti). North-Holl. Publ. Cy., Amsterdam. 235-256 (1975).
5. Radola, B. J.: Isoelectric focusing in layers of granulated gels; thin-layer isoelectric focusing of proteins. Biochim. Biophys. Acta 295, 412-428 (1973).
6. Bours, J., Doepfmer, K. and Hockwin, O.: Isoelectric focusing of crystallins from different parts of the bovine and the dog lens in dependence on age. Docum. Ophthal. Proc. Ser. (1976) (in press).
7 Bours, J. and Hockwin, O.: Artunterschiede bei Linsenproteinen nach Trennung mit Isoelektrofokussierung auf Polyacrylamid-Dünnschichtplatten. Berl. & Münch. Tierärztl. Wochenschr. (1976) (in press).

Separation of Erythrocyte Membrane Components in Electric Field. Polyacrylamide Gel Isoelectric Focusing: A Potential Method for the Diagnosis of Membrane Abnormality

P.K. Das, D. Graesslin and H.W. Goedde

ABSTRACT

The results of separation of human erythrocyte membrane components, such as proteins, glycoproteins and acetylcholinesterase (AChE) using polyacrylamide gel electrophoresis (PAGE), starch gel electrophoresis (SGE), SDS-PAGE and polyacrylamide gel isoelectric focusing (PAGIF) are reported. Although the known good separation qualities of SDS-PAGE with respect to proteins and glycoproteins could be confirmed, it is limited for identification of enzyme activity. Surprisingly, Triton-X-100 solubilized protein components could be better separated using SGE as compared to 5 % PAGE. In contrast, for acetylcholinesterase PAGE showed higher resolution than SGE. On the otherhand, PAGIF was found to be equally good as SDS-PAGE in separation qualities and in addition, clearly demontrated a gross abnormality in protein patterns of erythrocyte membrane from DMD, whereas no significant difference in glycoprotein and AChE patterns could be found. The present study indicated a new possibility for studying membrane polymorphism in health and disease using PAGIF. The conjectures of normally occuring heterogeneity of membrane components and its relationship to the genetic influence are presented.

INTRODUCTION

No aspect of membrane biology probably presents so much problem as is faced with the qualitative separation of membrane components and their subsequent relationship to disease. The reasons are difficulty in solubilising the membrane components in their native functional states and their subsequent separation by conventional procedures, like electrophoresis, ultracentrifugation. The critical point is to fractionate the membrane proteins, in presence of solubilizers, since membrane proteins tend to aggregate. In this respect, however, the most promissing approach appears to be dissolution of the membrane in SDS-containing polyacrylamide gels (SDS-PAGE) and this technique is widely used. Although several disadvantages of SDS-PAGE have been noted by various investigators (1, 2) the main disadvantage is that SDS denatures the functional activity of the membrane protein, like enzymatic antigenic properties. Moreover, this method is not neccessarily an absolute parameter for molecular distinction of membrane in situ, since the criterian in this case is subunit size. Adequate resolution, therefore, neccessiates the evaluation of other separation systems for the study of macromolecular segregation with the membrane in health and disease.

Our present investigations were motivated by the well excepted hypothesis that the primary cause of Duchenne muscular dystrophy (DMD, a fatal sex-linked hereditary disorder) may be the results of a generalised plasma membrane defect involving even non-muscle cells (3 - 7), such as red blood cell. Electronmicroscopic and physical experiments supported the view of a gross alteration of erythrocyte membrane in this disorder (6, 8). Erythrocyte membrane bound enzymes have also been reported to be different in DMD (3 - 6). In this respect, DAS (9) suggested a probable genetic polymorphism of erythrocyte membrane bound acetylcholinesterase (EAChE) associated that K_m values and percentage inhibition patterns (of EAChE from normal and Duchenne subjects) in presence of 9 - 10 mM Nialamide HCl, 50 nM eserine sulphate and 0.3 mM NaF at pH 7.4 in 20 mM phosphate buffer at 25° C are different.

It was clearly shown that inhibition values using 1 mM acetylthiocholine iodide (Ac) with Nialamide HCl (NialN) and those using 0.05 mM Ac with NaF (FN) for normal adult Caucasian EAChE are 17 - 28 and 19 - 27 respectively; and those values for the dystrophie enzyme are 46 - 55 and 40 - 58 respectively. Similarly, precentage remaining activities in the presence of eserine sulphate (EsN) for normal and dystrophic enzymes are 100 - 125 and 75 - 80 respectively. These results indicated that an altered type of EAChE is associated with DMD. This type of approach together with the familiar studies is well known to the investigators engaged in the polymorphic studies of soluble enzyme (10). Although this kind of interpretation involving membrane bound enzyme should be treated with caution, it is indicative that membrane abnormality associated with this genetic disease offers a convenient model to study human membrane polymorphism as a whole rather than only one component.

Present experiments arise from the well known fact that various investigators used electrophoretic methods in studying the individual variation in protein and enzyme pattern (11). We attempted to explore the possibility by investigating several electrophoretic and also isoelectrofocusing techniques, whether this latter approach could also be applied for such study involving membrane.

This paper reports the results of separation of human erythrocyte membrane components using starch gel electrophoresis, polyacrylamide gel electrophoresis, SDS-electrophoresis and isoelectrofocusing on slabs of polyacrylamide gel (PAGIF) from normal as well as Duchenne muscular dystrophic individuals.

MATERIALS AND METHODS

Erythrocyte Samples:

Human erythrocytes were obtained from either heparinized or citrate-dextrose treated blood samples of various individuals. Blood samples from Duchenne Patients were collected at the Kinderklinik of Freiburg (Dr. Beckmann). The samples were stored at 4^{0} C without delay not for more than three days. White erythrocyte ghosts was prepared by the modified method of MARCHASI & PALLADE (12) as reported by DAS & LO (13). Membrane was solubilized in 1 % Triton-X-100 solution in 0.02 M phosphate buffer pH 7.4 according to DAS & LO (13). This preparation was used for the investigations.

Classification of the Samples

The samples were then classified as the normal and DMD by measuring the acetylcholinesterase activity (EAChE) and the inhibition values e.g.

$Nial_N$, FN and EsN as reported by DAS (9). Samples having $Nual_N$, FN and EsN with the range of 17-28, 15-27 and 98-130 were also treated as normal although they might have some clinical syndrome of muscle weakness, in addition to those samples which had been collected from normal healthy volunteers. Samples from individuals clinically confirmed Duchenne dystrophy were also evaluated by the same test of EAChE types.

Separation Techniques

SGE eas carried out at 4^0 C according to DAS et al. (14). PAGE was performed by a modification of the method of DAVIS (15) as mentioned by DAS & LO (13). SDS-PAGE was performed by a modification of the procedure of WEBER & OSBORNE (16).

Concentrated stock solutions were mined in order and proportions given in the following table and added to 7.5 cm glass tubes of 5 mm i.d. that had been soaked in chronic-sulfuric acid or by washing with hot SDS. Each column hight was overlaid gently with 300 ml of the electrod buffer.

Formulas for Stock Solutions, Buffers and SDS-Gels:

A. Stock Solution:

Cone Ac Bis:

Acrylamide: 44.4 g

Bis : 1.2 g

H_2O to : 200 ml

B. Conc. Buffer:

1.0 M Tris: 400 ml

2.0 M Acetic Acid: 100 ml

0.2 M EDTA: 100 ml

Acetic Acid to: pH 7.4

H_2O to: 1000 ml

C. Electrophoresis Buffer:

200 ml of B + 20 g of SDS

H_2O to 1000 ml

D. Fresh Ammoniumpersulphate:

45 mg Ammoniumpersulphate +

3 ml H_2O

E. Preparation of Gels:

30 ml C

13.6 ml A

2.0 ml D

30.0 ml of TEMED

F. Electrode Buffer:

500 ml Solution C

H_2O to 1000 ml

G. Sample Preparation Soln:

5 ml Solution C
5 g Sucrose + 85 ml H_2O
3 g SDS
3 ml Mercapto-aethanol

PAGIF was performed in 0.8 mm thick polyacrylamide gel layers (5 % T and 2.6 % C) containing 0.25 % Triton-X-100, in the pH range between 3.5 - 10 using 1 % Ampholine from LKB-Produkter, Stockholm. The apparatus used was manufactured by one of the authors (D.G.): an aluminium box coated with plastics and equipped with electrographitated carbon electroces putted directly on the gel surface. Starting with 100 V not exceeding 7 mAmp for about 30', the final voltage for the run was held at 650 V by stepwise increasing and never exceeding 20 mA. The run was continued for 7 - 8 hours. Subsequent handling of the gels during staining - destaining was performed according to GRAESSLIN et al. (17).

Staining Techniques:

For both electrophoretic and focusing methods the following stainings were carried out:

Protein: Protein staining was done in a 0.1 % Coomassie Brillant Blue in water : ethanol : acetic acid (50 : 45 : 5), destaining was achieved by a solution of water : ethanol : acetic acid (55 : 40 : 5).

Glycoprotein staining using PAS reagent was carried by incubating the gels in a solution made of 15 g $NaJO_4$, 50 ml acetic acid, 100 ml ethanol and 350 ml water for 30' at 20^0 C under gentle agitation. Then the gels are washed in a solution containing 100 ml ethanol, 50 ml acetic acid and 350 ml H_2O at 4^0 C for 45'. Immediately the gels were then transferred into Schiff's reagent (Merck, West-Germany) over night at mechanical agitation at 4^0 C. The stained gels can be kept in this reagent in the dark at 4^0 C for several weeks.

EAChE activity was detected using 1 mmol/l Ac as substrate according to the method of JUUL (18) as described earlier (19).

RESULTS

Table 1 shows the range of values of EAChE obtained from 20 normals. 15 Duchenne and 5 carrier mothers. These values are compatable with those previously reported with the exception that the specific activities as against activity/ml of RBC of the diseased samples are in lower range of normals. These samples were subsequently examined by the different electrophoretic methods.

Table 1:

SAMPLES	*SPEC. ACTIVITY*	$NIAL_N$	E_SN	*FN*
Normals	*1.88 - 3.1*	*22 - 29*	*94 - 102*	*23 - 28*
Duchenne	*1.09 - 2.0*	*37 - 47*	*78 - 83*	*35 - 44*
Carrier	*2.2 - 3.0*	*32 - 37*	*89 - 93*	*31 - 34*

Table 1: Enzyme activities and inhibition values for EAChE preparations from normal and Duchenne individuals together with carrier mothers.

Fig. 1 shows the diagrammatic protein pattern of normal human erythrocyte membrane on SGE and PAGE respectively. At least 4 distinct protein bands could be detected on SGE and no differences could be found in disease in several parallel runs. In this a major portion of protein remain near the origin. The components staining for AChE activity have also been demonstrated. In a survey of 20 normals, altogether 5 bands in different proportions could be detected in various individuals. According to numbers of these bands, individuals may be divided into 3 - 4 groups.

The majority of these groups demonstrate either band 2 and 4 or band 2 and 3. In addition a very small percentage of individuals either only bands 3 and 5 or bands 1, 2, 4 and 5 were noted. This normally occuring heterogeneity of EAChE on SGE as compatible to earlier results of DAS (9) and makes it difficult task for recognising any differences in disease. However in majority cases of dystrophic only band 2 was detected although in several dystrophic patients both bands 2 and 4 like majority of normals were present. This indicates that no clear cut differences in EAChE pattern could be seen on SGE in normals and dystrophy.

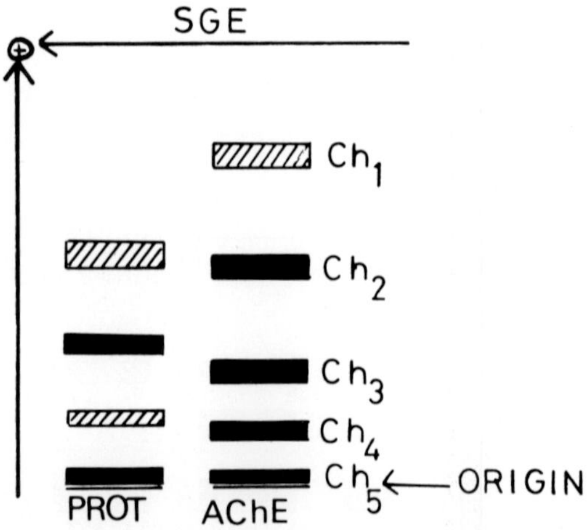

Fig. 1: Starch gel electrophoretic pattern of normal human erythrocyte ghosts. The gels were stained for proteins (Coomassie Brillant Blue) and for enzyme activity (AChE).

Fig. 2 shows the diagramatic patterns of protein, glycoprotein and EAChE of normal erythrocyte on 5 % PAGE. Surprisingly the resolution of protein pattern on PAGE is less convincing as compared to SGE. A major nonpenetrable aggregate was stained at the origin with a second distinct band. In addition about two more faint protein bands of faster mobilities could sometimes be observed. Two glycoprotein (PAS 1 and 2) bands are present of which one is stained near the origin and the other approximately at the similar position of 2nd major protein band. No differences could be obserbed in normals and disease in several parallel runs.

In the same figure, the diagramatic pattern of EAChE on PAGE are also given. In our experience, altogether 4 bands can be detected in differing proportion in various individuals of which band 1 is either present in trace amount or totally absent.

In majority of individuals only bands 3 and 4 and in some cases bands 2 and 4 are present. In rare cases bands 2, 3 and 4 were seen. Hereagain, because of such a normally occuring heterogeneity of EAChE pattern makes this technique hardly attractive for the diagnostic and polymorphic studies. Nevertheless, it is worth mentioning that at least in 3 dystrophic

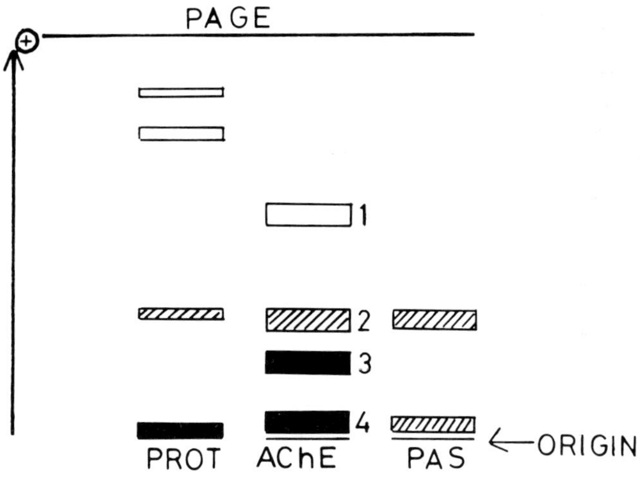

Fig. 2: Protein-Acetylcholinesterase-Glycoprotein pattern of normal erythrocyte ghosts on PAGE.

samples only bands 1 and 4 are present as against usually normal pattern with either bands 3 and 4 or 2 and 4. However, in several other parallel runs this difference could not be seen.

Fig. 3 (i-iii) shows types of protein pattern of normal samples on SDS-PAGE. In 3 (i) a typical pattern is usually seen in majority individuals. Band I and II (in range of average mol.weight 250 - 230 x 10^3) followed by band III (MW 90 000) and band IV (MW 78 000), bands V and VI (MW 42 000 and 35 000). In between those zones there are also bands with MG of 50 - 60 000 followed by several bands with MW between 12 - 25 000. In few exceptional cases, it was found that this typical normal pattern is not universal; examples of two such exceptions are illustrated in 3 (ii) and 3 (iii). Where band III and IV are present in only trace amount whereas bands V and VI are present proportionately in increased quantities as judged by the relative intensities of the bands, in addition a new band IIa of MW 125 000 could be found. In other exception an additional band of lower MW is present in relatively higher amount. Considering this normal heterogeneity, no conclusive differences between normal and Dystrophic samples could be expected. However, in several parallel runs no difference could be observed. Usual glycoprotein pattern demonstrating the pre-

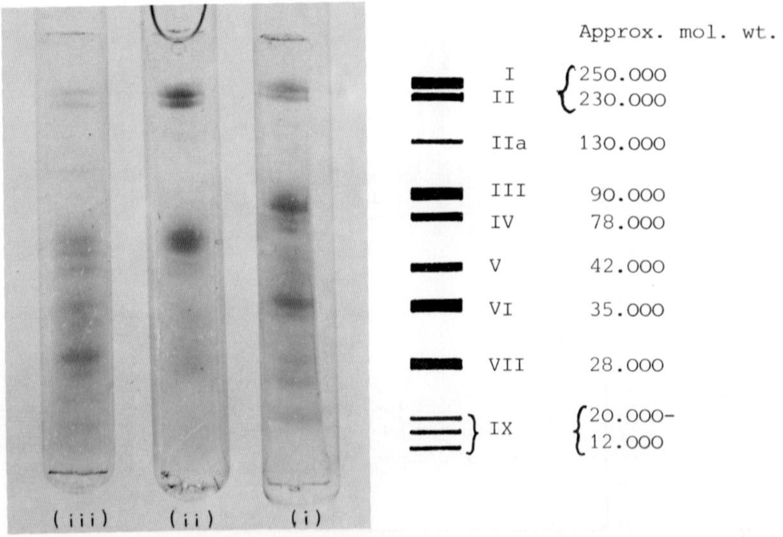

Fig. 3 (i-iii): Protein patterns of normal erythrocyte ghosts on SDS-PAGE. Pattern in (i) is usually seen and in rare cases, different banding pattern as in (ii) and (iii) have also been observed.

sence of PAS 1, PAS 2 and PAS 3 (20) has been observed.

Fig. 4 shows a typical diagramatic protein, glycoprotein and ChE pattern on PAGIF together with the pH gradient. About 20 protein bands could be resolved of which about 80 % as indicated by eye estimation are located between 3.8 - 4.5. About 5 to 6 of these bands, especially band 7 and 8 (arbitarily numbered) are present in all samples; bands in the range of pH 6 - 8 somewhat varied in their distinctness and sharpness in comparison to those in the acidic region. Similar was the case in extreme pH ranges of 3.0 - 3.3, where comparetively 3 or 4 faint zones can be seen. However about 6 to 8 protein bands near the main two bands have invariably been seen as distinct and sharp. The ratio of these bands appeared to be fairly constant.

In contrast only two distinct PAS zones in pI ranges of 4 - 4.3 and 7.8 - 7.4 for PAS can be observed of these the acidic zones appear to be split into two to three distinct sharp bands, whereas the alkaline zones are somewhat diffused.

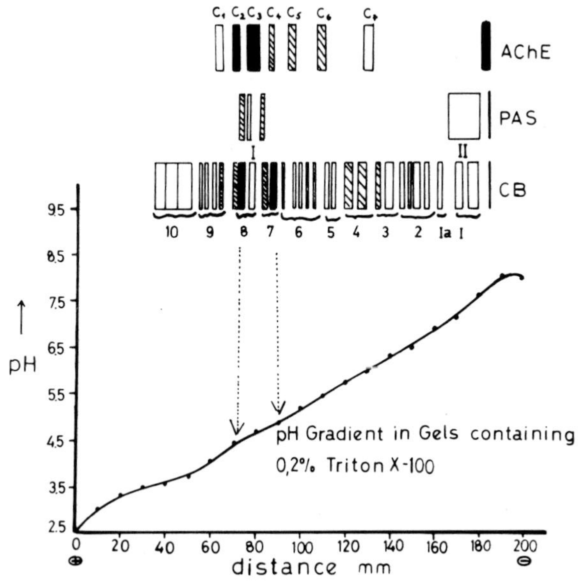

Fig. 4: A typical PAGIF pattern of Triton-X-100 solubilized erythrocyte ghosts from normal Caucasian individual. The gels were stained for protein (CB), glycoprotein (PAS) and acetylcholinesterase (AChE). pH gradient in gel is also shown. The arbitary band Nos. are given.

About four bands of AChE (C_2 - C_5) activity could be seen after four hours staining, of which C_2 and C_3 are the major components. And after overnight staining up to total of 7 bands are seen, unfortunately after such long staining major fast moving components become diffused to form a diffused broad zone. In addition, a sharp zone staining for AChE activity remains at the origin.

Fig.5A shows typical reproductions pointing out two major bands in several normal individuals, whereas in Fig.5B the protein pattern of DMD individuals appear to vary considerably from normal and as well as among themselves. Probably 4 types of variations could be seen.

Type I: Normally occuring bands 7 and 8 are present only in very small amounts whereas the only major band occurs in the region of pI 3.3 - 3.5 (sample a and f).

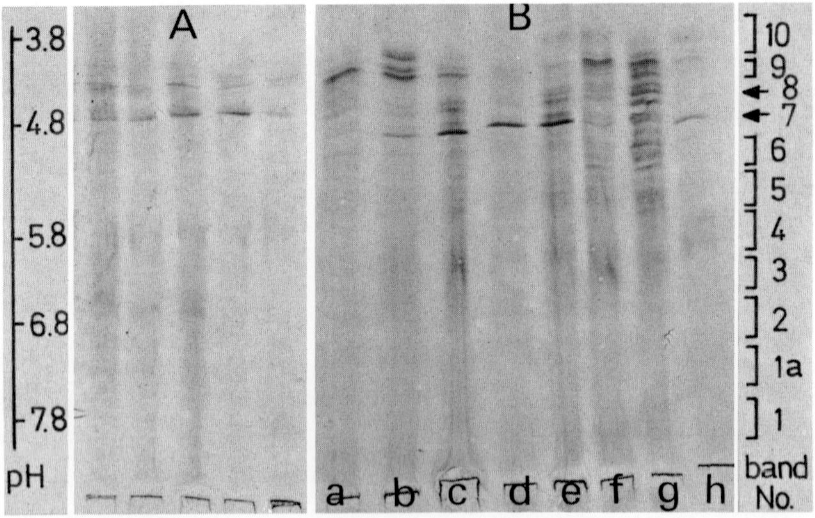

Fig. 5: Protein pattern of Triton X-100 solubilized erythrocyte ghosts after separation on PAGIF. A: 5 normal individuals (the 2 characteristic major bands 7 and 8 are also seen on Fig. 4. B: 8 patients with Duchenne muscular dystrophy (DMD).

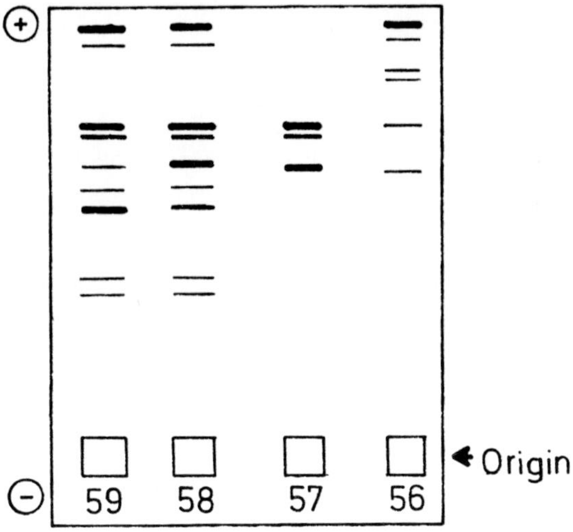

Fig. 6: Protein pattern on PAGIF of erythrocyte ghosts from a DMD patient together with that from carrier mother. No. 56-58 = family; 56 = affected son; 57 = father (normal); 58 = mother; 59 = unrelated Duchenne patient.

Type II: Again bands 7 and 8 are present in small amount and three faster moving zones are distinctly visible in the pI region of 3 - 3.5 (sample b) and one of these band correspond to type I major band.

Type III: Here band 8 was the major band together with a second main band in the area of the type I major band.

Type IV: Even distribution of at least 10 distinct bands in the pI regions between 5.5 and 3.2 (sample g) and hereagain one of these bands correspond to type I, major band.

According to the above classification we have investigated at least 16 Duchenne dystropic patients. Among these 5, 2, 5 and 4 individuals showed the patterns of type I, II, III and IV respectively.

Fig. 6 shows the potential use of this technique in studying membrane polymorphism. It is seen that the affected son belongs to type I pattern and his carrier mother also possess the similar characteristic band together with normally occuring bands.

DISCUSSION

The procedures described here are characterized by the following main features: 1. Treatment of the membrane sample with detergents which achieved 70 - 100 % solubilization of membrane proteins; 2. incorporation of the same detergents into separation media thus providing a constant environment of the solubilizing agent so that reaggregation during electrophoresis can be prevented. It turns out that Triton X-100, a non-ionic detergent which can conveniently be used, because a) they are less apt to alter protein structure than most commonly used SDS and permit recovery of enzymic activity; b) their lack of charge permits separation of protein according to their ionic behaviour within the nuall units of "lipid-protein-carbohydrats complex" by direct electrophoretic and isoelectrofocusing methods.

The above results indicate that plasma membrane can successfully be resolbed into their components by both electrophoretic and isoelectrofocusing methods without disturbing the functional properties like enzymatic activity.

This paper also describes the extension of PAGIF to high degree of resolution of membrane components which this technique affects. Several other workers (21, 22) applied this technique to human erythrocytes. However, these workers used 8 M urea which rendered membrane composition into subunits prior to separation and thus functional membrane activity could not be recovered. The present study shows that by simply using 0.25 % Triton X-100 in the gel without urea or any other denaturing agent equally good resolution can be achieved.

In addition, the extension of the high resolution technique of PAGIF to erythrocyte membrane units represents a possible method of fractionation for the study on membrane polymorphism using either protein, glycoprotein or enzymes as markers. Although the present paper demonstrated the abnormality in protein pattern in a genetic disease like DMD, it remains to be seen whether this technique will find the success in a similar way as the starch gel electrophoresis used in the polymorphic studies of soluble protein. Caution, however, in this respect, should be taken until extensive survey of normally occuring heterogeneity together with formal genetic studies are carried out.

ACKNOWLEDGMENTS

This work was supported by Alexander Humboldt Foundation (Senior Fellowship of P.K.D.) and Sonderforschungsbereich 34 (Endokrinologie).

REFERENCES

1. *Wallach, D.F.H., Winzler, R.J.: Membrane "Macromolecules". In: Evolving Strategies and Tactics in Membrane Research. New York, Berlin, Heidelberg: Springer-Verlag (1974), p 110-116.*
2. *Steck, T.L., Fox, C.F. (ed.): Membrane Proteins in Membrane Molecular Biology. Stamford, Conn. 27: Sinauer Assoc. Inc. (1972).*
3. *Brown, H.D., Chattopadhay, S.K., Patel, A.B.: Erythrocyte Abnormality in Human Myopathy. Science 157, 1577-1578 (1967.*

4. *Das, P.K., Watts, D.C., Cole, H.M.T.: Red Cell Cholinesterase with Altered Properties from Human in Duchenne Dystrophy. Abstr. Commun. Eur. Biochem. Soc. Meeting 7, 307 (1971).*

5. *Das, P.K., Williams, M.H., Watts, D.C.: Probable Use of Red Cell Cholinesterase Activity (RChE) for the Diagnosis of Carriers for Duchenne Dystrophy. Scand. J. Clin. Invest. Suppl. 126, 29, Abstr. 2351 (1972).*

6. *Roses, A.D., Herbstreith, M.H., Appel, S.H.: Membrane Protein Kinase Alteration in Duchenne Muscular Dystrophy. Nature (London) 254, 350-351 (1975).*

7. *Matheson, D.W., Howland, J.L.: Erythrocyte Deformation in Muscular Dystrophy. Science 184, 165-166 (1974).*

8. *Percy, A.K., Miller, M.E.: Reduced Deformability of Erythrocyte Membranes from Patients with Duchenne Muscular Dystrophy. Nature (London) 258, 147-149 (1975).*

9. *Das, P.K.: On Probable Red Cell Cholinesterase Variants and their Significance in Disease. In: Rapoport, S. (ed.): VII. Int. Symp. Struktur und Funktion der Erythrozyten. Abhandlungen der Akademie der Wissenschaften der DDR, Akademie-Verlag (1975), p 643-649.*

10. *Lehmann, H., Liddell, J.: Human Cholinesterase (Pseudocholinesterase). Genetic Variants and their Recognition. Brit. J. Anaesth. 41, 235-244 (1969).*

11. *Harris, H.(ed.): The Principles of Human Biochemical Genetics, North Holland: American Elsevier (1975).*

12. *Marchasi, V.T., Pallade, G.E.: Localisation of Mg-Na-K ATPase on Red Cell Ghost Membrane. J. Cell. Biol. 35, 385-403 (1967).*

13. *Das, P.K., Lo, E.H.M.: Heterogeneity of Human Erythrocyte Acetylcholinesterase. Individual Variation Revealed by DEAE-Cellulose Chromatography and Gel Disc Electrophoresis. Enzyme 21, 304-316 (1976).*

14. *Das, P.K., Watts, R.L., Watts, D.C., Dimelow, E.: Distribution, Specificity and Function of General Esterases, Proteases and Cholinesterases of Various Species of Star Fish. Comp. Biochem. Physiol. 39B, 979-997 (1971).*

15. *Davis, D.J.: Discelectrophoresis II. Method and Application to Human Proteins. Ann. N.Y. Acad. Sci. 121, 404-409 (1964).*

16. *Weber, K., Osborn, M.: The Reliability of Molecular Weight Determinations by Dodecyl Sulfate-Polyacrylamide Gel Electrophoresis. J. Biol. Chem. 244, 4406-4412 (1969).*

17. *Graesslin, D., Weise, H.C., Rick, M.: Preparative Isoelectric Focusing of Proteins on Slab of Polyacrylamide Gel. Anal. Biochem. 71,492-499 (1976).*

18. *Juul, P.: Human Plasma Cholinesterase Isozymes. Clin. Chim. Acta 19, 205-213 (1968).*

19. *Das, P.K.: Further Evidence on the Heterogeneity of "Silent" Serum Cholinesterase Variants. Hum. Hered. 13, 88-96 (1973).*

20. *Fairbanks, G., Steck, T.L., Wallach, D.F.H.: Electrophoretic Analysis of the Major Polypeptides of the Human Erythrocyte Membrane. Biochemistry 10, 2606-2617 (1971).*

21. *Bhakdi, S., Knüfermann, H., Hoelzl Wallach, D.F.: Separation of EDTA-Extractable Erythrocyte Membrane Proteins by Isoelectric Focusing Linked to Electrophoresis in Sodium Dodecyl Sulphate. Biochem. Biophys. Acta 345, 448-457 (1974).*

22. *Litmann, G.W., Litmann, R.T., Merz, D.L.: Analysis of Polypeptide Disposition in Human Erythrocyte Membranes Employing Membrane Inversion. Biochim. Biophys. Acta 394, 348-360 (1975).*

Microheterogeneity of Serum Glycoproteins as Revealed by Flat-Bed Gel Isoelectric Focusing

A. Hamann

ABSTRACT

The microheterogeneity of the serum glycoproteins fetuin and transferrin is investigated by flat bed gel isoelectric focusing and enzymatic treatments. Fetuin shows beside the sialic acid caused heterogeneity charge differences due to partial phosphorylation. In transferrin small fractions with higher sialic acid content as hitherto observed could be detected by isoelectric focusing. A subheterogeneity, probably caused by non-covalent interactions, can be observed in the transferrin bands.

INTRODUCTION

Isoelectric focusing in thin layer gels is outstanding for its resolution power and therefore has been used widely for the analysis of microheterogeneity of proteins. The carbohydrate part of glycoproteins and especially the terminal sialic acid was early recognized to be responsible for a great part of observed heterogeneities in otherwise homogenous glycoproteins, but some other causes were equally reported for various proteins (15). Therefore we applied isoelectric focusing in thin layer gels for the study of a number of serum glycoproteins. In this paper experiments on fetuin, a glycoprotein of fetal calf serum and on human serum transferrin are presented.

Several authors observed earlier by means of starch gel or polyacrylamide gel electrophoresis or by column focusing microheterogeneity in these proteins, which was due to the different content of sialic acid or, for transferrin, additional genetically determined variances in the peptid moiety (1, 5, 6, 7, 13).

Based on chemical analysis, detailed proposals for the structure of the carbohydrate part of fetuin and transferrin were made (4, 9, 11). Our experiments show that by flat bed gel isoelectric focusing a further, hitherto not described heterogeneity can be demonstrated, and, in combination with enzymatic treatment, a more detailed analysis of the molecular diversity can be achieved.

MATERIALS AND METHODS

Transferrin and Vibrio cholerae neuraminidase was purchased from the Behringwerke, fetuin was prepared by cold ethanol fractionation of fetal calf serum (GIBCO) (10).

For neuraminidase treatment the enzyme preparation (0.1 U/ml) was used either undiluted, concentrated by vacuum dialysis or diluted with the appropriate buffer. Enzyme activity was determined with fetuin as substrate (2). The incubation temperature was 37° C. Incubation with 10 U/ml phosphatase (alkaline phosphatase from calf intestine, Boehringer) was performed at 25° C (2 hours) after dialysis of fetuin + enzyme against 50 mM veronal buffer, pH 9.5. Two other phosphatases were used in a similar manner. The treated samples were stored frozen or applied directly without further treatment.

Isoelectric focusing was carried out in 5 % polyacrylamide gels, 0.6 or 1 mm thick, in a flat bed apparatus with carbon electrodes (3) with an additional thermostated cooling plate. 1.5 % Ampholine (LKB), pH range 4-6 + 0.5 % pH 2,5 - 4 was used for fetuin and 2 % pH 5-7 for transferrin. The samples (10 - 50 µg) were soaked in filterpaper pieces and placed on the surface of the prefocused gels, in the middle or near the cathode. After covering with a thin polyethylene sheet, the gels were focused at 4° C for 1 h with 200 V and 3 to 4 hours with 1000 V. Application of the samples at different positions in the gradient was used to prove equilibrium conditions.

For staining, Coomassie blue in a mixture described recently (8), but with higher alcohol content (48 %), was used. High alcohol concentrations during the entire staining procedure were necessary to prevent bleeding out of the highly glycosylated proteins. For preparative isolation, the visible bands of transferrin were cut out and electrophoretically eluted. pH determinations were carried out at 4° C in gel pieces of 0.5 x 2 cm, eluted with 200 µl 10 mM KCl.

RESULTS AND DISCUSSION

As shown in Fig. 1, fetuin shows about 10 bands with an isoelectric point between 4.0 and 4.5, and the pattern is virtually the same in the purified material and in fetal calf serum.

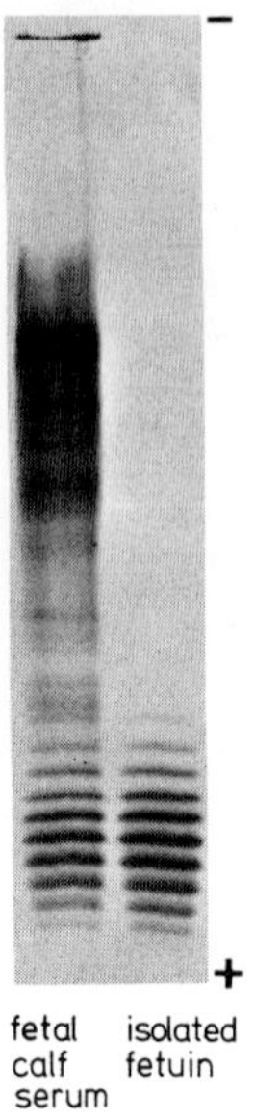

Fig. 1. Microheterogeneity of Fetuin.

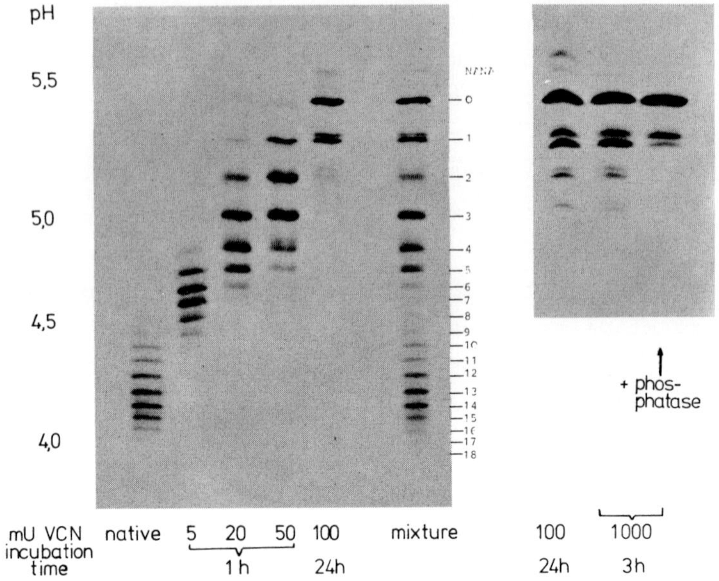

Fig. 2. Influence of Neuraminidase and Phosphatase on the Microheterogeneity of Fetuin.

Treatment with neuraminidase from Vibrio cholerae (Fig. 2) reveals that this heterogeneity is due to a differing number of sialic acid residues per molecule. Upon incubation with increasing concentrations of neuraminidase, the pattern shifts stepwise to more basic positions in the gradient. The stepwise enzymatical cleavage of sialic acids makes it possible to determine the actual number of sialic acid residues per protein molecule simply by isoelectric focusing. The value of 12 to 15 sialic acid residues in the most abundant fractions determined by isoelectric focusing agrees well with the average content of 13 to 14 residues determined chemically by several authors (5, 10). Detailed proposals for the structure and number of carbohydrate sequences in fetuin were published based on the average values for carbohydrate content (11, 12). However, the results presented with isoelectric focusing indicate that there is the possibility of a higher glycosylation as suggested in the models, at least with regard to sialic acid.

As can be seen from Fig. 2, material which is extensively treated with neuraminidase and which contains virtually no more sialic acid as revealed by colorimetric determination, still shows a heterogeneity. The material with a higher pI than the main fraction seems to be an artifact caused by the long incubation time. It disappears if higher enzyme concentrations together with shorter incubation times are used. However, beside the main fraction there remains a considerable part of material being more negatively charged. Treatment with high concentrations of phosphatases led to the partial disappearance of these bands (Fig. 2), while extensive dialysis against high salt concentrations, extraction with organic solvents, treatment with phospholipase A, RNase or DNase had no effect. This shows, that obviously covalently bound phosphate is present in a part of the fetuin molecules and is responsible for the residual heterogeneity. This result is confirmed by chemical analysis. In contrast to previous investigations (10), we could detect about 0.5 Mol. phosphorus per mol fetuin, 50 % can be cleaved by unspecific phosphatases.

In our experiments with transferrin, we were confronted with initial difficulties to obtain sharp bands with the ironfree material, perhaps due to a reversible binding of trace amounts of iron. Upon incorporation of 1 mM ferric chloride into the gel, sharp pictures of the iron-saturated transferrin are obtained. Native transferrin shows one main band, with an isoelectric point of 5.56 and can be shifted in 4 steps by neuraminidase to the band representing sialic acid free transferrin (Fig. 3).

Beside the main band, small fractions are present in the commercially obtained transferrin, which have either a lower or a higher sialic acid content. The fact that the hitherto not described fractions with a lower pI compared to the pre-

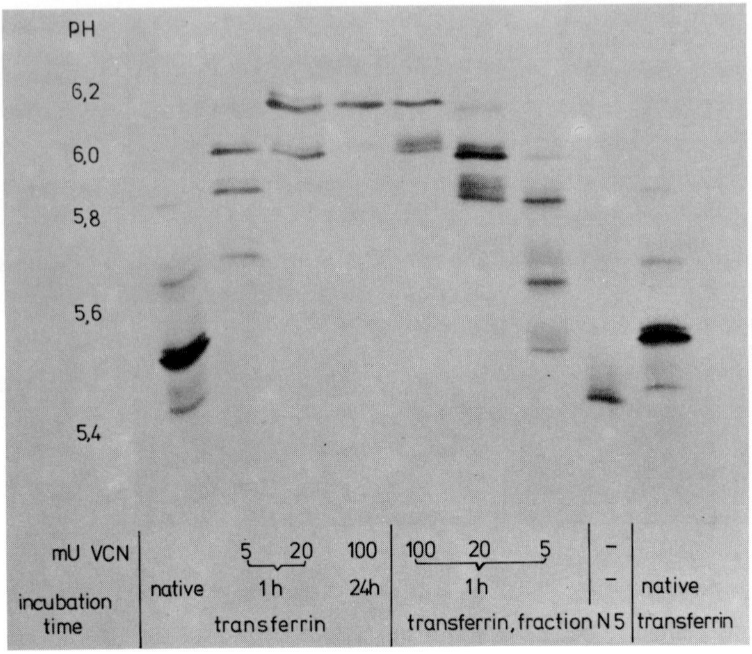

Fig. 3. Influence of Neuraminidase on Microheterogeneity of Transferrin.

dominant material really contain more sialic acid is seen by isolation and neuraminidase treatment. The same products are obtained from this fraction with 5 sialic acid residues as from the main fraction (Fig. 3). This shows that transferrin can obviously contain more than 2 carbohydrate units with maximally 2 sialic acid residues per unit, as was previously suggested based on chemical analysis (4, 9).

In addition to the heterogeneity caused by the sialic acid, transferrin shows a further heterogeneity in isoelectric focusing. Each main band is composed of up to three narrow spaced bands, lying about 0.02 pH units apart (Figs. 3 and 4). This subheterogeneity is not removed by the presence of mercaptoethanol nor seems to be due to differences in the position of the sialic acid, since it is also present in sialic acid free transferrin (Fig. 3). Two observations seem to signify that the subheterogeneity is not caused by variances in the primary structure: 1) the percentual distribution between the bands can obviously vary under certain conditions (Fig. 4). 2) Preparatively isolated transferrin shows a slightly altered pI, compared to the original bands (Fig. 3). Small conformational differences, binding of ampholytes or differences in the anions, which are bound together with iron (7, 13) may account for the subheterogeneity observed.

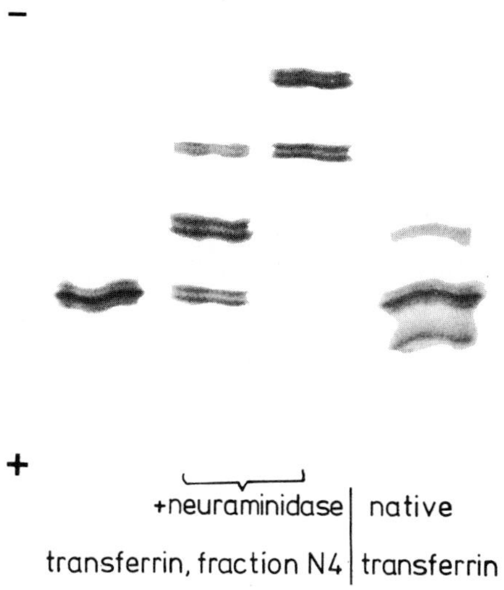

Fig. 4. Subheterogeneity of Transferrin

The results presented demonstrate that isoelectric focusing in flat bed gels, combined with enzymatic treatment, offers more insight into the microheterogeneity of proteins than the formerly used methods starch gel or polyacrylamide gel electrophoresis or column focusing.

Certain discrepancies were found in the determination of isoionic points. Our values for fetuin and transferrin are about 0.2 to 0.4 pH units higher than those measured by means of column focusing (1, 6, 14). Other authors (5, 10) measured the isoionic point of fetuin directly and obtained a value of 0.1 to 0.2 pH units lower than our values, unless they added salt to the protein solution. Since we have, as far as it is possible, proved the existence of equilibrium conditions, it remains to be proved, whether the differences are caused by measurement at different temperatures, influence of the sucrose used in column focusing or by additional factors.

ACKNOWLEDGEMENTS

This investigation was supported by the Deutsche Forschungsgemeinschaft.

REFERENCES

1. Begbie, R.: Studies on fetuin from foetal bovine serum. The composition and amino acid sequences of glycopeptides from fetuin. Biochim. Biophys. Acta 371, 549-567 (1974).

2. Drzeniek, R.: Viral and bacterial neuraminidases. Curr. Topics Microbiol. Immunol. 59, 35-74 (1972).

3. Graesslin, D., Weise, H.C. and Rick, M.: Preparative isoelectric focusing of proteins on slabs of polyacrylamide gels. Anal. Biochem. 71, 492-499 (1976).

4. Jamieson, G.A., Jett, M. and DeBernardo, S.L.: The carbohydrate sequence of the glycopeptide chains of human transferrin. J. Biol. Chem. 246, 3686-3693 (1971).

5. Oshiro, Y. and Eylar, E.H.: Physical and chemical studies on glycoproteins. III. The microheterogeneity of fetuin, a fetal calf serum glycoprotein. Arch. Biochem. Biophys. 127, 476-489 (1968).

6. Pagé, M.: Demonstration of the microheterogeneity of fetuin by electrofocusing. Biochim. Biophys. Acta 236, 571-577 (1971).

7. Putnam, F.W.: Transferrin. In: The Plasma Proteins, Vol. I, p. 265-316. Ed. F.W. Putnam. New York, San Francisco, London (1975).

8. Righetti, P.G. and Drysdale, J.W.: Isoelectric focusing in gels. J. Chromatogr. 98, 271-321 (1974).

9. Spik, G., Bayard, B., Fournet, B., Strecker, G., Bouquelet, S. and Montreuil, J.: Studies on glycoconjugates. LXIV. Complete structure of two carbohydrate units of human serotransferrin. FEBS-Letters 50, 296-299 (1975).

10. Spiro, R.G.: Studies on fetuin, a glycoprotein of fetal calf serum. J. Biol. Chem. 235, 2860-2869 (1960).

11. Spiro, R.G.: Periodate oxidation of the glycoprotein fetuin. J. Biol. Chem. 239, 567-573 (1964).

12. Spiro, R.G. and Bhoyroo, V.D.: Structure of the O-glycosidically linked carbohydrate units of fetuin. J. Biol. Chem. 249, 5704-5717 (1974).

13. Sutton, H.E. and Jamieson, G.A.: Transferrins, Haptoglobin and Ceruloplasmin. In: Glycoproteins, Part A. p. 653-698, Ed. A. Gottschalk. Amsterdam, London, New York (1972).

14. Wenn, R.V. and Williams, J.: The isoelectric fractionation of hen's-egg ovotransferrin. Biochem. J. 108, 69-74 (1968).

15. Williamson, A.R., Salaman, M.R. and Kreth, H.W.: Microheterogeneity and allomorphism of proteins. Ann. N.Y. Acad. Sci. 209, 210-224 (1973).

Diagnosis of Cystic Fibrosis (CF) – Purification and Distinction Between CF Protein and Ciliary Dyskinesia Activity in CF and Asthmatic Sera Using Isoelectric Focusing and the Rabbit Tracheal Ciliary Bioassay

G.B. Wilson, M.T. Monsher, P. Arnaud and H.H. Fudenberg

ABSTRACT

A cystic fibrosis protein (CFP) can be detected in whole sera from CF homozygotes and heterozygotes by isoelectric focusing (IEF). The detailed methodology using the LKB multiphor is described. Studies using whole sera indicated that CFP was correlated to a ciliary dyskinesia activity (CDA) found in individuals with the CF gene, and that several CDA's may exist in human serum. Chromatography, coupled with IEF, and disc gel electrophoresis allow for the partial purification and distinction between a CF-specific CDA found associated with CFP and a nonspecific CDA found in asthmatic sera.

INTRODUCTION

Previous studies have indicated that a CFP can be detected in sera from CF patients and heterozygote carriers of CF (HCF) by IEF. A standardized assay for the detection of CFP has been reported and although CFP was shown to be specific for the CF gene, its possible role in the pathophysiology of CF is unknown (9, 11, 12). We have recently perfected a modified version of the rabbit tracheal bioassay for the detection of CF ciliary dys-

+ *The research was supported in part by NTH Grant HDO 9938-02 and by the South Carolina State Appropriations for Research HR 524, A 705. This is publication 43 from the Department of Basic and Clinical Immunology and Microbiology of the Medical University of South Carolina. Presented in part at the 17th Annual Meeting of the Cystic Fibrosis Club, April 27, 1976, in St. Louis, Missouri. We thank Susan Berry and Leigh Williams for technical and typing assistance.*

kinesia factor (CDF). Studies using whole sera indicated complete concordance between CFP positivity and CDF activity in all sera tested except for asthmatic sera (patient controls, CDF+, CFP-). This finding suggested that CFP may be related to a CF-specific-CDF, and the possibility that the ciliary dyskinesia activity (CDA) found in patients with asthma differs from a CF-specific-CDF.

The present report describes the detailed methodology for detection of CFP in sera using the LKB multiphor as compared to the apparatus of Awdeh et al. (1), employed in previous studies. In addition, CFP and the CDA in sera from CF patients, HCF and asthmatics have been partially purified and demonstrated by several procedures to be distinct.

MATERIALS AND METHODS

Collection and Processings of Samples

Serum samples obtained from clotted venous blood, were collected and processed for electrofocusing and bioassay exactly as described previously (11, 12, 15).

Isoelectric Focusing

Isoelectric focusing in thin-layer polyacrylamide gels (IEF) (pH 5 - 10 gradient) was performed using the apparatus of Awdeh et al. (1) exactly as described previously (11, 12). Isoelectric focusing in disc gels (pH 3.5 - 10 gradient) was performed by the method of Eder (4) as described previously (9) with modifications: (A) Gels of 10 % acrylamide were used and N,N-methylene-bisacrylamide was replaced by N,N-diallyltartardiamide (DATD) at an acrylamide was replaced by N,N-diallyltartardiamide (DATD) at an acrylamide to DATD ratio of 1 : 10. (B) The anode and cathode solutions were 1 % phosphoric acid and 1 % TEMED to conform to the solutions used in thin-layer IEF (11, 12). Isoelectric focusing employing the LKB multiphor was performed using electrode solutions and a gel composition identical to those previously described except that a gradient of pH 2.5 - 10 was used. Three percent (W/V) ampholytes were used, and the gel was prefocused at 350 V (constant voltage) for 20 min. Samples were applied onto

6 x 8 mm pads of Whatman no. 3 MM chromatography paper placed with the lower edge of the pad 5 mm from the anodal strip. The gel was focused at 35 watts (constant power) for 1.5 hrs and the pads were not removed. In preliminary experiments, using the LKB multiphor, an acidic pH range of less than pH 5 was found to be essential for the demonstration of CFP (11) and a constant power source was necessary to obtain adequate resolution of the protein bands between pH 8 - 9. Disc gel and thin-layer electrofocusing (LKB multiphor) and disc gel electrophoresis were performed at 4° C.

Electrophoresis

Disc gel electrophoresis was performed in 15 % polyacrylamide gels with minor modifications of the procedure of Reisfeld et al. (7) employing ß-alanine-acetic acid buffer, pH 4.5. DATD was used for a cross-linker, and samples were electrophoresed for 1 hr at 6 mA/gel.

Ion Exchange Chromatography and Gel Filtration

DEAE-cellulose was performed using either .01 M phosphate (pH 8.0) or .005 m Tris-HCl (pH 8.6) and a discontinuous step gradient of NaCl (0.00 M, 0.05 M, 0.1 M, 0.2 M) as described previously (9) except that EDTA was not used. Gel filtration on Sephadex G 200 was performed as described (9) using .01 M phosphate, 0.15 M NaCl buffer, pH 8.0 (PBS, pH 8.0) for elution. Gel filtration on Bio Gel P 10 was performed using a 1 x 10 cm column at a flow rate of 5 ml per hr. Fractions of 2.0 ml were collected and PBS, pH 8.0 was used for elution. Sephadex G 200 and Bio Gel P 10 columns were calibrated with markers of known molecular weight as indicated. All column fractionations were performed at 4° C using siliconized glass columns and plastic tubes. Fractions were pooled, dialyzed using Spectrapor 3 dialysis membrane tubing (MW retention of 3500; Spectrum Medical Industries, Los Angeles CA) to equilibrium against 0.0025 M phosphate, pH 7.2, and lyophilized. Fractions were reconstituted in distilled water and redialyzed if found to contain excessive electrolytes.

For EDTA treatment, whole sera or fractions were mixed with solid EDTA (Fisher) to saturation (pH 3.7) and left at room temperature for 15 - 20 min. The protein concentration in different fractions was estimated by reading the absorbance at 280 nm and exact protein determinations were made using the procedure of Lowry et al. (6).

Bioassay

Determinations of ciliary dyskinesia activity in whole sera and column fractions were performed as described previously for whole sera (15). All fractions (in 0.01 M phosphate pH 7.2) were mixed 1 : 1 (v/v) with incomplete medium and warmed to 37^{0} C just prior to bioassay (15).

RESULTS AND DISCUSSION

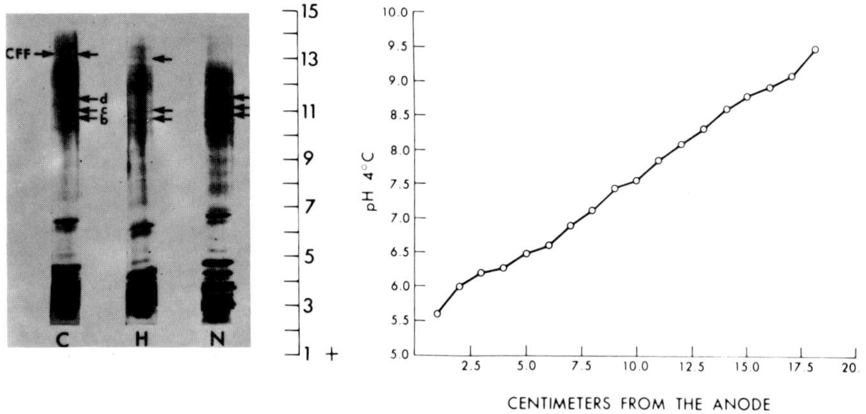

Fig. 1: Left: Results of electrofocusing sera using apparatus of Awdeh et al. (1), pH 5 - 10, 4 M urea. Scale: CM from anode. C: CF homozygote; H: obligate heterozygote carrier for CF. N: Normal individual. Right: pH gradient as measured on surface of gel at 4^{0} C.

Detection of CFP by Electrofocusing

Typical patterns obtained using the apparatus of Awdeh et al. (1) to electrofocus whole sera are shown in Fig. 1. Also shown is the pH gradient as recorded on the surface of the gel at 4^{0} C (12). CFP has a calculated pI value of 8.46 ± 0.05 and focuses at 13.3 ± 0.2 cm from the anode as determined from the analysis of over 300 positive samples (9, 12). Bands B, C, D, used to distinguish between CF homozygote and heterozygote serum samples (9, 12) are also shown (Fig. 1). Prolonged exposure of the serum samples to an acidic pH and 4 M urea are considered to favor the demonstration of CFP, possibly due to the dissolution of an IgG-CFP complex (12) or the alteration of another precursor which may be liberating CFP (11).

The Awdeh apparatus requires 18 - 24 hrs to focus (11, 12), however the LKB multiphor requires less than 3 hrs. To fulfill the possible requirement for acidic hydrolysis, it was found necessary to use a pH 2.5 - 10 gradient and to apply the sample 5 mm from the anodal strip. Typical patterns obtained using the LKB multiphor are shown in Fig. 2 as is the pH

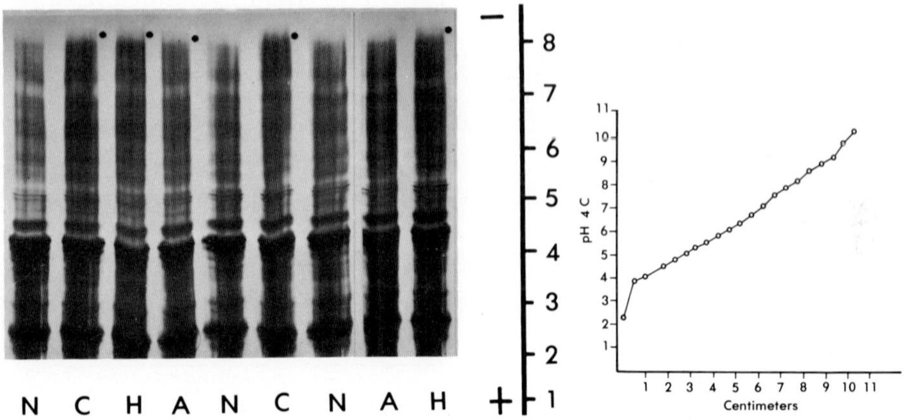

Fig. 2: Left: Results of electrofocusing sera using LKB multiphor, pH 2.5 - 10, 4 M urea. Scale: CM from bottom of sample pad (5 mm from anode), as in all LKB IEF Figures. Right: pH gradient at 4^{0} C, CM from anode.

gradient recorded at 4^{0} C. The location of a band doublet representing CFP is indicated with a pI of about 8.5 - 8.6. The serum protein patterns were found to extend to a maximum of pH 9.0. Since an extended gradient was used, it was found necessary to prefocus the gel at 350 V to generate the acid section of the gradient and focus at 35 W using constant power to increase the resolution. analysis of approximately 60 samples (roughly equal numbers of CF, heterozygote, and normal) using the LKB multiphor and our standardized assay conditions for CFP detection (e.g. samples contain 300 µg IgG) (9, 12) indicated complete concordance between the results using the LKB multiphor and the apparatus of Awdeh et al. (1). Problems in detection occurred primarily when greater than 50 µl of sera were required for analysis.

Table 1: *CDA and CFP Positivity in Human Sera*

Serum Type, Number	*CFP*[1]	*Onset of Ciliary Dyskinesia (min)*[2]
CF Homozygote		
(16)	+	*14 $\pm$ 8*
Obligate Heterozygote		
(15)	+	*25 $\pm$ 8*
Sibling of CF Patient		
(3)	-	*NR*[3]
4)	+	*19, 24, 29, 35*
Normal		
(13)	-	*NR*
(1)	+	*20*
Asthma		
(7)	-	*22 + 9*
Asthma		
(1)	+	*15*

1. Mean $\pm$ SD; 2. CFP: +, present; -, absent;
3. NR: No Reaction at 60 Min.

Table 1 showing time of onset of dyskinesia versus CFP positivity in whole sera.

CDA and CFP in Whole Sera

Comparative analyses of whole sera for CFP and CDA positively indicated complete agreement between the presence or absence of these two markers (16) for CF in all sera tested except for 7 or 8 patient controls with bronchial asthma (CDA+ and CFP-) (Table 1). The sequence of events which occur upon application of a CDA (or CFP) positive sample are shown in Fig. 3. The increased frequency of CFP + samples (carriers) in our asthmatic population (13 % vs 5 % in the Caucasian population) may be related to the observation of Warner et al. (10) that carriers of CF show an increased prevalence of allergic conditions and may suggest a genetic predisposition for allergy in carriers of CF, possibly related to a metabolic regulatory abnormality in the complement or kinin cascades (2, 10, 14).

The results of our comparative analyses of CFP and CDA indicated that CFP and a CF-specific CDF may be identical or closely related markers, and unlike the implication of Conover et al. (3), suggested that the activity detected by the rabbit tracheal bioassay in sera from patients with asthma is probably a substance different from CF-specific CDF. To test our hypothesis we purified CFP and the CDA's in CF and asthmatic sera.

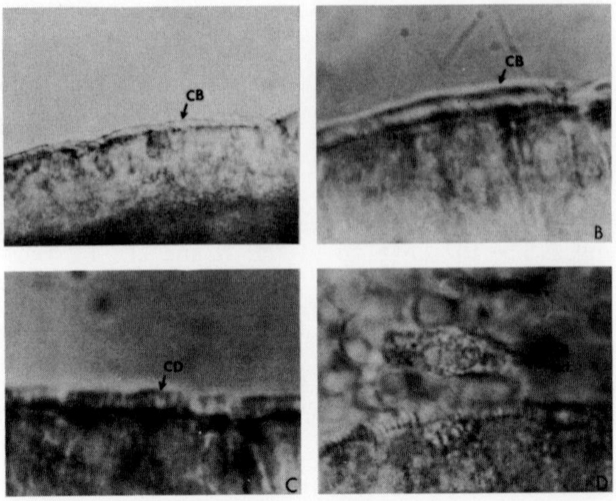

Fig. 3: Affect of CFP (CDA) positive serum on rabbit tracheal epithelium. A. Time equals 5 min (400 x) CB: ciliated boundry. B. Higher magnification of A (1000 x) C. Ciliary dyskinesis (CD), (1000 x). Time equals 15 min. D. Destruction of tissue. Time equals 35 min. cell ext.: Cell extrusion (1000 x).

Distinction Between CFP, CF-CDA and Asthmatic CDA

Fractionation of sera by DEAE-cellulose chromatography demonstrate that CFP was found only in the IgG containing fraction (F) (Table 2, Fig. 4). When phosphate buffer was used, the CDA's in CF and asthmatic sera were found in F I with CFP. However, when TRIS-HCl was used, CFP and the CF-CDA were separated from the asthmatic CDA; the latter being found in F II. The asthmatic-heterozygote carrier presented two activities, one with CFP in F I, the other in F II. The use of TRIS-HCl buffer thus afforded 1 method of distinguishing between CF-CDA and asthmatic CDA and suggested a close association of CFP with CF-CDA. Also, since the method clearly separates bands B, C, D from CFP, (F II vs F I) and CF DEAE-TRIS-HCl F II is CDA negative; we can infer that bands B, C, D do not have dyskinesia activity.

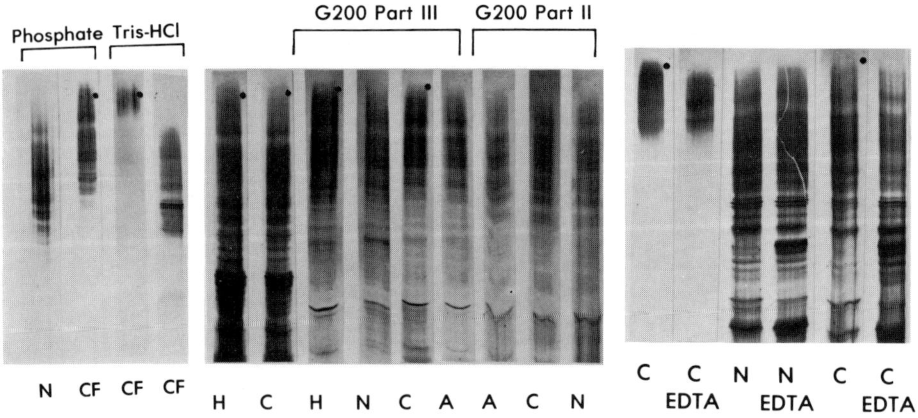

Fig. 4: Results of electrofocusing whole sera and column fractions (F) on LKB multiphor, pH 2.5 - 10. Left to right: N and CF: F1 from DEAE-cellulose, phosphate buffer. CF, FI and FII from DEAE-cellulose TRIS-HCl buffer H, C: whole sera. H, N, C, A: Sephadex G 200 FIII 40 µl (1 x). A, C. N: Sephadex G 200 FII 40 µl (1 x). CF DEAE-TRIS-HCl FI before (C) and after EDTA treatment (C EDTA). N, N-EDTA, C, C-EDTA: N or C G 200 F-III from sera pretreated with EDTA 40 µl (1 x). All IgG preps and whole sera run at levels of 300 µg IgG. Dot marks CFP doublet (lower one is CFP).

Table 2: *DEAE-Cellulose Chromatography of Human Serum*

Serum Type [1]	*Phosphate 0.01M, pH 8.0* Fraction				*TRIS-HCl, 0.005M, pH 8.6* Fraction			
	I 0.0	*II* 0.05	*III* 0.1	*IV* [2] 0.2	*I* 0.0	*II* 0.05	*III* 0.1	*IV* 0.2
Normal (CPF-, CDA-)	-	-	-	-	-	-	-	-
CF Homozygote (CFP+, CDA+)	*CFP/ CDA*	-	-	-	*CFP/ CDA*	-	-	-
Obligate Heterozygote carrier (CFP+, CDA+)	*CFP/ CDA*	-	-	-	*CFP/ CDA*	-	-	-
Asthmatic (CFP-, CDA+)	*CDA*	-	-	-	-	*CDA*	-	-
Asthmatic-heterozygote carrier	*CPF/ CDA*	-	-	-	*CFP/ CDA*	*CDA*	-	-
IgG	+	-	-	-	+	+	-	-

[1]Sera and fractions are indicated as being cystic fibrosis protein (CFP) positive as determined by electrofocusing, and as having ciliary dyskinesia activity (CDA) as determined by the rabbit tracheal bioassay. Sera and

fractions noted as CDA positive produced ciliary dyskinesia by 30 minutes.
[2]Fractions obtained from DEAE-cellulose column eluted with buffer and molariry of NaCl indicated. Presence of IgG in a fraction is indicated as + or -.

Fractionation of all serum types by gel filtration on Sephadex G 200 provided an additional method to partially purify and distinguish between the asthmatic and cystic CDA's. A typical chromatogram obtained by fractionating 2 ml of serum is shown in Fig. 5. Bioassay of all fractions (reconstituted 1 - 2 x) showed that the CDA in CF sera was contained in F III and IV (weak CDA) and was distinct from asthmatic CDA which was found totally in F II (Table 3). The asthmatic-heterozygote serum showed activity in F II - IV, while the normal control sera tested showed activity in no peak. Analysis of F I - V by electrofocusing indicated that CFP was found only in F III, (Fig. 4).

Table 3:

Serum Type	*CDA and CFP Positivity*					
	I	*II*	*III*	*IV*	*V*	*VI*
Normal	-	-	-	-	-	-
Normal-EDTA[1]	-	-	-	-	-	*CDA*
CF Homozygote	-	-	*CFP/ CDA*	*CDA*	-	-
CF Homozygote-EDTA	-	-	-	-	-	*CFP/ CDA*
Heterozygote Carrier	-	-	*CPF/ CDA*	*CDA*	-	-
Heterozygote Carrier-EDTA	-	-	-	-	-	*CFP/ CDA*
Asthmatic	-	*CDA*	-	-	-	-
Asthmatic-EDTA	-	-	-	-	-	*CDA*
Asthmatic-Heterozygote	-	*CDA*	*CFP/ CDA*	*CDA*	-	-
Asthmatic-Heterozygote-EDTA	-	-	-	-	-	*CFP/ CDA*

[1]*Serum pretreated by mixing with solid EDTA to saturation, final pH: 3.7.*

Table 3: Results of bioassay and electrofocusing of G 200 serum fractions before and after treatment with EDTA.

Treatment with EDTA prior to gel filtration routinely eliminated all CDA activity in F II - IV and generated a CDA found exclusively in F VI (MW: 1100 - 13 700) in all sera (Fig. 5). Analysis of F VI (1 mg/ml) (with or without treatment) by immunoelectrophoresis, against antiserum to whole human serum failed to show any reaction, however, analysis of F VI (25 ug) by disc gel electrophoresis, showed that presence of several cathodal bands and indicated obvious differences between the serum types (Fig. 6).

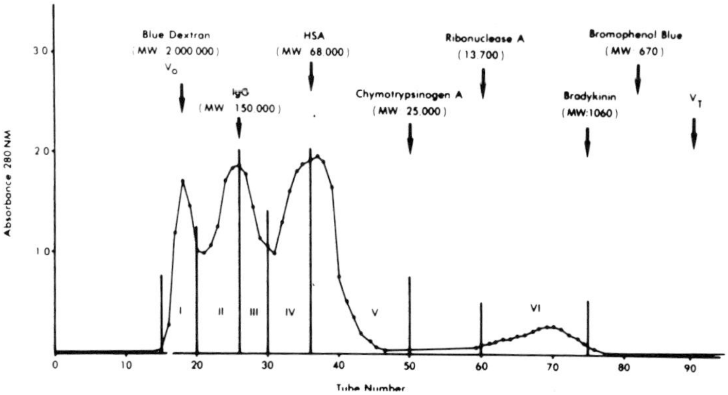

Fig. 5: Typical chromatogram of 2.0 ml whole serum on Sephadex G 200. HSA: Human serum albumin V_0; V_T; Void volume and total column volume respectively, for details see Ref. 9.

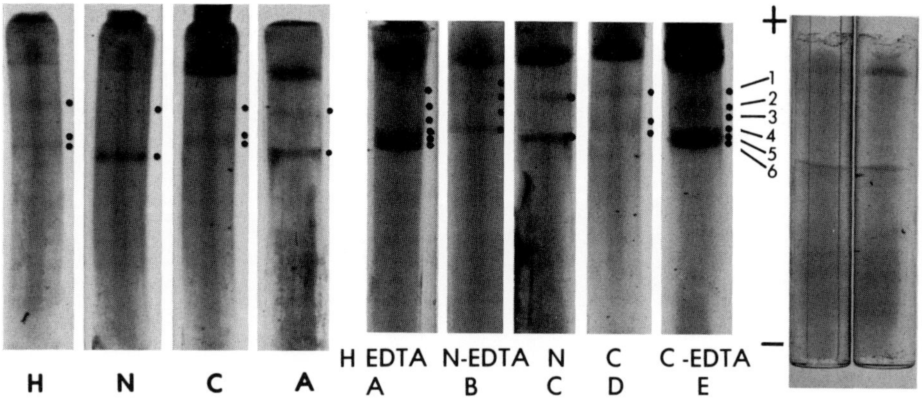

Fig. 6: Above left and center; results of disc gel electrophoresis, pH 4.5, acetic acid- alanine buffer. Left: H,N,C,A, G 200 F-VI. Center: A-E G 200 F-VI from untreated and EDTA treated sera. Dots mark cathodal migrating bands of interest. Far Right: Fig. 7: Results of disc gel electrofocusing, of G 200 F-VI from N (left) or CF EDTA treated sera, cathode below.

In particular, an extra cathodal band was noted in untreated CF and carrier samples and two additional cathodally migrating bands noted in EDTA treated CD and carrier samples, as compared to normal samples. Electrofocusing of F I - V from EDTA treated sera indicated that CFP was not found in any fraction (Fig. 4), showed that CFP was not generated in normal sera and that in lieu of the disc gel results, that CFP was shifted from F III to F VI in CF sera. Apparently, dissolution of a CFP-IgG (or other precursor complex) by acid treatment had occurred (9). Partial direct confirmation of the movement of CFP to F VI was obtained by disc gel IEF. As shown in Fig. 7, banding differences are evident when equal amounts (25 µg) of F VI from EDTA normal and CF sera are analyzed. Bands near pH 8.5 are evident in the CF sample. Disc gel IEF and electrophoresis add evidence for the association of CFP with the activity in F VI from CF or carrier EDTA treated sera. However, these data do not serve to explain the generation of activity in normal sera or the movement of the activity from F II to VI in asthmatic sera.

Table 4: *P10 Gel Filtration of Active Fractions*

Sample	*Fraction* A	*I*	*II*	*III*	*IV*	*V*	*VI*
CF G 200-EDTA P6	*NP*[1]	*CDA*	-	*CDA*	*CDA*	*CDA*	-
N G 200-EDTA P6	*NP*	*CDA* (±)	-	-	-	*CDA*	-
H G 200-EDTA P6	*NP*	*CDA*	-	*CDA*	*CDA*	*CDA*	-
A G 200 P2	*CDA*	-	-	-	-	-	-
A G 200 P2+EDTA	-	*CDA*	-	-	-	-	-
C G 200 P3, P4	*CDA*	-	-	-	-	-	-
CF G 200 P3, P4+EDTA	-	*CDA*	-	*CDA*	*CDA*	-	-
CF IgG	*CDA*	-	-	*CDA* (±)	-	-	-
CF IgG + EDTA	-	*CDA*	-	*CDA*	*CDA*	-	-

1. NP: no peak; CDA (±): weak activity

Table 4: Results of bioassay for CDA activity of fractions from Bio Gel P10. CDA positive: active at 30 min.

Chromatography of G 200 F-VI- (EDTA) samples (from 2 ml serum) on Bio Gel P10 routinely resulted in a chromatogram as shown in Table 4. Bioassay of the fractions indicated the existence of 3 separate activities (peak I, III - IV, V, Fig. 8), in CF and heterozygote F-VI but only two in normal. The activity in F III - IV is felt to represent a unique CF-CDA and has a MW of around 5000. Chromatography of CF IgG (TRIS-HCl F-I) and CF G 200 F III - IV with and without EDTA treatment, indicated that the activity in P10 F III - IV is derived from an IgG complex or precusor found in G 200 F III - IV. The activity in P10 F I (MW 9 000 - 10 000) apparently is created upon EDTA treatment since it is found in normal G 200 F-VI-EDTA and derived from asthma G 200 F-II. However, the amount of protein in P10-FI is approximately 5 - 10 times less in normal as compared to CF or carrier G 200 F-VI-EDTA. Preliminary experiments using monospecific antiserum to human C3a, indicate that the activity in P10-F-I is C3a (anaphylatoxin). Apparently, EDTA treatment at room temperature serves to strip CF-CDA from IgG and generates C3a from C3 (found in G 200 F-II) in a manner similar to treatment of serum with E-amino caproic acid (5). From the results obtained with CF IgG, apparently both a C3a derivative and a CF-specific CDA are weakly ionically complexed to IgG. Removal of CFP by EDTA treatment of CF-IgG is clearly shown in Fig. 4. Bulk purification and further characterization of the activities in P10 FI, III - IV and V is currently in progress. At present, it would appear that either quantification of possiblyC3a levels or definitely of the component in P10-F III - IV, could serve as a new diagnostic test for CF.

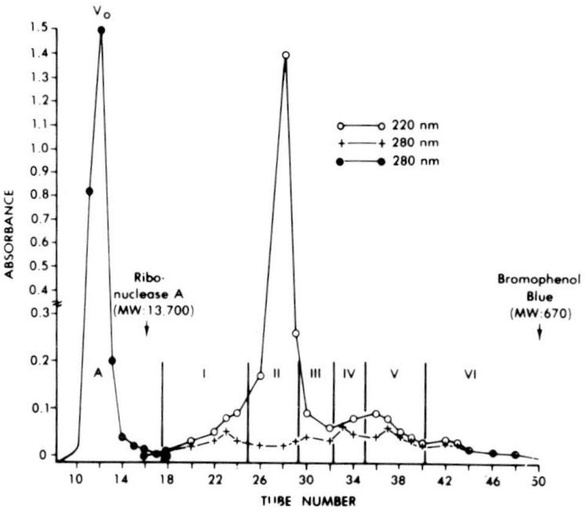

Fig. 8: Chromatogram for elution of G 200 or DEAE TRIS-HCl fract. on Bio Gel P10 with or without prior EDTA treatment or from sera pretreated with EDTA (g 200-EDTA). A: Void vol. peak found in fract. other than G 200 P6.

The finding that one CDA activity is C3a suggests a defect in an enzyme or enzyme inhibitor found associated with the complement and kinin generating systems since the amount of normal P10-FI is much less than CF or carrier P10-FI. The mechanism responsible for increased C3a is unknown, however it may be related to a defect in a_2-macroglobulin as proposed by Wilson and Fudenberg (10) rather than carboxypeptidase B as proposed by Conover et al. (3). Alternatively, a structural mutation in C3a (in CF) leading to its decreased catabolism and increased binding affinity for IgG could be operable. Assuming one of the above mechanisms explains increased C3a levels, then we speculate that the activity in P10-F III - IV (CF-specific CDF) is generated through a related defect in metabolism (10).

REFERENCES

1. Awdeh, Z.L., Williamson, A.R., Askonas, B.A.: Isoelectric Focusing in Polyacrylamide Gel and its Application to Immunoglobulins. Nature (London) 219, 66-67 (1968).
2. Conover, J.H., Conod, E.J., Hirschhorn, K.: Ciliary Dyskinesia Factor in Immunological and Pulmonary Disease. Lancet p. 1194 (1973/I).
3. Conover, J.H., Conod, E.J., Hirschhorn, K.: Studies on Ciliary Dyskinesia Factor in Cystic Fibrosis, IV. Life Sci 14, 253-266 (1974).
4. Eder, J.: Isoelectric Focusing of Antibodies in Polyacrylamide Gels. J. Immunol. Meth. 2, 67-74 (1972).
5. Hugli, T.E., Vallota, E.H., Muller-Eberhard, H.J.: Purification and Partial Characterization of Human and Porcine C3a Anaphylatoxin. J. Biol. Chem. 250, 1472-1478 (1975).
6. Lowry, O.H., Rosenbrough, N.J., Farr, A.C., Randall, R.J.: Protein Measurement with the Folin Phenol Reagent. J. Biol. Chem. 193, 265-271 (1951).
7. Reisfeld, R.A., Lewis, V.J., Williams, D.A.: Disk Electrophoresis of Basic Proteins and Peptides on Polyacrylamide Gels. Nature (London) 195, 281-283 (1962).
8. Warner, J.V., Norman, A.P., Soothill, J.F.: Cystic Fibrosis Heterozygosity in the Pathogenesis of Allergy. Lancet p. 990-991 (1976/I).
9. Wilson, G.B., Fudenberg, H.H.: Further Purification and Characterization of Serum Proteins Used to Detect Cystic Fibrosis Genotypes by Isoelectric Focusing. Tex. Rep. Biol. Med. 34, in press (1976).
10. Wilson, G.B., Fudenberg, H.H.: Studies on Cystic Fibrosis Using Isoelectric Focusing. II. Pediat. Res. 10, 87-96 (1976).
11. Wilson, G.B., Arnaud, P., Monsher, M.T., Fudenberg, H.H.: Detection of Cystic Fibrosis Protein by Electrofocusing. Pediat. Res. in press (1976).

12. Wilson, G.B., Fudenberg, H.H., Jahn, T.L.: Studies on Cystic Fibrosis Using Isoelectric Focusing. I. Pediat. Res. 9, 635-640 (1975).

13. Wilson, G.B., Jahn, T.L., Fonseca, J.R.: Demonstration of Serum Protein Differences in Cystic Fibrosis by Isoelectric Focusing in Thin-Layer Polyacrylamide Gels. Clin. Chim. Acta 49, 79-91 (1973).

14. Wilson, G.B., Jahn, T.L.: A Preliminary Report on the Kinetics and Mode of Action of Cystic Fibrosis Serum on Colpidium striatum. Cystic Fibrosis Club Abstr. 14, 59 (1973).

15. Wilson, G.B., Monsher, M.T., Fudenberg, H.H.: Studies on Cystic Fibrosis Using Isoelectric Focusing. III. Correlation Between Cystic Fibrosis Protein and Ciliary Dyskinesia Activity in Serum Shown by a Modified Rabbit Tracheal Bioassay. Pediat. Res. in press (1976).

16. Wilson, G.B., Monsher, M.T., Fudenberg, H.H.: Cystic Fibrosis Protein Factor Detected by Isoelectric Focusing and Ciliary Dyskinesis. Clin. Res. 24, 295 (1976).

Microheterogeneity of Human Kininogen, Alpha-2-HS and Alpha-1-Acid Glycoproteins by Thin-Layer Isoelectric Focusing

U. Hamberg, U. Turpeinen and U. Knuutinen

Summary

Kininogen and alpha-2-HS-glycoprotein were isolated from pooled human plasma or plasma taken from single donors under controlled conditions to prevent the activation of Factor XII. Proteolytic inhibitors were added during preparative steps to minimize increase of heterogeneity.

A main fraction of alpha-2-HS-glycoprotein could be separated from the main fraction of kininogen by linear gradient chromatography on DEAE-Sephadex. About 20 % alpha-2-HS-glycoprotein found with the kininogen fraction is believed to be due to variants causing overlapping of the pI-ranges of the proteins between pH 4.5-5.1 $\pm$ 0.2.

Other interferring proteins, albumin at pI 5.o-5.2 and alpha-1-acid glycoprotein due to relatively high concentration, were removed using specific immunoadsorption. Molecular variants (4-5) of alpha-1-acid glycoprotein were regularly detected at pI 4.1-4.2. Purified kininogen exposed several stained immunoreactive bands which apparently are antigenically identical multiple forms of the low molecular weight kininogen.

Introduction

Kininogen is a plasma glycoprotein containing a vasoactive peptide segment which can be released by proteinase action related to the activation of the blood coagulation and fibrinolytic systems. Kininogens of varying molecular weights have been described (1,2). Earlier studies (3) showed that the low molecular weight human kininogen is heterogenous and apparently subject to various types of structural polymorfism. This presents problems as concerns purification of the kininogen antigen. The present study aims to investigate in more detail the interference of some related proteins by thin layer focusing in polyacrylamide gel. Molecular variants of alpha-1-acid glycoprotein were regularly detected at pI 4.1-4.2. With alpha-2-HS-glycoprotein it was found that 18-28 % of the protein in part represents molecular forms with a pI-range over lapping that of the heterogenous kininogen.

Materials and methods

Materials. Fresh human plasma was obtained from the Finnish Red Cross Blood Transfusion Center. Pooled plasma contained citrate-phosphate-dextrose 67 ml per 500 ml blood in polyvinylchloride bags (30 % phtalate) and was stored for a maximum of 6 month at $-20^{o}C$ (CPD-plasma). Plasma from single donors was taken into polyvinyl bags containing 750 mg disodium ethylenediamine tetra-acetic acid (Merck p.a.), 160 mg polybrene[R] (hexadimethrine bromide, Lot 10,768-9, Aldrich-Europe, Belgium) and 20 mg 1, 10-phenantrolin (Merck p.a.), molecular weight 198.23, dissolved in 10 ml saline. Red cells were removed by centrifugation 20 min at 900 x g, $4^{o}C$, and the plasma was used directly to gel filtration (PBR-plasma).

All reagents unless specified were of analytical grade (p.a.). Distilled water was deionized and twice distilled in quartz apparatus. Highly purified trypsin inhibitor from bovine lung, Trasylol[R], was a gift from Bayer AG; 6-aminocaproic acid (EACA), highly purified human albumin and transferrin were gifts from AM Kabi.

Specific commercial antisera to human plasma proteins were purchased from Behringwerke AG and Dakopatts A/S, Denmark (Dako), Monospecific anti-human kininogen serum was raised in rabbits immunized with 50-200 ug immunologically pure kininogen (3).

Isolation of kininogen and alpha-2-HS-glycoprotein. Low molecular weight kininogen was collected after Sephadex G-200 gel filtration of whole plasma at K_{av} 0.34 (3). The preparation followed the previously given flow scheme (3) using DEAE-Sephadex chromatography in the linear gradient between 0.06 to 0.40 M NaCl at pH 5.5. The alpha-2-HS-fraction (I) was collected between 0.16-0.21 M NaCl and the kininogen fraction (II) between 0.21-0.25 M NaCl (see preparations no. 2-5, Table 1). Preparation no. 6 (not listed in the table) corresponded to 1400 ml pooled plasma; after DEAE-chromatography protein was collected at 0.11-0.18 (I) and 0.18-0.22 (II) M NaCl respectively. A large scale preparation was a preparation of ammonium sulphate precipitated plasma (appr. 5 l) (no. 1, Table 1).

Purification with Sepharose-bound antibody. Anti-albumin serum raised in rabbits (titre 1:32) 32 ml was precipitated at 50 % saturation with ammonium sulphate pH 6.0, 4°C. The precipitate was collected after 2 h by centrifugation 20 min. at 10.400 x g, 4°C, washed with 40 % ammonium sulphate and redissolved in 7.5 ml distilled water. The procedure was repeated at 40 % saturation. The immunoglobulin (331 mg) was redissolved in 3 ml water and dialysed 1 h against water and 16 h against 0.1 M $NaHCO_3$, 0.5 M MaCl, pH 8.0, 4°C. The anti-alpha-1-acid glycoprotein serum (Dako, no. 10-011) 30 ml was precipitated once by addition of solid ammonium sulphate to 40 % saturation. The immunoglobulin (892 mg) was collected and dialysed as above.

Sepharose 4B (Pharmacia) was activated with CNBr (Eastman Kodak Co., Rochester, N.Y.) immediately followed by coupling of the antibody in accordance with the instructions of the manufacturer (4).

Pool II material (no. 6 Table 2), 98 mg was gel filtred through Sephadex G-200 (2.5 cm x 182 cm, upward flow 7 ml/h, 4°C). The homogenous main protein fraction (V_e 464 to 608 ml, K_{av} 0.45) was collected (3). After desalting and lyophilization the protein (4.1 mg kininogen/34 mg) was stored in a desiccator. Albumin and alpha-1-acid glycoprotein were removed by passing throug the Sepharose-antibody columns (0.8 cm x 42 cm, 13 ml/h and 1.6 cm x 40 cm, 16 ml/h, respectively)., using 0.1 M Tris-HCl, 0.5 M NaCl, pH 8.0, buffer.

Thin layer gel electrofocusing in polyacrylamide gel was performed following the LKB Multiphor method with some modifications. Gels 110 x 250 x 1 mm were prepared (5) using 1 % Ampholine (w/v) pH 3-6 and pH 4-6. The electrode solutions were 1 M H_3PO_4 (anode) and 1 M NaOH (cathode). The voltage was allowed to rise slowly (40 min.) from 300 V to 900 V kept constant during the 4 h run at 10°C on an aluminum cooling block connected with the KT 33 Refrigirating Bath Circulator (Haake, Berlin, GFR). The gels were sliced without delay. The pH was measured using the Ingold Lot 403-30-M8 and the Radiometer pH-meter PHM 22. Gels were sliced

(5 x 10 mm) immediately and stored at -20°C. Gels were stained for protein with Coomassie Brilliant Blue R-250 according to the instructions.

Immunodiffusion analyses were performed according to Ouchterlony (3) using purified agar Difco (7.5 ml, 1 %) on 5 x 7.5 cm glass plates. Pieces of the gel sliced after focusing were applied directly for identification of protein.

Radial immunodiffusion (6) was used for immunochemical protein determination with M-Partigen® plates (Behringwerke AG, Marburg-Lahn) or glass plates prepared with monospecific antisera raised in rabbits.

Protein determinations were performed at $A^{1\,\%}_{280\ nm} = 10.0$; ceruloplasmin at $A^{1\,\%}_{610\ nm} = 6.8$ (3).

Results and discussion

Table 1. summarizes the results of thin layer focusing between pH 3-6 of fractions I and II obtained after DEAE-chromatography (3). The main fraction (I) of alpha-2-HS focused between pH 4.7-5.1. Regardless of plasma type about 20 % was obtained in the main kininogen fraction (II). Overlapping to some extent of the fraction pools may in part explain this regularity. However, the lower pI-range obtained in fraction II suggests that true variants of alpha-2HS account for the interference with the pI-range of kininogen. This was sustained with a kininogen preparation focused between pH 4-6 (Table 2). The focusing patterns are shown in Figs. 1-2. As demonstrated in Figs. 3-4 alpha-2HS was detected by immunodiffusion at pI 4.4-4.6. Only after 48 h reaction time small amounts of alpha-2HS could be detected at pI 4.7-4.8 and pI 4.2-4.4. Alpha-2HS from different types of plasma were antigenically identical (Fig. 5).

FOCUSING RANGE OF a_2HS IN FRACTIONS ELUTED AFTER DEAE-SEPHADEX CHROMATOGRAPHY IN A LINEAR SALT GRADIENT FROM VARIOUS TYPES OF HUMAN PLASMA MATERIALS

table 1

plasma type	0.160-0.210 M NaCl (I)				0.210-0.245 M NaCl (II)				% distribution	
	mg protein	mg a_2HS	% a_2HS	pI-range	mg protein	mg a_2HS	% a_2HS	pI-range	I	II
1.	290.8	91.9	31.5	5.0-5.1	81.8	26.0	31.8	4.5-5.0	78	22
2.	75.0	10.3	13.7	4.9-5.1	23.6	2.2	9.3	4.5-4.8	82	18
3.	164.4	31.2	21.4	4.7-4.8	88.3	11.9	13.5	4.6-5.0	72	28
4.	193.0	40.6	21.0	4.7-4.8	67.4	9.5	14.1	4.6-5.0	81	19
5.	50.4	5.8	11.5	4.9-5.0	22.1	1.7	7.5	-	78	22

ISOELECTRIC FOCUSING IN GEL OF KININOGEN COLLECTED AFTER DEAE-SEPHADEX CHROMATOGRAPHY IN THE LINEAR GRADIENT BETWEEN 0.18-0.22 M NaCl

table 2

PROTEIN BY MANCINI TEST	pI-RANGE	% OF TOTAL PROTEIN A_{280}	PROTEINS BY IMMUNODIFFUSION
KININOGEN	4.7-5.1	12	ceruloplasmin
a_2HS-GLYCOPROTEIN	4.4-4.9	17	Gc-globulin β_1C/β_1A-globulin
ALBUMIN	4.9-5.1	7	haptoglobin
a_1-ACID GLYCOPROTEIN	4.1-4.2	-x)	

x) protein amount uncertain; focusing range differs from pI 2.7 according to Putnam 1975 (see The Plasma Proteins, Vol. I)

After gel filtration (G-200) of the kininogen preparation (material 6, Table 2) the homogenous protein fraction contained 12 % albumin, 4 % alpha-2HS, 63% alpha-1-acid and traces of Gc-globulin (not determined). Albumin and alpha-1-acid were removed with the respective Sepharose-antibody (in that order). The result is shown in Fig. 2, 4. The protein responded by immunodiffusion only to the three proteins seen in Fig. 4. Immunodiffusion analysis of slices of the gel after thin layer focusing (Fig. 2) revealed that Gc-globulin can be removed by focusing (Fig. 4), however, as expected the alpha-2HS appeared still within the focusing range of kininogen. Details are given in Fig. 3. In some runs the unusually low range of pI 4.2 ($\pm$ 0.2) for alpha-2HS is explained by the bending of the protein zones during focusing. The focusing pattern seen in Fig. 2 corresponds to the immunoreaction against the monospecific anti-kininogen serum (Fig. 3a) obtained with the 0.5 cm slices of the gel between pH 4.8-5.1 $\pm$ 0.2 and the single precipitin lines seen in the immunodiffusion analysis against the polyvalent antinormal human serum (Fig. 3b).

As shown before (3) the immunologically pure kininogen displayed several protein zones in disc gel electrophoresis. Although alpha-2HS due to the low content was not removed present findings would suggest a higher degree of heterogeneity. The mean value of the focusing range of the antigenically active kininogen obtained in ten separate runs was pI 4.7-5.0$\pm$0.2. As judged by the results in Table 1 there are no striking differences between the pI-ranges of alpha-2HS from various plasma sources. A somewhat different focusing pattern occurred with alpha-2HS from a large scale ammonium sulphate precipitated plasma preparation (Fig. 1, type 1). Similarly no great variations were observed with the focusing patterns of kininogen.

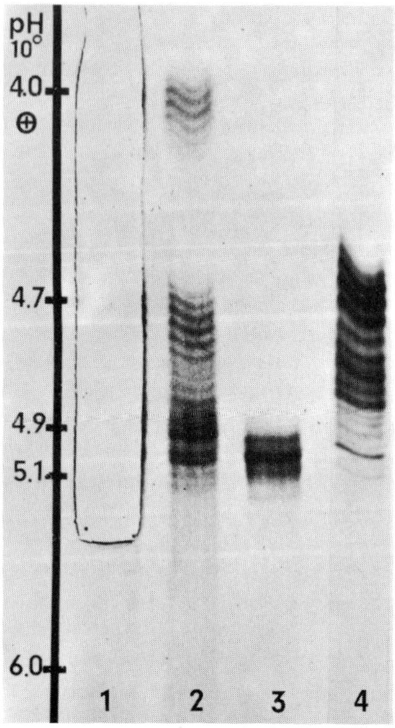

Fig. 2. Electrofocusing (pH 4-6) of fraction II material (no. 6, Table 2) 50 ug (1,2) before and after the removal of albumin and alpha-1-acid, 128 ug (4); control run with albumin, 10 ug (proteins by weight). Technique used for identification shown in gel (1) (for details see methods)

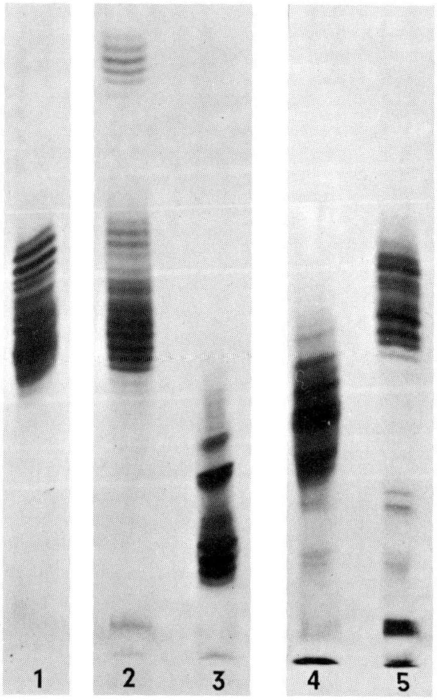

Fig. 1. Electrofocusing (pH 3-6) of fraction I materials (Table 1): type 2 (=3,4 two donors), homogenous fraction from Ultrogel ACA 44 (LKB), 19 ug (1); 20 ug transferrin (3); CPD-plasma 1:40, 10 ul (4); type 1 (large scale, am. sulph. ppted, several donors), 29 ug (5). For pI-ranges see Table 1

Fig. 5. Immunodiffusion analysis with 5 ug protein (clockwise) of CPD-material fr.I and fr.II, PBR-material fr.I and fr.II, am.sulph.ppted material fr.II and PBR-material fr.I (for fractions see Table 1). In center well 30 ul A-alpha-2HS serum. Precipitin reaction after 24 h.

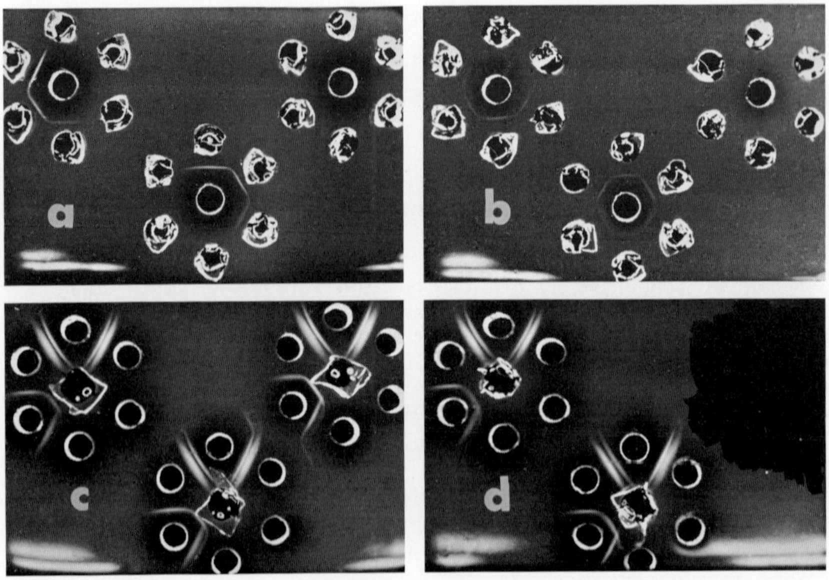

Fig. 3. Immunodiffusion analysis with gel slices (half a slice respectively) (a,b) placed clockwise from pH 5.4 to 4.3 (patterns from left to right); precipitin reactions in (a) against monospecific anti-kininogen serum, (b) polyvalent anti-normal human serum (30 ul, center wells)

(c,d): Identification of alpha-1-acid glycoprotein between pH 4.3 to 4.0 (gel slices in center wells) clockwise against monospecific plasma protein antisera: A-normal human, A-ceruloplasmin, A-kininogen, A-haptoglobin, A-alpha-1-acid, A-alpha-2HS (30 ul). Precipitin reaction is seen only against alpha-1-acid glycoprotein

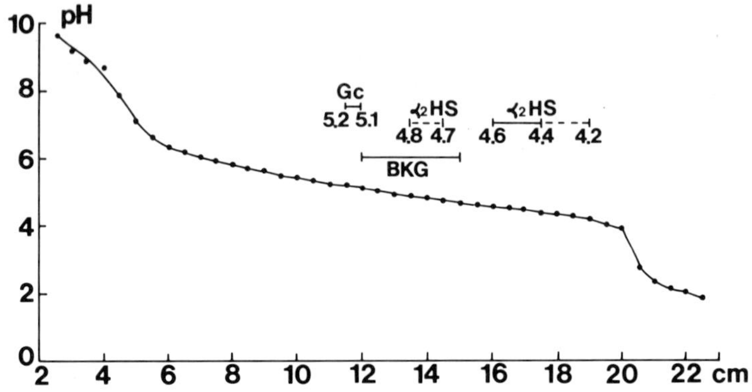

Fig. 4. pH-ranges (·) of focused kininogen (BKG), alpha-2HS and Gc-globulin (pH4-6) full line: precipitin reaction after 18 h, dotted: 48 h. pH is given as the mean of two parallell measurements at different places along the length of the gel

With alpha-1-acid glycoprotein five to six protein bands were regularly obtained at pI 4.1-4.2$\pm$0.2 (mean of nine runs) similarly between pH 3-6 and pH 4-6 (Table 2, Figs. 1-2). This heterogeneity may depend to some extent on the ampholine buffers or lower content of sialyl residues compared with the native protein of pI 2.7. Several polymorfic patterns of alpha-1-acid glycoprotein have been described (7) depending in part on various degrees of desialyzation.

Acknowledgements. Supported by grants from the Science Research Council (no. 413-2551-3 01036680-6) and the Signe and Ane Gyllenberg Foundation (to U.H.). Thanks are due to Prof.V.Mutt,Karolinska Institutet, Drs. H. Björling and R.Lundén, AB Kabi, Stockholm, for use of laboratory and fractionation facilities for large scale; to Miss H.-L.Kauppinen, M.Sc., Finnish Red Cross Blood Center, for providing activated Sepharose 4B.

References

1) Bradykinin, Kallidin and Kallikrein, Handbook of Experimental Pharmacology Vol. XXV pp. 1-768 (E.G. Erdös, ed.). Springer-Verlag, Berlin, Heidelberg, New York (1970)

2) International conference on chemistry and biology of the kallikrein-kinin system in health and disease (J.J. Pisano and K.F. Austen, eds.) Life Sciences, 16, 781-832 (1975)

3) Hamberg, U., Elg, P., Nissinen, E. and Stelwagen, P.: Purification and heterogeneity of human kininogen. Use of DEAE-chromatography, molecular sieving and antibody specific immunosorbents. Int. J. Peptide Protein Res. 7, 261-280 (1975)

4) CNBr- activated Sepharose 4 B, Pharmacia Fine Chemicals, Sweden, Upplands Grafiska AB, pp. 3-17 (1974)

5) Jeppson, J.O. and Berglund, S.: Thin layer isoelectric focusing for haemoglobin screening and its application to haemoglobin Malmö. Clinica Chimica Acta 40, 153-158 (1972)

6) Mancini, G., Carbonara, A.O. and Heremans, J.F.: Immunochemical quantitation of antigens by single radial immunodiffusion. Immunochemistry 2, 235-254 (1965)

7) Schmid, K: α_1-acid glycoprotein, pp. 183-228 in The Plasma proteins (F.W. Putnam, ed.). Academic Press, New York, San Francisco, London (1975)

A Genetic Variant of Amylase from Human Parotid Saliva Detected by Isoelectric Focusing

J.C. Pronk

ABSTRACT

Three different protein patterns of human parotid α-amylase were obtained by isoelectric focusing in a pH-gradient from 3.5-10.0. Family data confirm the hypothesis that these patterns are phenotypic expressions of one autosomal locus with two co-dominant alleles. An estimate of the frequency of the variant allele in the Dutch population gives a value of 0.065. The possibility that this variant is identical with one of the variants described with electrophoresis is discussed.

INTRODUCTION

Genetic variants of human parotid α-amylase are described by means of agar (1) and polyacrylamide electrophoresis (2, 3). Kamaryt and Laxova (1) found a double α-amylase band in saliva from 2.7% of the individuals studied; 92.8% showed a single band and 4.5% showed no activity at all. With polyacrylamide-electrophoresis at pH 8.6 the common pattern (Amy 1 A) shows 6-8 anodally migrating isozyme bands (Sa 1 ... 6). Boettcher and de la Lande (2) found genetic variability in the intensity as well as in the electrophoretic mobility of the amylase isozymes. They suggested that two different loci were involved in saliva α-amylase synthesis. Ward and Merrit (3) summarized the genetic variants described with this method. They all showed lower electrophoretic mobility (Amy 1 B ... G). Except for one variant (Sa Amy 1 E), which might be relatively common in negroes, the frequency in the populations studied is very low. It was recently recognized (4) that isoelectric focusing in thin layer polyacrylamide gels in a pH range from 3-10 reveals patterns that differ from individual to individual. However, no details were given, nor were family studies performed. This communication presents a comparison of individual α-amylase isoelectric focusing patterns. A (genetic) hypothesis is advanced that the variant patterns observed are phenotypic expressions of one autosomal locus with two co-dominant alleles.

MATERIAL AND METHODS

Parotid saliva (10 ml) was stimulated by sour lemon candy and collected directly from the ductus by use of a special device (5). After centrifugation for 10 minutes at 3,000 r.p.m. 5 ml samples were concentrated 10 times by use of a B-15 minicon microconcentrator (Amicon N.V. Oosterhout, N.B. Holland) or desalted on a sephadex G-25 column (18 x 2.5 cm) and subsequent lyophilized. Thin layer isoelectric focusing was essentially performed as described by Davies (6) on the LKB 2117 multiphor. The contents of a 5.4% polyacrylamide gel are: N,N,N',N'-tetramethylethylenediamine 0.043 ml; acrylamide 3.25 g N,N'-methylenebisacrylamide, 0.088 g; carrier ampholines (LKB) pH- range 3.5-10.0, 3 ml; riboflavine 0.34 mg, 60 ml.
As electrode solutions 5% ethylenediamine was used on the cathode and 1 M phosphoric acid on the anode. Samples were applied near the anode using Whatman 3 M filterpapers (5.0 x 7.5 mm). Gels were cooled with tap water (14° C) and run across the width in about 1.5 hour. 40 Watt was the upper limit for heat production. The pH gradient was determined with the gel on the cooling plate with a surface elctrode (type LOT 403-30-48, Ingold, Zürich, Switzerland). Gels were stained as described in detail (7) and stored by overlaying of a solution containing 7% gelatine and 5% glycerine and drying in a dustfree room. Discontinuous horizontal sheet polyacrylamide electrophoresis at pH 8.6 was employed, essentially as described by Ward and Merrit (3). Gels were stained with KI-starch. Photographs were taken immediately after staining since the patterns

become less distinct after 10-15 minutes. Gels were scanned with a Quick scan Jr. (Helena Laboratories Corp. Beaumont, Texas, U.S.A.) with an automatic integrator. Chemicals were obtained from Merck A. G. Darmstadt, Germany.

RESULTS AND DISCUSSION

On staining with coomassie blue R 250 the most frequent pattern shows three dense staining bands with isoelectric points 6.4, 6.0 and 5.6 respectively and 4-8 additional minor bands (pattern I). This pattern was present in 37 out of 43 randomly selected nonrelated individuals. Five individuals showed a pattern with two additional major bands with pI 6.2 and 5.7 (pattern II). One individual missed the three major bands seen in the common pattern. Her pattern showed two bands with pI 6.2 and 5.7 (pattern III). The three different patterns are shown in fig. 1. From three individuals with different patterns samples were taken on four consecutive days. The patterns appeared to be reproducable from day-to-day. Four successive 5 minute samples were taken from these individuals at

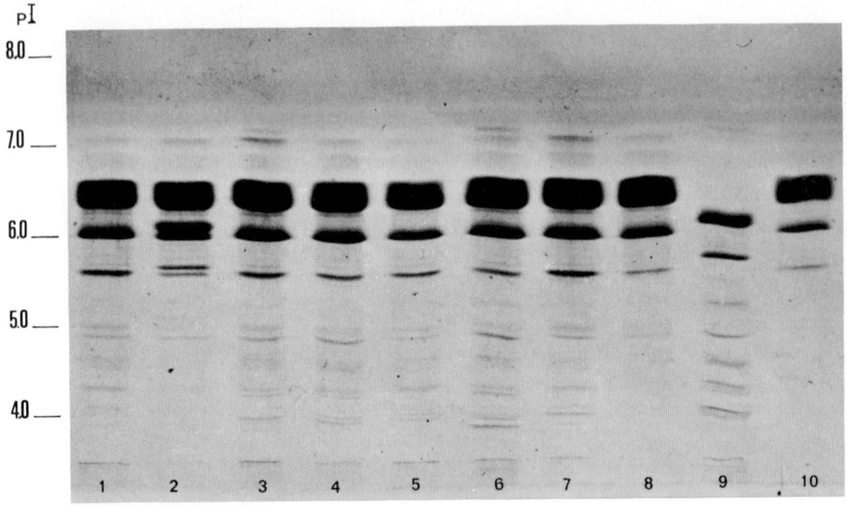

Fig. 1. Thin layer polyacrylamide isoelectrofocusing of parotid saliva from 10 different individuals. No. 1, 3 up until no. 8: pattern I, no. 2: pattern II, no. 9: pattern III. Staining: coomassie blue R 250.

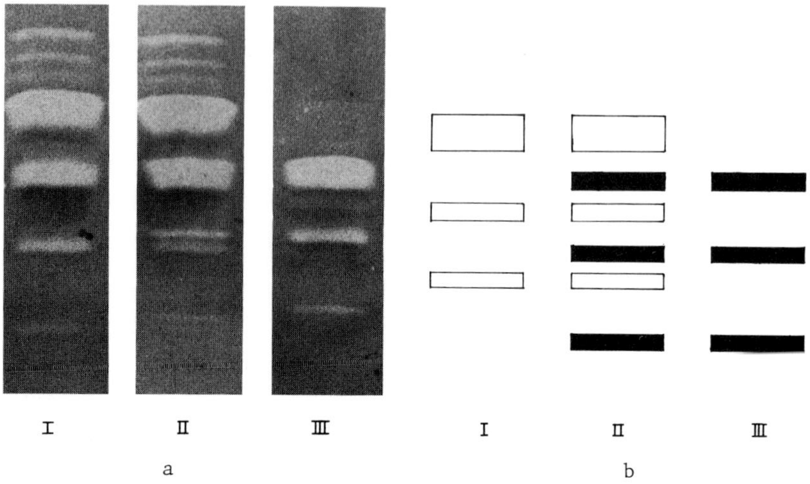

Fig. 2 a. Thin layer isoelectric focusing in a pH 3.5-10.0 gradient of three different patterns stained on α-amylase activity. b. Diagrammatic representation of the genetic hypothesis. I: common homozygote, II: heterozygote, III: homozygote for the variant protein.

three different times a day. The patterns were not dependent on the length of the stimulation nor on the time of the day. Autoincubation of two samples from individuals with pattern I for 4 hours at 37° C did not alter the pattern. From this it is concluded that these patterns are characteristic for particular individuals.

Fig. 2 a shows the result of isoelectric focusing from samples with different patterns after coomassie blue staining, stained for α-amylase activity with KI-starch. By this more sensitive and specific staining method it is shown that pattern I and pattern III are very much alike but shifted along the pH-gradient for about 0.2 pH units. We tried to explain these patterns with a genetic hypothesis. In this hypothesis pattern I represents the common homozygote and pattern III the homozygote for the variant allele. Pattern II represents the heterozygote for the variant allele. This is schematically drawn in fig. 2 b.

Although we did not yet compare the activities of α-amylase in fresh saliva from individuals with the three different patterns the zymogram indicates that the activity of the variant protein is similar to the

activity of the common protein.
In table 1 it is shown that (in the coomassie stained patterns) the relative intensity of the bands with pI 6.0 and 5.6 is diminished in individuals with the additional bands. This is also compatible with a genetic hypothesis but it does not absolutely prove it as one cannot exclude the possibility that the proteins are altered post-translational within the parotid gland cells or during the excretion process.
Quantitative observations on the KI-starch stained isoelectric focusing gels are difficult to interprete because the bands are rather diffuse.
Fig. 3 shows the result of family studies. It is shown that the variant protein is inherited as if it was coded for by an autosomal co-dominant gene. These family studies confirm our hypothesis that the variation observed has a genetic background and is not caused by exogenous factors.
We also studied four families in which both parents showed pattern I. All fifteen children in these families showed pattern I as well.
A calculation of the frequency of the variant allele in the Dutch population by gene counting gives a value of 0.065. This is a high frequency for a genetic variant which makes it very suitable as a genetic marker for family and population studies.
Fig. 4 shows the result of polyacrylamide electrophoresis of three saliva samples from individuals with different phenotypes with isoelectric focusing.

Isoelectric point	5.6	5.7	6.0	6.2	6.4
Pattern I (n=8 mean ± s.d.)	4.8 ± 0.9		22.3 ± 3.7		71.5 ± 4.4
Pattern II (n=3 individual values)	2.0 1.0 1.0	3.0 3.0 3.0	12.0 9.0 9.0	16.0 22.0 12.0	67.0 65.0 74.0
Pattern III (n=1)		24.0		76.0	

Table 1. Relative intensities as per cent of total area of major α-amylase peaks in individuals with different isofocusing patterns.

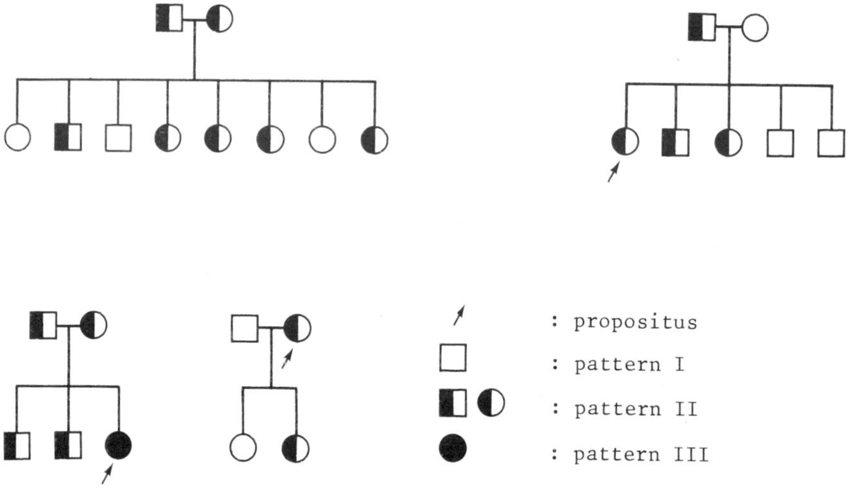

Fig. 3. Pedigrees of families with the variant protein.

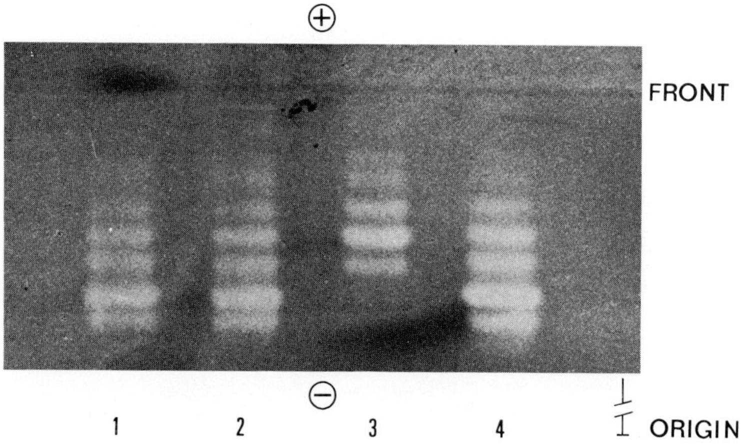

Fig. 4. Polyacrylamide electrophoresis from parotid saliva from individuals with different isoelectric focusing patterns.
1: common homozygote, 2 and 4: heterozygote, 3: homozygote for the variant protein. Staining KI-starch. Only the one half of the gel most distal from the origin is shown.

No difference can be observed between the patterns from the common homozygote and the heterozygote. The pattern from the homozygote for the variant protein however is anodal to the normal pattern. A difficulty in the interpretation of the patterns in polyacrylamide electrophoresis is that the two slowest bands in the pattern from the homozygote for the variant protein have the same electrophoretic mobility as the third and fourth band in the common pattern. This should one make to believe that these proteins are identical. We believe that this is not so unless this polymorphic system is dependent on the absence or precense of an enzyme which converts the fourth band in the common pattern in the second band. This second band is until now believed to be the primary gene product for the α-amylase from which all other bands are derived by post-translational glycosidation and deamidation (8).
The variant described here is not identical with one of the Sa Amy 1 B ... G variants (3) which are all cathodal to the normal pattern.
It cannot yet be concluded however whether these variants are identical or not with the two anodal variants described (1, 2).

ACKNOWLEDGEMENTS

The author wishes to thank Mrs. Nanda van Empelen for her technical assistence and Mrs. Leny Dufay for the collection of the samples.
Drs. Bertie Reijnders, Drs. Rune Franz and Mr. Gerard Pals are thanked for their stimulating discussions. Prof. Dr. J. Huizinga is thanked for reading and improvement of the manuscript.

REFERENCES

1. Kamaryt, J., Laxova, R.: Amylase heterogeneity: variants in man. Hum. Genet. 3, 41-45 (1966).

2. Boettcher, B., de la Lande, F.A.: Electrophoresis of human saliva and identification of inherited variants of amylase iso-enzyms. Aust. J. exp. Biol. med. Sci. 47, 97-103 (1969).

3. Ward, J.C., Merrit, A.D., Bixter, D.: Human salivary amylase: genetics of electrophoretic variants. Am. J. Hum. Genet. 23, 403-409 (1971).

4. Beeley, J.A.: Fractionation of human salivary proteins by isoelectric focusing. In: Isoelectric Focusing ed. by Arbuthnott and Beeley, Butterworths & Co. Ltd. London. Pag. 240-247 (1975).

5. Curby, W.A.: Device for collection of human parotid saliva. J. Lab. Clin. Med. 41, 493-496 (1953).

6. Davies, H.: Thin-layer gel isoelectric focusing. In: Isoelectric Focusing ed. by Arbuthnott and Beeley, Butterworths & Co. Ltd. London. Pag. 97-113 (1975).

7. Vesterberg, O.: Isoelectric focusing of proteins in polyacrylamide gels. Biochem. Biophys. Acta. 257, 11-19 (1972).

8. Karn, R.C., Rosenblum, B.B., Ward, J.C., Merrit, A.D.: Genetic and post-translational mechanisms determining human amylase isozyme heterogeneity. Isozymes 4, 745-761 (1975).

6. Two-Dimensional Techniques

Biological Applications of Two-Dimensional Gel Electrophoresis

R.D. Ivarie, D.H.Gelfand, P.P. Jones, P.Z. O'Farrell, B.A. Polisky, R.A. Steinberg and P.H. O'Farrell

ABSTRACT

Two-dimensional gel electrophoresis described by O'Farrell [1] is used to display the electrophoretic properties of several identified gene products in the total gel patterns from bacterial and mammalian cells. Several known proteins yield only one spot after two-dimensional separation. Thus, the method does not generate polypeptide microheterogeneity. Several proteins from animal cells, however, are heterogeneous in both charge and molecular weight. In two cases, the isoelectric point heterogeneity is caused by protein phosphorylation; in two others, by the addition of sialic residues. Examples are given demonstrating the use of the technique in detecting missense mutations in structural genes and in monitoring changes in gene expression in response to environmental signals. These data illustrate the application of the method to many different biological problems.

INTRODUCTION

Two-dimensional gel electrophoresis described by O'Farrell [1] combines isoelectric focusing with sodium dodecylsulfate (SDS) stacking gel electrophoresis to give a sensitive analytical method for the high

---

Abbreviations. SDS, sodium dodecyl sulfate; TAT, tyrosine amino transferase (E.C. 2.6.1.5); R, regulatory subunit of cyclic AMP-dependent protein kinase; C, catalytic subunit of cyclic AMP-dependent protein kinase.

resolution separation of proteins. The samples are run under denaturing and reducing conditions; thus, the displayed spots represent single polypeptide chains. Because resolution is based on isoelectric point in the first dimension and molecular weight in the second, the positions of spots provide useful physical parameters. Very small changes in charge or molecular weight of a protein are readily detected. The resolving power generated by the combination of the two separating methods allows the detection of at least a thousand radioactively labeled cell proteins. Although similar attempts have been made in the past [2,3,4], this method has much higher resolution of individual components and is particularly suitable for analyzing complex mixtures of proteins.

In this paper, application of the two-dimensional gel method to several different biological problems is presented to illustrate the versatility of the technique and the use of a variety of isotopic labeling and protein identification procedures. In each case, the electrophoretic properties of identified gene products are displayed. These gene products include: (1) the enzyme β-galactosidase of the bacterium *Escherichia coli*; (2) the major viral capsid protein, VP-1, of simian virus 40; (3) the regulatory subunit of cyclic AMP-dependent protein kinase from mouse lymphoma cells; (4) the surface antigens specified by the major histocompatibility gene complex of the mouse; and (5) a set of steroid inducible proteins in rat hepatoma cells of which one is the enzyme tyrosine aminotransferase (TAT). Less attention is given here to the details of the technique itself since these have been thoroughly described [1].

To show that the method does not artifactually generate polypeptide heterogeneity, examples are presented of proteins which give rise to single spots on two-dimensional gel patterns. Nevertheless, many gene products do show microheterogeneity in the gel system and for several of these heterogeneity is the result of *in vivo* protein modification reactions including phosphorylation and glycosylation. Based on these examples, at least, it appears that intracellular modification of proteins may be quite extensive in mammalian cells.

RESULTS

Identification of Several E. coli Proteins on Two-Dimensional Gel Autoradiograms

As described below, many proteins of mammalian cells show microheterogeneity; that is, the product of a single gene produces more than one electrophoretically identifiable spot. To emphasize that not all proteins are microheterogeneous, we show that the *E. coli* enzyme, β-galactosidase, electrophoreses as a single spot, and describe several other identified gene products of *E. coli*. The demonstration that a number of proteins produce single spots indicates that the technique for sample preparation and electrophoresis as described [1] does not introduce artifactual heterogeneity.

For a series of experiments involving isolation of elements of the lactose operon [5], it was necessary to determine the position of the spot corresponding to the β-galactosidase subunit. Figure 1 shows the pattern obtained after electrophoresing about 1 μg of unlabeled purified β-galactosidase with ^{35}S-methionine labeled extracts from wild type *E. coli* (Figure 1a) or from the same strain grown in the presence of cyclic AMP and isopropylthiogalactoside (Figure 1b), effectors known to induce maximal synthesis of β-galactosidase. Purified β-galactosidase, seen only on the stained patterns (data not shown), migrates as a single spot to the position indicated by the arrow in the autoradiograms shown in Figure 1 . Clearly, the induced extract contains a major labeled protein that coelectrophoreses with the carrier enzyme; this labeled spot is absent from the uninduced control. The large size of the enzyme spot is due to the large amount of protein contributed by the carrier (ca. 1 μg). Since the labeled extracts contain very little protein, all other spots in the patterns approach minimal size [1]. Further evidence for identifying this inducible spot as β-galactosidase was obtained by showing that *E. coli* strains harboring deletions in the structural gene for β-galactosidase also lack this single protein spot. The observation that β-galactosidase migrates as a single spot was used to show that an altered form of the enzyme lacking the terminal 15 amino acids was synthesized from a portion of the lactose operon contained in a plasmid.

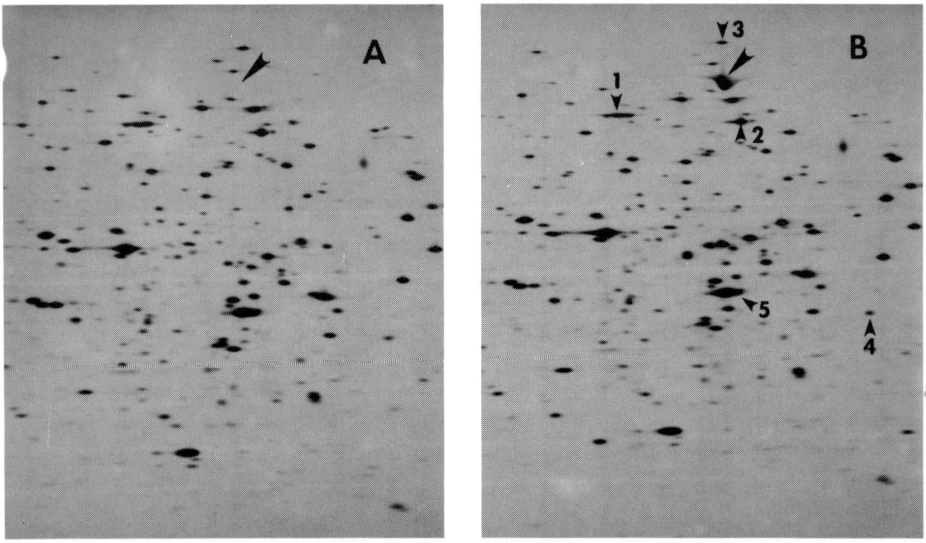

Figure 1. Identification of β-galactosidase by induction and comigration. a) An autoradiogram depicting the ^{35}S-methionine labeled proteins made in *E. coli* in the absence of induction. The large arrow indicates the position of unlabeled purified β-galactosidase (ca. 1 μg) as detected on the stained gel. b) An autoradiogram depicting the labeled proteins made after induction of β-galactosidase by cyclic AMP and isopropylthio-galactoside. The large arrow indicates the position of added carrier enzyme as above, and the small arrows indicate the position of (1) the met E gene product; (2) EF-G [6]; (3) β-subunit of RNA polymerase; (4) α-subunit of RNA polymerase; and (5) EF-Tu [7]. A number of other spots are induced in this gel by cyclic AMP [8].

Using a similar genetic, biological and/or electrophoretic criteria, a number of proteins have been identified in two-dimensional patterns from *E. coli*. The positions of the following proteins have been determined and shown to correspond to a single spot: (1) methyl transferase, the product of the met E gene (O'Farrell, *unpublished observation*), (2) translational elongation factor G [6], (3) the β-subunit of RNA polymerase (O'Farrell, *unpublished observation*), and (4) the α-subunit of RNA polymerase (O'Farrell, *unpublished observation*). In addition, translational elongation factor Tu has been identified as a single spot even though two separate genes code for it [7].

The Resolution of the Major Capsid Protein of Simian Virus 40

The high resolution afforded by two-dimensional electrophoresis allows the synthesis of viral specific proteins to be monitored directly in whole cell extracts of infected cells without purification of viral specific components. Here we illustrate the resolution of the major capsid protein of SV40, VP1 (for details of these analyses, see 9). Figure 2 compares a portion of the two-dimensional gel pattern for uninfected monkey cells (Figure 2a) and for monkey cells infected with either wild type (Figure 2b) or a temperature sensitive mutant of SV40 (Figure 2c). In the region of the infected cell pattern corresponding to the molecular weight expected for VP1 (47,000 daltons; 10), there is a striking array of protein spots (Figure 2b). These protein spots are the products of a single gene coding for the major capsid protein of SV40, VP1; a temperature sensitive mutation in the region of the genome coding for VP1 results in an altered isoelectrophoretic migration of all six protein spots (Figure 2c).

Thus the two-dimensional separation of VP1 reveals that it is heterogeneous, consisting of one major protein (molecular weight 47,000 daltons and isoelectric point of approximately pH 6.8) and five minor protein components. The minor forms of VP1 are 10 percent of the total VP1 and differ from the major form of VP1 in both molecular weight (by approximately 500 daltons) and isoelectric point (ranging from approximately pH 6.7 to pH 6.9). The various forms of VP1 are synthesized in an ordered and reproducible fashion during SV40 infection. The kinetics of production of the various forms of VP1 suggest that they are the result of modification reactions which occur within minutes of the primary translation of VP1 messenger RNA. Two of the forms of VP1 are phosphorylated derivatives (the two rightmost minor protein spots). The molecular weight differences between several of the protein spots suggest that they may be the result of the addition of one or two glycosyl residues. To date it has not been possible to ascribe a function to the modifications of VP1. The data illustrate that a single gene product may be subject to rapid post translational modification resulting in the

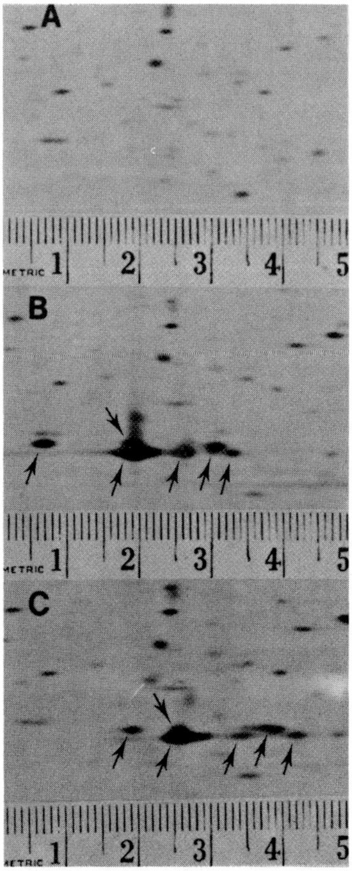

Figure 2. The viral origin of the various forms of VP1 of SV40. CV-1 monkey cells were infected with wild type and temperature sensitive mutant BC11 virus. Uninfected cells, labeled for 3 hours at 68 hours post infection at 33°C as described [9]. The VP1 region of the two-dimensional separation from total extracts of (a) uninfected, (b) wild type infected, and (c) temperature sensitive mutant infected cells are shown. (Published with the kind permission of Cell, MIT Press.)

appearance of several protein spots in two-dimensional electrophoretic separations.

Regulatory Subunits from Cyclic AMP-Dependent Protein Kinase: Induction of Charge Heterogeneity by Protein Phosphorylation

As the intracellular transducer for cAMP mediated effects in higher eukaryotic cells, cAMP-dependent protein kinase plays a central role in many hormone mediated responses. S49 mouse lymphoma cells, which are normally killed by either exogenous dibutyryl cAMP or hormone inducers of cAMP, provide a system in which protein kinase can be functionally altered by mutation [11]. The two-dimensional gel technique provides a powerful tool for investigating some of these alterations. In order to pursue these studies, however, it was necessary to identify and characterize the wild type gene products.

Cyclic AMP-dependent protein kinase is composed of cAMP binding regulatory (R) subunits and catalytic (C) subunits which are activated by the cAMP induced dissociation of R's and C's. R subunits purified to near homogeneity from ^{35}S-methionine labeled cells give two spots in the two-dimensional gel system with an SDS mobility corresponding to a molecular weight of about 50,000 daltons and separation in the isoelectric focusing dimension equivalent to about a single charge difference [12]. Figure 3a shows the gel pattern of purified R's (indicated by arrows) which were mixed with a small amount of ^{35}S-methionine labeled protein from a whole cell extract for orientation. The heterogeneity in the isoelectric point of R is attributable to phosphorylation as shown in Figure 3. Figure 3b shows the gel pattern of a total soluble extract from S49 cells labeled for 3 hr with $^{32}PO_4$. When kinase R's are purified from this extract by affinity chromatography on a cAMP-agarose derivative [12], a much simpler pattern is seen with only one prominent spot (Figure 3c). When this purified $^{32}PO_4$ labeled material is co-electrophoresed with the ^{35}S-methionine labeled preparation shown in Figure 3a, it is apparent that the prominent $^{32}PO_4$ labeled spot is identical to the more acidic of the ^{35}S-methionine labeled R spots. Confirmation that these spots both are derived from the same gene product has been obtained from studies of labeling kinetics (Steinberg et al., *unpublished observations*) and from missense mutations causing charge shifts in the two R spots [12].

Fig. 3

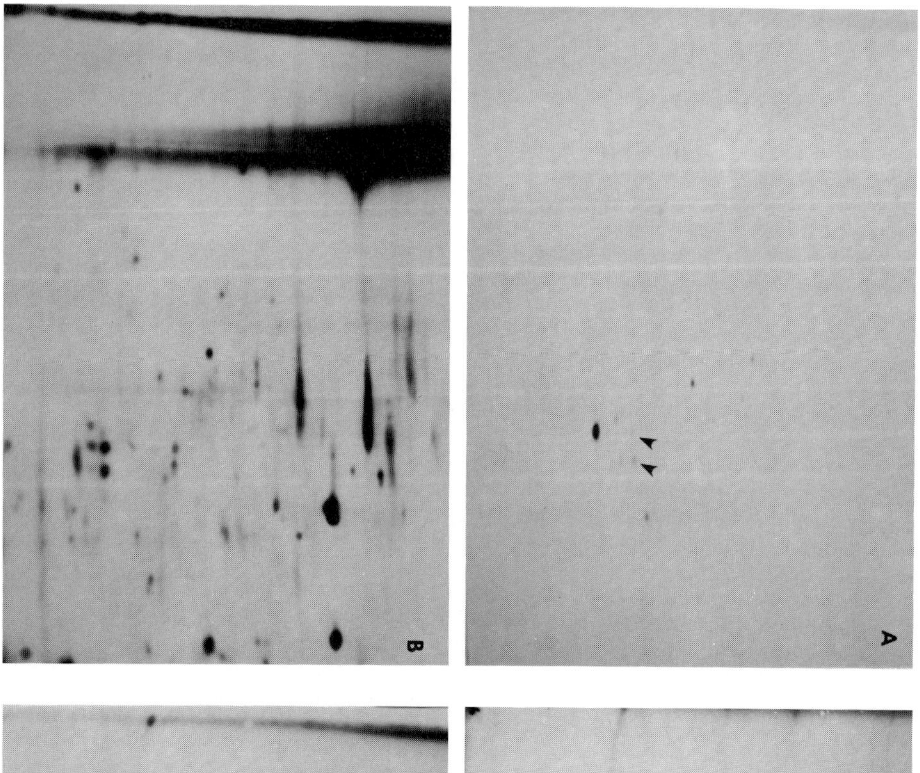

Two-Dimensional Gel Analysis of Lymphocyte Surface Antigens

The H-2 gene complex of the mouse codes for a number of traits important to the immune system [14]. Included among these are genes which determine H-2K and H-2D, the major histocompatibility antigens. In addition, genes mapping between H-2K and H-2D control the expression of the Ia antigens. The Ia proteins may actually be the products of the H-2-linked immune response genes which regulate immune responsiveness to certain antigens [13].

Knowledge of the molecular structure of H-2 and Ia antigens should facilitate understanding both the functions of these molecules and the complex genetics which control them. Specific immunoprecipitation of H-2 and Ia from radiolabeled lymphocyte extracts has revealed that H-2 and Ia are glycoproteins with molecular weights of about 45,000 daltons [16] and 25,000-35,000 daltons [17], respectively. However, such preparations clearly contain heterogeneous populations of H-2 and Ia molecules. With the two-dimensional gel electrophoresis technique it has become possible to resolve all the molecular species of H-2 and Ia into discrete spots, and, in combination with several different radiolabeling procedures, to elucidate the basis for some of the molecular heterogeneity of these antigens [18].

Figure 3. Phosphorylation of the regulatory subunit from S49 cell cyclic AMP-dependent protein kinase. Cultures of S49 mouse lymphoma cells were labeled with ^{35}S-methionine or $^{32}PO_4$, extracts were prepared, and cyclic AMP-dependent protein kinase regulatory subunits were i isolated as described [12]. Samples were run on two-dimensional gels [1] except that the second dimension gels were 7.5% polyacrylamide. (a) Highly purified ^{35}S-methionine labeled regulatory subunit (1500 cts/min) with ^{35}S-methionine labeled crude cell extract (15,000 cts/-min). (b) $^{32}PO_4$ labeled cell extract (2 X 10^6 cts/min). (c) $^{32}PO_4$ labeled regulatory subunit after step affinity purification (12,000 cts/min). (d) Mixture of samples as for (a) and (c). Autoradiograms of (a), (c), and (d) were exposed for 15 days; that for (b) for 3 days.

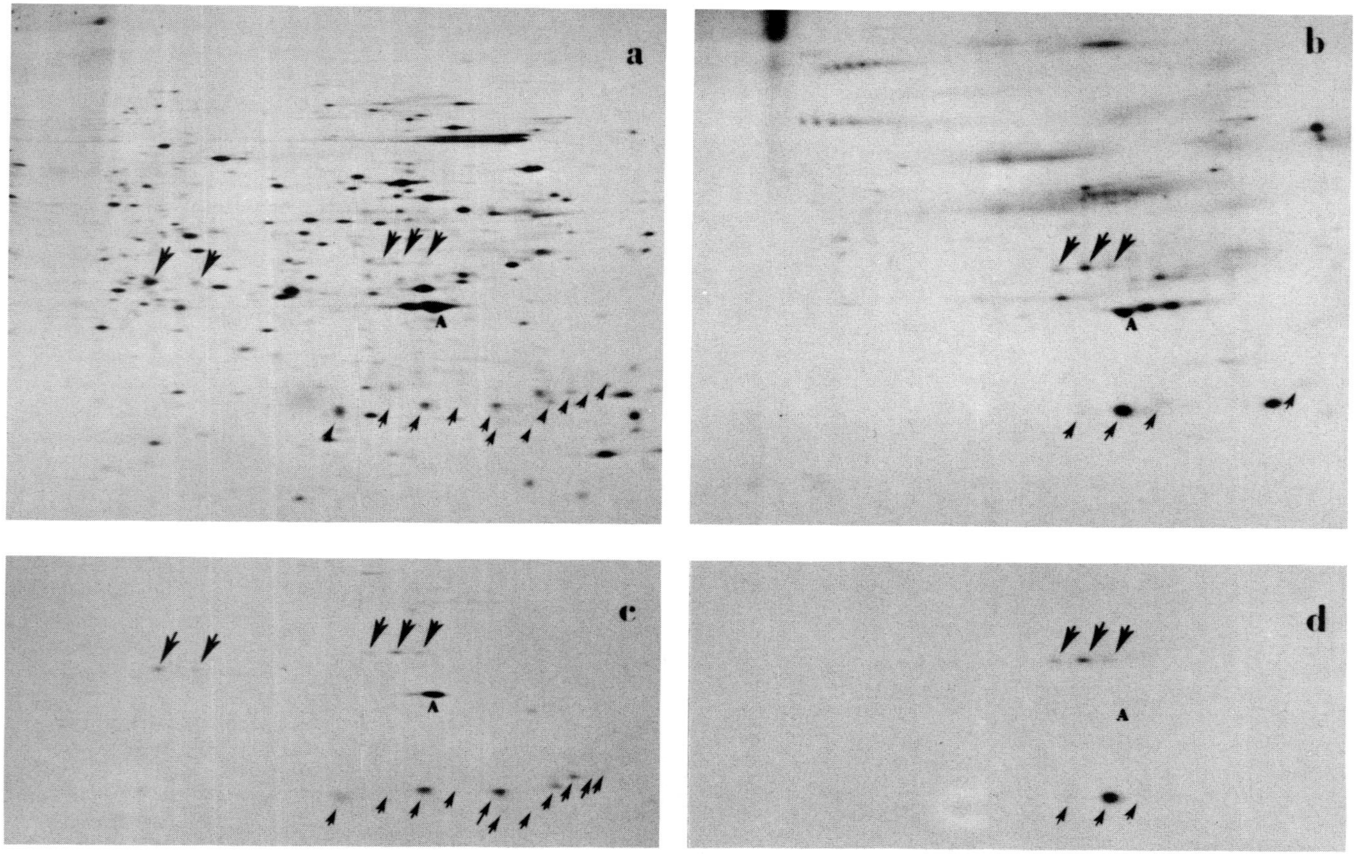

Fig. 4

Autoradiograms of two-dimensional gels prepared from detergent extracts of BALB/c spleen lymphocytes internally labeled with ^{35}S-methionine and surface labeled with ^{125}I are presented in Figure 4a and b, respectively. From these two gels it can be seen that only a small portion of the total cell proteins are on the surface and are capable of being radio-iodinated. H-2K and Ia were specifically isolated from these complex mixtures of proteins by immunoprecipitation; the two-dimensional gel patterns of these precipitates show that only subsets of the total ^{35}S-methionine labeled spots (Figure 4c) can be surface labeled with ^{125}I (Figure 4d).

A dominant feature of the H-2K and Ia spot patterns is their marked heterogeneity. Several lines of evidence suggest that some of the heterogeneity is due to sequential glycosylation of the polypeptide chains as they migrate to the cell surface. In some cases the mobility differences between spots represent minor changes in size alone, as would be expected from the addition of neutral sugars. There are also combined shifts in isoelectric point (equivalent to increments of unit negative charge; 1,12) with small increases in size, consistent with the addition of a sialic acid residue. For both H-2K and Ia, the spots which are iodinated tend to be those which are larger and more acidic, as expected for membrane proteins which should have their full complement of carbohydrate. Preliminary experiments have shown that neuraminidase treatment of radioiodinated cells chases the ^{125}I label into smaller, more basic spots, confirming the role of sialic acid residues in generating some of the observed heterogeneity.

Figure 4. ^{35}S-methionine and ^{125}I labeled lymphocyte proteins. BALB/c spleen lymphocytes were labeled either with ^{35}S-methionine for 4 hours or with ^{125}I by surface iodination. As has been described [17], the cells were lysed with the detergent NonIdet P-40, and portions of the extracts were immunoprecipitated with antiserum (B10 X A)F_1 anti-B10.D2, which reacts with H-2K and Ia antigens. (a) ^{35}S-methionine NP-40 extract, (b) ^{125}I NP-40 extract, (c) ^{35}S immunoprecipitate, (d) ^{125}I immunoprecipitate. H-2K and Ia molecules are indicated by large and small arrows, respectively. Other spots were present in the normal mouse serum control precipitate and thus represent nonspecifically precipitated proteins. Actin is indicated by the letter A.

Steroid-Inducible Proteins in Rat Hepatoma Cells

Because so many gene products are resolved, a useful application of two-dimensional gel electrophoresis is the measurement of changes in gene expression in response to environmental stimuli. Specific questions can be asked concerning the number of polypeptide chains affected by a given signal (domain of response; 8,18), the direction of the response (induction vs. repression), the kinetics of the response, and the dose-response relationships among affected proteins. Such measurements have important implications for the mechanism of action of the effector.

Steroid hormones alter the intracellular levels of specific proteins by controlling their rates of synthesis [19]. Shown in Figure 5 are two-dimensional gel patterns from rat hepatoma cells labeled for 30 min with ^{35}S-methionine following treatment with (Figure 5a) or without (Figure 5b) dexamethasone, a synthetic glucocorticosteroid. As marked by arrows, only a limited number of polypeptides have increased rates of synthesis in response to the hormone. One is the steroid-inducible enzyme tyrosine aminotransferase identified by its reaction with anti-TAT serum. Five of the induced polypeptides form a series, termed "belt I," which is heterogeneous in both charge and molecular weight much like the surface antigens on mouse lymphocytes discussed in the previous section. Belt I is a surface protein and presumably represents a single gene product which has undergone extensive glycosylation and sialation. Two major conclusions can be drawn from an experiment of this kind: (1) dexamethasone affects an extremely limited number of hepatoma cell proteins and (2) dexamethasone in general increases the rate of synthesis of affected gene products.

Ideally, assaying a domain of hormonal response by two-dimensional gels measures the number of genes expressed in a given cell that are under hormonal control. Implicit in such a measurement, however, is that each gene gives rise to only one spot on a two-dimensional gel pattern. Such an assumption is clearly unwarranted because, as described in earlier sections, eukaryotic proteins can be structurally modified to generate spot microheterogeneity. While the number of modified proteins is

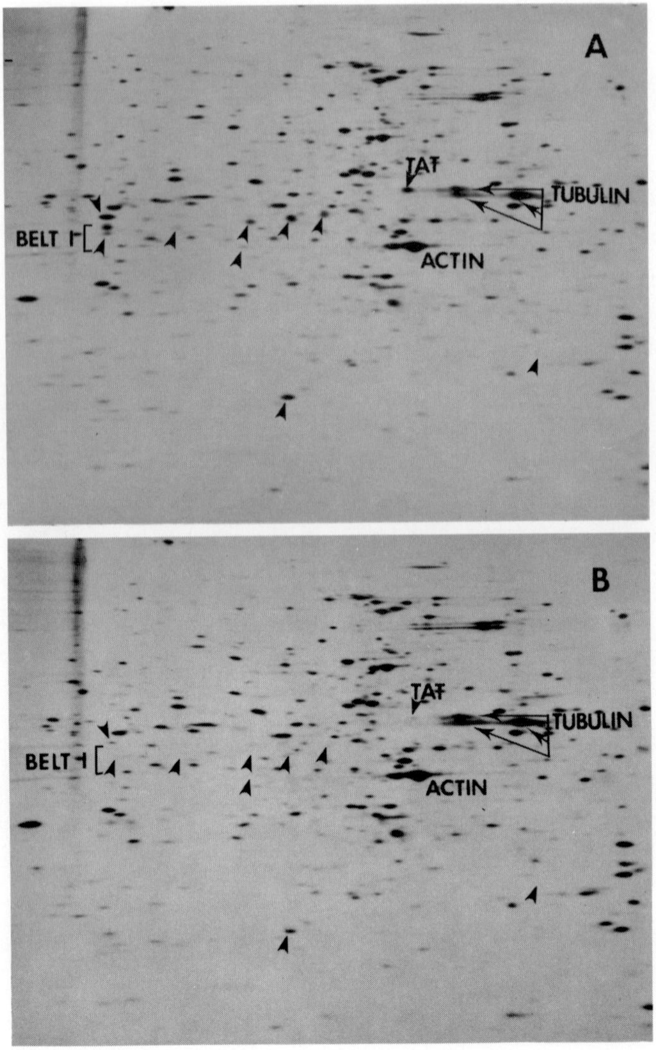

Figure 5. Induction of rat hepatoma cell proteins by dexamethasone. Rat hepatoma cells were plated in multiwell dishes, grown for three days, and treated for 9 hr with or without dexamethasone (10^{-6} M). Following induction, cells were labeled for 30 min with ^{35}S-methionine and lysed directly in lysis buffer [1]. Samples were electrophoresed in two dimensions [1] and the gels were stained, dried, and exposed to X-ray film (Kodak NS-2T). Small arrows mark the position of induced polypeptides; the large arrow, the position of TAT. The positions of actin and tubulin are also indicated. (a) induced and (b) uninduced.

unknown, modification reactions introduce a degree of uncertainty in measuring the size of a domain.

DISCUSSION

The work presented here confirms earlier conclusions that the procedures described for sample preparation and two-dimensional electrophoresis [1] do not induce artifactual heterogeneity in protein samples. Proteins which migrate as single spots (Figure 1) show that there is no generalized protein modification. The demonstration that some modified forms of a protein are produced by phosphorylation or sialation--known *in vivo* enzymatic modifications--shows that observed microheterogeneity has a biological origin. Since intracellular modification of proteins might be quite common, it cannot be concluded that every spot is the product of a different gene. Accordingly, considerable caution must be observed in interpretation. As a further word of caution, investigators should be aware that artifactual modification gives rise to distinctive electrophoretic patterns which can be easily distinguished from biological modification [1,12].

The determination of the position of different proteins in two-dimensional gel patterns illustrates the criteria and approaches which can be utilized in identification--immunological, genetic, biological, and biochemical. In addition, the applications described here point to the usefulness of the technique in the study of diverse biological problems.

ACKNOWLEDGEMENTS

This work was funded by grants from the National Institutes of Health and the National Science Foundation. RDI and BAP were supported by training grants from the National Institutes of Health. PPJ is a fellow of the Arthritis Foundation; PHO'F, a fellow of the Jane Coffin Childs Memorial Fund for Medical Research; PZO'F, a fellow of the American Cancer Society, California Division; and RAS, a fellow of the Cardiovascular Research Institute, University of California - San Francisco.

REFERENCES

1. O'Farrell, P.H.: High resolution two-dimensional electrophoresis of proteins. J. Biol. Chem. 250, 4007-4021 (1975).

2. Raymond, S.: Acrylamide gel electrophoresis. Ann. N.Y. Acad. Sci. 121, 350-365 (1964).

3. Macko, V. and Stegemann, J.: Mapping of potato proteins by combined electrofocusing and electrophoresis: Identification of varieties. Hoppe-Seyler's Z. Physiological Chem. 350, 917-919 (1969).

4. Kaltschmidt, E. and Wittmann, H.G.: Ribosomal proteins VII. Two-dimensional polyacrylamide gel electrophoresis for fingerprinting of ribosomal proteins. Anal. Biochem. 36, 401-412 (1970).

5. Polisky, B.A., Bishop, R.J., and Gelfand, D.H.: A plasmid cloning vehicle allowing regulated expression of eukaryotic DNA in bacteria. Proc. Nat. Acad. Sci. (U.S.A.), in press (1976).

6. Furano, A.V. and Wittel, F.P.: Synthesis of elongation factors Tu and G are under stringent control in Escherichia coli. J. Biol. Chem. 251 898-901 (1976).

7. Pederson, S., Blumenthal, R.M., Rech, S., Parker, J., Lemaus, P., Laursen, R.A., Hagarkatti, S., and Friesen, I.D.: A mutant of Escherichia coli with an altered elongation factor Tu. Proc. Nat. Acad. Sci. (U.S.A.) 73, 1698-1701 (1976).

8. O'Farrell, P.H.: Catabolite repression as a pleiotropic response. J. Biochem. 79, 33 (1976).

9. O'Farrell, P.Z. and Goodman, H.M: Resolution of simian virus 40 proteins in whole cell extracts by two-dimensional electrophoresis: Heterogeneity of the major capsid protein. Cell 9, 289-298 (1976).

10. Estes, M.K., Huang, E.S., and Pagano, J.S.: Structural polypeptides of simian virus 40. J. Virol. 7, 635-641 (1971).

11. Coffino, P., Bourne, H.R., Friedrich, U., Hochman, J., Insel, P.S., LeMaire, I., Melmon, K.L., and Tomkins, G.M.: Molecular mechanisms of cyclic AMP action: A genetic approach. Recent Progress in Hormone Res. 32, 669-684 (1976).

12. Steinberg, R.A., O'Farrell, P.H., Friedrich, U., and Coffino, P.: Mutations causing charge alterations in regulatory subunits of the cyclic AMP-dependent protein kinase culture S49 lymphoma cells. Submitted for publication (1976).

13. Shreffler, D.C. and David, C.S.: The H-2 major histocompatibility complex and the I immune response region: Genetic variation, function, and organization. Adv. Immunol. 20, 125-195 (1975).

14. Benacerraf, B. and McDevitt, H.O.: Histocompatibility-linked immune response genes. Sci. 175, 273-279 (1972).

15. Nathenson, S.G. and Cullen S.E.: Biochemical properties and immunochemical-genetic relationships of mouse H-2 alloantigens. Biochim. Biophys. Acta 344, 1-25 (1974).

16. Cullen, S.E., David, C.S., Shreffler, D.C., and Nathenson, S.G.: Membrane molecules determined by the H-2 associated immune response region: isolation and some properties. Proc. Nat. Acad. Sci. (U.S.A.) 71, 648-656 (1974).

17. Jones, P.P.: Analysis of H-2 and Ia molecules by two-dimensional gel electrophoresis. Submitted for publication (1976).

18. Tomkins, G.M.: The metabolic cod. Sci. 189, 760-763.

19. Yamamoto, K.R. and Alberts, B.M.: Steroid receptors: Elements for modulation of eukaryotic transcription. Ann. Rev. Biochem. 45, 722-745 (1976).

Plant Proteins Evaluated by Two-Dimensional Methods

H. Stegemann

The best ways to overcome difficulties in a tedious process and secondary alteration of proteins are 1) separation methods taking care of crude samples to shorten the working time and 2) a sensitive detection of the components to handle minute amounts. In addition, plant proteins are generally less soluble then animal proteins, they belong more often to the glycoprotein group and their handling is rather likely interfered by tanning due to polyphenols formed in plant tissues.

Gels are effective carriers in this respect and suitable for microestimation of enzymes especially if made from acrylamide (1) (2). Isoelectric focusing in ampholytes in one dimension (3) is a powerful tool if done in tubes and gels (4). If done in slabs, the method is "allergic" against many ions which have to be removed. It is recommended to use focusing among the very first steps in any protein work, not to determine the mixture's complexity but to learn something about the distribution of the isoelectric points for further handling. Application of 2-dimensional methods and taking advantage of physical and/or physiological parameters will give the most satisfying results.

METHODS

All the different methods of electrophoresis and focusing described later were performed in one little apparatus, PANTA-PHOR, Labor-Müller, D-3510 Hann.Münden (5). Technical details of the methods mentioned are available from the author. Ask for the student's leaflet "Instructions for Electrophoresis and Focusing with the PANTA-PHOR", where we have compiled our experiences with polyacrylamide as carrier since 1960.

PROPERTIES OF SOME PLANT PROTEINS

Before discussing 2-dimensional techniques, mainly applied by us to proteins of potato tubers and maize seeds or seedlings, I like to summarize the work done earlier in our institute. It may be fairly well known from our and other colleagues work in the early sixties, that the distribution patterns of proteins, esterases and peroxidases are characteristic for varieties and cultivars. The latest result of these findings is an index of all potato varieties of Europe based on electrophoretic patterns (6). Fig. 1 gives the corresponding patterns of esterases for 18 cultivars and cv. MARITTA as an internal standard. We have found that the pattern is not influenced by environmental factors and that the size distribution of protein monomers in potato tubers was less changed during evolution and breeding compared with the change in charge (7). Therefore, focusing would be the method of choice also for the "Index" if not the current price of various ampholytes would be a limiting factor.

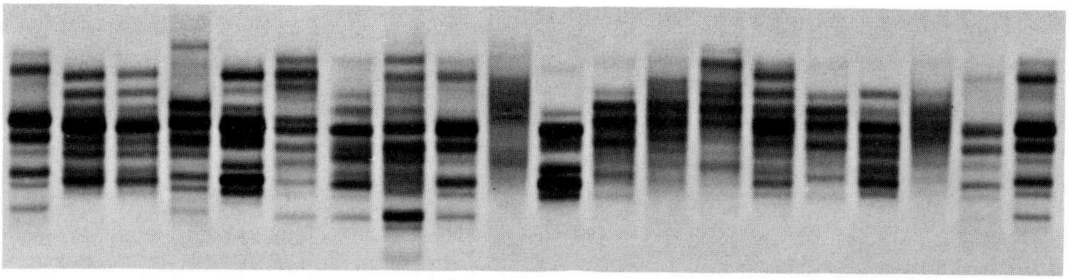

Fig. 1. Cultivar depending patterns of esterases in the raw sap from potato tubers. PAGE with 6% cyanogum in buffer (Tris/borate) pH 8.9 at 2°C. Stained with naphthylacetate and Fast blue RR-salt. From left to right: MARITTA, RENATA, GLORIA, PORTA, WEIMA, MONI, REGINA, AGNES, VINETA, NORDSTERN, GUNDA, ERSTLING, BELLADONNA, TAIGA, BISON, FINA, ASTRID, SOMMERSTÄRKE, REMEDY, MARITTA.

2-D-TECHNIQUES, PHYSICAL PARAMETERS. CAUTIONS.

More individual protein bands or spots and hence more detailed genetic information can be obtained using two variables, for example charge and size, which we have called "mapping technique" (7) (8). After focusing in one dimension, electrophoresis with or without pretreatment with Na-dodecylsulfate (SDS) is applied in the rectangular direction. Especially worthwhile is this technique if done in a block of 16 mm thickness with samples of about 400 mg (sometimes kept in solution with ampholytes pH 7 - 10). The block can be cut into 8 slabs and stained for different enzymes. In addition it is one of the fastest procedures for getting pure proteins on a mg-scale.
A disadvantage of cutting out spots from a gel is the contamination with oligomeric acrylamides. This is of no importance if the sample is used for further electrophoresis or enzymatic essay. However, for fingerprints and so on the elimination of

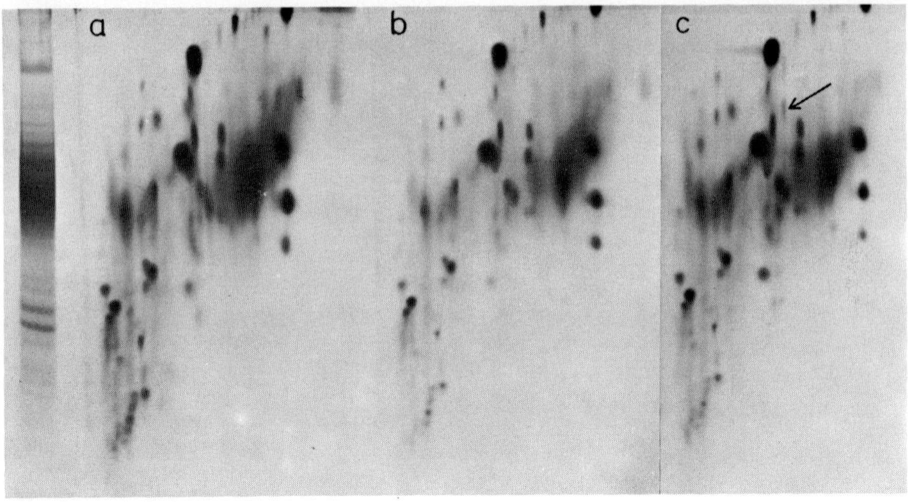

Fig. 2. "Mapping" of seed proteins of maize after buffer extraction. First dimension (left to right): Focusing with 1% Servalyt pH 6 - 8 in 6% acrylamide / 0.16% BIS. Second dimension (top to bottom): PAGE with 5% Cyanogum in Tris/. borate pH 8.9 at 2 °C.
a and b: different runs of the same extract; c: extract of a closely related inbred line. Stained with Supranolcyanin (Bayer) for proteins.

oligomeres can be overcome by preparative techniques with intermittent elution during electrophoresis (5).

An example for mapping seed proteins from maize is given using the extract from dry defatted seeds of the same cultivar (Fig. 2a and 2b) to show the highly reproducible pattern, and of a closely related cultivar (Fig. 2c) to show the slight difference of the protein distribution not to detect in either of the separation methods alone.

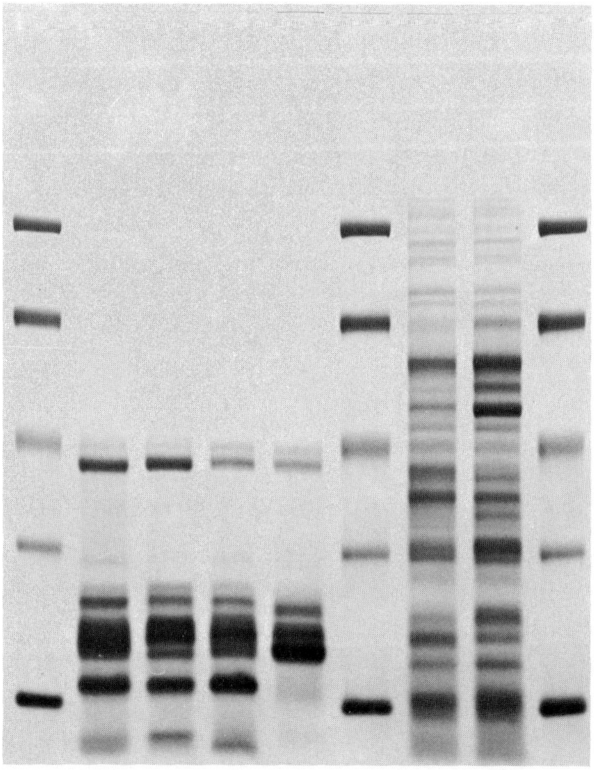

Fig. 3. Comparison of tuber proteins of potatoes, cv. MARITTA (2), SIEGLINDE (3), VORAN (4,5) and seed proteins of maize (7,8). (See Fig. 2). PAGE in Tris/borate/0.1% SDS pH 7.1 and 5% Cyanogum at 2 °C. Markerproteins (1,6,9) MW 1000, 670, 370, 257, 143 x 10^2. Supranolcyanin stain. All samples reduced with 2-mercaptoethanol, except No. 5. (Numbers from left to right).

In contrast to the rather small range of molecular weights of protein monomers in the potato (7) the distribution of the main proteins in maize seeds is found to be between 10^4 and 10^5 daltons (Fig. 3). - Two-dimensional techniques give a sound and quick insight whether there was a shift of the propagated protein or enzyme pattern after passage through a meristemic culture (9).

Another related 2-D-method introduced by us (7) which is less versatile in respect to enzyme differentiation (due to the lack of enzyme activity after treatment with SDS but more precise in respect to physical parameters) is the "SDS-mapping".

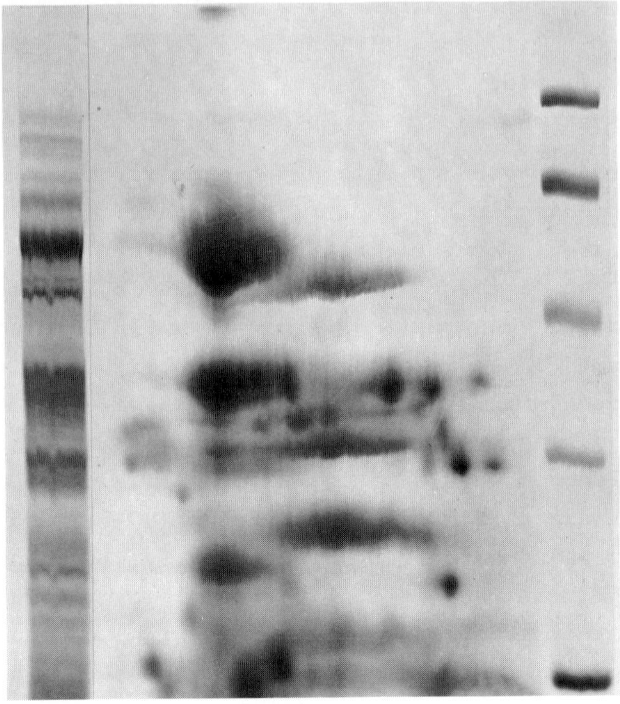

Fig. 4. "SDS-Mapping" of seed proteins (maize). First dimension (left to right): Focusing with 1% Servalyt pH 6-8 in 6.0% acrylamide/0.16% N,N'-methylene-bisacrylamide. Incubation at 20/50/100 °C for 10/20/5 min. in 2% SDS/2% ME, followed by PAGE in Tris/borate/0.1% SDS pH 7.1 and 7.5% Cyanogum at 2°. Note the uneven bands of the electrophoretic pattern on the left (for comparison) due to uneven polymerization. Marker-proteins as in Fig. 3.

After focusing in gels the separated proteins are not reduced, or reduced by mercaptoethanol, transformed to negatively charged rod-like particles with SDS according to (10) (11) and than forced into a gel slab containing SDS with or without a porosity gradient. Marker proteins for the determination of molecular weights are run besides the focused sample. SDS-mapping of maize (seed) proteins show - in contrast to those of potatoes - a very reproducible smearing in some parts of the slab due to carbohydrate moieties with a size distribution not as discrete as in pure proteins (Fig. 4). The most promising but somewhat tedious method seems to be focusing in the first dimension and a porosity gradient electrophoresis without SDS in the second. Here proteins stay in the original conformation and enzyme activity can still be observed which makes possible a very specific identification of multiple forms caused by genetic variation.

Before using protein- or enzyme-patterns for chemotaxonomy or genetic screening in the bacteria-, fungi-, plant- and animal field, one should be sure that there is a certain time interval where the patterns are not influenced by environmental or physiological changes. A warning example is a complex of proteins found in the immature potato tuber obscuring the cultivar-dependent pattern (7). After standard electrophoresis we see a single protein band, which was isolated by preparative electrophoresis taking samples of 1 - 3 g. In a porosity gradient (Fig. 5) two proteins (MW 2.7 and 4.9 x 10^5) can be observed disappearing with maturity. It is noteworthy that this protein band is disintegrated during focusing or in urea but not by addition of ampholytes without focusing. If "mapped" in a 16 mm thick block taking cuts 2 mm thick and stained for different enzyme activity, a considerable complexity is seen. We wondered whether this complex is variety-dependent or characteristic of the immature state among all the potato cultivars since the position of this band in standard electrophoresis is the same with all samples tested. There-

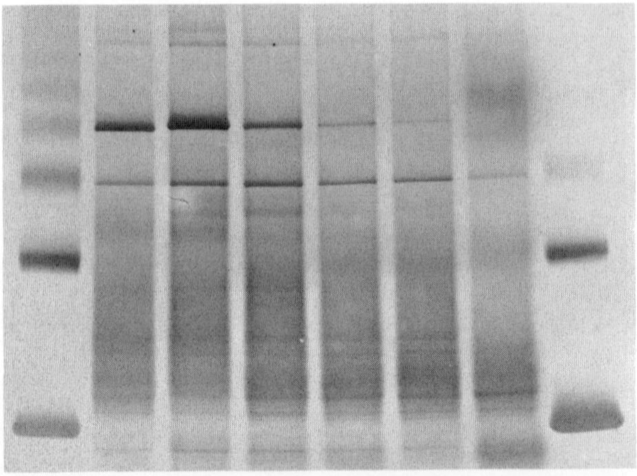

Fig. 5. Porosity gradient (PAA 6-26%) electrophoresis of proteins from tubers cv. MARITTA of different maturity. Left to right: harvest June to October. 400 volt for 27 hours in Tris/borate pH 8.9. References: oligomeric serum albumin (bovine).

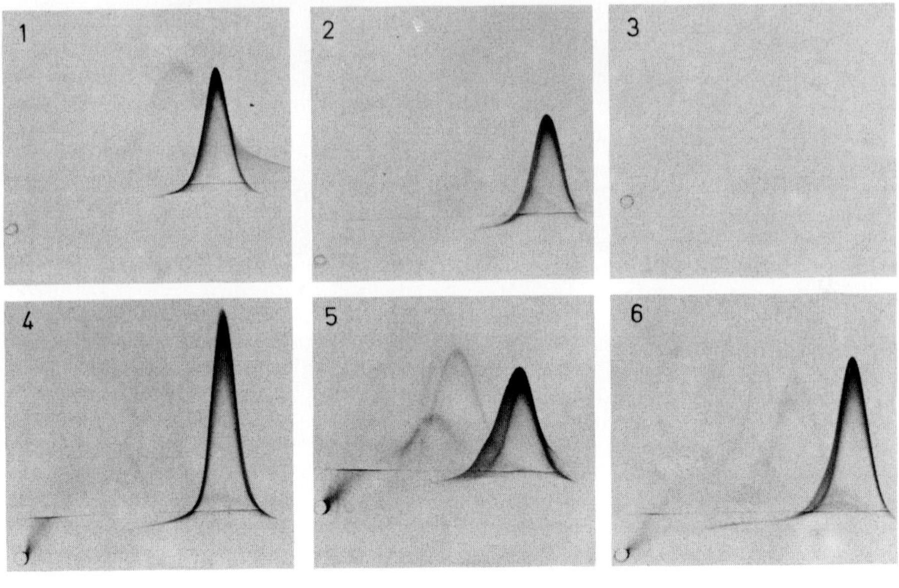

Fig. 6. 2-D-immuno electrophoresis (Clarke-Freeman-Technique). First dimension (left to right): electrophoresis in 1% agarose in buffer Tris/borate pH 8.9, anode at the right side. Samples: isolated fraction from immature potato tuber, cv. GRATA (1); raw sap from immature tubers, cv. SIEGLINDE (2), HANSA (4), GRATA (5), MARITTA (6). From mature tuber, cv. GRATA (3). Second dimension: electro-diffusion in the same sytem, containing antiserum (rabbit) against sample (1). Anode on top.

fore we applied 2-dimensional immuno-techniques. The protein-complex from the immature cultivar GRATA was crossreacted with other cultivars (Fig. 6). It turned out that the immunoresponse is identical among all cultivars. Furthermore the interrelationships of other proteins in mature tubers were checked by immuno techniques either taking isolated fractions after preparative electrophoresis or using a mixture separated in a gel by focusing (12). The almost absent antigenicity of the main fractions and the close relationship among the minor bands in the anodic range is seen in fig. 7.

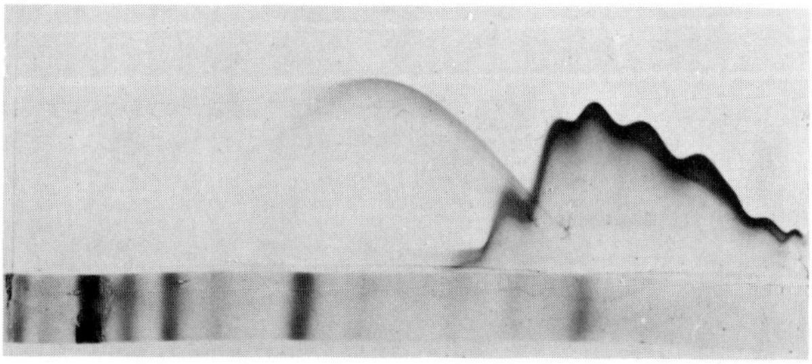

Fig. 7. Immun-mapping of tuber proteins cv. VORAN. First dimension (lower gel, anode at the right): focusing in 1% Servalyt pH 4 - 6 and 4% PAA for 3 hours at 200-300 volt. Second dimension: electrodiffusion into 1% agarose containing rabbit antiserum against the totality of proteins of cv. VORAN. Buffer Tris/borate pH 8.9, anode on top.

2-D-TECHNIQUES, PHYSIOLOGICAL PARAMETERS

A physiological dimension may be used as well, for instance the growth of a mycelium to evaluate the age-dependent enzyme synthesis or the enzyme breakdown. You simply incubate an agar-solidified nutrient in the center of a petri dish with an inoculum of a fungus, let it grow almost to the edges. Then a strip is cut out including the center, the strip is placed horizontally - frozen or unfrozen - into the PANTA-PHOR. Then acrylamide is jelled around it and the vertical focusing or electrophoresis is started (13). Depending on the fungus or the enzyme tested the activity is more pronounced in the young or in the old mycelium, sometimes the synthesis is not age dependent or is enhanced in the medium range. Finally, in gels with small pores and 20 mm thick even a 3-dimensional separation is possible (to be published).

REFERENCES

1. Raymond, S., Wang, Y.J.: Preparation and properties of acrylamide gel for use in electrophoresis. Anal. Biochem. [New York] 1, 391-396 (1960).
2. Raymond, S., Weintraub, L.: Acrylamide gel as a supporting medium for zone electrophoresis. Science 130, 711 (1959).
3. Vesterberg, O., Svensson, H.: Isoelectric fractionation, analysis and characterization of ampholytes in natural pH gradients. IV. Further studies on the resolving power in connection with separation of myoglobins. Acta chem. scand. 20, 820-834 (1966).
4. Fawcett, J.S.: Isoelectric fractionation of proteins on polyacrylamide gels. FEBS Letters 1, 81-82 (1960) and other authors.
5. Stegemann, H.: Apparatur zur thermokonstanten Elektrophorese oder Fokussierung und ihre Zusatzteile. Zeitschr. analyt. Chemie 261, 388-391 (1972).
6. Stegemann, H., Loeschcke, V.: Index Europäischer Kartoffelsorten / Index of European Potato Varieties (bilingual), based on electrophoretic spectra. Mitt. Biol. Bundesanstalt, Berlin, Heft 168, p. 1-215 (1976).
7. Stegemann, H., Francksen, H., Macko, V.: Potato proteins; genetic and physiological changes, evaluated by one- and two-dimensional PAA-gel-techniques. Zeitschr. Naturforsch. 28 b, 722-732 (1973).
8. Macko, V., Stegemann, H.: Mapping of potato proteins by combined electrofocusing and electrophoresis. Identification of varieties. Hoppe-Seyler's Zeitschr. physiol. Chemie 350, 917-919 (1969).
9. Stegemann, H., Loeschcke, V., Bode, O., Huth, H.: Gelelektrophoretische Untersuchungen von Kartoffelknollen nach Aufzucht aus Meristem-Kultur. Jahresberichte Biol. Bundesanstalt A 65 (1970) and P 62 (1971).
10. Shapiro, A.L., Viñuela, E., Maizel, J.V.: Molecular weight estimation of polypeptide chains by electrophoresis in SDS-polyacrylamide gels. Biochem. Biophys. Res. Commun. 28, 815-820 (1967).

11. Koenig, R., Stegemann, H., Francksen, H., Paul, H.L.: Protein subunits in the potato virus X group. Determination of the molecular weights by polyacrylamide electrophoresis. Biochim. Biophys. Acta 207, 184-189 (1970).

12. Stegemann, H.: Electrophoretic characterization of potato varieties. International Symposium on "The Biology and Taxonomy of the Solanaceae", Birmingham, England. Edited by J.G. Hawkes and R.N. Lester, in print.

13. Roeb, L.: Eine neue Technik zur in-situ-Bestimmung von Enzymen in Pilzkulturen, dargestellt an Polygalakturonasen. Phytopath. Zeitschr. 79, 359-363 (1974).

Two-Dimensional Separation of Lymphocyte Microsomal Membrane Proteins: Isoelectric Focusing Linked to SDS-Polyacrylamide Gel Electrophoresis

H. Knüfermann

ABSTRACT

Microsomal membrane fractions from calf lymphnode lymphocytes were solubilized by treatment with SDS, Triton X-100 or Empigen BB. The membrane proteins were separated by SDS-polyacrylamide gradient gelelectrophoresis and by isoelectric focusing. These two separation methods were combined: first dimensional separation was done by isoelectric focusing; gels were then dialysed against the SDS buffer system and SDS electrophoresis was performed subsequently at a right angle. These two-dimensional separations provide a far higher resolution of lymphocyte membrane proteins than obtained with either of the both methods alone.

INTRODUCTION

Solubilization of membrane proteins in sodium dodecylsulfate followed by electrophoresis in detergent containing polyacrylamide gels (SDS-PAGE) has become one of the most powerfull tools in membrane protein analysis. Among a lot of other applications this technique was used intensively to investigate erythrocyte (1) and thymocyte (2) plasma membranes. The method of isoelectric focusing of membrane extracts with non ionic detergents was introduced by Merz et al. (3). For the erythrocyte membrane both separation parameters, i.e. molecular weight and protein charge, were combined to a two dimensional separation technique (4,5,6) resulting in a far higher membrane protein resolution and prooving the heterogeneity of induvidual bands of the SDS-PAGE. Lymphocyte membranes have a far more complex protein composition than erythrocytes even in the simple SDS-PAGE; thus it was attempted to improve the resolution by introducing polyacrylamide gradient gels in combination with the two-dimensional separation technique.

MATERIALS AND METHODS

Chemicals

Unless otherwise stated all chemicals and biochemicals were obtained from Serva (Heidelberg), Boehringer (Mannheim) and Merck (Darmstadt). As carrier ampholytes the wide range Servalyt type pH 2 to 11 was used. The amphipathic detergent Empigen BB, an alkylbetaine ($R-N^{+}-(CH_3)_2-CH_2.CO_2^{-}$ ($R = C_{10}-C_{16}$)) was provided by Albright and Wilson Ltd. Marchon Division, Whitehaven, Cumberland, England (Lot. No. 62604).

Membrane Preparation and Solubilization

Lymphocyte suspensions from calf mediastinal lymph nodes were prepared according to Ferber et al. (7), as well as the cell disruption by the nitrogen cavitation method, differential centrifugation and washing of the crude microsomal membrane fraction to release the cytoplasmatic proteins. The resulting washed microsomal membrane fractions were solubilized by adding SDS up to 1%, by adding the non-ionic detergent Triton x-100 up to 2% plus dimethylsulfoxide (DMSO) up to 10% or by adding the amphipathic detergent Empigen BB up to 2% plus DMSO up to 10%. The detergent extracts were shaken at room temperature for 15 min. and centrifuged at 7.4×10^6 $g_{av.}$ x min. in a Beckmann centrifuge type L5-75 B (rotor type SW 50.1 with 0.8 ml adaptors). The clear supernatant of these detergent extracts were separated from the gelatineous pellet and were directly used for isoelectric focusing experiments after adding glycerol up to 10%. For electrophoresis the supernatants were dialysed against two changes of a 5 mM phosphate buffer pH 7.4 containing 0.1% SDS at a temperature of +10°C; no precipitation of membrane proteins occured during this dialysing procedure. Before electrophoresis samples were concentrated to a protein concentration of 10 to 20 mg/ml by means of a Amicon micro pressure filtration cell using PM 10 Diaflo membranes.

SDS-Gradient-PAGE

For analytical SDS-PAGE the Pharmacia apparatus GE-4 and the commercially available gradient gels (4% to 30% polyacrylamide) were used. A pre-electrophoresis was done for 15 min. at 25 mA per gradient gel slab. The sample applicator was fixed to the gels by a small amount of hot 1% agarose solution containing 1% SDS and the marker dye. Electrophoresis of about 0.1 mg membrane protein containing samples was done for about 4 h at 50 mA per gel slab using the continous buffer system as in (1).

Isoelectric Focusing

Isoelectric focusing of the detergent extracts in detergent containing gels without incorporation of urea was done as described in principle in (4,5,6) with the following modifications: Servalyt pH range 2-11 up to 2% was used; gel dimensions were 6.5 cm length and 2.5 mm diameter for two-dimensional separations or 10 cm length for one dimensional focusing. Gels were prefocused at constant current of 0.2 mA/gel up to 25 V/cm. Focusing was done at +4°C over night (16-18 h)at this constant voltage. Determination of the pH was done directly on the gels using a micro surface glass electrode (Ingold Type 273-M2) in combination with a reference electrode (Ingold Type 374-M8).

Two Dimensional Separation

Gels from the isoelectric focusing experiment were dialysed against the SDS buffer system for 20 min. at room temperature and placed on top of the gradient gel slabs. They were fixed to the SDS-gel slabs by 1% agarose containing the marker dye. Electrophoretic conditions were the same as in SDS-PAGE.

Staining Procedures

Analytical SDS-PAGE gels were stained in a 0.05% Coomassie brilliant blue solution (25%isopropanol/10% acetic acid) for 24 h. Diffusion destaining was done for about 20 h with several changes in 10% acetic acid. Gels from isoelectric focusing and two-dimensional gel slabs were first fixed in 25% isopropanol/10% acetic acid to remove the carrier ampholytes and then stained as described above.

RESULTS AND DISCUSSION

SDS-Gradient-PAGE

Electrophoresis of solubilized microsomal membrane proteins

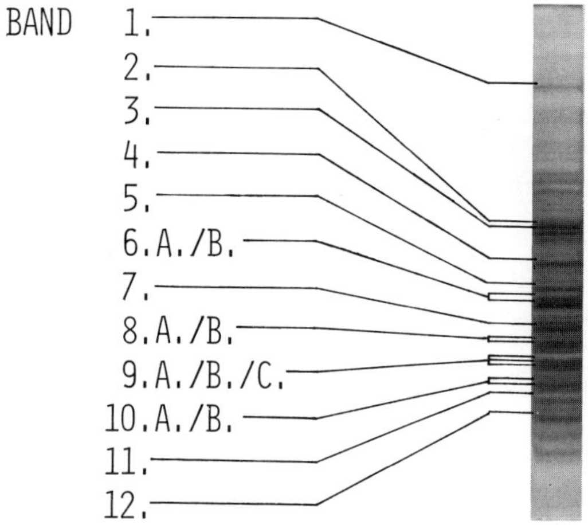

Fig. 1.: SDS-Gradient-PAGE: Membrane protein pattern of calf lymph node microsomal membranes.

Table 1. Maior Proteins of Lymphocyte Microsomal Membranes

Band	Apparent molecular weight	Standart Deviation
1.	~ 232.000	± 4.3 %
2.	80.900	± 3.5 %
3.	77.100	± 3.2 %
4.	59.600	± 3.7 %
5.	48.800	± 2.9 %
6.A.	45.500	± 2.1 %
B.	42.700	± 3.7 %
7.	35.000	± 3.7 %
8.A.	31.300	± 3.5 %
B.	30.300	± 3.6 %
9.A.	27.100	± 4.1 %
B.	26.400	± 4.2 %
C.	25.300	± 4.0 %
10.A.	22.400	± 4.0 %
B.	21.700	± 3.7 %
11.	19.800	± 3.5 %
12.	16.900	± 4.0 %
Number of determinations n = 9		

using polyacrylamide gradient gels greatly improves the resolution of the protein pattern in comparison to the normal polyacrylamide gels. Up to 51 different bands can be detected, demonstrating the very complex composition of this lymphocyte membranes. Fig. 1. shows this intricate separation pattern; Table 1. lists the apparent molecular weights of the maior protein components. Triton- and Empigen-extracts yield an identical electrophoretic separation pattern, i.e. all membrane proteins are solubilized. Nevertheless some proteins are extracted to a lower extend (protein No. 7. and No. 10.).

Isoelectric Focusing

The one step solubilization procedure of microsomal membranes of a fairly high protein concentration (about 20 mg/ml) yields a solubilization of about 50% of membrane proteins if Triton

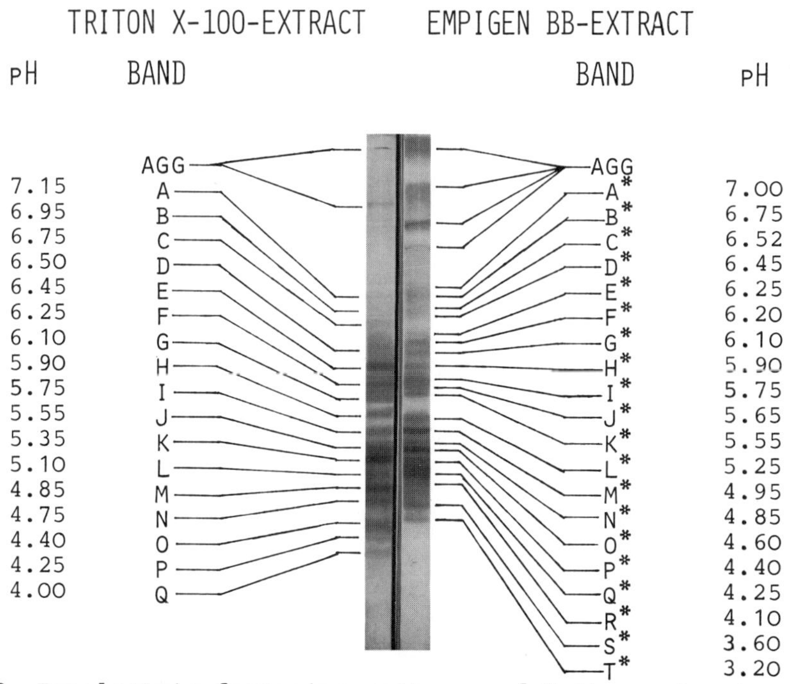

Fig. 2. Isoelectric focussing patterns of Triton-extracts and Empigen-extracts of lymphocyte microsomal membranes on polyacrylamide gels containing the same detergents. (AGG = aggregates).

X-100 is used and a solubilization of up to 40% in the case of Empigen BB.

Both extracts can be separated by isoelectric focusing on polyacrylamide gels with incorporation of the same detergent into the gel. In both cases some precipitation occurs on top of the gel and on the gel surface due to a shrinking process during prefocusing which allows part of the sample to migrate in between the glass tube wall and the gel. A stable and linear pH gradient is built up and maintained; the pH gradient in the Empigen containing gels is flattened at the alkaline end and is somewhat steeper than the gradient in the Triton gels. As shown in Fig. 2. and 3. most of the soluble membrane proteins focus in the pH range from 6.5 to 4.0 (Triton) and 6.25 to 3.2 (Empigen) respectively; 17 protein bands can be detected in

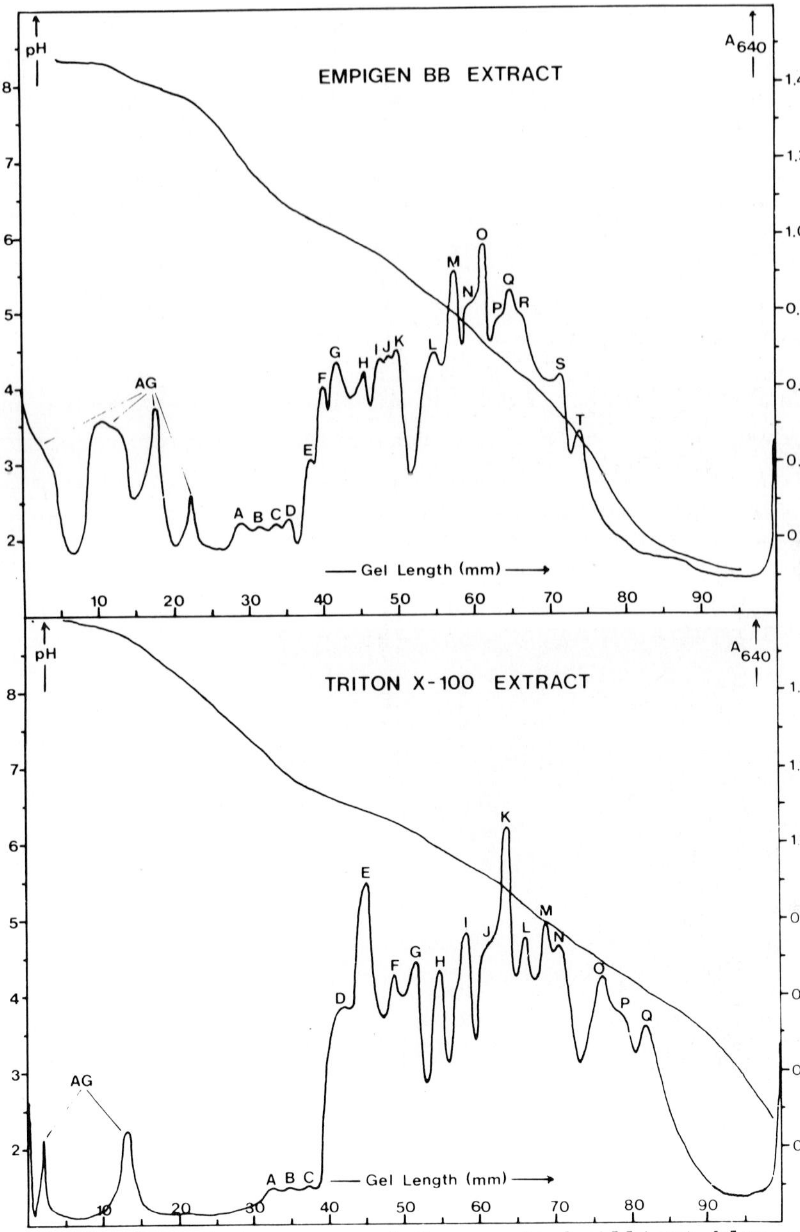

Fig. 3. Densitometric scans of Coomassie brilliant blue stained isoelectric focusing gels with the corresponding pH gradients. The upper part shows the Empigen extract the lower part shows the Triton extract of lymphocyte microsomal membranes. (AG = aggregates).

the Triton extracts, whereas the Empigen extracts exhibit 20 proteins. The focusing patterns differ from each other considerably. This clearly reflects a different solubilization behaviour of the membrane protein complexes in the different dissociating media used. Thus, for comparison different isoelectric focusing patterns, the use of identical dissociating media is a essential prerequisite.

Two-Dimensional Separation

Combining isoelectric focusing of Triton-X-100-extracts and SDS-gradient-PAGE to a two-dimensional separation results in a highly complex separation pattern; up to 95 microsomal membrane proteins can be distinguished (Fig. 4.). Importantly one

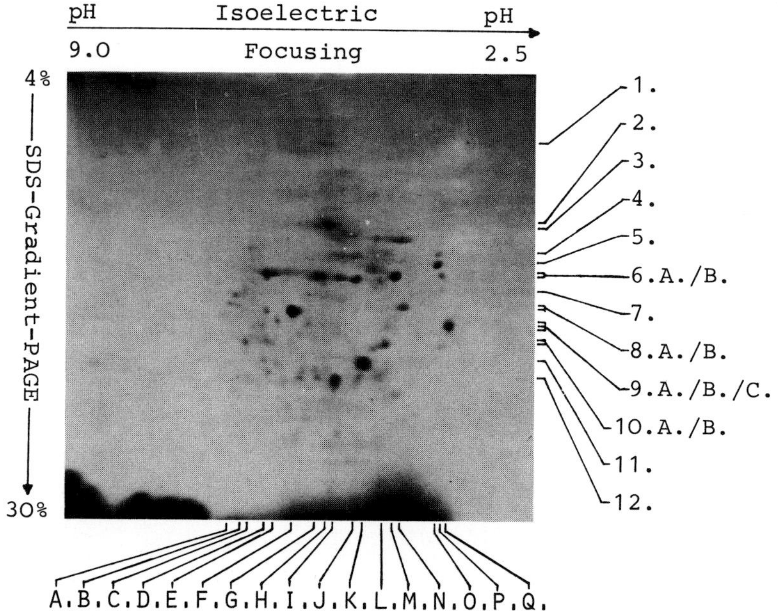

Fig. 4. Two-dimensional separation pattern of a Triton X-100 extract from lymphocyte microsomal membranes. First dimension: isoelectric focusing; second dimension: SDS-gradient-PAGE. Numbers correspond to the electrophoretic separation pattern of Fig. 1., letters to the focusing pattern of Figs. 2. and 3.

can recognize, that none of the both separation methods alone separates the molecular protein subunits. This ist most clearly demonstrated in the case of SDS-bands 6.A. and 6.B., which focus at a variety of pH values, or the focusing band I., which splits into the bands 2., 3., 6.B. and 12. upon SDS-gradient PAGE. Thus, as has been shown for the erythrocyte membrane proteins, this two-dimensional separation technique defenitely prooves the heterogeneity of the single SDS-PAGE protein bands and the focusing bands. It clearly demonstrates too that preparative isolation of induvidual membrane components cannot be achieved with a single step procedure. At an analytical scale this communication points out, that for this purpose one has to apply a combination of different electrophoretic and/or chromatographic methods, just as in the case of erythrocyte membrane proteins (8,9,10).

AKNOWLEDGEMENTS

I thank Mrs. France Pressler for skillfull technical assistance, Prof. D.F.H. Wallach and Prof. H. Fischer for continous encouragement and advice and Dr. S. Bhakdi for critical discussions. This work was supported by the Deutsche Forschungsgemeinschaft.

REFERENCES

1. Fairbanks,G., Steck,T.L. and Wallach, D.F.H.: Electrophoretic analysis of the maior polypeptides of the human erythrocyte membrane. Biochemistry 10 2606-2617 (1971).
2. Schmidt-Ullrich,R., Ferber,E., Knüfermann,H., Fischer,H. and Wallach,D.F.H.: Analysis of the proteins in thymocyte plasma membrane and smooth endoplasmatic reticulum by sodium dodecylsulfate gel electrophoresis. Biochim. Biophys. Acta. 332 175-191 (1974).
3. Merz,D.C., Good,R.A. and Litman,G.W.: Segregation of membrane components using isoelectric focusing in polyacryl-

amide gels. Biochim.Biophys. Res. Comm. 49 84-91 (1972).

4. Bhakdi,S., Knüfermann,H. and Wallach,D.F.H.: Separation of EDTA-extractable erythrocyte membrane proteins by isoelectric focusing, linked to electrophoresis in sodium dodecylsulfate. Biochim.Biophys. Acta 345 448-457 (1974).
5. Bhakdi,S., Knüfermann,H. and Wallach,D.F.H.: Separation of hydrophobic membrane proteins by isoelectric focusing linked to sodium dodecylsulfate gel electrophoresis. In: Righetti,P.G. (Ed.) Progress in Isoelectric Focusing and Isotachophoresis. pp. 281-291. North-Holland Publishing Company (1975).
6. Bhakdi,S., Knüfermann,H. and Wallach,D.F.H.: Two-dimensional separation of erythrocyte membrane proteins. Biochim. Biophys. Acta 394 550-557 (1975).
7. Ferber,E., Resch,K., Wallach,D.F.H. and Imm,W.: Isolation and characterization of lymphocyte plasma membranes. Biochim.Biophys. Acta 266 494-504 (1972).
8. Knüfermann,H., Bhakdi,S. and Wallach,D.F.H.: Rapid preparative isolation of maior erythrocyte membrane proteins using polyacrylamide gel electrophoresis in sodium dodecylsulfate. Biochim.Biophys. Acta 389 464-476 (1975).
9. Bjerrum,O.J., Bhakdi,S., Bøg-Hansen,T.C., Knüfermann,H and Wallach,D.F.H.: Quantitative immunoelectrophoresis of proteins in human erythrocyte membranes. Analysis of protein bands obtained by sodium dodecylsulfate gel electrophoresis. Biochim.Biophys. Acta 406 489-504 (1975).
10. Bhakdi,S., Bjerrum,O.J. and Knüfermann,H.: The maior "intrinsic" membrane protein of human erythrocytes: Preparation, isolation and immunoelectrophoretic analyses. Biochim. Biophys. Acta (in press) (1976).

Mammalian Mitochondrial Ribosomes: Two-Dimensional Gel Electrophoresis of the 55 S- and the Corresponding Subunit Proteins

W. Czempiel

ABSTRACT

The combination of isoelectric focusing (*first dimension*) and slab gel electrophoresis (*second dimension*) in a two-dimensional electrophoretic system allows separation of the small amounts of protein (80 µg), which are available from mammalian 55S mitochondrial ribosomes and their subunits. The protein pattern of the small 28S ribosomal subunit revealed 27 dark and 16 weak spots. The pattern of the large 39S subunit displays 45 dark and 18 weak spots. In contrast 60 dark and 47 weak spots were found for the 55S monosome protein pattern. Isoelectric points of the majority of proteins ranged from pH 7 to pH 9. Molecular weights of 20 000 to 45 000 daltons were calculated for the bulk of ribosomal proteins by preliminary experiments with SDS electrophoresis in the second dimension.

INTRODUCTION

The elucidation of the structural components of mitochondrial ribosomes is a prerequisite for understanding their function in protein biosynthesis. The characterization of mammalian mitochondrial ribosomes in respect to their physico-chemical properties have been attempted previously (1). The S-values of this class of ribosomes were found to be 55S to 60S (2). From the RNA percentage of about 30% a high protein content of approximately 2×10^6 daltons was calculated (3).

Mapping of the ribosomal proteins by two-dimensional polyacrylamide gel electrophoresis was expected to reveal the number of proteins of the 55S monosome and of the 28S and 39S subunits. The originally used two-dimensional polyacrylamide gel electrophoresis developed for mapping ribosomal proteins from bacteria (4) had to be modified in our studies for two reasons: Firstly, mitochondrial ribosomal proteins exhibit a lower basicity than bacterial ribosomal proteins and, therefore, hardly migrate cathodically in a basic 1-D gel system. Secondly, proteins from mitochondrial ribosomes are only available in minute quantities. Two-dimensional separation procedures combining isoelectric focusing (IEF) in the first dimension either with gel electrophoresis or SDS electrophoresis in the second dimension have been described for a variety of proteins (5,6,7,8). A similar system allowed the two-dimensional separation of the proteins from 55S rat liver mitochondrial ribosomes (9). Additional information on these proteins was obtained by dissociation of the 55S particle into subunits. The separation of the proteins from the small 28S and the large 39S ribosomal subunits by the afore mentioned electrophoretic procedures is presented in this paper.

MATERIAL AND METHODS

Preparation of Mitochondrial Ribosomes

Mitochondria were isolated from homogenized rat livers by differential centrifugation as described elsewhere (9). The mitochondrial lysate that pelleted through a 1 M sucrose cushion yielded 'crude' ribosomes. After further purification on a 10% - 30% convex sucrose gradient 'pure' 55S ribosomes were obtained. The gradient was fractionated by monitoring at 254 nm. The fraction containing the 55S ribosomes was pelleted again. Ribosomal proteins were subsequently extracted by 66% acetic acid, dialysed stepwise against 10% to 0.2% acetic acid and lyophilized. The lyophilized proteins dissolved in 8 M urea and mixed with Sephadex were directly layered on top of the 1-D focusing gel (9).

Preparation of 28S and 39S Ribosomal Subunits

The 55S ribosomes were resuspended in a buffer containing 0.5 M KCl, 3 mM $MgCl_2$, 0.1 mM EDTA, 1 mM dithioerythrit, 2 mM TES-buffer, pH 7.6, and incubated in 0.1 mM puromycin for 10 min. at $37^{o}C$. After overnight centrifugation on a 10% - 30% sucrose density gradient made up in the same buffer, the two ribosomal subunits were obtained. The UV_{254} profile of the fractionated density gradient is given in Fig.1. The dissociation was quantitative; no 55S material could be detected. The proteins of the 28S and 39S subunits were prepared as described for the proteins from the 55S particles.

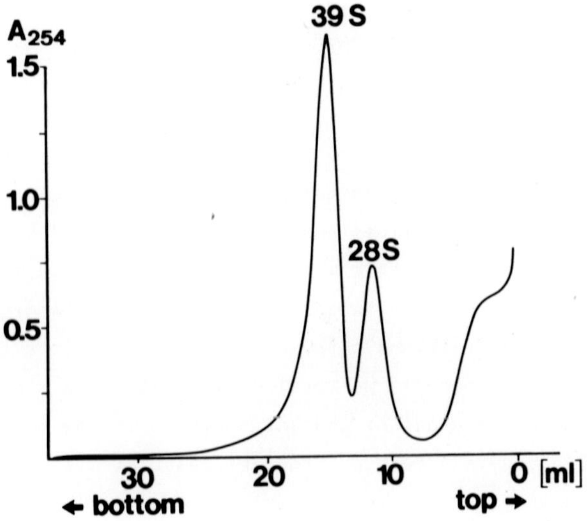

Fig. 1. Sucrose density gradient profile of subunits prepared from purified 55S mitochondrial ribosomes (rotor SW 27 Beckman, 14 hrs., 52 000x g).

Electrophoretic Separation of Mitochondrial Proteins

Isoelectric focusing in the first dimension was performed in 6 M urea in a 4.8% T, 3.8% C polyacrylamide gel according to KLOSE (9) supported by LKB ampholines stabilizing a pH gradient from pH 3.5 to pH 10. Isoelectric focusing proceeded for 5 hrs.

Separation in the second dimension was then carried out as described by KALTSCHMIDT and WITTMANN (4), however, the dimensions of the gel chamber for the 2-D gel were 10 x 10 x 0.3 cm. The monomer concentrations in the gel were 19,2% T, and 2.7% C. The 2-D gel again contained 6 M urea. The run proceeded for 17 hrs. at 60 Volts. The SDS gel that was used in additional experiments for separation in the second dimension had the following composition: 15.2% T, 1.3% C, 0.1% SDS in separation- and spacer gels, and 0.2% SDS in embedding gel and buffer as reported by KLOSE (10). This run was carried out in a Pharmacia electrophoresis apparatus for 1 hr. 20 min. with increasing voltage; final voltage: 400 Volts (10).

After fixation of the slab gels in an aqueous solution of 50% methanol, 10% acetic acid, the gels were stained overnight in Coomassie Brilliant Blue R 250 in methanol, acetic acid, water at a ratio of 4:1:5. Destaining was performed as described (9).

RESULTS AND DISCUSSION

The determination of the pH gradient in the 1-D focusing gel was performed by cutting gels in 0.5 cm slices and elution with 0.5 ml boiled distilled water for pH measurements with a micro glass electrode. After 5 hrs. basic marker proteins with known pI values as e.g. myoglobin (horse), RNAse, and cytochrome c reached positions corresponding to their isoelectric points. However, when electrophoresis was prolonged to 10 hrs.a slow cathodic migration of both the pH gradient and the position of the marker proteins occured. Fig.2 presents a mean pH gradient calculated from 3 single gels of the same run after a focusing time of 5 hrs. The slope of the gradient is nearly linear over a wide pH range.

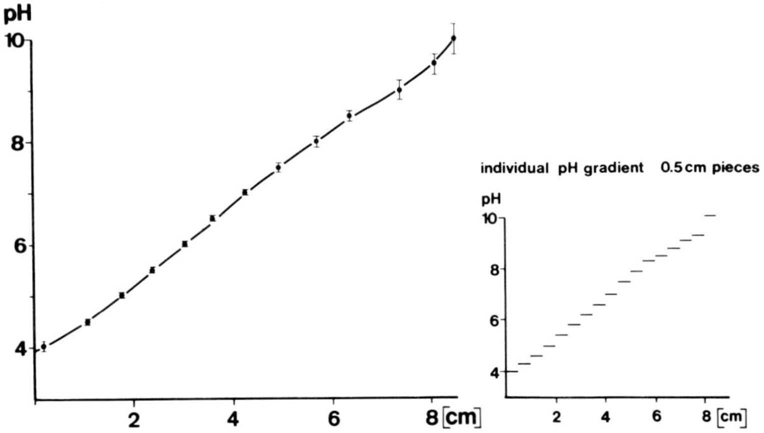

Fig. 2. pH gradient calculated from unstained 1-D gels, ampholines 3.5-10 (gel slices: 0.5 cm; elution: 0.5 ml H_2O)

The two-dimensional protein pattern of the small 28S ribosomal subunit proteins was obtained by isoelectric focusing in the first dimension and slab gel electrophoresis in the second dimension (Fig.3). 27 dark spots and 16 weak spots could be detected. 20 of the dark spots are identical with the dark spots in the previously described 55S protein pattern (Fig.5) (9).

Fig. 3. Two-dimensional protein pattern of the 28S mitochondrial ribosomal subunit showing 43 spots (27 dark and 16 weak spots). First dimension: IEF, anode at left; second dimension: slab gel electrophoresis, pH 4.0, anode on top of the gel.

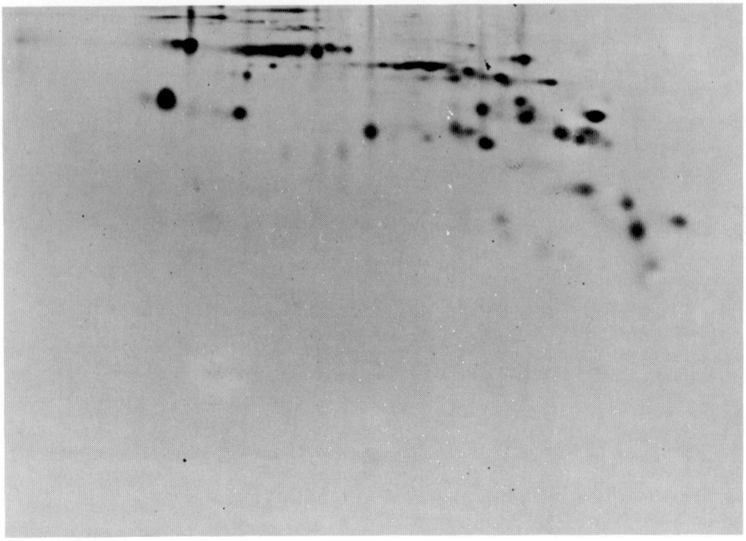

Fig. 4. Two-dimensional protein pattern of the 39S mitochondrial ribosomal subunit showing 63 spots (45 dark and 18 weak spots). The conditions are the same as in Fig.3.

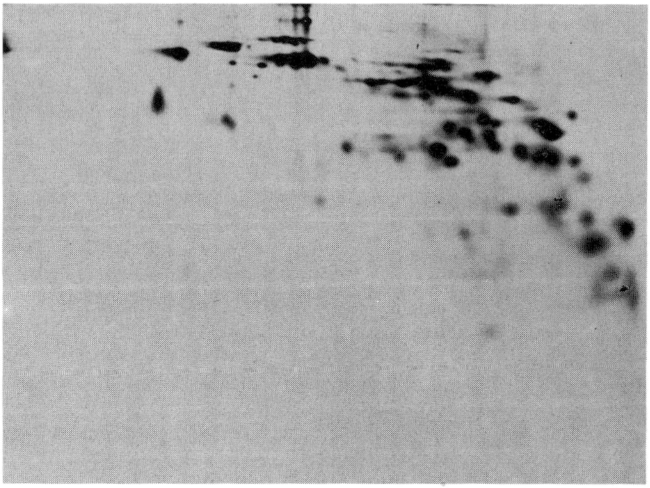

Fig. 5. Two-dimensional protein pattern of the 55S mitochondrial ribosomes showing 107 spots (60 dark and 47 weak spots). The conditions are the same as in Fig. 3.

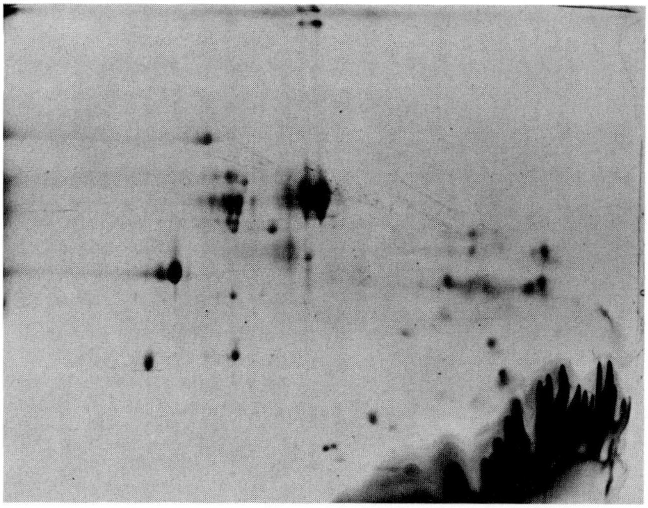

Fig. 6. Two-dimensional protein pattern of the 55S mitochondrial ribosomes. First dimension: IEF; second dimension: SDS gel electrophoresis, pH 8.3, cathode on top of the gel.

Fig.4 represents the 2-D protein pattern of the large 39S ribosomal subunit proteins. 45 dark spots and 18 weak spots were found. Compared to the 55S pattern 36 dark spots exhibit similar staining intensities and are located in identical positions (Fig.5). A few spots are identical in the protein patterns of the two ribosomal subunits. The question if whether these proteins are cross-contaminations caused by the other subunit or if whether they exist on both subunits cannot be decided by the protein mapping experiments.

The total number of spots in the two subunit patterns is approximately the same as in the 55S pattern. Differences may be caused by the salt washing conditions during the dissociation procedure. A number of about 72 mitochondrial ribosomal proteins is plausible, when only the dark spots of the 28S and the 39S subunits are taken into account.

Preliminary experiments combining IEF and SDS gel electrophoresis were performed to determine the molecular weights of the 55S proteins (Fig.6). The molecular weights of the bulk of proteins from the 55S ribosomes range from 20 000 to 45 000 daltons. Final results will be obtained when the calibration of SDS gels by marker proteins shows a linear relationship in the Ferguson plot. Another problem is the removal of the heavily stained ampholine-SDS complex in the lower right corner of the slab gel, since it may cause a distortion of the protein pattern in this gel region. Some spots may also be covered by the heavily stained material. Finally, the summation of the molecular weights of our 72 ribosomal proteins will prove whether it correlates to the postulated total of 2×10^6 daltons.

ACKNOWLEDGEMENTS

This work was supported by grants from the Deutsche Forschungsgemeinschaft awarded to Sfb 29 (Embryonalpharmakologie). I would like to express my appreciation to Mrs. Inge Schütte for her expert technical assistance. I'm indebted to my colleagues Drs. J. Klose and H. Spielmann for their critical appraisal.

REFERENCES

1. De Vries, H., van der Koogh-Schuuring, R.: Physicochemical characteristics of isolated 55-S mitochondrial ribosomes from rat-liver. Biochem. Biophys. Res. Comm. 54, 308-314 (1973).

2. O'Brien, T.W., Kalf, G.F.: Ribosomes from rat liver mitochondria I. Isolation procedure and contamination studies. J. Biol. Chem. 242, 2172-2179 (1967).

3. Hamilton, M.G., O'Brien, T.W.: Ultracentrifugal characterization of the mitochondrial ribosome and subribosomal particles of bovine liver: Molecular size and composition. Biochem. 13, 5400-5403 (1974).

4. Kaltschmidt, E., Wittmann, H.G.: Ribosomal Proteins VII Two-dimensional polyacrylamide gel electrophoresis for fingerprinting of ribosomal proteins. Anal. Biochem. 36, 401-412 (1970).

5. Dale, G., Latner, A.L.: Isoelectric focusing of serum proteins in acrylamide gels followed by electrophoresis. Clin. Chim. Acta., 24, 61-68 (1969).

6. Macko, V., Stegemann, H.: Mapping of potato proteins by combined electrofocusing and electrophoresis Identification of varieties. Hoppe-Seyler's Z. physiol. Chem. 350, 917-919 (1969).

7. Klose, J.: Protein mapping by combined isoelectric focusing and electrophoresis of mouse tissues. A novel approach to testing for induced point mutations in mammals. Humangenetik 26, 231-243 (1975).

8. O'Farrell, P.H.: High resolution two-dimensional electrophoresis of protein. J. Biol. Chem. 250, 4007-4021 (1975).

9. Czempiel, W., Klose, J., Bass, R.: Mammalian mitochondrial ribosomes: Characterization of ribosomal proteins by two-dimensional gel electrophoresis. FEBS Letters 62, 259-262 (1976).

10. Klose, J.: Protein mapping as a tool for investigating mutagenic and teratogenic effects in mouse embryos in: New Approaches to the Evaluation of Abnormal Embryonic Development (Neubert, D. and Merker, H.J., eds.), pp. 375-387, Thieme, Stuttgart (1975).

Crossed Immunoelectrofocusing for Standardization and Characterization of Fungal Antigens

K. Holmberg and T. Wadström

INTRODUCTION

Analytical disc gel electrophoresis has been frequently used in studies on microbial taxonomy and for characterization of microbial antigens. As yet, very few studies have been devoted to the use of newer methods with higher resolving power, e.g. isoelectric focusing (IFPAG) and gradient acrylamide electrophoresis for such a purpose (1, 2). Pathogenic fungi such as Candida albicans contain very complex antigen mixtures and patients show antibody responses to many different antigens during infections (3). However, a considerable variation in different antigen preparation and in the individual immune responses makes immunodiagnosis a complicated task. The appearance of yeast and mycelial forms of fungi, the age of cultures, the method for cell disintegration make characterization of standard antigen mixtures very important for taxonomic studies and diagnostic purposes. Up to recently, crossed immunoelectrophoresis has been the method of choice for such studies of reference systems for Candida, Aspergillus, Actinomyces and Nocardia (3-5). However, it has been emphasized that identification based solely on position electrophoretic mobility and morphology of single rocket immunoprecipitates for individual antigens in complex mixtures is often problematic. Separation of antigens by isoelectric focusing in polyacrylamide gel compared to agarose electrophoresis can increase their resolution (6). The recent development of a "laying on" method for the electrophoresis of antigens into the antibody-containing agarose gel has also made crossed disc immunoelectrophoresis, crossed immunoelectrofocusing and crossed isotachophoresis potential tools for such analyses (6-8). The present study has been performed to compare the use of crossed immunoelectrofocusing with crossed immunoelectrophoresis for the standardization of Candida and Aspergillus antigens for diagnostic purposes.

MATERIALS AND METHODS

Rabbit antiserum to Candida albicans (lot no 076) was obtained from Dakopatts A/S, Copenhagen, Denmark; anti-Aspergillus fumigatus serum (lot no SBL 3B/76) from our own laboratory; cytoplasmic antigen preparations of C.albicans were obtained from Institute Pasteur, Paris, France, (lot no 03-5297) and Mycological Ref. Lab., London School of Tropical Medicine and Hygiene, London, England (PHLS 2109 kindly supplied by Dr. J. McKenzie).Antigens were also prepared in our own laboratory (SBL lots no 2/76 and 1/3/4/76, respectively). Fungal cells were disrupted by using a Braun homogenizer (Braun, Melsungen, Germany) and centrifuged at 104.000 x g, for 60 min and the supernatant used as antigen (3). Extracellular antigens from Asp. fumigatus were concentrated filtrates of liquid cultures containing antigens secreted during growth and/or released upon cell autolysis; batch 33,34/7 and 35/7 from Bencard, Brentford, England; antigen batch 90828 from Institute Pasteur, Paris, France, and lots no 4/76 and 5/76 from our own laboratory.

Polyacrylamide gels containing Ampoline ^{R} (PAG plates, lot no 026, T=5%: C=3%) were obtained from Aminkemi, Bromma, Sweden; and agarose (batch no 685B) from Miles Lab., Stokes Poges, England. All chemicals used were of analytical grade. Isoelectric focusing was performed as previously described under conditions of constant wattage (2-3 MW/m^3 for 60 min) (1, 6). Samples were applied 15 mm from the cathode on strips of Whatman 3 MM chromatography paper (5 x 40 mm). The amount of antigens applied to each strip varied from 50 μl to 200 μl depending on protein content. The IFPAG gels were sliced into 5 mm broad polyacrylamide strips and transferred to agarose gels for electrophoresis (2). Agarose 1% (w/v) was dissolved in barbital buffer pH = 8.6, I = 0.01 containing sodium azide (200 mg/ml) (6). Agarose gels (1.5 x 100 x 100 mm) were prepared on glass plates (0.8 x 100 x 100 mm). Different amounts of antiserum (3-10 $\mu l/cm^2$) were added to the agarose solution immediately before moulding at 55°C. Electrophoresis was performed for 3 h with a field strength of 10V/cm or for 18 h with a field strength of 2V/cm. The gels were then pressed for 10 minutes, dried, and then stained (6). Crossed immunoelectrophoresis and protein determination were performed as previously described (2, 4).

RESULTS AND DISCUSSION

Analysis of different batches of C. albicans cytoplasmic and extracellular antigens from Asp. fumigatus yielded complex immunoprecipitate patterns against the homologous reference antisera (Fig 1A and B). The patterns showed great qualitative and quantitative differences with regard to rocket position and morphology depended on electrophoretic mobility at pH 8.6. The separation of the proteins in isoelectroc focusing revealed different complex protein band patterns. The individual bands in the polyacrylamide gel could be identified after crossed immunoelectrofocusing and referred to by their pI values. This facilitates the identification of the individual precipitin lines in a complex pattern (Figs 2A-C). A close study of these patterns revealed dissimilarities in term of individual antigens with different pI values between the antigen preparations, e.g. concerning peak height, and morphology and sharpness or strength of the immunoprecipitates.

It should be noted that antigens with very similar pI´s not resolved in isoelectric focusing were identifiable as immunoprecipitate rockets of different height and morphology.

Previous studies have shown the importance of a standardized method for sample application and for the performance of electrofocusing to obtain optimal resolution (1, 6). The highest resolution with distinct and reproducible immunoprecipitate pattern was obtained in experiments at the low fieldstrength (2V/cm) run for 18 h (Fig 3). A careful procedure for slicing the acrylamide gel and transferring the slices was found to be the critical step in the whole process as pointed out for analysis of diphtheria toxin (6).

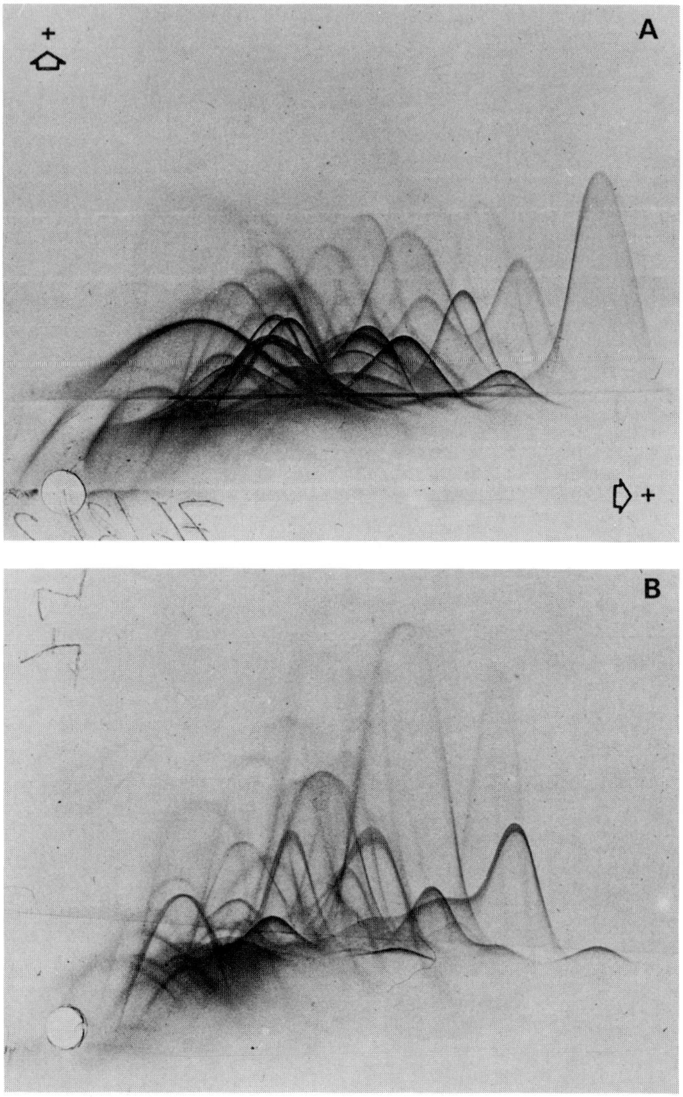

Figure 1. Crossed immunoelectrophoresis of 5 µl of extracellular antigens from Aspergillus fumigatus (A) SBL lot no 4/76 and (B) batch no 90828 from Pasteur Institute, Paris. The second dimension gel contained 5 µl of a rabbit standard antiserum to Asp.fumigatus (lot no SBL 3B/76), per cm^2. The first-dimension electrophoretic run was done at 10 V/cm for 50 min, and the second run was done at 2 V/cm overnight.

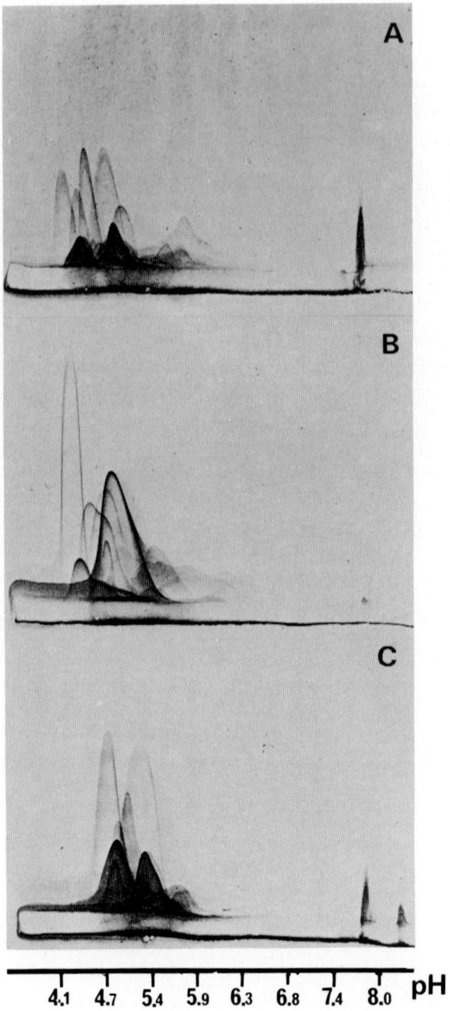

Figure 2. Comparison of three preparations of cytoplasmic antigens from Candida albicans (A) SBL lot no 2/76 (B) Institute Pasteur lot no 03-5297 and (C) SBL lot no 1/3/4/76 by crossed immunoelectrofocusing. Isoelectric focusing of the antigens was performed in pH 3.5-9.5 gradient. Polyacrylamide gel slices used for electrophoresis into antibody-containing agarose gel each contained approximately 100 µg protein. Electrophoresis was performed into an agarose gel containing 10 µl of reference anti-Candida albicans-serum (Dakopatts, Copenhagen, Denmark) per cm^2 for 18 h with a field strength of 3 V/cm. The anode edges of the acrylamide strips are to the left of the figure and the anode during electrophoresis is at the top.

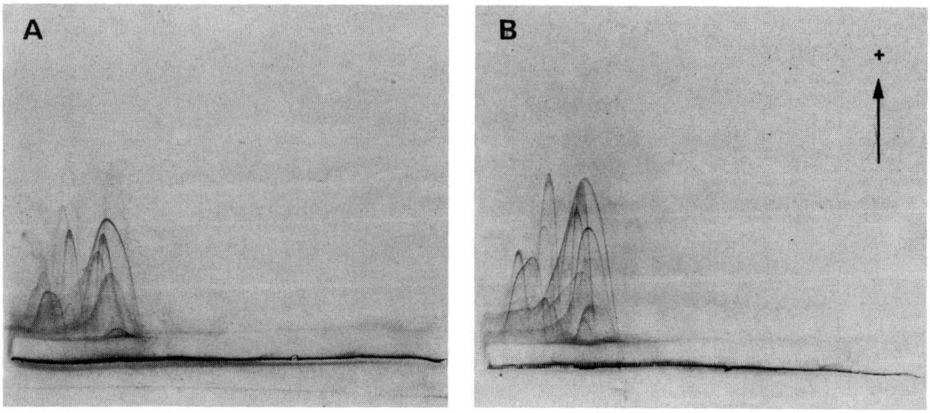

Figure 3. Comparison of two crossed immunoelectrofocusing experiments of an extracellular antigen preparation of Aspergillus fumigatus (SBL 4/76) Isoelectric focusing of the antigen preparation was run on the same electrofocusing acrylamide gel in a pH 3.3 - 9.0 gradient. Electrophoresis was performed (A) for 3 h with a field strength of 10V/cm, and (B) for 18 h with a field strength of 3V/cm, into an agarose gel containing 5 µl anti-Asp. fumigatus-serum (SBL 3B/76) per cm^2.

The potential use of crossed immunoelectrofocusing for analysis of protein antigens from animal, plant and microbial origin is obvious. The recent development of the "laying on" method has provided us with a simple and accurate method for obtaining reproducible results upon transfer of slices of acrylamide to agarose gels. One advantage of crossed immunoelectrofocusing over other high resolving immuno-gel methods, e.g. crossed disc electrophoresis (7) and crossed isotachophoresis (9), is identification of individual antigens by determination of pI values such a defined physical property of individual antigens permits better control of intra-and inter-laboratory variations. The pI value is also an important property when preparative methods for purification of individual antigens in the crude mixtures are to be chosen. However, technical difficulties (a) to separate very basic and acidic antigens in suitable pH gradients, (b) to get basic antigens to migrate into the agarose gel in the second step, remain important problems for further studies. For antigens in these extreme pH ranges, the other two high resolution methods are probably to be preferred at the present moment.

REFERENCES

1. Wadström, T., Smyth, C.J.: Isoelectric focusing in polyacrylamide gel for analysis of cell proteins in bacterial taxonomy. A methodological study. In Progress in Isoelectric Focusing and Isotachophoresis (Ed., P.G. Righetti) North-Holland Publishing Co., Amsterdam, pp. 149-163 (1975).

2. Axelsen, N.H. (Ed.): Quantitative immunoelectrophoresis, New developments and applications. Scand. J. Immunol. 4 Suppl. 2, (1975).

3. Axelsen, N.H.: Quantitative immunoelectrophoretic methods as tools for a polyvalent approach to standardization in the immunochemistry of Candida albicans. Infect. Immunity 7, 949-960 (1973).

4. Holmberg, K., Nord, C-E., Wadström, T.: Serological study of Actinomyces israelii by crossed immunoelectrophoresis I. Standard antigen-antibody reference system for A.israelii. Infect. Immunity 12, 387-397 (1975).

5. Holmberg, K., Nord, C-E., Wadström, T.: Serological study of Actinomyces israelii by crossed immunoelectrophoresis II. Diagnostic and taxonomic applications. Infect. Immunity 12, 398-403 (1975).

6. Söderholm , J., Smyth, C.J., Wadström, T.: A simple and reproducible method for crossed immunoelectrofocusing. Comparative studies on microheterogeneity of diphtheria toxin from different preparations. Scand. J. Immunol. Suppl. 2, 107-113.

7. Ekwall, K., Söderholm, J., Wadström, T.: Disc-crossed immunoelectrophoresis. A simple "laying on" technique permitting the use of commercially available agarose. J. Immunol. Methods 12, 103-115 (1976).

8. Brogren, C.H.: Crossed immunoisotachophoresis - a new method for analysis of protein. Isoelectric Focusing and Isotachophoresis

7. Preparative Separations

Isoelectric Focusing in Free Water

W.D. Denckla

Due to the advantages of focusing in free water an extensive attempt was made to develope instruments for this purpose. There appear to be three major problems with free water: heat, electroosmotic flow and solvent drag secondary to electroosmotic flow. All three problems can be resolved with basically the same solution: relatively narrow channels that are well-cooled (1).

It may be useful to present a tentative, new model for the effects of the Helmholtz layer in order that the design solution might be better understood. In the past, a two layer model has been generally presented: the dipole-oriented water (Helmholtz layer) near a charged stationary surface and the bulk water. Such a model, however, could not explain a number of observations, notably the fact that colored indicators several milimeters from the wall were observed to move at a rate and in a direction under the influence of the Helmholtz layer. A tentative model shown in Fig. 1 evolved. Adjacent to the stationary boundary layer of water, the Helmholtz layer moves proportional to the zeta potential and the electric field. However, as this layer moves, bulk water adjacent to it must also move as a consequence of the viscous forces between the bulk and dipole oriented water. The water moving as a consequence of viscous drag may move relatively slowly but the volume moved far exceeds that of the Helmholtz layer and can cause severe distortions of the pH gradient. In the extreme case, as shown in Fig. 1, ampholytes contained within this layer can be dragged to positions away from their appropriate place and zones of the same pH can be found at different distances from the electrodes within the instrument. The small volume of the Helmholtz layer can be appreciated from an analysis of the weal forces that generate it; dipole orientation is based on charge attraction and cannot exceed a layer of a few Angstroms as the force of the surface charge falls off as the square of the distance. Therefore, the Helmholtz layer cannot be cumulative as it moves through the instrument. Any attempt of the electrically oriented molecules to build up a layer thicker than that permissible by the zeta potential of the wall would result in the water returning to random orientation of bulk water. There is virtually nothing one can do to stop the thin, rapidly moving Helmholtz layer during its progress through an instrument; one can however, prevent its re-entry into the instrument as pure water by forcing it to re-enter through narrow channels and be refilled with ampholytes driven into it by the electric field and diffusion.

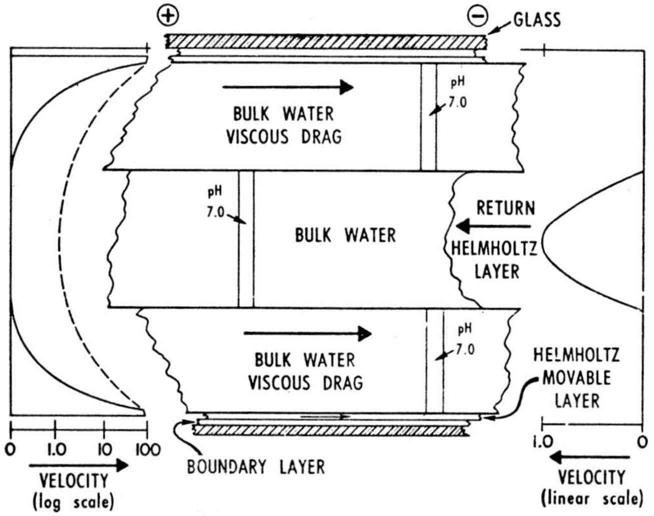

Fig. 1: Model of water movements primarily (Helmholtz layer) and secondarily (bulk water viscous drag) caused by electric field. Approximate relative velocities shown. Dashed lines indicate velocity with walls closer together.

The movement of the water secondary to viscous drag can also be limited by narrow channels. As the channels approach closer to each other, the velocity on either walls and the velocity of the water in the middle of the channels become more nearly the same (Fig. 1). This tends to prevent leaving ampholytes of a given pH elsewhere in the instrument. In addition all the instruments are made from a series of U tubes (1) that are open at the top. As the water driven by viscous drag reaches the top of a U tube, it cannot be driven into the next U tube in appreciable amounts due to the weak forces involved. The Helmholtz layer can, of course, be driven through the length of the instrument. This admittedly hypothetical analysis would suggest that with only a few U tubes per instrument, the legs of the U tubes would have to be relatively long; conversely if one had many U tubes and multiple junctions to break up the water moved by viscous drag, one could have considerably shorter legs on the U tubes. In tests with over 40 instruments, these design limits appear to be reasonably effective.

Drift: Two very different effects can account for most but not all of the drift of the pH gradient observed in free water: CO_2 and chemical decomposition of ampholines. Exclusion of CO_2 arrests the first and ascorbic acid arrests the second cause of drift. It must be emphasized that these two have only been tested in free water instruments that are unaffected by electroosmotic flow; drift in gels may have other causes as well. Furthermore, the amounts of ascorbic acid recommended will only permit operation approximately twice as long as is needed to gain adequate separation. Atmospheric CO_2 is excluded by pumping air (small aquarium pump) through soda lime and then into the instrument. A filter is required to trap fine particles of soda lime.

Apparently some product of the positive electrode moves out through the instrument and causes decomposition of the ampholines and proteins. Ascorbic acid added to the electrode solution (0.02 M H_3PO_4) destroys this compound and will not itself migrate into the instrument appreciably as it is a weak acid. However, ascorbic acid is itself destroyed rapidly unless kept at a very acidic pH. Consequently spacers of pH 2.5 - 4 ampholines must also be added to the electrode solution at a concentration of 0.05 to 0.1 % of the concentration of the main ampholines. Because of the wide number of times, types of instruments and pH ranges that might be used, only an approximate value for the amount of ascorbic acid can be given. The requirement for ascorbic acid is proportional to the time and the current only and is independent of the voltage. Consequently the amount used should be matched to the ampholine concentration as the latter limits the current at a given voltage. Twenty to 40 % (w/total volume ampholines) is used. Failure to use enough ascorbic acid is indicated by: drift of gradient, formation of yellow colored compounds in the acid region (? ascorbic acid ampholine complexes), increase in OD 280 in acid region, increase in OD 280 of proteins, decomposition of proteins such as hemoglobin. The proteins complexed with partly decomposed ampholines have up to 3 x higher OD than the protein itself and can be disassociated largely by dialysis.

Coacervates: There have been two major criticisms of IF: a re-run of an apparently single protein results in finding nearly as many proteins as in the original mixture and a given protein will have a broad peak of distribution, far from the narrow zones expected with purified proteins. Reichert, for example, found partially purified equine LH distributed from pH 3.78 to 8.9 (3). Both of these difficulties have been traced to the existence of coacervates in partially purified or crude mixtures of proteins. Coacervates are defined as heterogenous aggregates of proteins or proteins and other molecules such as lipids. As the techniques for separating coacervates are new, only a few general guidelines can be offered. Following these guidelines examples will be given of separations under coacervating and decoacervating conditions.

Alkali and dilution: Only these two factors have been found to routinely decoacervate all proteins so far examined. The concentration of protein used depends entirely on the nature of the coacervates present. With very crude extracts only 50 - 500 ugm/ml total volume of the instrument can be used. With proteins from serum that do not as readily form coacervates, 10 to 50 times these amounts have been used. In practise the proteins are placed in a volume 5 - 10 % the total volume of the instrument in Tris base adjusted to as close to pH 9.5 as the proteins can tolerate. This solution is placed near the negative electrode and Tris serves as the electrolyte. Due to the varied amount of acidic proteins that may be present in the mixture, the Tris base is added until the desired pH is reached; no single molarity can be recommended for all protein mixtures. It should be noted that any given protein is never diluted throughout the instrument with this method, a feature of some importance with proteins that are unstable when diluted. Since IF is a concentrating method, the concentration of a given protein should not decrease below that found in the sample application zone.

The success of the above method may be attributed, in part, to its being a combination of electrophoresis and IF. As pH 9.5 is above the pI of most proteins they will all move rapidly into the instrument according to their electrical mobilities. Separation of one protein from another thus starts immediately in the alkaline region and discourages the reformation of coacervates favored by neutral and mildly acidic conditions. As the proteins move through the instrument, there is a continuous simplification of the protein mixture because the more alkaline proteins are left behind at their pI's. When the acid proteins reach their pI's, no neutral or alkaline proteins remain.

Application in acid: Theoretically acid should disassociate coacervates as readily as base. However, in a wide variety of protein mixtures application of the sample in acid near the positive electrode was found undesireable. It was found that when the alkaline proteins move through the acidic proteins they form very stable coacervates before ever reaching the appropriate pH region. For the rare neutral or alkaline protein that does not form coacervates under these conditions, this method might be useful as it could yield a great deal of purification. The stability of coacervates formed under these conditions can be appreciated from the observation that growth hormone and hemoglobin once coacervated in this fashion, could not be removed in appreciable amounts from the precipitated acid protein coacervate with either pH 10 Tris base alone or in combination with 8 M urea.

Urea, detergents, solvents: When urea at high concentrations, 68 M, is used throughout the instrument, unusual coacervates form with certain proteins. The same observation was made with either non - ionic detergent (Brij) or ionic detergents (sulfobetaine type). Glycerol or ethelene glycol did not appreciably help decoacervation and, again, with some proteins unusual coacervates were formed. However, lower concentrations of urea might be helpful and it has been used at high concentrations in the sample compartement.

Dialysis: Due to the requirement for salt-free solutions, dialysis is frequently used. However, due to the carbonate ion, unless dialysis is against a weak base such as Tris, stable coacervates form when the pH within the dialysis bag falls below pH 6.0. A considerable non-recoverable loss was observed with certain water soluble proteins under these conditions (growth hormone, hemoglobin).

In the subsequent figures the protein solutions were first focussed in free water on instruments that consisted of 10 U tubes in series (1). Samples were then taken of the 10 tubes and these were applied to a PAG plate (LKB) in the alkaline region. Fixation, staining and destaining followed the instructions of the manufacturer. In this manner the nature of proteins found at a given pH could be studied to determine the presence or absence of coacervates. A large amount of proteins with pI's, as determined by the PAG plate, that were inappropriate for the average pH of a given tube indicated the presence of coacervates. In Fig. 2 are shown the results of the re-run of a pituitary extract in which coacervates were present because too high a concentration of protein was used. The original

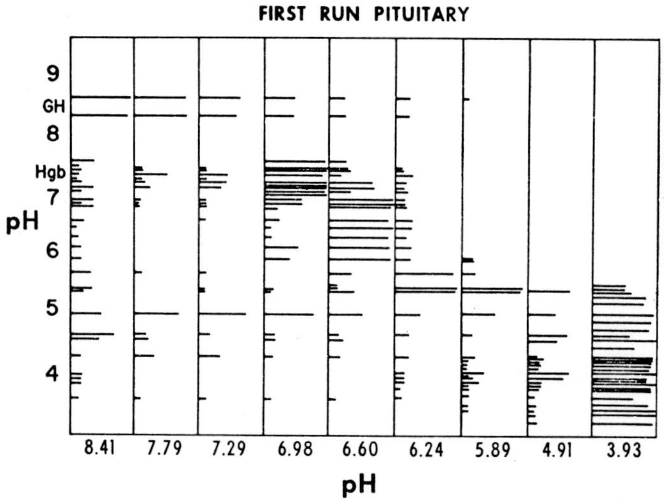

Fig. 2: Drawing of a PAG plate (LKB) of a pituitary extract after focusing on a 10 U tube, 22 liter, instrument. Abcissa: pH of the tube from the instrument; ordinate: approximate pH on PAG plate. The samples all applied in base near cathode. Length of line approximately proportional to concentration of protein. Too much protein applied to original run. Note presence of growth hormone (GH) and hemoglobin (Hgb) in tubes with inappropriate pH's. Note presence of some acidic proteins in each tube.

sample was a high speed supernatant of a Tris extract (pH 9.0) of homogenized fresh frozen bovine pituitaries (2.5 mg/ml total volume). The sample was placed in alkali (pH 9.5, Tris base) in the alkaline region but clearly each tube contained a rich variety of proteins with inappropriate pI's. As might be expected if tubes from a run such as that shown in Fig. 2 were pooled and re-run on the same instrument, little purification would be accomplished as one would be simply re-focusing coacervates with a broad spectra of pI's. Fig. 3 is an example of such a re-run.

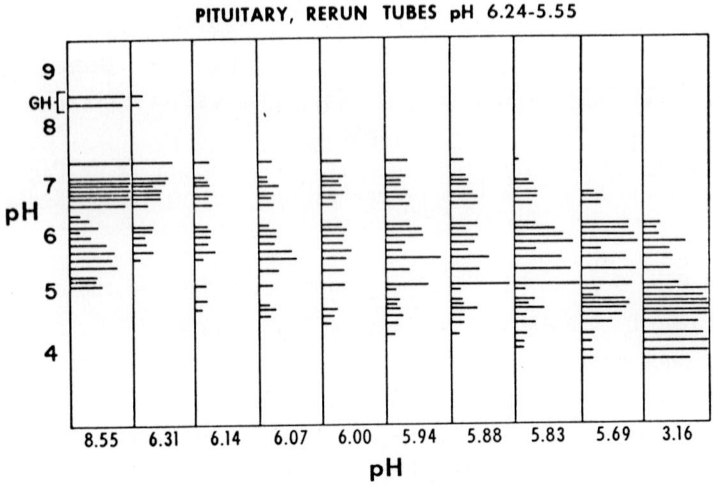

Fig. 3: Drawing of PAG plate of a second run with pooled tubes from a run like that shown in Fig. 2. Failure of separation is apparent.

However, in Fig. 4 one can see that with a 5 fold dilution the same pituitary extract now can be divided into discrete groups of proteins according to their pI's. Thymic extract (fraction F 5,2) was focused under coacervating and decoacervating conditions. This extract is used as an example to emphasize that even relatively small proteins can form stable coacervates. All proteins in this extract are small enough to have passed through a PM 10 membrane. In addition the inadequacy of DISC electrophoresis to decoacervate proteins can be appreciated from the fact that the F 5 fraction yields only approximately 5 major bands with this method. DISC electrophoresis is not alkaline enough and the proteins must be excessively concentrated for application. In order to demonstrate the effects of application of the sample in acid, the same concentration of F 5 extract

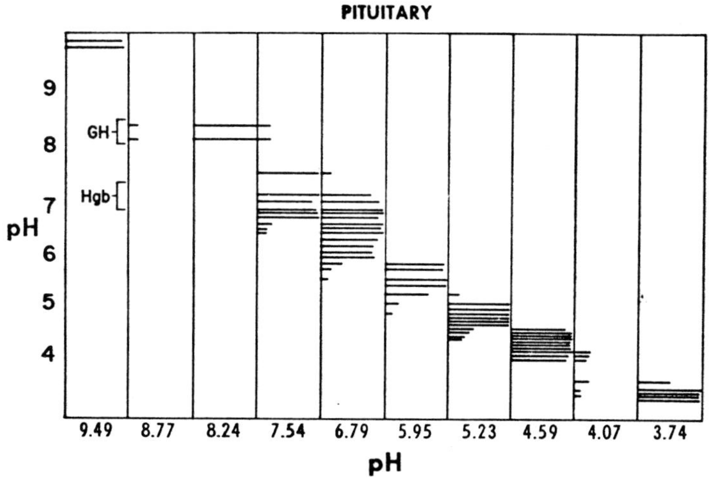

Fig. 4: Drawing of PAG plate of same pituitary extract as in Fig. 2 but put on alkaline side at 5 x less concentration. Separation markedly improved.

was used in Fig. 5 as in Fig. 6. The phenomenon described above is clearly demonstrated by the lack of protein in tubes above pH 7.5, despite the presence of such proteins in the coacervates found in the acid region. The effectiveness of decoacervation by base is shown in Fig. 6, wherein considerably more bands are apparent than DISC electrophoresis would have in-

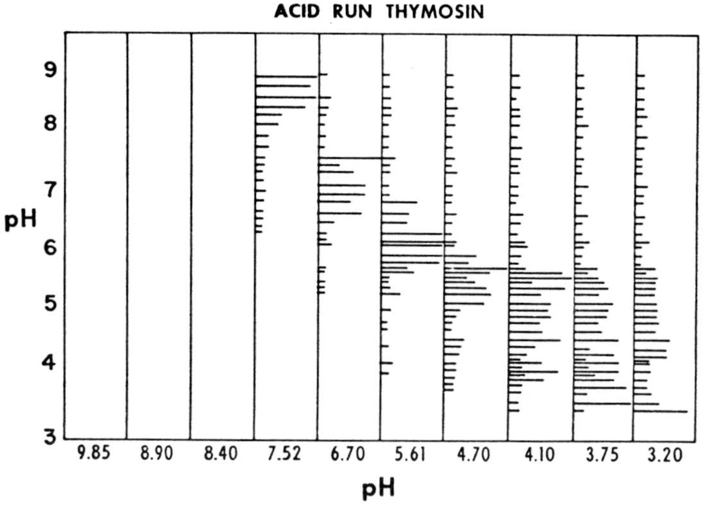

Fig. 5: PAG plate of run of Hooper)s fraction F 5 (2) at correct concentration but placed in acid region before focusing. Note rich variety of proteins in every tube and lack of proteins above pH 7.52.

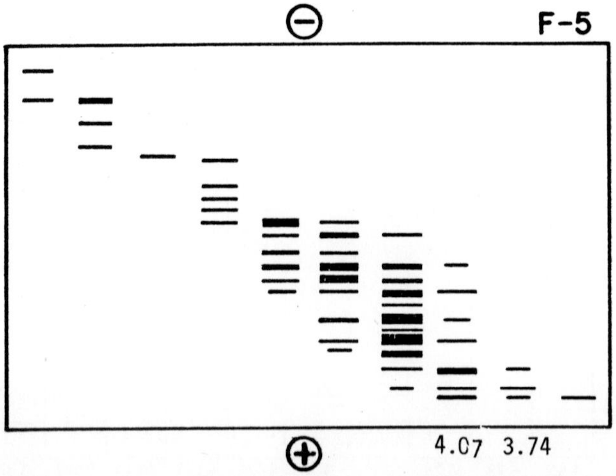

Fig. 6: Drawing of Pag plate of samples from a thymosin run of the same concentration as in Fig. 5 but placed in alkaline region. Thickness of lines indicates relative concentration as judged by intensity of stain.

dicated. The effectiveness of the method is further indicated by the re-run of two tubes on smaller 10 U tube instruments to resolve the contents of the tubes shown in Fig. 6. Fig. 7 indicates the proteins are now distributed in a manner one would expect when the proteins were individually in solution. The bioassay performed in Dr. Alan Goldstein's laboratory, indicated that 90 % of the activity was contained in one tube that also contained approximately 10 % of the total protein. Therefore, one would have expected only a 10 fold increase in specific activity. The disproportionately large increase in specific activity (1000 x) suggested the possibility that the active material had previously been coacervated with an inhibitor. The location of the active material largely in one tube, despite a shallow pH gradient also indicated absence of coacervates. Previously the biological activity has focussed on an LKB column in a broad zone around pH 4 with an ill-defined peak (Dr. Alan Goldstein, personal communication).

It should be noted that the requirement for dilution and alkali for decoacervation has been tested on the LKB column and on a 1.000 ml acrylamide gel slab and in both cases coacervates were found unless the rules were followed that were derived from experiments in free water. Ampholine concentration appears to have little effect on coacervates within the range

THYMOSIN, RERUN

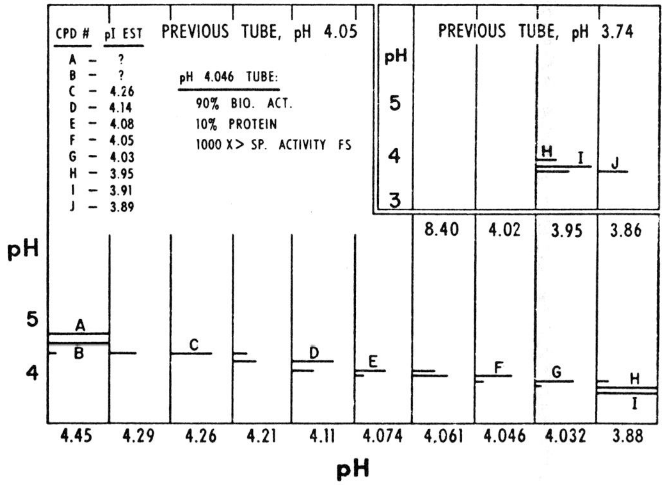

Fig. 7: Drawing of PAG plate with samples from a re-run of tube pH 4.07 and tube pH 3.74 shown in Fig. 6. Absence of coacervates indicated by relatively narrow distribution of proteins into narrow pH zones also by lack of contaminating protein.

In conclusion, it is hoped the above findings may be of use to others who will be able to add other ways of dealing with coacervates.

REFERENCES

1. Denckla, W.D., U.S. Patents, 3.901.780 : 3.951.777.
2. Hooper, J.A., McDaniels, M.C., Thurman, G.B., Cohen, G.H., Shulof, R.S. Goldstein, A.L.: Purification and properties of bovine thymosin. Ann. Y.Y. Acad. Sci 249, 125-144 (1975).
3. Reichert, L.E., Jr.: Electrophoretic properties of pituitary gonadotropins as studied by electrofocusing. Endocrinology 88, 1029-1044 (1971).

This work was completed while the author was at the Roche Institute of Molecular Biology, Nutlex, N.J., U.S.A.

Preparative Electrofocusing in Flat-Beds of Granulated Gel – Methodological Aspects

A. Winter

ABSTRACT

The use of a granulated gel as the stabilizing medium in preparative electrofocusing allows gram-quantities of sample to be separated. However, good results require a standardized practical procedure. The pretreatment of the gel and a way of controlling the final water content of the gel bed are important parts of such a procedure.

INTRODUCTION

Preparative electrofocusing in flat beds of a granulated gel (PEGG) (1,2) constitutes an alternative to electrofocusing in a density gradient. A number of properties specific for PEGG make it especially well suited for preparative work. The loading capacity of the method is high and gram-quantities may be applied when dealing with complex protein samples e.g. at an early stage of a preparative purification procedure. PEGG is a method relatively insensitive to precipitation. Any precipitate that may be formed during the focusing will be trapped within the gel bed at the place of its formation and will have very little or no influence on the focusing of the other sample components. The accessibility of an open flat bed technique like PEGG enables the use of prints for detection of separated zones. These are stained either for protein in general or by a zymogram technique. Another advantage of PEGG is that the final fractions do not contain any stabilizing material which makes it very easy to use the cascade principle (3) where the interesting part of a focusing experiment is re-focused in a second run.

As with all other separation techniques the successful use of PEGG relies on a standardized practical procedure and three points have been found to be of vital importance namely

1. the choice of stabilizing gel
2. the pretreatment of the gel, and
3. the final water contents of the gel bed

The stabilizing gel should be hydrophilic and show no or very weak electroendosmotic properties. Radola (1) showed that beaded polyacrylamide or Sephadex gels of the superfine grade meets these requirements optimum being G-75 or G-100.

Recently it was shown at the LKB Application Laboratory that the Sephadex gels as they are delivered from the manufacturer contain soluble charged components thatinterfere with pH-gradient formation in electrofocusing (2). These contaminants are readily washed out of the gel by distilled water. This pre-treatment of the gel changes its swelling properties in terms of final gel volume per gram of dry gel.

The final water-content is important. If the gel bed is too dry it will crack. On the other hand if the gel bed is too wet the focused protein zones will tend to sediment and make it difficult to visualize the zones by the print technique. In order to prepare a gel bed with properties optimal for electrofocusing a procedure has been developed that takes into account the change in swelling properties due to the pre-treatment of the gel (2).

MATERIAL AND METHODS

All experiments were carried out in the Multiphor (LKB 2117-301) equipped for PEGG (LKB 2117-101 and LKB 2117-501), and with a constant wattage power-supply (LKB 2103) as the electrical sources.

Gel beds were prepared as described in the LKB Application Note 198 (1975). The sample was added to the initial gel slurry which contained 2% (w/v) Ampholine® carrier ampholytes unless otherwise stated. Untreated Sephadex G-75 superfine was used as 7% (w/v) slurries before evaporation, while pre-washed Sephadex G-75 superfine was used as 5% (w/v) slurries. Some of the experiments were carried out with ULTRODEX (LKB 2117-510) a commercially available dextran gel specially prepared and purified for PEGG.

pH was measured directly in the gel bed with the aid of a surface glass electrode (LOT 403-30-M8, Ingold, Zürich).

Pre-washing of the gel

Swollen Sephadex G-75 superfine was washed on a glass filter funnel with 10 gel volumes of distilled water. The gel was then dehydrated by repeated washings in absolute ethanol and dried under vacuum.

Determination of the optimal water content of the gel bed

100 ml of a 5% (w/v) gel slurry in distilled water was poured onto the gel tray and evaporated until minor cracks were visible in the gel bed. The water loss was determined by weighing and expressed as percentage of the initial slurry weight. It has been found empirically that if the evaporation is stopped at a water loss corresponding to 75% of this figure, the bed properties will be suitable for electrofocusing.

RESULTS AND DISCUSSION

Electrofocusing in general is sensitive to higher concentrations of salt ions which tend to interfere with pH-gradient formation. As mentioned above Sephadex gels contain water-soluble charged contaminants which severely distort the pH-gradient. Fig. 1 shows the results of an experi-

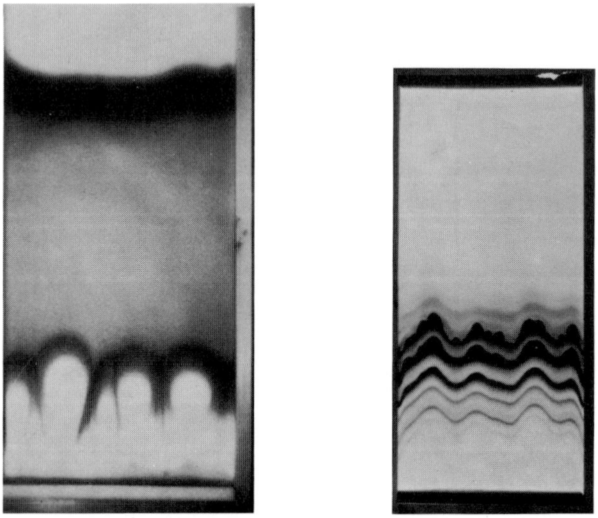

a b

Fig. 1. The pH-gradient distortion obtained in untreated gel.

ment where 100 mg of human hemoglobin (Hb) was focused in a pH-range of 6-8.5 and with untreated Sephadex G-75 superfine as the stabilizing gel. Already after a couple of hours severe disturbances were noted at the cathodic end of the gel bed (Fig. 1a). These disturbances persisted even in the fully focused pH-gradient (Fig. 1b). The effect of washing the gel before electrofocusing is dramatic as is illustrated in Fig. 2.

In this experiment Hb was run under the same conditions as in Fig. 1 but for the pre-treatment of the gel. The focused zones now appeared as straight lines perpendicular to the electrical field and transversal sectioning of the gel bed can be performed in order to collect the separated proteins without any re-mixing caused by irregularly shaped zones. At both ends of the focused zones slight edge-effects are seen. These were found to develop only after the steady state was reached in which the conductivity of the system is at its lowest.

Fig. 2. The effect of pre-washing the stabilizing gel.

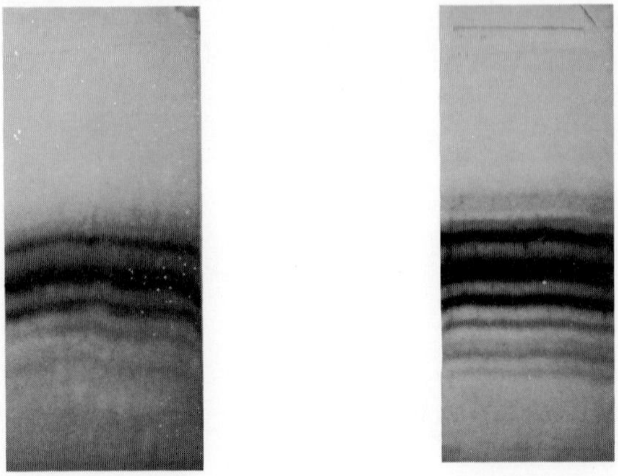

a b

Fig. 3. The effect of the water content of the gel bed on the "printability". a) To high a water content. b) Normal water content.

Figure 3 illustrates the adverse effect on the "printability" of the gel bed that results from a too high water content of the gel. In one experiment the evaporation was stopped at half of the normal water-loss. As seen in Fig. 3a this resulted in a much fainter print than that of the control experiment (Fig. 3b).

In preparative electrofocusing one should always use as narrow pH-ranges

Table 1

Ampholine Carrier Ampholytes \ pH-range	SYSTEM NO 1 2.5–4.5	2 3.5–5.5	3 5–7.5	4 6–8.5	5 7–10
2.5–4	3.75 ml				
3.5–5	1.25 ml	2.5 ml			
4–6		2.5 ml			
5–8			5 ml		
6–8				2.5 ml	
7–9				2.5 ml	2.5 ml
9–11					2.5 ml
Distilled water + sample	95 ml	95 ml	95 ml	95 ml	95 ml

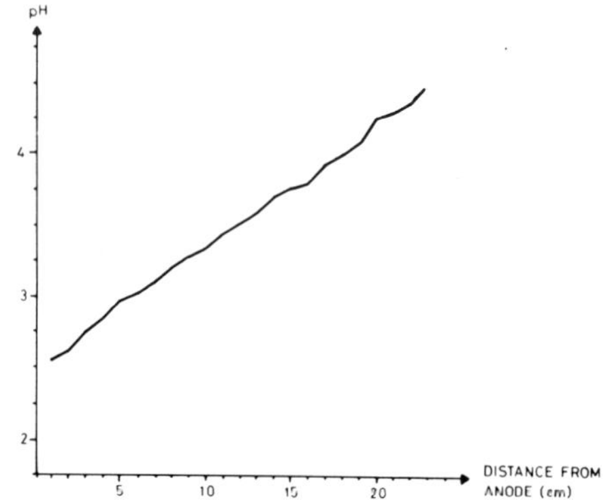

a
Ampholine System No. 1, pH 2,5–4,5

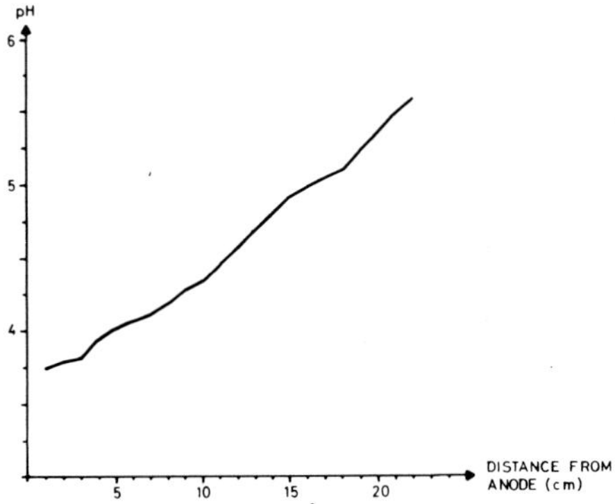

b
Ampholine System No. 2. pH 3,5–5.5

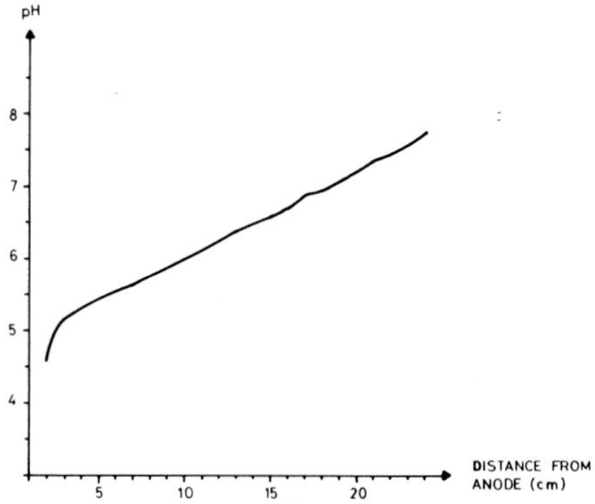

c
Ampholine System No. 3. pH 5–7,5

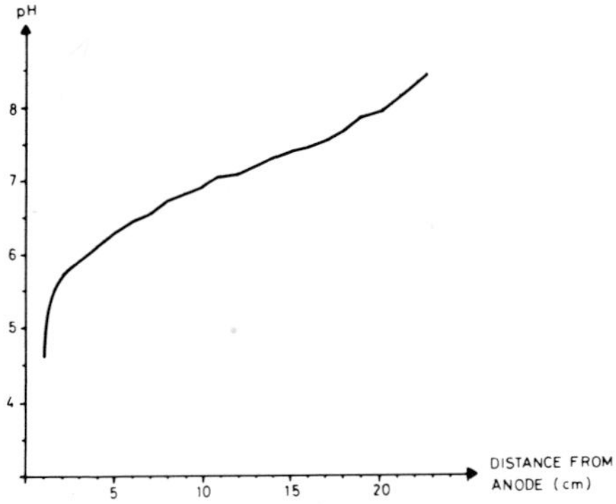

d
Ampholine System No. 4. pH 6–8,5

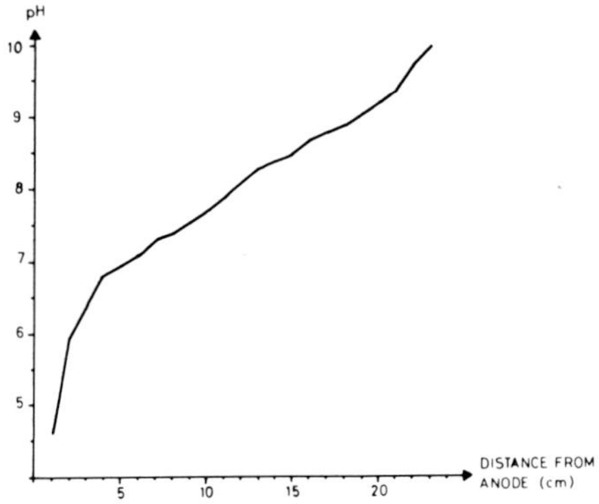

e
Ampholine System No. 5. pH 7–10

Fig. 4. Resulting pH-gradients of the Ampholine systems in Table 1.

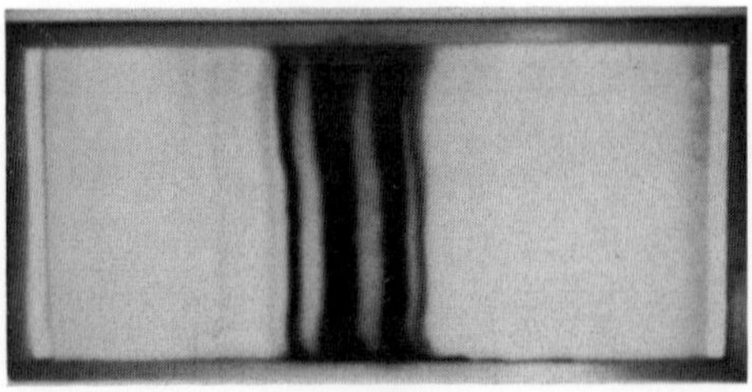

Fig. 5. Electrofocusing of 100 mg Hb at 40W constant power. pH-gradient 6-8.5; running time: 4 hours.

as possible in order to maintain loading capacity and resolution as high as possible (4). Table 1 gives the composition of five overlapping narrow pH-ranges. The figures assume an initial gel slurry of 100 ml and an Ampholine concentration of 2%. The resulting pH-gradients are shown in Fig. 4a-e. In Fig. 4c a "knee" appears at the alkaline end of the pH-gradient and in Figures 4d-e this is increasingly pronounced. This phenomenon is the result of a slight cathodic drift of the whole pH-gradient, which in part may be attributed to the weak electroendosmotic properties of the gel. Thus the nominal pH-range (that stated on the Ampholine bottle) does not always agree with the actual pH-range in PEGG.

At an applied constant power of 8-10 W the normal focusing time ranges from 14 to 20 hours depending on the Ampholine concentration, the pH-range, the composition of the sample etc. This power is certainly not the highest allowable and the result shown in Fig. 5 was obtained at 40W constant power (cooling temperature $4^{o}C$) and with the voltage limited to maximum 1.5 kV. The focusing time was as short as four hours. However, taking into account the time needed for preparation of the gel bed, pH-measurements, harvesting etc, the total experimental time will exceed a normal working day. Hence it is more convenient to adjust the focusing time so as to fit over-night runs.

When looking for one component out of a complex protein mixture which often is the case at an early stage of a protein purification procedure,

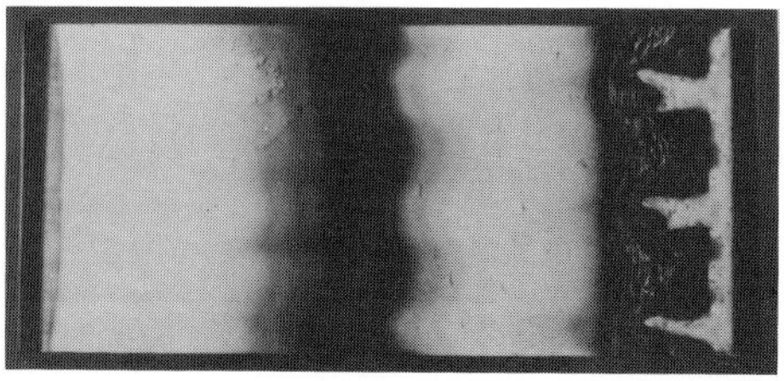

Fig. 6. The result of a PEGG-experiment with 1 g of serum proteins and 100 mg Hb. pH-gradient: 6-8.5.

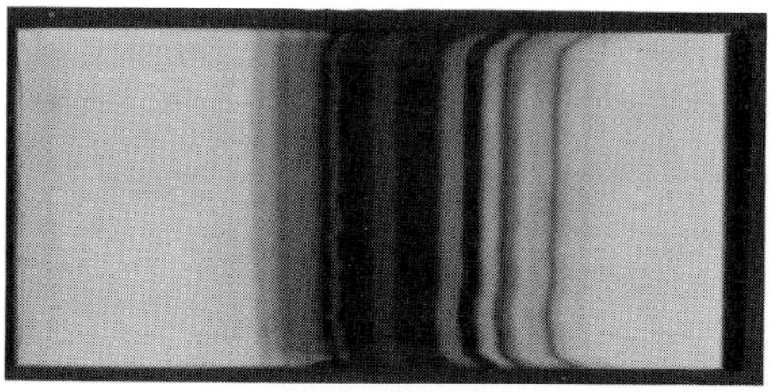

Fig. 7. The result of refocusing the Hb-containing part of the experiment in Fig. 6.

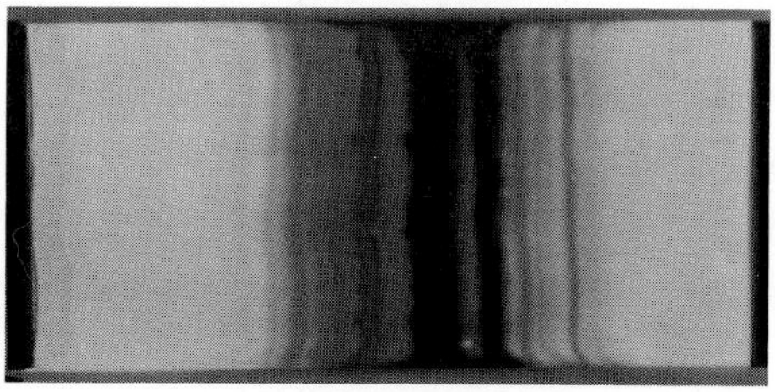

Fig. 8. The result of electrofocusing 1000 mg Hb. Ampholyte-concentration 4% (w/v); pH-gradient: 6-8.5.

up to gram-quantities may be applied in PEGG. In order to mimic such an "early stage" situation 1 g of the proteins of whole human serum was mixed with 100 mg Hb and focused in a pH-gradient of 6-8.5 pH. The initial slurry volume was 150 ml and the concentration of Ampholine was 2.5%. The result is shown in Fig. 6. No pre-treatment of the sample was carried out and the salt ions in the serum sample produced the usual wavy appearance of protein zones. A large number of the serum proteins have isoelectric points below pH 6 and these ended up close to the anode. They are seen as the dark irregular band at the top. In fact the protein concentration within this band was that high that it caused a slight shrinkage of the gel. The Hb part was then scraped out of the gel tray and applied to a second run in the pH-range 6-8.5 according to the cascade concept. The resulting pattern is shown in Fig. 7. Now the Hb zones appeared as straight lines and the whole pattern resembles very much that of Fig. 2. Sometimes even a single-component sample can be focused in very large amounts as exemplified in Fig. 8. Here 1 g of Hb was focused in a pH-gradient of 6-8.5. The Ampholine concentration was 4% and

Fig. 9. An example of the print technique.

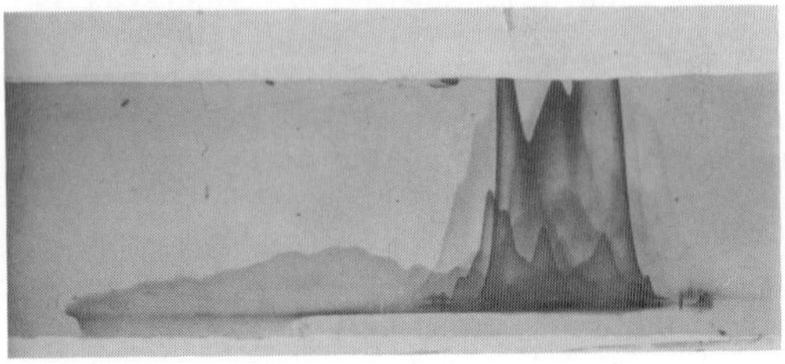

Fig. 10. Immunoelectrophoretic analysis of a PEGG-experiment with 2 ml human serum in a pH-gradient of 3.5-10.

volume of the initial slurry was 150 ml.

Figures 9 and 10 represent two ways of monitoring separation prophiles in PEGG. The result in Fig. 9 is taken from an experiment with an extract from human muscle. After focusing a dry chromatography paper (Whatman No 1) was rolled onto the gel bed surface and left there for two minutes thus allowing small amounts of the separated proteins to be absorbed by the paper. This print was then dried, fixed and stained with Coomassie Brilliant Blue R 250. The result in Fig. 10 was obtained by taking a 0.5 cm wide print. The unfixed print was then transferred onto an antibody-containing agarose gel and analysed according to the rocket technique of Laurell. In this experiment 2 ml of human serum was focused in a pH-gradient of 3.5-10.

Up to now PEGG has not been a widely spread purification technique. One of the reasons for this is probably the unawareness of the necessary pre-treatment of the stabilizing gel. It is my hope that the now available standardized practical procedure will contribute to a wider acceptance of this powerful purification method.

REFERENCES

1. Radola, B.J.: Isoelectric focusing in layers of granulated gels. II Preparative isoelectric focusing. Biochim. Biophys. Acta 386, 181 (1975).

2. Winter, A. et. al.: Preparative flat-bed electrofocusing in a granulated gel with the LKB 2117 Multiphor. LKB Application Note 198 (1975).

3. Fawcett, J.S.: Some recent developments in preparative isoelectric focusing. p. 23 in Isoelectric Focusing, Butterworths, England (1975). Editors: Arbuthnott, J.P. and Beeley, J.A.

4. Rilbe, H. and Pettersson, S.: Preparative isoelectric focusing in short density gradient columns with vertical cooling. p. 44 in Isoelectric Focusing, Butterworths, England (1975). Editors: Arbuthnott, J.P. and Beeley, J.A.

Preparative Scale Purification of Bacterial Enzymes and Toxins by Isoelectric Focusing and Isotachophoresis

T. Wadström, R. Möllby, B. Olsson, J. Söderholm and C.J. Smyth

Introduction

The significant contribution that isoelectric focusing in density gradients has made to the purification and characterization of bacterial toxins and enzymes has been previously reviewed (1,2). However, there are only a few instances where the technique has been used preparatively at an early stage in purification (see 3 for refs.). Furthermore there are few cases where density gradient electrofocusing has been used as virtually the sole purification step to yield material of demonstrated high purity (4-6). Recently Morris and Morris (7) made the following assertion: "The following general principles can be stated which at least avoid waste of time and effort. High resolving power methods should not be used in the early stages of the purification." Our experience over the past decade with isoelectric focusing in density gradients for the purification of a number of bacterial proteins (as opposed to idealized separations with highly purified proteins e.g. myoglobin) does not substantiate their claim, although it is true to say that some proteins have presented more difficulties than others, e.g., Escherichia coli enterotoxin (8). In this report unpublished data concerning the purification of one bacterial enzyme and two toxins will be discussed in relation to our comparative experience with density gradient and granulated gel electrofocusing (Sephadex G-75) and with isotachophoresis.

Escherichia coli enterotoxin

E.coli like Vibrio cholerae, has been shown to cause acute diarrhoea by producing enterotoxins, which act on the intestinal epithelium, one of which causes fluid accumulation in the lumen through activation of adenyl cyclase (9). The purification of this heat-labile enterotoxin has been the subject of intensive effort in many laboratories (10). Our preliminary studies with ion exchange chromatography and isoelectric focusing showed that the heat-labile enterotoxin had a pI of 4.5 and that it was unstable at low pH (8). Moreover preparative scale electrofocusing in density gradients gave unpromisingly low yields of active material. However, more recent findings have indicated ways of alleviating such problems and also some interesting observations regarding pI and biological activity of this enterotoxin.

Bacterial cell lysates obtained by freeze-pressing cells were treated with $(NH_4)_2SO_4$, $MnCl_2$ and by negative adsorption with CM and QEAE-Sephadex (8). Isoelectric focusing of such material in density gradients of glycerol (8) using several batches of enterotoxin resulted in a pI of 4.5-4.9 with 10-15% recovery of input activity. With the use of concentrated crude extracellular enterotoxin and loading of the sample into the alkaline part of a preformed 3.5-10 pH gradient, a second peak of enterotoxin activity with a pI around 7 (cf. pI of 6.9, ref. 12) and a third component with a pI around 6.5 were observed with good reproducibility (Fig. 1). However, the elution profiles of enterotoxin activity as measured by the adrenal cell test and the rabbit intestinal loop assay (9, 10) were not exactly superimposable suggesting partial separation of adenylcyclase-activating and fluid accumulation-promoting activities, respectively, of the enterotoxin. Increased recoveries of input activity by both assays were obtained (up to 60%). Fractionation of partially purified enterotoxin by Sepharose 2B chromatography has also revealed differences in the elution profiles of these biological activities (Fig. 2). Similar observations have been noted by Dorner (personal communication). Isoelectric focusing in layers of granulated gel have also indicated that application of the sample around neutrality in a preformed pH gradient will increase the recoveries of activity dramatically. Preliminary results indicate no differences in the stability properties between fractions with different relative biological activity.

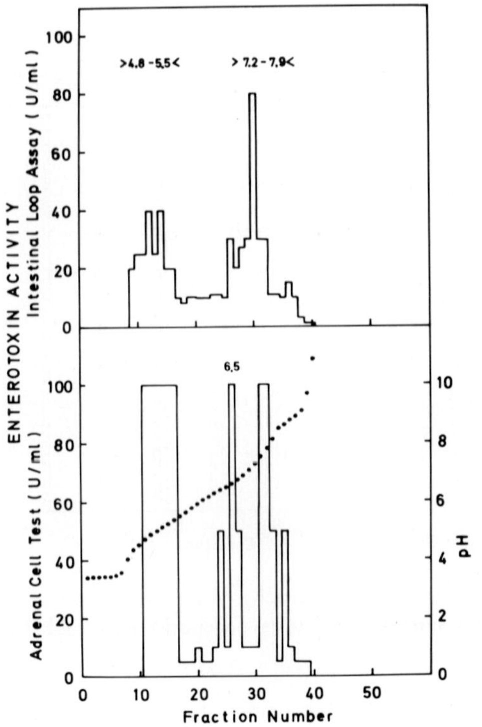

Figure 1. Isoelectric focusing of E.coli enterotoxin in a density gradient of glycerol. The pH gradient comprised Ampholine pH range 3.5-10. (•) pH gradient. Enterotoxin activity was measured in the rabbit intestinal loop assay (upper part of Fig.) and in the adrenal cell test (lower part). The pIs of activity peaks are inserted.

Isotachophoresis seemed an attractive alternative to isoelectric focusing for preparative scale purification of enterotoxin because the sample need not be exposed to unfavourable pH conditions during loading or separation (11). Moreover, the resolving power and sample capacity of the two techniques are similar. Isotachophoresis of E.coli enterotoxin in the LKB Uniphor represents the first application of this technique to the purification of bacterial toxins and enzymes (11,12). Dorner (12) reported large scale purification of the enterotoxin from a porcine strain. From 50 litres of starting material, 0.3 mg of purified heat-labile enterotoxin was obtained with an

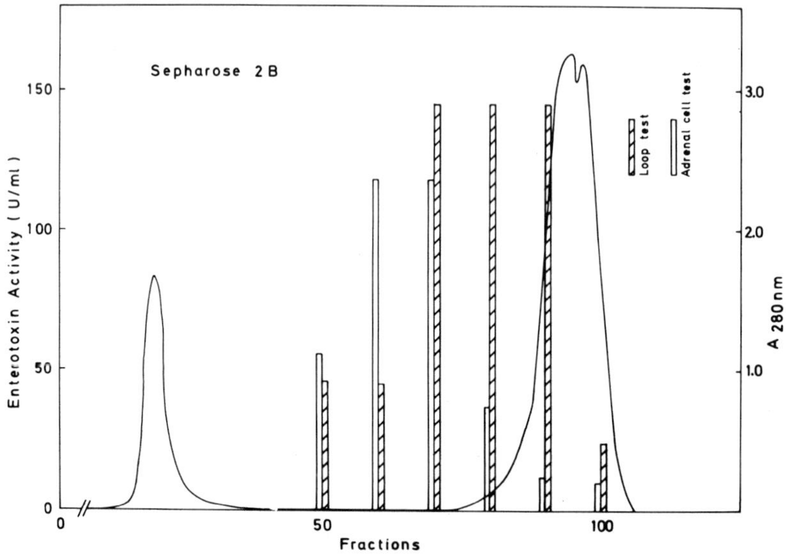

Figure 2. Sepharose 2B chromatography of E.coli enterotoxin. —— Absorbance at 280 nm; open bars - adrenal cell test; hatched bars - rabbit intestinal loop assay.

overall recovery of 58%. In contrast the yields reported by Möllby et al. (11) were lower with material released from cells (5-15%), but about equal with prepurified extracellular material (25-65%). In summary the most remarkable new observation from our current investigations has been the differences in relative biological activity between fractions obtained both by isoelectric focusing and gel chromatography.

Deoxyribonuclease B(DNAase B) from Streptococcus pyogenes

Four immunologically distinct deoxyribonucleases designated A, B, C and D have been shown to be elaborated by group A streptococci (13). The human immune response to DNAase B is diagnostically indicative of streptococcal skin infection in patients with glomerulonephritis. Electrofocusing studies on the haemolysins and enzymes produced by streptococci of groups A, C and G (14) acted as the trigger for efforts to devise a simple purification of DNAase B on preparative scale for use as a diagnostic reagent.

Extracellular proteins from Strep. pyogenes, strain C203S were precipitated with $(NH_4)_2SO_4$, dialysed, adsorbed with DEAE-Sephadex and finally dialysed against glycine prior to electrofocusing in narrow alkaline pH gradients (Figure 3). Columns were usually eluted upwards to avoid contamination of active fractions with precipitate which formed in the anodic region. Overall yields of DNAase B are difficult to calculate as at present there is no way of specifically measuring DNAase activity attributable to DNAase B in crude starting material. However, sufficient DNAase B for 50-100 diagnostic tests per day for up to 6-8 months can be obtained from 10 l of culture supernatant fluid. Peak fraction pools stored as small aliquots at -70C without further processing have been stable for up to two years.

Attempts have been made to purify DNAase B using a commercial trough for granulated gel electrofocusing (15). Very clear cut separation was obtained between DNAase B and residual DNAase D. Moreover yields of activity from such electrofocusing experiments were approximately 85-95% of that applied. However, a disturbing discrepancy was noted between the pI values observed in density gradient and Sephadex trough electrofocusing studies (Table 1). Electrofocusing of peak fractions of DNAases B and D from both procedures was also performed in thin layer polyacrylamide gel. DNAase activity was subsequently detected using an agarose overlayer zymogram technique (16, and Figure 4). The DNAases obtained by the Sephadex gel technique were not cross-contaminated with each other even after prolonged incubation times for the development of the zymogram. Such clear-cut separations were not obtained with fractions from density gradient columns. The latter DNAase B contained some DNAase D and was also resolved into two components with pIs of approximately 9.0 and 8.3 corresponding to the values observed in density gradient and

Table 1. Isoelectric points (pI values) of DNAases from Streptococcus pyogenes C203S

DNAase	pI Density gradient	Sephadex gel
D	7.9	7.4
B	9.0	8.3

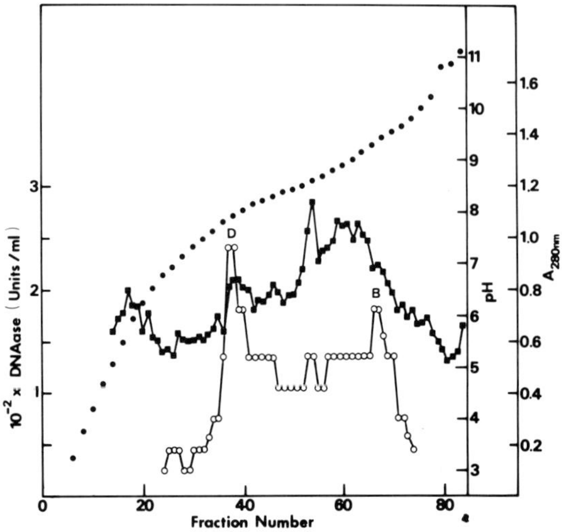

Figure 3. Preparative scale isoelectric focusing of partially purified DNAase B from group A streptococci. Focusing was performed in the LKB 440 ml column. The pH gradient was from 7-10. (•), pH gradient, (■), absorbance at 280 nm, (o), DNAase activity as measured by a radial diffusion techniques (see Ref. 14).

granulated gel experiments. The nature of these anomalous findings is unclear. Horseradish peroxidase also a basic protein with a pI of 9.0 in density gradient experiments, exhibits similar anomalous behaviour on electrofocusing in these different media. Delincée and Radola (17) discussed different explanations for this phenomenon. Clearly, however, these two instances illustrate that apparently different pIs may be determined using the density gradient and granulated gel techniques.

Diphtheria toxin from Corynebacterium diphtheriae

Diphtheria toxin was one of the first bacterial toxins to be recognised as the cause of disease. It is now known to catalyze ADP ribosylation of elongation factor 2 (18). Using high voltage isoelectric focusing in thin layer polyacrylamide gel at constant power, preparations of diphtheria toxin from different reference laboratories were shown to exhibit complex microheterogeneities (19 and Figure 5). It was suggested

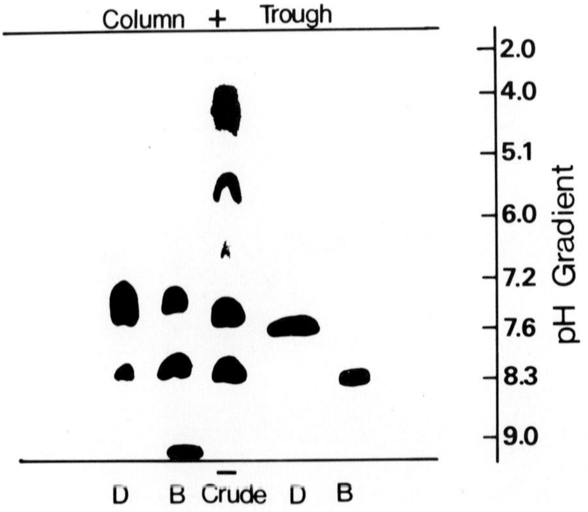

Figure 4. DNAase zymogram of streptococcal DNAases separated by isoelectric focusing in thin layer polyacrylamide gel. Peak fractions of activity of DNAases B and D from density gradient column and granulated trough experiments were electrofocused in a pH 3.5-10 gradient in parallel with crude starting material (centre). DNAase bands were developed by an agarose overlay technique (16). Zones of clearing against a dark background are diagramatically represented.

that these could have resulted from partial nicking of the toxin. This microheterogeneity has been further studied by crossed immunoelectrofocusing (19). It was thus of interest to develop a procedure allowing isolation of these individual multiple forms to permit further investigation of the nature of the microheterogeneity. However, in density gradient experiments heavy precipitation of diphtheria toxin occurs resulting in poor resolution. Isoelectric focusing in layers of granulated gel was thus tried as an alternative approach.

Several practical points must be stressed regarding the use of granulated gel techniques as a result of our experiences with diphtheria toxin. The Sephadex gel should be washed thoroughly with distilled water to remove soluble dextran and other impurities which will otherwise contaminate the separated fractions (15). However, washing increases the swelling properties of the gel. Moreover, the swelling of individual gel batches vary considerably. Both of these factors are important in relation to the

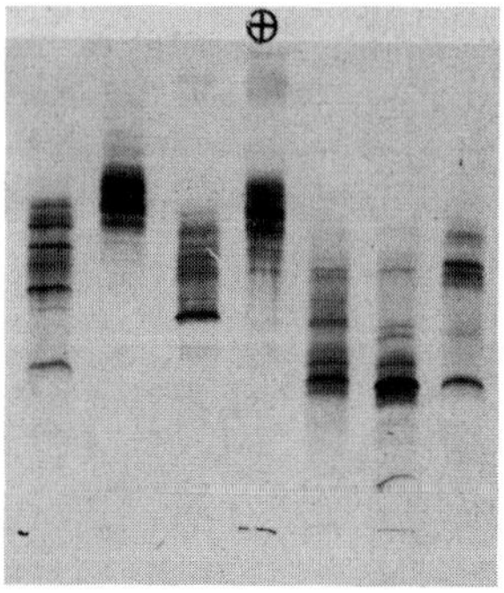

Figure 5. Isoelectric focusing in thin layer polyacrylamide gel of seven diphtheria toxin preparations from several laboratories in a pH 4–6 gradient. Protein bands stained with Coomassie brilliant blue.

determination of the cracking point of the gel bed (15). Sample application as a zone is best done on a levelling table to aid even spreading. Any observed binding affinity of sample components for the gel mitigates against application of the sample close to the expected pI in prerun gradients or incorporation of the sample throughout the gel bed. Such problems were noted with pigmented components in crude diphtheria toxin.

During initial experiments it was noted that diphtheria toxin precipitated when Sephadex fractions were being eluted. However, incorporation of 20% (w/v) glycerol into the gel bed and elution of gel fractions with buffer containing glycerol prevented this from occurring. The stabilizing properties of glycerol speak for its wider use. A paper print of a gel in which diphtheria toxin had been electrofocused in a narrow pH 4–6 gradient, revealed the separation of discrete protein bands reflecting the previously recorded microheterogeneity. These fractions were insufficiently well separated to allow maintenance of the resolution using the LKB separator for harvesting gel fractions. However, fused rocket immunoelectrophoresis indicated that partial fractionation was retained (Figure 6). Preparative scale fractionation will only be possible with the use of narrower pH gradients prepared from commercially available carrier ampholytes (20) or by the use of specific amino acid additives (21).

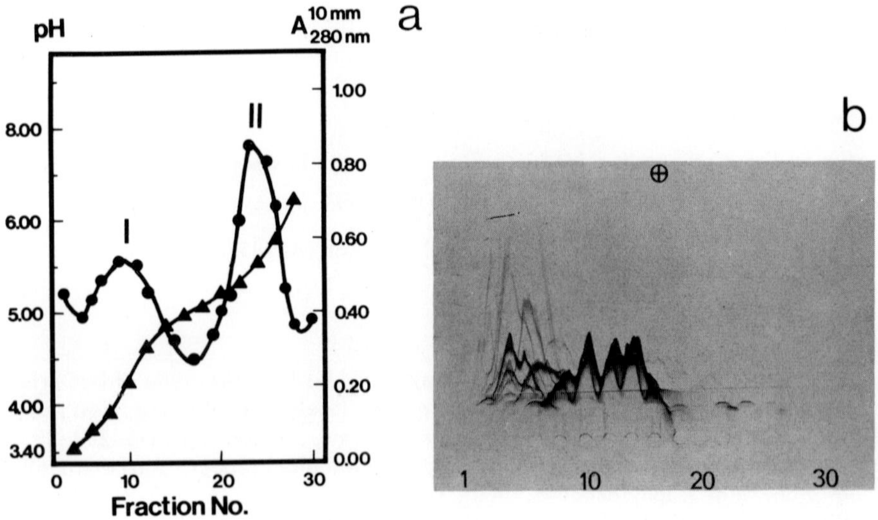

Figure 6. Isoelectric focusing of diphtheria toxin in flat bed of granulated gel. (a) Elution profile of protein from gel fractions obtained with the LKB separator. (•) Absorbance at 280 nm, (▲) pH values of gel fractions measured with a surface electrode. (b) Fused rocket immunoelectrophoresis of 10 µl aliquots of the eluted fractions. Note that peak II in (a) at the sample application zone contains pigmented UV-absorbing material without antigenic reactivity by the fused rocket procedure.

Concluding remarks

Biochemical separation methods such as isoelectric focusing have provided the means for obtaining bacterial toxins and enzymes of the purity required for meaningful biological studies. An important point with regard to resolution has been the maintenance of the separation achieved <u>in situ</u> during electrofocusing on elution of fractions. The granulated gel technique offers many advantages over columns in this respect. However, observations such as those reported herein for DNAase B and by Delincée and Radola for horseradish peroxidase emphasize the need for careful comparative studies to resolve anomalous findings before one switches over to granulated gel electrofocusing on a preparative scale. Present apparatus available for isotachophoresis necessitates the use of a concentrated sample of small volume. Isotachophoresis in granulated gel beds may prove to be a simpler method with high load capacity (Brogren, personal communication).

Acknowledgements

The studies discussed herein relating to E.coli enterotoxin were supported by grants from the Swedish Medical Research Council (16X-4723 and 40P-4576) and the Swedish Council for Forestry and Agricultural Research (A3985/B2974).

References

1. Arbuthnott, J.P., McNiven, A.C., Smyth, C.J.: Multiple forms of bacterial toxins in preparative isoelectric focusing. In, Isoelectric Focusing, Eds., Arbuthnott, J.P. and Beeley, J.A., Butterworths, London, pp. 212-239 (1975).
2. Wadström, T., Thelestam, M., Möllby, R.: Biological properties of extracellular proteins of Staphylococcus. Ann. N.Y. Acad. Sci. 236, 343-361 (1974).
3. Winter, A., Karlsson, C.: Preparative electrofocusing in density gradients. LKB Application Note 219, LKB Produkter, Bromma, Sweden, April 1976.
4. McNiven, A.C., Owen, P., Arbuthnott, J.P.: Multiple forms of staphylococcal alpha toxin. J. Med. Microbiol. 5, 113-122 (1972).
5. Smyth, C.J.: The identification and purification of multiple forms of θ-haemolysin (θ-toxin) of Clostridium perfringens. J. Gen. Microbiol. 87, 219-238 (1975).
6. Möllby, R., Wadström, T.: Purification of phospholipase C (alpha toxin) from Clostridium perfringens. Biochim. Biophys. Acta 321, 569-584 (1973).
7. Morris, C.J.O.R., Morris, P.: Separation Methods in Biochemistry, 2nd Edit., Pitman Publishing Co., pp. 992-1005 (1976).
8. Söderlind, O., Möllby, R., Wadström, T.: Purification and properties of heat-labile enterotoxin from Escherichia coli. Zentralbl. Bakteriol. Parasitenkd. Infektionskr. Hyg. I. Abt. Orig. A 229, 190-204 (1974).
9. Finkelstein, R.A.: Progress in the study of cholera and related enterotoxins. In, Mechanisms in Bacterial Toxinology, Ed. A.W. Bernheimer, John Wiley & Sons, New York, pp. 53-84 (1976).
10. Craig, J.P. and co-editors: The structure and functions of enterotoxins. J. Infect. Dis. 133, Supplement, S1-S156 (1976).
11. Möllby, R., Hjalmarsson, S.G., Wadström, T.: Separation of E.coli heat-labile enterotoxin by preparative isotachophoresis. FEBS Letts. 56, 30-33 (1975).
12. Dorner, F., Jaksche, H., Stöckl, W.: Escherichia coli enterotoxin: Purification, partial characterization and immunological observations. J. Infect. Dis. 133, S142-S156 (1976).

13. Gray, E.D.: Nuclease of group A streptococci. In, The Streptococci and Streptococcal Diseases: Recognition, Understanding and Management, Eds., Wannamaker, L.W. and Matsen, J.M., Academic Press, New York, pp. 143-155 (1973).
14. Smyth, C.J., Fehrenbach, F.J.: Isoelectric analysis of haemolysins and enzymes from streptococci of groups A, C and G. Acta Pathol. Microbiol. Scand. 82, 860-870 (1974).
15. Winter, A., Perlmutter, H., Davies, H.: Preparative flat-bed electrofocusing in a granulated gel with the LKB 2117 Multiphor. LKB Application Note 198, LKB Produkter, Bromma, Sweden, September 1975.
16. Wadström, T., Smyth, C.J.: Zymogram methods applied to thin-layer isoelectric focusing in polyacrylamide gel. Sci. Tools 20, 17-21 (1973).
17. Delincée, H., Radola, B.J.: Anomalous behaviour of horseradish peroxidase in isoelectric focusing. In, Electrofocusing and Isotachophoresis, Eds., Radola, B.J. and Graesslin, D., Walter de Gruyter (1976).
18. Collier, R.J.: Diphtheria toxin: Mode of action and structure. Bact. Reviews 39, 54-85 (1975).
19. Söderholm, J., Smyth, C.J., Wadström, T.: A simple and reproducible method for crossed immunoelectrofocusing. Comparative studies on the microheterogeneity of diphtheria toxin from different preparations. Scand. J. Immunol. 4 (Suppl. 2), 107-114 (1975).
20. Haglund, H.: Isoelectric focusing in pH gradients. Methods Biochem. Anal. 19, 1-104 (1971).
21. Brown, R.K., Caspers, M.L., Vinogradov, N.S.: Carrier ampholyte distribution in isoelectric focusing. In, Electrofocusing and Isotachophoresis, Eds., Radola, B.J. and Graesslin, D., Walter de Gruyter (1977).

Preparative Zone Convection Electrofocusing in Helical Glass Tubes

R. Quast

Inspite of the great potential inherent in the electroconvection principle of Valmet (1) for preparative application of electrofocusing in a pH gradient, the technique was not further developed into commercially available equipment. Instead, electrofocusing in granulated gels (2), in short, density gradient columns (3,4), and in continuous flow apparatus (5) have dominated the interest of scientists working with isoelectric focusing as a preparative technique. I first commented on preparative electrofocusing in helical glass tubes in 1974 (6). Crosscontamination of zones during fractionation observed in the first design is successfully prevented in the present instrument (7) which in some aspects resembles appartus employed by Macko & Stegemann (8) and by Bours (9) for analytical separations, but it differs from it in other important aspects, namely reusability of the helix, direct access to each of the separated fractions, and sample capacity in the range of grams. The glass helix is inexpensive, and the design described in the following may be taken as an example which can be easily modified with regard to fraction number and total volume.

A helical glass tube (numbers in quotation marks referring to fig. 1-3, and 7-10) with turns "1" of equal diameter "6" carries on top of each turn a straight glass tube "4". During separation and fractionation,the helix is immersed in a thermostated bath of a fluorocarbon liquid (liquid Teflon) with its axis "11" kept horizontally, the straight tubes "4"

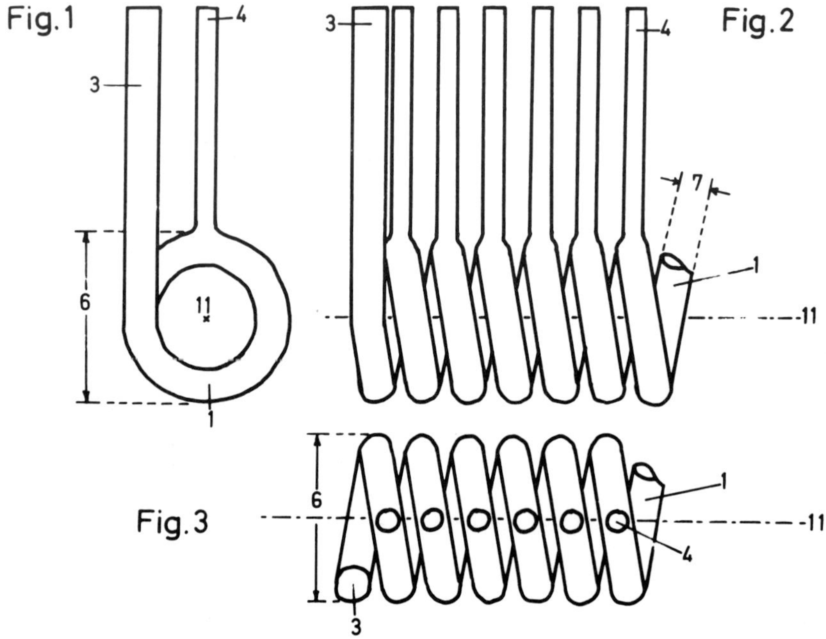

Fig. 1. End-on view of the glass helix.
Fig. 2. Side-on view of the glass helix.
Fig. 3. Top view of the glass helix. For explanation see text.

and the ends of the helix "3" with the electrodes sticking out of the coolant. Sample, solvent, and ampholytes fill the turns of the helix completely up to the base of the straight tubes. The platinum wire electrodes are surrounded by the respective electrode protective solution. The potential applied induces a pH gradient, and separation proceeds as described by Valmet (1). The liquid in the upper part of the turns of the helix is withdrawn simultaneously after the run by small plastic tubings fitted to syringes fixed to a frame (see fig. 4 b). Air bubbles thus separate the liquid of neighbouring turns, and the helix can be regarded as a series of discrete U-tubes similar to the sea-serpent shaped apparatus of Rose & Harboe (10). It is then easy to recover the separated fractions, one for each turn (fig. 5).

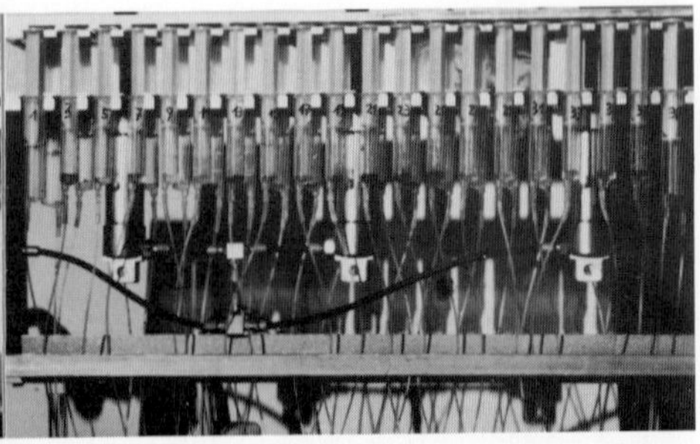

Fig. 4 a Fig. 4 b

Fig. 4 a. Helix after focusing a mixture of rabbit-haemolysate and human serumproteins in a wide-range pH gradient; results are shown in fig. 5; note small air bubbles at the bases of the straight tubes.

Fig. 4 b. Pneumatically driven frame with plastic syringes and capillary tubings in place; note small volume aspirated by the syringes.

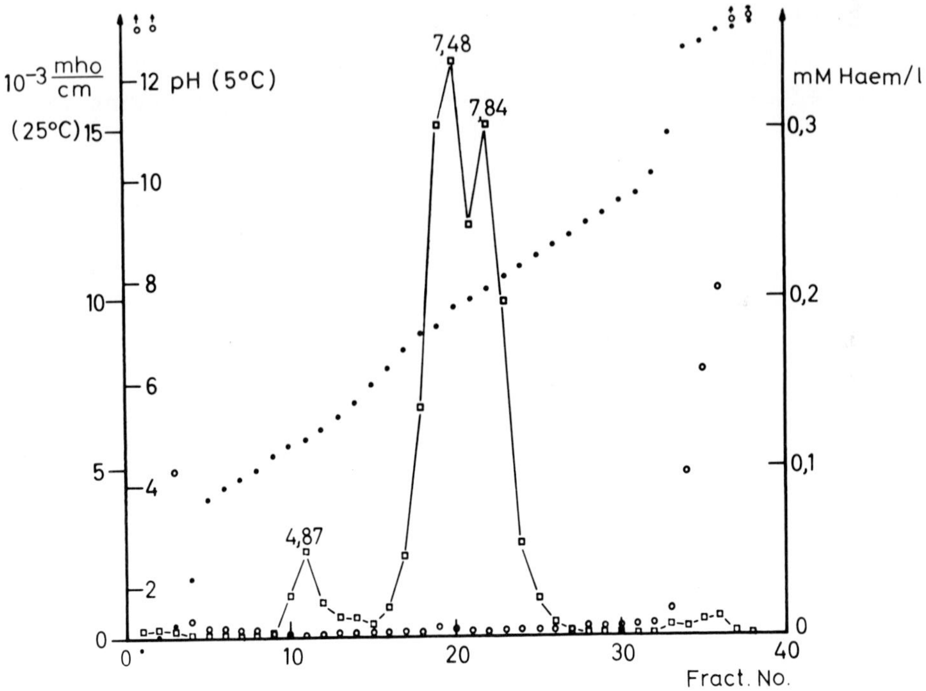

Fig. 5. Results of experiment shown in fig. 4 a. Dots = pH, open circles = conductance, open squares = concentration of haeme as reduced pyridine haemochromogen; the peak at pH 4.87 corresponds to Haeme-albumin.

The prototype helix in use in our laboratory consists of 37 turns, each of 4.5 ml volume, corresponding to a resolution of 37 fractions. The internal diameter "7" of the glass tube is about 6 mm, the outer diameter 9 mm. The diameter of the turns of the helix "6" - 4.5 cm - is determined by two limiting conditions: first, it is desirable to have as many fractions per volume of a given helix as possible, implying as many turns per length of glass tube, and second, the internal cross-section of the glass tube must be kept constant over the length of the helix during the manufacturing process which imposes severe restrictions to the glassblower with regard to a very small radius of the turns of the helix. The total length of the tube measured between the electrodes is about 3.6 m. Very high voltage must be applied to achieve a field strength approximating that used in other equipment, for instance 50 to 100 V/cm. Although the field strength is not an effective variable to increase separation power in electrofocusing (5,6,

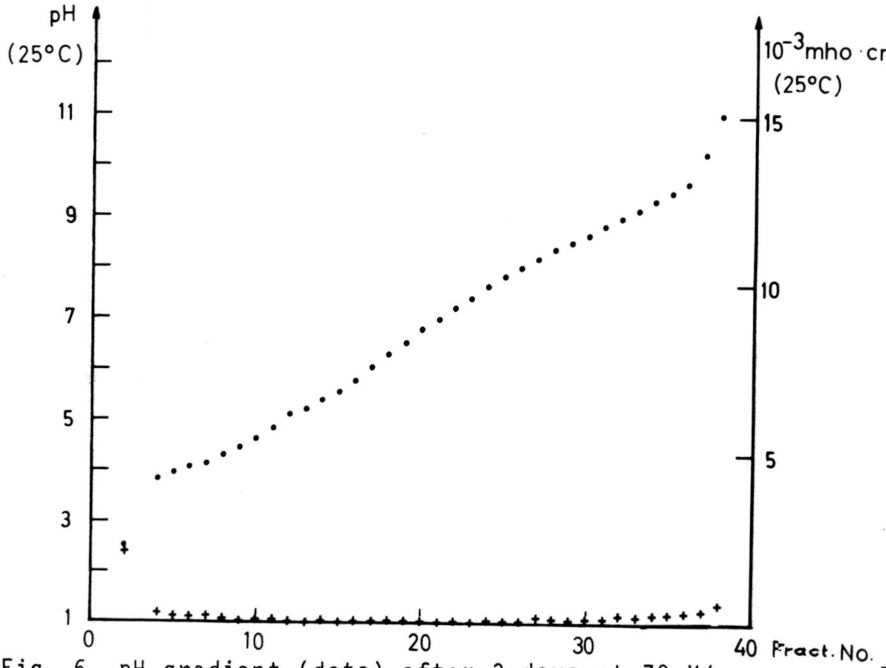

Fig. 6. pH gradient (dots) after 3 days at 70 V/cm; no sample; 1% Ampholine 3.5-10; crosses = conductance.

11), it helps shorten the time necessary for establishing the pH gradient and for focusing the sample molecules. The former is demonstrated by results of two experiments, one run at a maximum of 7.5 V/cm for about 13 days (fig. 5), the other run for only 3 days at about 70 V/cm (fig. 6). Of course, it is not necessary to have the sample present in the solution from the start. Instead, it is advisable in many cases to interrupt the focusing, to separate the volumes of the turns as described above for fractionation, to dissolve the sample in one or more fractions having a suitable pH, and to refill the turns with the liquid withdrawn into the syringes.

Fig. 7

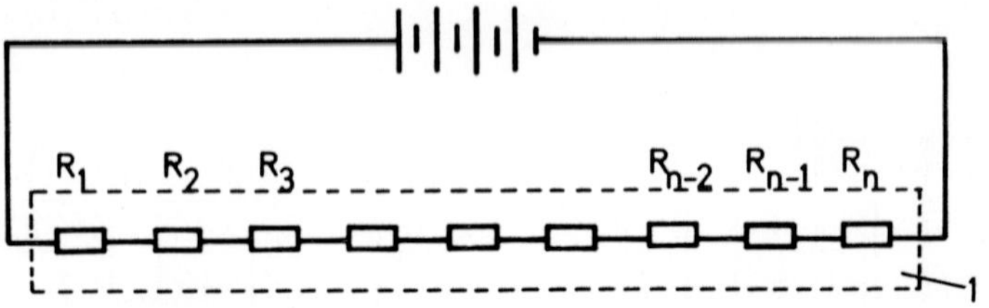

Fig. 8

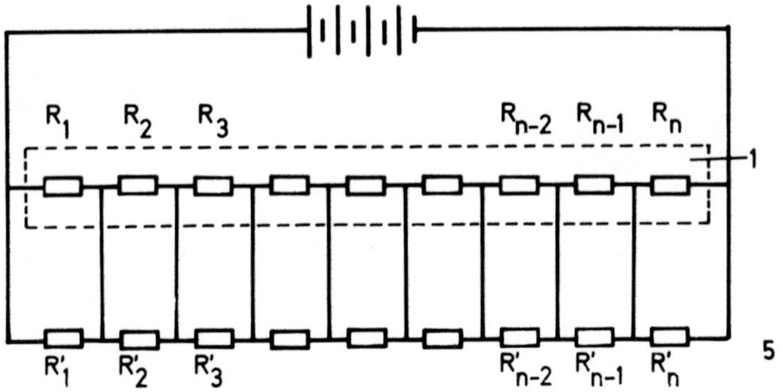

Fig. 7. Circuit diagram of helix (broken line) with power supply; R = resistance of the liquid in one turn.
Fig. 8. Circuit diagram of helix (broken line) with power supply and external voltage divider; R = resistance in one turn; R' = resistance of one power resistor; n = number of elements; the vertical lines connecting the series of power resistors with the helix symbolize the platinum wire dynodes; in first experiments R' was 28 kohms, the sum of 37 power resistors was about 1 Mohm.

The conductance along a pH gradient changes considerably (12) (see fig. 5 and 6). This leads to great differences in power distribution and hence of heat production in various turns of the helix. An attempt to smoothen out these differences by adding a voltage divider circuit "5", outlined in fig. 7 and 8, resulted in a pH gradient comparable to that of the experiments shown in fig. 5 and 6. The electrochemical effects of the platinum wire dynodes connecting the external power resistors - R' -through the straight tubes with the internal resistors - R - represented by the fluid in the turns of the helix are yet to be determined.

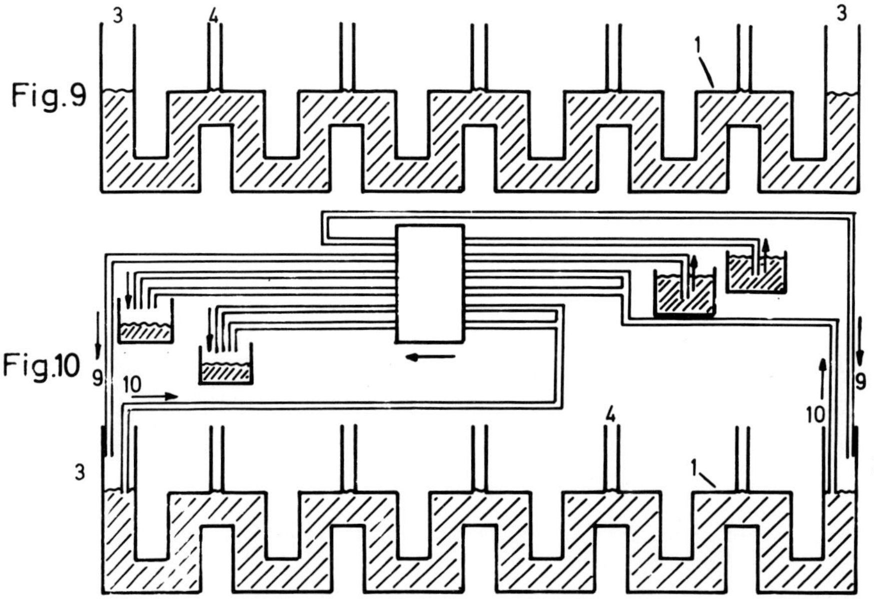

Fig. 9. Diagram of liquid distribution in helix (hatched area)
Fig. 10. Experimental setup for continuous, equal, and constant liquid exchange on both sides of the helix; storage tanks at right, waste tanks at the left, metering pump in the middle.

Another line of experiments is in progress in order to investigate the feasibility of changing continuously the electrolyte solutions near both electrodes during the run (fig. 9 and 10). Preliminary results show that hydrodynamic problems of a constant and equal rate of fluid exchange on both sides of the helix have to be solved before experiments under power can be conducted in which both anions and cations of suitable type and concentration may be sent over the pH gradient from the respective counter-electrode compartments at a constant rate thus leading to a constant stream of ions of a neutral salt through the pH gradient. Eventually, this will result in a steady state concentration of neutral salt in the vicinity of macromolecules helping to retain them in their native configuration when being focussed at their isoelectric point.

Acknowledgement: This work was supported in part by the Deutsche Forschungsgemeinschaft through a gift of analytical instruments. The skilful technical assistence of Mrs. B.Snaidero is greatfully acknowledged.

References:

1. Valmet,E.: Zone convection electrofocusing - a new technique for fractionating ampholytes in free solution. Science Tools 16, 8-13, (1969)

2. Radola,B.J.: Analytical and preparative isoelectric focusing of proteins in Sephadex and Bio-Gel layers. Prot.Biol.Fluids, Proc. 18th coll., Brugge, (1970) Pergamon Press, Oxford, 1971, pp. 487-491

3. Rilbe,H.: Rapid isoelectric focusing in density gradient columns. Ann.N.Y.Acad.Sci. 209, 80-93, (1973)

4. Rilbe,H. & Petterson,S.: Preparative isoelectric focusing in short density gradient columns with vertical cooling. In: Isoelectric focusing, J.P.Arbuthnot & J.A.Beeley, eds., Butterworths, London, 1975, pp. 44-57

5. Fawcett,J.S.: Some recent developments in preparative isoelectric focusing. ibidem, pp. 23-43

6. Quast,R.: Preparative isoelectric focusing of plasma proteins.
In: Plasma protein turnover, Bianchi,R., Mariani,G., & McFarlane,A.S.,eds., MacMillan Press Ltd., London, 1976, pp. 3-14
7. DBpat. applied for
8. Macko,V., & Stegemann,H.: Free electrofocusing in a coil of polyethylene tubing. Anal.Biochem. 37, 186-190, (1970)
9. Bours,J.: Free isoelectric focusing of bovine lens gamma-crystallins. Exp. Eye Res. 16, 501-515, (1973)
10. Rose,C., & Harboe,N.M.G.: Preparative isoelectric focusing in a sea-serpent shaped apparatus.
Prot.Biol.Fluids, Proc. 17th coll., Brugge, (1969)
Pergamon Press, Oxford, 1970, pp. 397-400
11. Vesterberg,O.: Isoelectric focusing of proteins.
Thesis, Stockholm, (1968)
12. Haglund,H.: Isoelectric focusing in pH gradients - a technique for fractionation and characterization of ampholytes.
Methods in Biochem.Anal. 19, 1-104, (1970)

Chromatofocusing: Isoelectric Focusing on Ion Exchangers in the Absence of an Externally Applied Potential

L.A.AE. Sluyterman and J. Wijdenes

SUMMARY

A pH gradient, slowly moving down a column, is produced inside a column of an ion exchanger, previously adjusted to one pH, by running through an elution buffer of another pH. Proteins dissolved in the first few ml of the elution buffer pass slowly through the column and emerge in the order of their pI values. It is shown that focusing effects occur. Results are comparable to those obtained with the electrophoretic method in a density gradient.

INTRODUCTION

Electrofocusing depends on protein charges and an internally produced pH gradient. Ion exchange chromatography of proteins also depends on protein charges and can be carried out with a pH gradient. We therefore wondered whether focusing effects could be produced in the latter system.

THEORETICAL CONSIDERATIONS

A pH gradient can be developed inside a column in the following manner. A buffer of say pH 7 is run through a column of an ion exchanger, for example an anion exchanger previously adjusted to pH 11. The pH of the buffer increases as the solution moves down the column, owing to the base content of the exchanger. Therefore a pH gradient in the range of pH 7 to 11 is established inside the column, which moves down the column as the elution with the buffer of pH 7 continues. Proteins with isoelectric points within this pH range, dissolved in the first few ml of elution buffer, are carried along by the buffer as long as the pH of that part of the buffer is below the isoelectric point of each particular protein and the charge of the proteins is positive. As soon as the pH of the buffer fraction containing the proteins is increased to a value slightly above the pI of a protein, that protein is negatively charged and retained by the positive anion exchanger, to be released again by a subsequent buffer fraction of slightly lower pH. This process of adsorption and desorption is repeated continuously as the pH gradient moves down the column. In this way proteins emerge in the order of their pI values.

The focusing action was demonstrated by applying a sample of myoglobin to the column and starting the elution. When the sample had proceeded along one fifth of the column length, a second sample of myoglobin was applied. The elution being continued, the second sample ran down the column at a higher rate than the first sample until the second sample caught up and joined the first one. Therefore the two samples were focused into one peak before emerging from the column.

The theoretical bandwidth in pH units, as far as it depends on axial diffusion, is proportional to

$$(1) \quad \left\{ \left(D \frac{dpH}{dV} \right) / \left(-\frac{dV}{dt} \frac{F\Psi}{RT} \frac{dZ}{dpH} \right) \right\}^{\frac{1}{2}}$$

in which D denotes the diffusion constant of the protein, $\frac{dpH}{dV}$ the pH gradient, $\frac{dV}{dt}$ the flow rate, Ψ the Donnan potential of the ion exchanger and $\frac{dZ}{dpH}$ the slope of the titration curve of the protein around its pI. This equation is similar to the equation derived for the electrophoretic method, indicating the resemblance of the two methods. Furthermore it is apparent from equation (1) that it is advantageous to utilize a low buffer concentration, causing a low pH gradient and, owing to low ionic strength, a high Donnan potential.

EXPERIMENTAL RESULTS

Papain, cytochroom C and myoglobin were successfully run on a 1.6x12 cm bed of DEAE-Biogel A (Bio-Rad Laboratories)contained in a glass column of Pharmacia (Sweden) type K 16/40. The bed material was washed with 1 M ammonia. The column bed was poured and equilibrated with 0.01 M acetate-ammonia buffer pH 10.0. Successively 4 ml water, 2 ml elution buffer containing 10-25 mg protein and the elution buffer were applied. The latter buffer was 0.15 % Ampholine range 8-9.5 (L.K.B.) adjusted to pH 7.5 by addition of acetic acid. The flow rate was 5-10 ml/h. For elution of papain the buffer contained 5 % dimethylsulphoxide for increasing the solubility of the protein. The elution pattern of commercial myoglobin (Serva) showed one main band and four minor coloured components

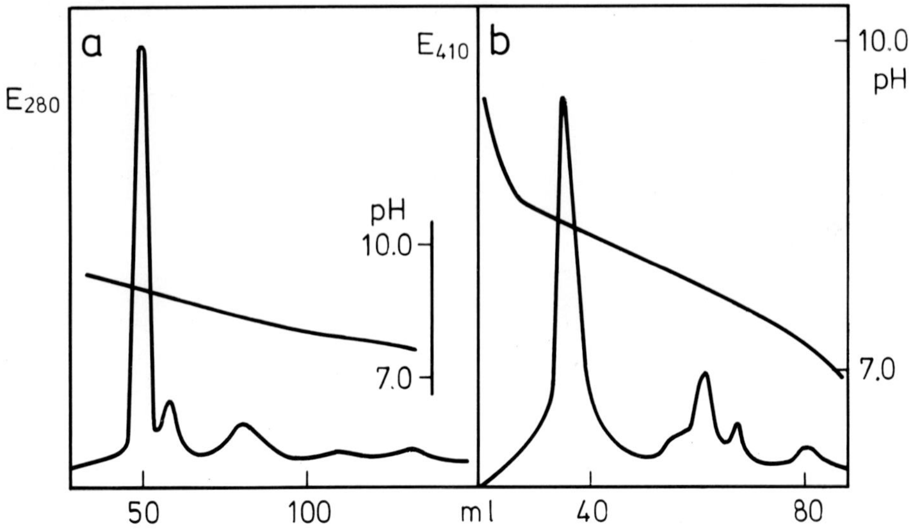

Fig. 1. Isoelectric fractionation of commercial myoglobin, a) by chromatography, b) by electrophoresis in a sucrose gradient, 0.25 % Ampholine 7-9, 3 days 400-1000 V.

(fig. 1a). This is comparable to the results obtained with the electrophoretic method in a density gradient (fig. 1b).

A few experiments were carried out in the acid range with bovine serum albumin on a column of CM-Bio-Gel A, previously adjusted to pH 4.0 with 0.1 M acetate-ammonia buffer. Elution buffer was 0.08 % Ampholine range 3-10, adjusted to pH 7 with ammonia. Both the chromatographic and the electrophoretic method gave a double peak.

DISCUSSION

The chromatographic method of focusing can be carried out with ordinary chromatographic equipment. Any size of column can be utilized, including large ones because there is no cooling problem. The use of Ampholine buffers, though convenient for obtaining smooth pH gradients, is not essential.

8. Separation of Cells

New Instrumentation and Procedures for the Analysis of Mammalian Cells by Electrophoretic Techniques

N. Catsimpoolas and A.L. Griffith

ABSTRACT

A new instrumentation system (TRANS-Analyzer) has been developed for the separation and characterization of mammalian cells by electrophoretic techniques. The supporting medium is provided by an iso-osmolar density gradient throughout the separation path. Monitoring of the cell migration in the electric field is performed _in situ_ and repetitively by axial light loss (ALL) and scattering measurements. Data acquisition, processing and display is performed by a computer. Kinetic analysis of the cell distribution provides precise information on mobilities and dispersion of the population as a function of time. Applications of the method to the analysis of erythrocytes and lymphocytes by continuous and pH gradient (isoelectric focusing) techniques is presented.

INTRODUCTION

We have recently described an instrument and procedures (1,2) for the separation and analysis of mammalian cells based on the principles of electrophoresis in an iso-osmolar density gradient (3,4) and velocity sedimentation at unit gravity (5) and on earlier instrumental developments for transient state isoelectric focusing (TRANS-IF) (6,7). In this paper, we report the construction of a more advanced apparatus (TRANS-Analyzer) which incorporates larger columns, multiple column scanning, a two-channel photometer and associated electronics for control of the scanner and processing of the data. Applications of the TRANS-Analyzer to the analysis of mammalian cells by electrophoresis and isoelectric focusing is presented.

THE TRANS-ANALYZER

A general schematic diagram of the TRANS-Analyzer is shown in Fig. 1. Cell separation takes place in temperature regulated (4°C) quartz columns which bear electrode reservoirs for electrophoresis. An electronically controlled X-Y scanner moves the quartz columns sequentially in the path of the light beam (X-positioning) where each column is scanned by vertical movement (Y-scanning) using a narrow (100μ) slit of light. The resulting "absorbance" (axial light loss) and angle scattering is detected by two photomultipliers equipped with appropriate filters, measured by two photometers and recorded on a two-channel strip-chart recorder. In addition, the signals are digitized and processed by a computer. The liquid flow unit includes a cooler-circulator to provide temperature control for the columns, a density gradient former, syringe pumps for circulation of the buffers through the electrode reservoirs, semi-automatic sample injection loops, valves and associated tubing.

MATERIALS AND METHODS

Human and rabbit erythrocytes and mouse (C57BL/6J) spleen lymphocytes were prepared as described previously (2,4). The preparation of the iso-osmolar density gradient consisting of Ficoll-sucrose and the electrophoresis buffer were made as reported in a previous paper (1).

The isoelectric focusing experiments were performed in the Ficoll-sucrose gradient by using carrier ampholytes instead of the buffer. An 1:1 mixture of pH 2.5-4.0 Ampholine (LKB) and pH 5.0-8.0 Bio-lytes (Bio-Rad) in 1% dextrose were used to form the pH gradient which covered the pH range 2.5 to 8.0. The cell sample solution (placed on top of the density gradient) contained 1% carrier ampholytes (see above) in 2.0% Ficoll, 6.44% sucrose and 1% dextrose mixture. A "protecting" solution of 1% carrier ampholytes in 1.0% Ficoll, 6.62% sucrose and 1% dextrose was overlayed above the sample. The upper electrolyte was 0.02 M NaOH in 6.8% sucrose and the lower 5% phosphoric acid in 5.1% sucrose and 10% Ficoll. Estimation of the statistical moments of the cell distribution and determination mobilities were performed as described previously (1,7).

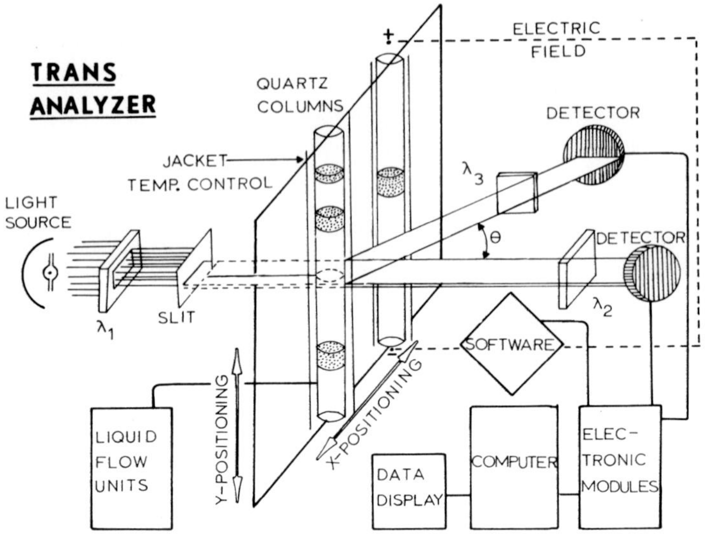

Fig. 1. Schematic diagram of the TRANS-Analyzer instrumental assembly.

RESULTS AND DISCUSSION

Typical electrophoretic scans of the separation of a model mixture of man and rabbit erythrocytes ($3x10^6$ cells from each species) as a function of the electrophoresis time are shown in Fig. 2. The mean electrophoretic velocity ($\bar{v}$) of each cell distribution is estimated from a plot of the first statistical moment (m'_1) i.e. centroid of the peak versus elapsed electrophoresis time (t) so that $\bar{v} = d\,m'_1/dt$ in cm/sec (7). Knowledge of the current, buffer conductance and cross sectional area of the tube allows the determination of the electric field strength (E) and therefore the electrophoretic mobility (M) in TU (Tiselius Unit, TU = $1x10^{-5}$ $cm^2 . V^{-1} . sec^{-1}$ (8)). At 4°C and under the conditions of the experiment in respect to the viscosity of the gradient and ionic strength of the buffer the mobility of man erythrocytes

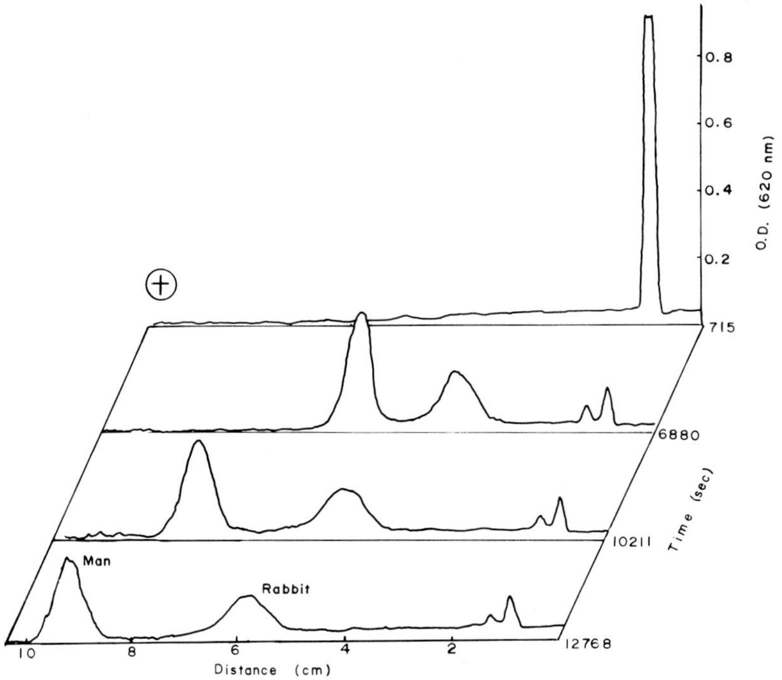

Fig. 2. Electrophoretic separation of a model mixture of rabbit and man erythrocytes. Axial light loss (ALL) at 620 nm scans of the cell distributions are shown as a function of electrophoresis time. The two small peaks at the sample origin indicate the interfaces of gradient/sample solution and sample solution/upper buffer.

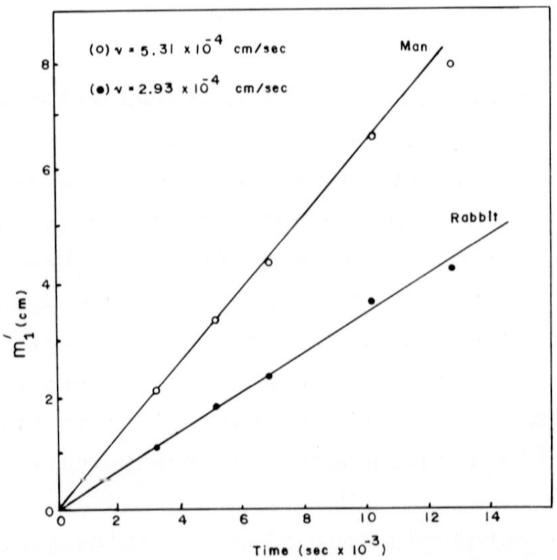

Fig. 3. Plot of the first statistical moment, m'_1 (i.e. centroid of peak position), versus electrophoresis time for man and rabbit erythrocytes from the scans shown in Fig. 2. The slope of the lines represent the electrophoretic velocity of the cells.

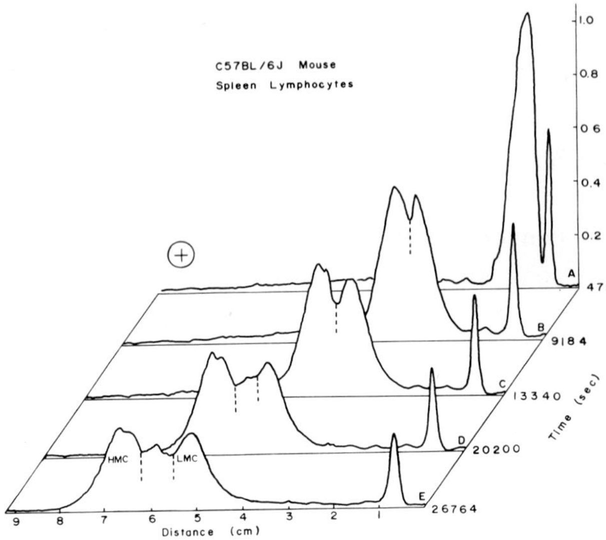

Fig. 4. Electrophoretic scans of C57BL/6J mouse spleen lymphocytes as a function of time. HMC and LMC denote the high and low mobility cells, respectively. HMC represent primarily T lymphocytes and LMC the B cells.

was 2.86 TU and that of rabbit 1.58 TU. These values can be used only for comparative purposes in estimating relative mobility (R_f) changes among various populations of cells. It should be noted that although the viscosity becomes progressively higher (3) as the cells migrate into the denser part of the gradient, the electrophoretic velocity did not change appreciably. This particular behaviour cannot be adequately explained at present. It is possible that an interplay of alterations in the dielectric constant of the medium and of progressive increase of the zeta potential of the cell membrane due to the possible binding of Ficoll contribute to the maintenance of a steady-state velocity.

Mouse spleen lymphocytes from C57BL/6J animals are separated electrophoretically into more than two peaks (Fig. 4). The high mobility (HMC) and low mobility (LMC) cells (Fig. 5) exhibited values of 2.52 TU and 1.72 TU, respectively. In preparative density gradient electrophoresis experiments (9), the HMC cells have been identified as typical T cells and the LMC as B lymphocytes.

The two representative experiments discussed above demonstrate the usefulness of the TRANS-Analyzer in following the electrophoretic separation of cells optically and measuring their mobilities. Consequently, we were interested in applying the same methodology in exploring the possibility of separating cells by isoelectric focusing. A preliminary report (10) utilizing free flow isoelectric focusing techniques seemed promising in this respect. Transient State Isoelectric Focusing (TRANS-IF) scans in the pH 2.5-8.0 range as a function of time for man erythrocytes and mouse (C57BL/6J) spleen lymphocytes are shown in Figs. 6 and 8, respectively. Both cell populations appeared to be initially focused and subsequently "unfocused" and finally lysed. Figs. 7 and 8 show the kinetics of focusing and unfocusing by following the m'_1 and m_2 values of each distribution with time. In the

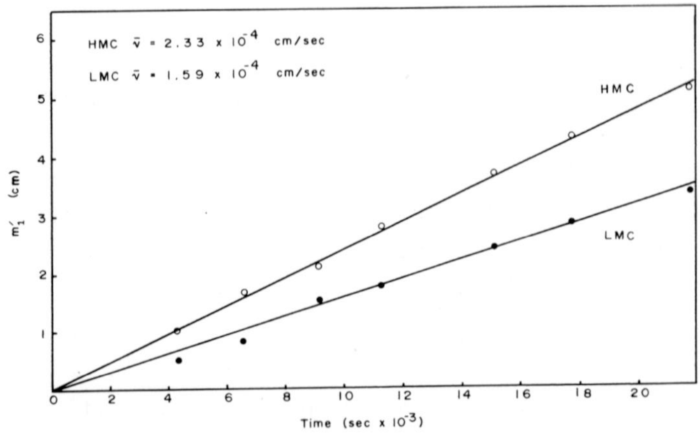

Fig. 5. Plot of the centroid of peak position of HMC and LMC C57BL/6J mouse spleen lymphocytes as a function of electrophoresis time . $\bar{v}$ denotes mean electrophoretic velocity.

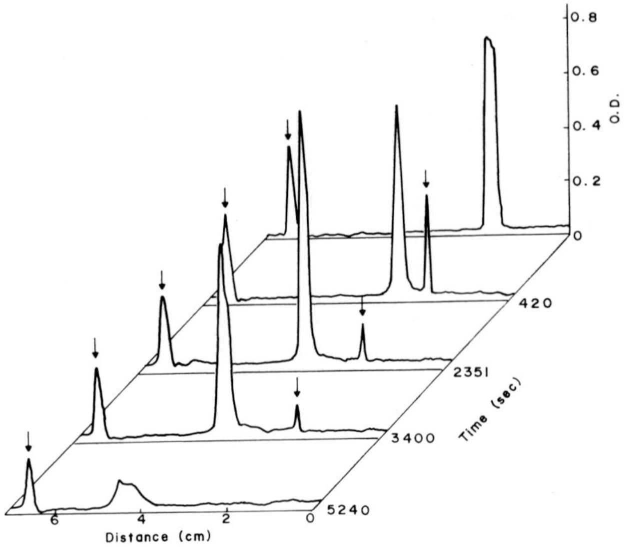

Fig. 6. Isoelectric focusing scans of man erythrocytes in the pH 2.5-8.0 range (2% carrier ampholytes). The peaks marked with arrows are interfaces indicating the start and end of the density gradient. Note the sharply focused peak at the 2351 sec scan and subsequent unfocusing. Migration is toward the positive electrode (acidic pH).

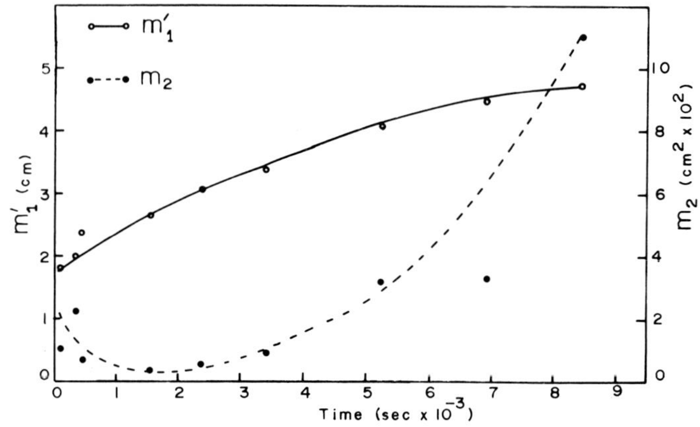

Fig. 7. Plot of the first (m'_1) and second (m_2) statistical moments of the cell distribution shown in Fig. 6 as a function of electrofocusing time.

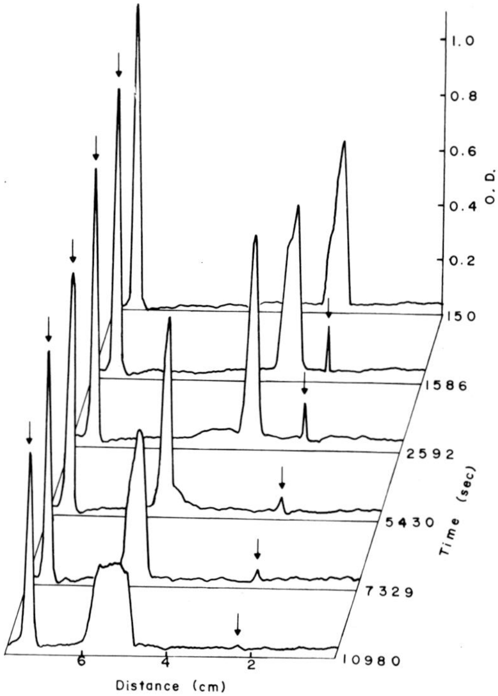

Fig. 8. Isoelectric focusing scans of C57BL/6J mouse spleen lymphocytes in the pH 2.5-8.0 range (2% carrier ampholytes). The peaks marked with arrows are interfaces. Migration is toward the positive electrode. Note the unfocusing of the distribution at the 7329 sec and 10980 sec scans.

case of proteins, the m'_1 reaches steady-state upon completion of focusing and the m_2 (variance of peak) reaches a minimum value and remains unchanged (7). When man erythrocytes were used for focusing the m_2 value reached a minimum (Fig. 7) which indicates "focusing", but then increased again indicating "unfocusing". The m'_1 value never reached the steady-state. Similar results were obtained with the mouse lymphocytes (Fig. 9) except that the cells remained "focused" (i.e. minimum m_2) for a longer period of time. The m'_1 value was not stabilized. The above results lead to the suggestion that as the cells reach an apparent pI they appear to be "focused" and exhibit minimum peak variance (m_2). At the region of the isoelectric pH, changes in the membrane may cause alteration of the pI toward more acidic values and therefore the m'_1 value is not stabilized since the cells are now "refocused" in a new position. This may occur repetitively to the point where cells with a wide range of pI's exist and therefore appear to be "unfocused" (i.e. m_2 values are increasing). Finally, extensive damage to the membrane causes lysis of the cells. When mixtures of rabbit and man erythrocytes were focused together the two types of cells appeared to be separated initially. The separation, however,

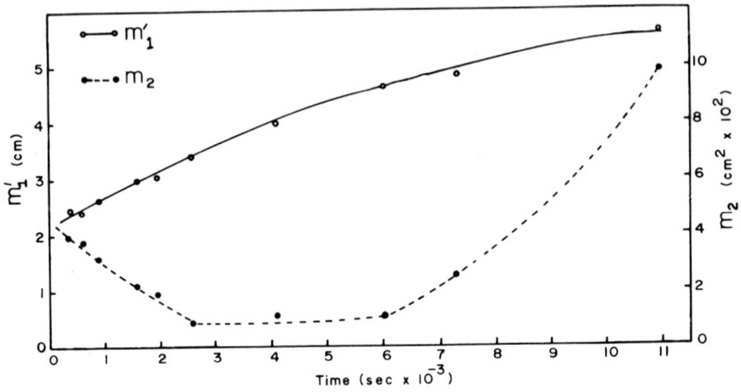

Fig. 9. Plot of the first (m'_1) and second (m_2) statistical moments of the cell distribution shown in Fig. 8 as a function of electrofocusing time.

was not due to the focusing of the cells but rather due to differences in mobility resulting from electrophoretic migration in a pH gradient, before the steady-state is achieved. From the above experiments, we have concluded that at the present state of development electrophoresis represents a milder and practically useful technique for the separation of mammalian cells than isoelectric focusing.

ACKNOWLEDGEMENT

This work was supported by NCI contract No. NO1-CB-43928 and NSF grant No. MPS74-19830.

REFERENCES

1. Catsimpoolas, N., Griffith, A.L., Williams, J.M., Chrambach, A., Rodbard, D.: Electrophoresis with continuous scanning desitometry: separation of cells in a density gradient. Analyt. Biochem. 69, 378-384 (1975).

2. Catsimpoolas, N., Rossi, R., Griffith, A.L.: Transient velocity sedimentation of cells at unit gravity. Life Sciences 18, 481-488 (1976).

3. Boltz, R.C., Todd, P., Streibel, M.J., Louie, M.K.: Preparative electrophoresis of living mammalian cells in a stationary Ficoll gradient. Prep. Biochem. 3, 383-401 (1973).

4. Griffith, A.L., Catsimpoolas, N., Wortis, H.H.: Electrophoretic separation of cells in density gradient. Life Sciences 16, 1693-1702 (1975).

5. Mel, H.C.: Sedimentation properties of nucleated and non-nucleated cells in normal rat bone marrow. Nature 200, 423-425 (1963).

6. Catsimpoolas, N.: Analytical scanning isoelectrofocusing. In "Isoelectric focusing and isotachophoresis" (N. Catsimpoolas, ed.), Ann. N.Y. Acad. Sci. 209, 65-79 (1973).

7. Catsimpoolas, N.: TRANS electrophoresis. In "Methods of Protein Separation", Vol. 1 (N. Catsimpoolas, ed.), pp. 27-67, Plenum, New York (1975).

8. Catsimpoolas, N., Hjertén, S., Kolin, A., Porath, J.K.: Tiselius Unit. Nature(London) 259, 264 (1976).

9. Platsoucas, C.D., Griffith, A.L., Catsimpoolas, N.: Density gradient electrophoresis of mouse spleen lymphocytes: separation of T and B cell fractions. J. Immunol. Methods (in press).

10. Just, W.W., Leon-V., J.O., Werner, G.: Isoelectric focusing in continuous-flow electrophoresis. 1. Separation of mixed red blood cells of different species. Analyt. Biochem. 67, 590 (1975).

Continuous-Flow Isoelectric Focusing: Further Studies on Human Red Blood Cells and on the Separation of Rat Liver Light Mitochondrial Fraction

W.W. Just and G. Werner

In the different areas of biochemical research, such as immunology, membrane biochemistry, cancer research, neurochemistry as well as others, the development of new methods for the isolation of purified cell or cell particle fractions is becoming more and more a basic necessity. Thus isoelectric focusing (IF), as described by Svensson (1)and Vesterberg (2) and known as a high resolving technique for the fractionation of amphoteric macromolecules, has been applied in different modifications by several authors for the fractionation or characterization of cells or subcellular particles (3-8). Up to now the question of whether steady state IF could be a useful method in the field of cell or cell particle fractionation has not been clearly answered. Without doubt, the application of this electrophoretic technique could pose several difficulties which may be briefly mentioned. First, is it at all possible to electrofocus cells or cell particles? If so, would the cells remain intact after the fractionation procedure? We have to consider in this connection that the biological material will be subjected for certain periods to a medium of very low "ionic strength" (see Discussion) and quite possibly outside the physiological pH range. Which criteria are suitable ones to varify the functional integrity of the cells after IF? It may also be of importance whether carrier ampholytes adsorbed on the cell surface may be removed after IF?

Of course the final aim of such studies is to obtain better results in the field of cell or cell particle fractionation. In this respect, one would first look at the quality of the separations and secondly at some of the above questions which however may be of theoretical importance. In that case it will be of interest whether the use of IF or electrophoresis in a pH gradient may provide some advantages over free-flow electrophoresis utilizing normal buffer systems. The latter technique, mainly elaborated by Hannig and his co-workers has been repeatedly demonstrated in the last years to be a valuable method for the fractionation of cells and subcellular particles (9,10).

Human red blood cells (RBCs) and the light mitochondrial fraction (LMF) of rat liver served as models to obtain, in the first case, new information on the application of IF to cell separation and, secondly, to test this separation technique on the isolation of a purified lysosomal fraction, a problem which has not been satisfactorily solved using centrifugation techniques (11,12). Furthermore, since damage of the lysosomal membrane immediately results in an increase of free lysosomal enzyme activity, as has been demonstrated by Appelmans and deDuve (13), the ratio of free to total lysosomal enzyme activity, often called latency, allows a good reflection of the state of preservation of these organelles.

MATERIALS AND METHODS

Since the description of the apparatus used for these studies (14), several modifications were made, of which the most important is the construction of a thermostat-controled cooling box around the equipment. This allows for precooling of all parts of the apparatus, especially the thick plexiglas front, and for keeping the biological material at low temperature at all times during the electrophoretic run.

Ampholine carrier ampholytes (LKB, Bromma) pI range 3.5-10 or 5-7 or, in one experiment, self-made ampholytes, pI range 4-9.5 and whose physico-chemical properties are briefly described following Results, were used.

Biological Materials

Human RBC suspensions were prepared by washing the cells three times in 0.9% saline and another three times in 0.25M sucrose, both containing 1mM EDTA adjusted to pH 7.0. With respect to native blood the final cell suspension was about 1:10 diluted.

The LMF was obtained from 4 rat livers by differential centrifugation, as previously reported (15,16). Homogenization and all centrifugation steps were carried out in the presence of 0.01mg/ml heparin and 0.01mg/ml RNA. Finally the LMF was suspended in 4-5ml of 0.25M sucrose containing 0.1mg/ml heparin and 0.1mg/ml RNA.

Electrophoretic Conditions

RBC experiments: Separation cell solution = 0.25M sucrose, 1% ampholytes; flow-through volume = 1.1-1.5ml/min; rate of RBC injection = 1.3ml/h; anolyte = 0.1M H_3PO_4; catholyte = 0.1M NaOH; field strength = 90V/cm; temperature = 4°C.

LMF experiments: Separation cell solution = 0.25M sucrose, 1% Ampholine pI 5-7; flow-through volume = 1.5ml/min; rate of LMF injection = 1.0-1.5ml/h; field strength = 100V/cm; anolyte = 0.1M H_3PO_4; catholyte = 0.1M NaOH; temperature = 4°C.

Biochemical Assays

Acid phosphatase activity was measured using both ß-glycerophosphate (17) and p-nitrophenyl phosphate (15) as a substrate. Aryl sulfatase, ß-glucuronidase and N-acetyl-ß-glucosaminidase activities were determined as described by Bowers et al.(17). Cytochrome c oxidase activity was determined as reported by Sottocasa et al.(18), using a Clark oxigen electrode. Glucose-6-phosphatase activity was measured in a double assay as outlined by Beaufay and deDuve (19). Inorganic phosphate set free by the enzyme acid phosphatase was determined as for the glucose-6-phosphatase assay (15). Uricase activity as well as protein was measured as previously described (6). Free and total activities of the lysosomal enzymes were determined as recently reported (16) except the use of 0.1M instead of 0.01M acetate buffer pH 5.0 at 37°C. Enzyme measurements performed after electrophoretic separations refer to enzyme activities which may be sedimented (particle-bound).

Electron micrographs were obtained using a general procedure (6) after having washed the fractions in 0.25M sucrose at 11700 x g for 20min.

RESULTS

The recent experiments were concerned with the separation of mixed RBCs from different species by continuous-flow IF (5,6). During electrophoresis, the migration of the cells was consistent with the principle of IF. The pH values at which the cells were collected after the separations were nearly the same regardless of whether ampholytes of pI 3.5-10 or 5-7 were used. In spite of this, there were critical doubts as to whether steady state IF was achieved in these experiments. With this in minde, an experiment of the type previously performed by Fawcett (20), but using human RBCs instead of hemoglobin, was performed to demonstrate if continuous-flow IF could be achieved with RBCs.

Fig.1 demonstrates the result of these experiments. After injection of RBCs at two different positions within the pH gradient, the cells fuse after a certain migration distance into one zone and were then eluted by following the liquid flow perpendicularly to the electric field. Fig.1a and 1b show identical experiments using Ampholine pI 3.5-10 and self-made ampholytes pI 4-9.5, respectively. The injection positions visible in Fig.1 do not represent the top of the separation cell but about the middle of it, which means that the RBCs are loaded into a gradient which is to some extent already

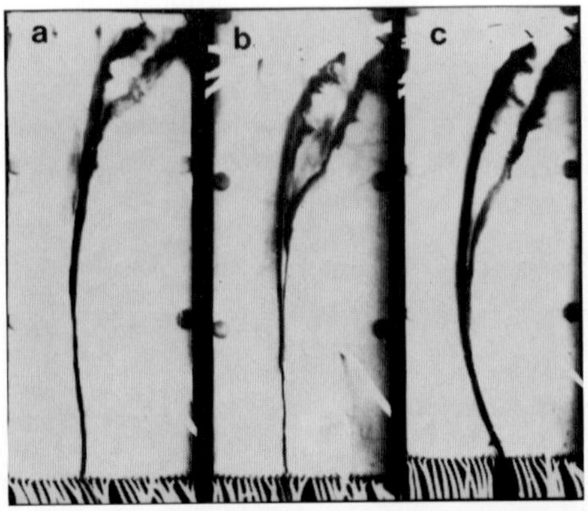

Fig.1. Human RBCs injected into a preestablished pH gradient simultaneously at two different positions. After a certain time of migration the two zones combine into one and with minimal further changing of position within the gradient the cells are eluted into 2-3 fractions (a,b). Under quite similar conditions, however, the RBCs sometimes start to migrate back cathodically (c), after having reached about the same peak acidic position within the gradient as in a and b. Self-made ampholytes pI 4-9.5 were used in experiment b.

established. The presence of 1mM EDTA in the RBC suspension prevents clumping during electrophoresis but causes severe disturbances, however, in the zones at the beginning of the migration which were not observed when, in addition to EDTA, 1% Ampholine was present in the cell suspension (5). The presence of ampholytes, on the other hand, decreased the migration rate of the cells so that the two initial zones failed or at least needed a longer time to combine. With sufficient time, the initial disturbances are completely balanced. For these experiments the pH at which the main fractions of the RBCs were collected was between 6.5-6.8, very close to the pI of hemoglobin. In the phase contrast microscope the cells look intact. After centrifugation of the focused RBCs the supernatant reveals a faint coloration from free hemoglobin.

Very often a migration pattern like that shown in Fig.1c was observed, especially at somewhat higher field strengths or at lower flow-through time. After reaching about the same position within the pH gradient as in the experiments shown in Fig.1a

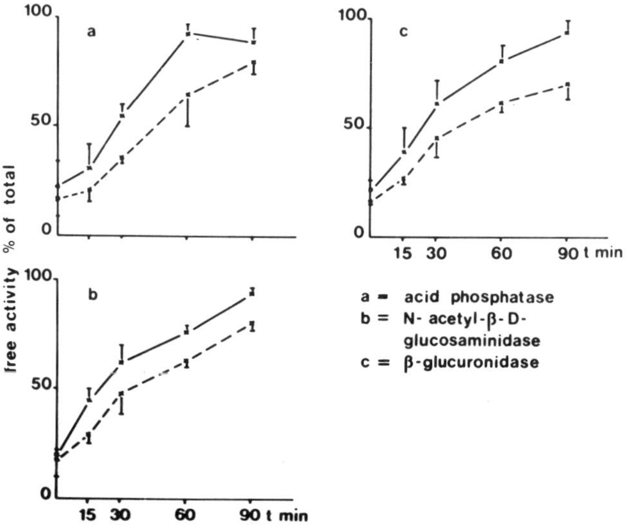

Fig.2. Stabilization of rat liver lysosomes by 1mg/ml RNA (---) by following the time dependent increase of the free activity of three different lysosomal enzymes after incubation at pH 5.0 and 37°C. The values are means of 4-6 experiments. The bars represent s.d. which are drawn to one side only.

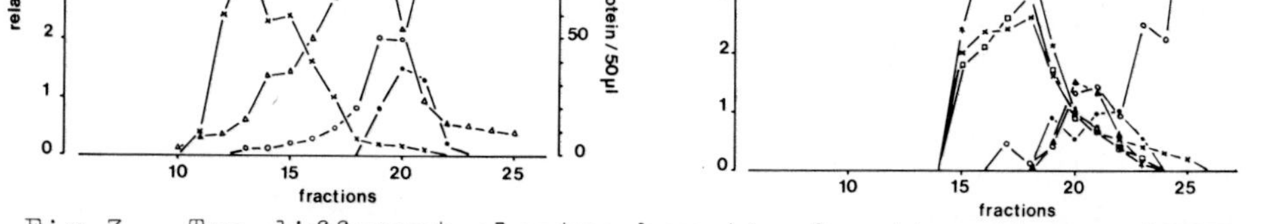

Fig.3. Two different electrophoretic fractionations of LMF by pH gradient electrophoresis using Ampholine pI 5-7 as carrier ampholytes. Specific enzyme activities are to be understood relative to the LMF. Acid phosphatase (—x—), aryl sulfatase (—□—), N-acetyl-ß-glucosaminidase (—✱—), glucose-6-phosphatase (—•—), uricase (—o—), protein (—△—) and cytochrom c oxidase (—▲—).

and 1b, the RBCs start to migrate back in the cathodic direction.

In addition to previously reported findings on the effects of polyanions on lysosomes and lysosomal enzymes (15,16,21), recent data have shown that RNA exhibited similar effects (22). Fig.2 shows the stabilization of rat liver lysosomes by 1mg/ml

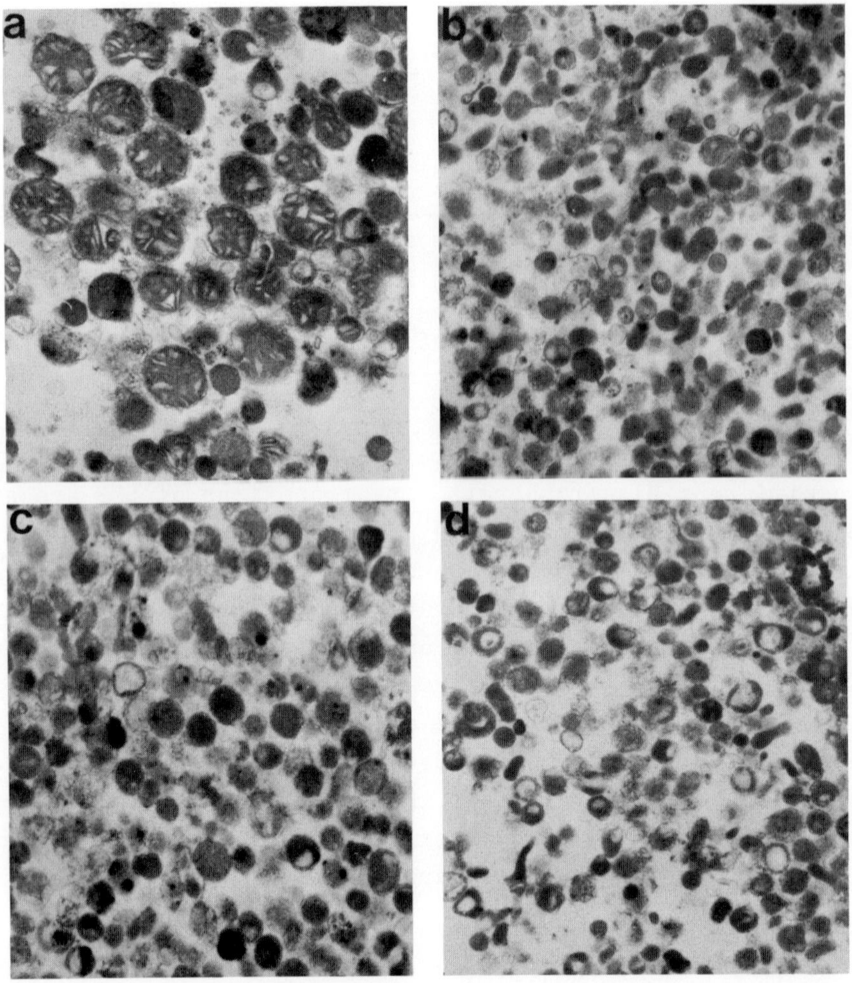

Fig.4. Electron micrographs of LMF (a) and of fractions 11 (b), 13 (c) and 14 (d) obtained by the pH gradient electrophoretic fractionation of LMF which is demonstrated in Fig.3a. A fair degree of purification of well-preserved lysosomes is observed, exhibiting the heterogeneous appearance known for these subcellular population.(x 14500).

RNA with respect to three different lysosomal enzymes. The presence of 0.1mg/ml heparin and 0.1mg/ml RNA in the particle suspension most effectively prevented clumping during electrophoresis. These findings essentially contribute to the possibility of LMF fractionation by pH gradient electropho -

	crude homogenate	LMF	fraction of maximal act.
Separation I*	1.0	12.3	47.9
Separation II**	-	1.0	3.4 (41.8)
Separation III**	1.0	14.1	36.7

Tab.1. Enrichment of lysosomes over the crude rat liver homogenate with respect to the relative specific activity of acid phosphatase. The enzyme activity was determined using ß-glycerophosphate (*) or p-nitrophenyl phosphate (**) as a substrate. The final enrichment factor of separation II was calculated by assuming that the LMF exhibited the same degree of purification as that of separation I.

resis.

Two different electrophoretic fractionations of LMF are shown in Fig.3 which were performed under quite identical conditions. The marker enzymes measured in the different fractions may provide the evidence that by this fractionation technique a purified lysosomal as well as peroxisomal fraction could be isolated. Out of the fractions 11,13 and 14 (Fig.3a) electron micrographs were prepared and are shown in Fig.4b, 4c and 4d, respectively. They demonstrate the high degree of purification and the good preservation of the lysosomal fractions. Enrichment factors of the lysosomes over the crude homogenate with respect to acid phosphatase are listed in Tab.1.

The properties of two different synthesis batches of self-made carrier ampholytes are presented in Fig.5, demonstrating the reproducibility of the synthesis method and the quite uniform distribution of buffering capacity and conductivity throughout the gradient. Although the product represents pure amino-carboxylic acid ampholytes of the type primarily synthesized by Vesterberg (2) and recently by Vinogradov et al.(24 and Righetti et al. (25), the route by which these ampholytes are obtained is different from that previously reported. The advantage of the synthesis procedure, which will be reported in detail elswhere, may be rapidity (maximal time for the complete synthesis 2 days), economy (extremely inexpensive

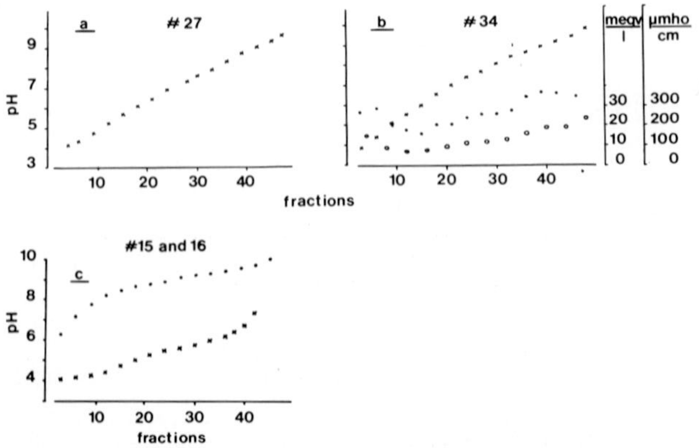

Fig.5. pH-Gradient profile, buffering capacity (-•-), conductivity of self-made amino-carboxylic acid carrier ampholytes suitable for isoelectric focusing. Applied on a test protein mixture, the ampholytes reveale similar properties to Ampholine but demonstrated, however, a different molecular weight profile and different properties upon the visualization reactions of various ampholyte species worked out by Radola (23) (personal communication Dr.Radola).

starting materials) and, with restrictions, the possibility of snthesizing by the same procedure pH gradients of narrower range, such as that shown in Fig.5c.

DISCUSSION

Two experiments were performed in this study of which the first one was to provide further information on the validity of IF of RBCs and the second on the advantageous use of pH gradient forming ampholytes for the electrophoretic fractionation of subcellular particles without the achievment of steady state IF.

Several facts have to be considered in explaining the electrokinetic behaviour of RBCs in that special medium. As shown by Heard and Seaman (26) there exists a strong relationship between stability of the erythrocyte membrane and ionic strength and pH of the suspending medium. These authors reported the occurence of membrane instabilities resulting in hemolysis and hemolysate,adsorption particularly with decreasing ionic strength and decreasing pH. This may also be occuring during IF of RBCs and may to some extent account for the pH at which the cells are apparently focused in the experiments shown in Fig. 1a and 1b. It may, however, not explain both the different pH found in previously reported experiments (5) and the cathodic migration demonstrated in Fig.1c.

The pH value at which the RBCs can be considered isopotential strongly depends on the ionic strength of the medium (26). This pH increases exponentially with decreasing ionic strength. Ionic strength (although this term in the concept of Debey and Hückel refers only to strong electrolytes) attributed to an ampholyte mixture decreases with focusing time and therefore will shift a proposed isopotential point concurrently towards higher pH as long as the pH gradient is in what Catsimpoolas called the transient state of IF (27). Assuming that under the applied experimental conditions of 90-100V/cm and flow-through times of about 15-20min somewhat different transient states of IF were achieved, this may explain both why the RBCs were eluted in previously reported experiments at a lower pH and the cathodic migration in the experiment depicted in Fig.1c. However, this explanation has to be examined experimentally.

The preparative character of pH gradient electrophoresis in the transient state of IF is well demonstrated in the fractionation of rat liver LMF, which leads to a properly purified and well-preserved lysosomal fraction within about 3 hours after isolation of the LMF. A very similar lysosomal fraction was obtained by Stahn et al. (28) using the Hannig type free-flow electrophoresis equipment. However, several differences to the results obtained by pH gradient electrophoresis may be pointed out. The lysosomal fractions obtained by pH gradient electrophoresis shows well-balanced enrichment factors whether acid phosphatase, aryl sulfatase or N-acety-ß-glucosaminidase are used as reference enzymes (Fig.3b). The 240-fold enrichment of lysosomes with respect to aryl sulfatase obtained by

Stahn et al. could not be confirmed. In the free-flow electrophoretic separation, the peroxisomal enzyme was not separated from the mitochondrial and microsomal ones. The fractionation of LMF in the free-flow electrophoretic system failed when heparin was added to the particle suspension (personal communication Dr.Stahn). This may indicate the possibility of using specific labelling materials in pH gradient electrophoresis which may influence the electrophoretic behaviour of a certain component of a particle mixture and therefore contribute to new electrophoretic fractionations.

ACKNOWLEDGEMENTS

The authors gratefully thank Dr.H.Fischer and especially V. Würfl for preparing the electron micrographs and W.Seyl for excellent mechanical assistance. We further wish to express our gratitude to Prof.Dr.K.Hannig, Max-Planck-Institut für Biochemie, Martinsried, for his critical comments during the present work.

REFERENCES

1. *Svensson, H.: Isoelectric Fractionation, Analysis and Characterization of Ampholytes in Natural pH Gradients. I. The Differential Equation of Solute Concentrations at a Steady State and its Solution for Simple Cases. Acta Chem. Scand. 15, 325-341 (1961).*

2. *Vesterberg, O.: Synthesis and Isoelectric Fractionation of Carrier Ampholytes. Acta Chem. Scand. 23, 2653-2666 (1969).*

3. *Sherbet, G.V., Lakshmi, M.S., Rao, K.V.: Characterization of Ionogenic Groups and Estimation of the Net Negative Electric Charge on the Surface of Cells Using Natural pH Gradients. Exp. Cell Res. 70, 113-123 (1970).*

4. *Ericson, I.: Determination of Isoelectric Point of Rat Liver Mitochondria by Cross Partition. Biochim. Biophys. Acta 356, 100-107 (1974).*

5. *Just, W.W., Leon-V., J.O., Werner, G.: Isoelectric Focusing in Continuous-flow Electrophoresis. I. Separation of Mixed Red Blood Cells of Different Species. Anal. Biochem. 67, 590-601 (1975).*

6. *Just, W.W., Leon-V., J.O., Werner, G.: Continuous-flow isoelectric Focusing: Studies on the Separation of Mixed Red Blood Cells of Different Species and of Light Mitochondrial Fraction of Rat Liver. In: Progress in Isoelectric Focusing and Isotachophoresis*

Righetti P.G. (ed), North-Holland Publishing Comp. (1975).

7. Catsimpoolas, N., Griffith, A.L.: New Instrumentation and Procedures for the Analysis of Mammalian Cells by Electrophoretic Techniques. This Volume.

8. Manske, W., Bohn, B., Brossmer, R.: Preparative Electrofocusing of Living Mammalian Cells in a Stationary Ficoll-sucrose Gradient. This Volume.

9. Hannig, K.: Free-Flow Electrophoresis. In: Noris, J.R., Ribbons, D. W. (ed.): Methods in Microbiology, London: Academic Press (1971).

10. Hannig, K., Heidrich, H.G.: The Use of Continuous Preparative Electrophoresis for Dissociating Cell Fractions and Isolation of Membraneous Components. In: Fleischer, S., Packer, L. (ed.): Methods of Enzymology, New York: Academic Press Inc. (1974).

11. Beaufay, H.: Methods for the Isolation of Lysosomes. In: Dingle, J.T. (ed.): A Laboratory Handbook, Amsterdam, London: North-Holland Publishing Comp. (1972).

12. Leighton, F., Poole, B., Beaufay, H., Baudhuin, P., Coffey, J.W., Fowler, S., DeDuve, C.: The Large Scale Separation of Peroxisomes, Mitochondria and Lysosomes from the Livers of Rats Injected with Triton WR-1339. J. Cell Biol. 37, 482-513 (1968).

13. Appelmans, F., DeDuve, C.: Tissue Fractionation Studies. 3. Further Observation on the Binding of Acid Phosphatase by Rat Liver Particles. Biochem. J. 59, 426-433 (1955).

14. Seiler, N., Thobe, J., Werner, G.: Elektrophorese im trägerfreien Pufferstrom. II. Eine modifizierte Trennkammer. Z. Anal. Chem. 252, 179-182 (1970).

15. Just, W.W., Leon-V., J.O., Werner, G.: Protective Effects of polyanions on the Constituents of Rat Liver Lysosomal Fraction in Acidic pH. Histochemie 41, 161-166 (1974).

16. Just, W.W., Leon-V., J.O., Werner, G.: Effect of Polyanions on Lysosomes and Lysosomal Enzymes of Rat Liver in Vitro. Hoppe-Seylers Z. Physiol. Chem. 355, 1565-1568 (1974).

17. Bowers, W.E., Finkenstaedt, J.T., DeDuve, C.: Lysosomes in Lymphoid Tissue. I. The Measurements of Hydrolytic Activities in Whole Homogenates. J. Cell Biol. 32, 325-327 (1967).

18. Sottocasa, G.L., Kuylenstierna, B., Ernster, L., Bergstrand, A.: An Electron Transport System Associated with the Outer Membrane of Liver Mitochondria. J. Cell Biol. 32, 415-438 (1967).

19. Beaufay, H., DeDuve, C.: Le system hexose-phosphatique. IV. Specifite de la glucose-6-phosphatase. Bull. Soc. Chim. Biol. 36, 1525-1537 (1954).

20. Fawcett, J.S.: Continuous-flow Isoelectric Focusing and Isotachophoresis. An.. N.Y. Acad. Sci. 209, 112-126 (1973).

21. *Trigaux, J.P., Ronvauy-Dupal, M.F., Wattiaux, R.: Stabilization of Rat Liver Lysosomes by Heparin in Vitro. Biochem. Pharmacol. 21, 429-431 (1972).*

22. *Just, W.W., Werner, G., Leon-V., J.O.: Effect of Polyanions on Lysosomes and Lysosomal Enzymes of Rat Liver in Vitro. Hoppe-Seylers Physiol. Chem. 356, 242 (1975).*

23. *Radola, B.J., Tschesche, H., Schuricht, H.: Characterization of Carrier Ampholytes by New Visualization Reactions and Chromatographic Methods. This Volume.*

24. *Vinogradov, S.N., Lowenkron, S., Andonian, M.R., Bagshaw, J., Felgenhauer, K., Pak, S.J.: Synthetic Ampholytes for the Isoelectric Focusing of Proteins. Biochem. Biophys. Res. Commun. 54, 501-506 (1973).*

25. *Righetti, P.G., Pagani, M., Gianazza, E.: Characterization of Synthetic Carrier Ampholytes for Isoelectric Focusing. J. Chromatogr. 109, 341-356 (1975).*

26. *Heard, D.H., Seaman, G.V.F.: The Influence of pH and Ionic Strength on the Electrokinetic Stability of Human Erythrocyte Membrane. J. Gen. Physiol. 43, 635-654 (1960).*

27. *Catsimpoolas, N.: Transient State Isoelectric Focusing. Digital Measurement of Zone Position, Zone Area, Segmental pH Gradient, and Isoelectric Point as a Function of Time. Anal. Biochem. 54,66-78 (1973).*

28. *Stahn, R., Maier, K.P., Hannig, K.: A New Method for the Preparation of Rat Liver Lysosomes. J. Cell Biol. 46, 576-591 (1970).*

Preparative Electrofocusing of Native and Modified Living Mammalian Cells in a Stationary Ficoll/ Sucrose Gradient

W. Manske, B. Bohn and R. Brossmer

ABSTRACT

An easy and reproducible method is described, by which living cells can be separated in a Ficoll/sucrose/Ampholine gradient according to their surface pK's. The main advantages of the procedure are: 1) standard column equipment can be used; 2) the focusing time can be kept rather short (not more than 4 to 5 h); 3) the supporting linear Ficoll gradient, ranging from 20 to 5% (w/v), is made isotonic over the whole range by addition of sucrose and allows for high migration velocity; 4) as a consequence of the latter two points, and in contrast to other methods in the sparse literature(<u>cf.</u> (1)), cell viability can largely be preserved.

INTRODUCTION

The surface charge of cells has become increasingly important in cell biology, especially with regard to malignancy, ever since Ambrose and co-workers (2) advanced thehypothesis that malignant cells exhibit a more negative net surface charge than their normal counterparts. Much controversy has arisen about this subject, since many experiments failed to show correlations between malignancy and surface charge alterations (for review: ref.3). It is generally believed that the bulk of the net negative surface charge of mammalian cells is generated by sialic acid residues which are bound to membrane glycoproteins and glycolipids.

The standard method for estimation of the cell surface charge is the determination of electrophoretic mobility, usually measured in a cytopherometer.This method allows to detect ionogenic groups which contribute to the so-called zeta-potential which is assumed to be generated by charged groups lying within 10 Å from the hydrodynamic plane of shear. Since we, like others (for review: ref. 3), have often found discrepancies between biochemical parameters and changes in the zeta-potential, especially with chemically or enzymatically modified cells, we thought it might be useful to get more information about possible charge carriers located too far away from the hydrodynamic plane of shear to be detected by cytophoresis. The most promising approach seemed to be adaptation of isoelectric focusing for the preparative separation of intact cells.

MATERIALS AND METHODS

Cells

Two strains of the Ehrlich-Lettré mouse ascites tumor(EAT) were used: the glycogen-storing(G+) and the glycogen-free(G-) sublines. The cells were harvested in the log phase of growth and washed twice in HBBS, pH 7.4. Furthermore, we used rat hepatocytes which were dissociated with a perfusion technique which essentially follows the directions given by Seglen(4).

Modification of the Cell Surface (see Fig.1)

Neuraminidase treatment (fig.1A): $5x10^6$ cells/ml were incubated with 10 U/ml vibrio cholerae neuraminidase(Behring-Werke, Marburg) in maleate buffer, pH 8.0, for 30 min at 37°C, while the cell suspensions were gently shaken. After incubation, the cells were pelleted and resuspended in HBSS. Sialic acid in the supernatant was determined by the thiobarbituric acid method, as described by Warren(5).

Abbreviations. EAT G+, Ehrlich-Lettré mouse ascites tumor, glycogen-storing; EAT G-, Ehrlich-Lettré mouse ascites tumor, glycogen-free; HBSS, Hank's balanced salt solution; PBS, phosphate-buffered saline; NANA, N-acetyl-D-neuraminic acid; NANA benzyl ketos., 4-diazo-benzyl-α-ketoside of NANA; NAP-taurine, N-(4-azido-2-nitrophenyl)-2-aminoethyl sulfonate; FDA, fluorescein diacetate.

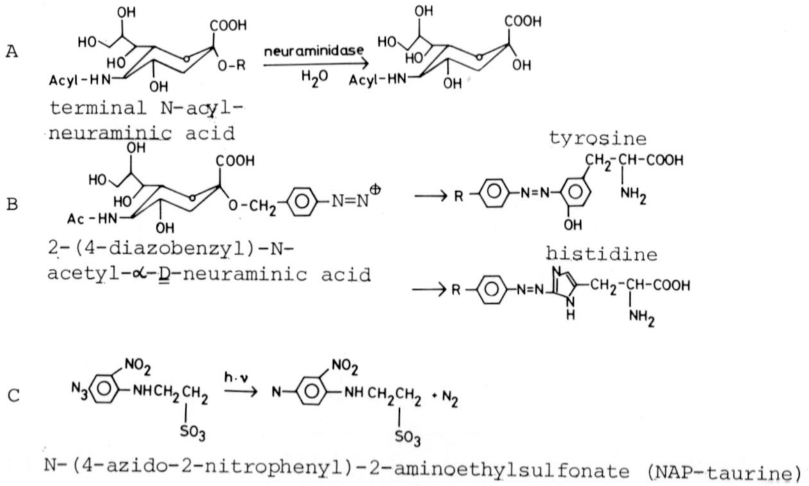

Fig.1. Enzymatic and chemical modification of the cell surface. A. Removal of terminal sialic acid residues from glycosides; B. Covalent attachment of additional sialic acid residues to protein moieties; C. Photolytic labeling with NAP-taurine.

"Grafting" of the 4-diazo-benzyl-α-ketoside of NANA to cell surfaces was carried out in the following manner (fig.1B): 4-amino-benzyl-N-acetyl-α-D-neuraminic acid, synthesized in our laboratories (6), was diazotized with sodium nitrite. Although we never detected unreacted nitrite under the conditions emploed, we prophylactically added small amounts of urea after neutralization of the solution. The cells were suspended in PBS, pH 8.0, containing 3.8×10^{-3}M diazo compound. During incubation at room temperature, the cells were gently shaken in Warburg flasks. For details cf. refs. (10,11). NAP-taurine, prepared in our laboratories, was coupled to outer cell surface components by photolysis at 4°C, as described in (7).

Determination of Electrophoretic Mobility of Cells

The anodic electrophoretic mobility of cells was measured in a Zeiss cytopherometer, as desribed by Fuhrmann and Ruhenstroth-Bauer (8). We used the following composition for the measuring fluid: 7.5ml isotonic sucrose solution, adjusted to 300 mOsm, 2.0 ml phosphate buffer, pH 7.18, made up to 10.0 ml with 0.5 ml cell suspension, diluted 1:4 with PBS.

Isoelectric Focusing of Cells

We used a 110ml LKB column, a fraction collector "Golden Retriever" (ISCO), equiped with an ISCO UA-5 absorbance monitor. Isotonic gradient solutions were prepared as shown in fig.2.

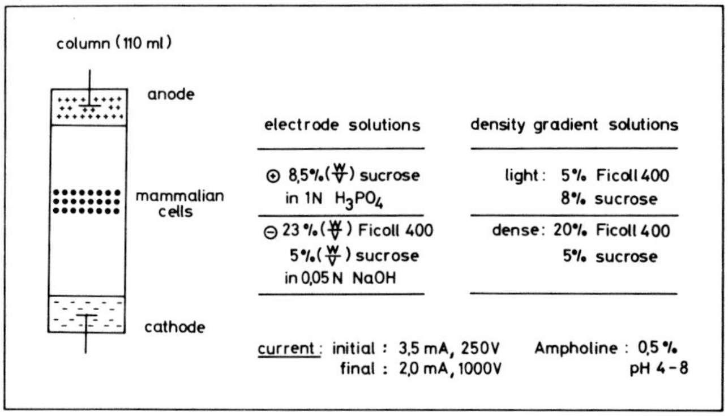

Fig.2. Materials for isoelectric focusing of living cells. Loading time: 40 min; cell number: $20x10^6$; loading zone: middle of column; elution volume per tube: 2.5 ml; elution speed: 2.5 ml/min.

Determination of Cell Viability

Cell viability was routinely determined with two methods: a) Fluorescent dye exclusion with Erythrosine B (Merck, Darmstadt). Cell suspensions were treated with 0.4% dye, and the percentage of fluorescent cells was determined in an automatic flow cytofluorometer ("Cytofluorograf 4800A", Bio/Physics, Mahopac, N.Y.). b) Application of the "fluorogenic substrate" FDA (Serva, Heidelberg), using the protocol for cytofluorometric viability determination described by Bohn (9).

RESULTS AND DISCUSSION

Our method of electrofocusing has some special features which have been designed -and certainly might be improved- to keep the viability rate of the cells to be separated as high as possible:

In the first place, we use Ampholine concentrations of only 0.5%. This certainly may cause a somewhat decreased resolution. On the other hand, it offers the advantage to keep the focusing time as short as 4 to 5 hrs. This is essential for the survival of the cells. Secondly, the gradient as well as the electrode solutions are made isotonic to prevent cell damage.

Fig.3 shows the results obtained with 3 cell types after focusing times of 4 hours (left) and 20 hours (right). It can readily be seen that focusing times of 20 hours make the bands more distinct. Furthermore, a small second peak can be detected. The main peaks, however, are located at essentially the same pH_I-values.

In fig. 4, a summary of surface pK-values, the viability rates of cells before focusing, and the relative decrease in viability after focusing for 4 to 5 hours is given for EAT G+, EAT G- and rat liver parenchyma cells. It should be emphasized that focusing times of 20 hours bring about a dramatic drop of the viability rates by more than 80%! It therefore seems unwise to focus cells over such a long period, unless purely analytical purposes are being pursued. The main peaks may also be obtained after 4 hrs.

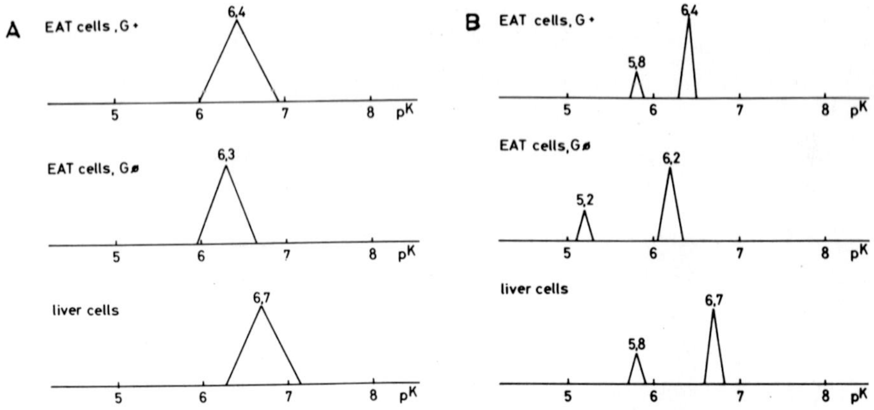

Fig.3. Cell distribution of EAT G+, EAT G- cells and rat hepatocytes after focusing for 4 hours (A) and 20 hours (B). The peaks represent the mean cell surface pK-values, whereas the distribution on the abscissae means the standard deviation from these values. Number of experiments as in fig.4.

cell type	pK	n	viability(%)	decrease of viability(%)
EAT G+	6.4±0.5	10	92	5
EAT Gø	6.3±0.4	8	92	5
liver cells	6.7±0.6	7	80	8

Fig.4. Mean cell surface pK-values, viability rates before electrofosusing, and decrease in viability after focusing between 4 and 5 hours.

A very useful application of cell electrofocusing concerns chemical and enzymatic modifications of cells with concomitant changes in surface charge. We modified tumor cells with various chemical reagents as well as with neuraminidase, and compared the shifts in the mean pK-values with the corresponding changes in electrophoretic mobility, and in the case of neuraminidase, with the amount of sialic acids liberated by the enzyme (see figs. 5 and 6).

reagent	pK	n	electrophoretic mobility $cm^2 \times v^{-1} \times sec^{-1} \times 10^{-4}$	n
none	6.4 ± 0.5	10	1.71 ± 0.25	16
NAP-taurine	6.6 ± 0.3	9	2.48 ± 0.15	20
NANA, benzyl ketos.	5.5 ± 0.3	9	1.90 ± 0.20	10
neuraminidase	7.1 ± 0.4	5	1.45 ± 0.14	12

Fig. 5. pK-values and electrophoretic mobilities of native and modified EAT G+ cells. The cells were treated with NANA benzyl ketos., NAP-taurine, and vibrio cholerae neuraminidase (cf. MATERIALS AND METHODS).

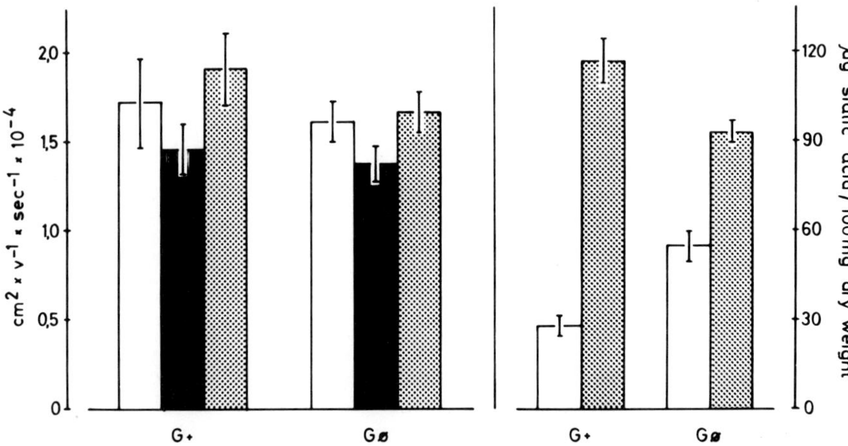

Fig. 6. (from (10)) Electrophoretic mobility (left) and sialic acid content (right) of EAT G+ and EAT G- cells.

 native cells;  cells treated with neuraminidase;

 cells treated with NANA benzyl ketos.

It can be seen that covalent attachment of the NANA ketoside to the cell surface brings about a decrease in the cell surface pK-values, with concomitant increase in anodic electrophoretic mobility. Enzymatic removal of sialic acids causes an increase in pK-values and a decrease in cytophoretic mobility. We could show (10, 11) that the electrophoretic mobility of both native and modified EAT cells is not necessarily correlated with the

sialic acid content, as determined with the Warren assay after neuraminidase treatment. Possible explanations ("cryptic" sialic acids, substituted neuraminic acids which may not serve as substrates for vibrio cholerae neuraminidase, etc.),are discussed in refs. 10 and 11.
From the data in fig.5, it appears that the pK-shifts caused by both modification procedures are more pronounced than the respective changes in electrophoretic mobility.
The sulfonate group of NAP-taurine, too, contributes an additional negative charge to the cell surface. As can be seen, here the difference in the pK-values is less marked than the change in electrophoretic mobility.

In conclusion, we think that the method of electrofocusing of intact cells should be further elaborated and refined. We believe that at least some of the apparentcontradictions, arising from cytopherometric data, may be reconciled by comparison of the results, obtained with both methods which seem to reflect charge properties in different planes of the cell surface.

ACKNOWLEDGMENTS

The expert assistance of Mrs. A. Sauer, Miss B. Löser and Mr. G. Krüger is gratefully acknowledged.
Our investigations were generously sponsored by the Deutsche Forschungsgemeinschaft, Sonderforschungsbereich 136 (Cancer Research).

REFERENCES

1. Sherbet,G.V., Lakshmi, M.S., Rao, K.V.: Characterisation of ionogenic groups and estimation of the net negative electric charge on the surface of cells using natural pH-gradients. Exptl. Cell Res. 70, 113-123 (1972).

2. Ambrose, E.J., James, A.M., Lowick, J.H.: Differences between electrical charge carried by normal and homologous tumour cells. Nature (London) 177, 576-577 (1956).

3. Weiss, L.: Neuraminidase, sialic acids, and cell interactions. J. Natl. Canc. Inst. 50, 3-19 (1973).

4.Seglen, P.: Preparation of rat liver cells. Exptl. Cell Res. 74, 450-454 (1972).

5. Warren, L.: The thiobarbituric acid assay of sialic acids. J. Biol. Chem. 234, 1971-1975 (1959).

6. Neumann, B.: Untersuchungen zur Chemie und Biologie der Sialinsäure. Synthese und Eigenschaften von 5-Acetamino- 3,5-didesoxy-D-glycero-D-galacto-nonulosidarsäuren. Verknüpfung von N-Acetyl-D-neuraminsäure mit Aminosäuren und Proteinen. Diss.Chem., Heidelberg 1974.

7. Staros, V. Richards, M.F.: Photochemical labelling of the surface proteins of human erythrocytes. Biochemistry 13, 2720-2726 (1974).

8. Fuhrmann, G.F., Ruhenstroth-Bauer, G.: Cell electrophoresis employing a rectangular measuring cuvette. In: Cell Electrophoresis (Ambrose, E.J., ed.), Little, Brown and Company, Boston (1965) 22-25.

9. Bohn, B.: High-sensitivity cytofluorometric quantitation of lectin and hormone binding to surfaces of living cells. Exptl. Cell Res., in press.

10. Brossmer, R., Bohn, B., Thies, C.: Modification of tumour cells by covalent attachment of N-acetyl-D-neuraminic acid to the cell surface. Submitted for publication in Eur. J. Cancer.

11. Bohn,B., Thies, C., Brossmer, R.: Cell surface charge, sialic acid content and metabolic behaviour of two tumour sublines. Submitted for publication in Eur. J. Cancer.

Part II. Isotachophoresis

1. Analytical and Preparative Methodology

The Principle of Preparative Capillary Isotachophoresis

L. Arlinger

INTRODUCTION

Analytical isotachophoresis in capillary tubes has shown to be a method of great separation potential for solving a variety of analysis problems. There has, however, been a restriction in the use of this technique because it has not been possible to investigate the highly resolved samples with other than the usual built-in detectors. Nor has it been possible to use capillary separation as an integral step in preparation procedures.

It was therefore considered of great value to use capillary isotachophoresis as a preparative procedure without losing the rapidity and resolving power inherent in the method.

PRINCIPLE

The Capillary Elution System

In the isotachophoretic capillary equipment[1] shown in Fig. 1(a), the leading electrolyte reservoir is separated from the capillary compartment by a semipermeable membrane, $\underline{m_1}$, to prevent hydrodynamic flows. The tension applied over the electrodes can be very high, up to 30 kV with a capillary length of 25 cm, thereby assuring rapid and highly resolving separation of the sample ions. When the electric current is switched off, the ions in the zones will quickly diffuse across the zone boundaries

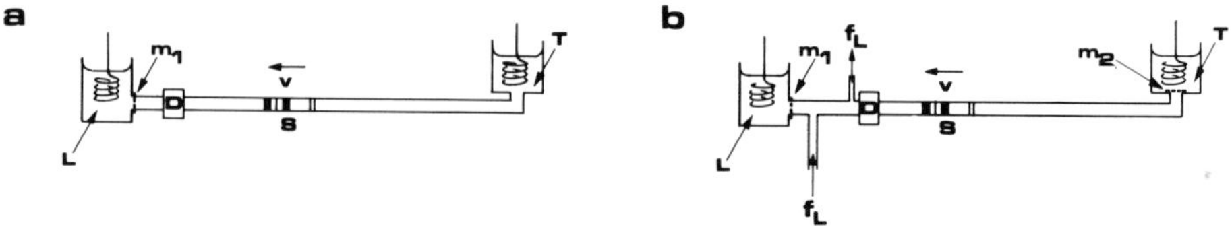

Fig. 1. The principle of preparative capillary isotachophoresis. Fig. (a) shows schematically the basic analytical equipment, and (b) the preparative version with the additional flow of leading electrolyte. L = leading electrolyte, T = terminating electrolyte, D = UV detector site, S = sample ions, v = migration velocity, m_1, m_2 = semipermeable membranes, f_L = flow of L to elute the sample zones.

and destroy the resolution within a few seconds, since no gel or other stabilizing medium is present in the liquid phase. The resolution is also immediately decreased by applying a flow along the capillary. Therefore, sample collection has to be performed with an electric field present and without moving the sample zones by flow along the direction of the capillary.

The technique chosen for the sample collection is to use a T-connection through which the sample is swept out by a flow of leading electrolyte. This is shown schematically in Fig. 1(b) where m_2 is another membrane inserted in the terminator reservoir to prevent counter-flow through the whole capillary. After passing the UV detector the zones will be swept out through the T-tube, provided the liquid flow rate from the pump is greater than the corresponding migration rate of the ions. Since the liquid flow will dilute the sample, the flow rate is set only a few percent higher than the migration rate. The transport situation for the different ions is described more in detail elsewhere (2).

The Fraction Collection Device

The flow rates through the T-tube in the present work were 2.75 µl/min or about 45 nl/s. Sample zones may have volumes as low as 10-20 nanoliter[3] and traditional fraction collection is not easily used. The sample is collected on a strip passing by the outlet of the T-tube. Since the high electric voltage is on during the collection of samples, there is an electric force present with which to transfer liquid from the T-tube outlet to the grounded strip. When the distance from the end of the T-tube to the strip is small, very small drops will pass over to the strip at a frequency of several Hz. This frequency can be increased by diminishing the distance until a continuous stream is obtained, and the electric tension will allow proper collection on the strip.

The identification of sample position on the strip and recorder chart is achieved by simultaneous, automatic marking of the strip and chart. There is also a constant time difference between the measurement of UV-absorbance of an ion and the moment the ion reaches the strip. This constant needs to be established only once for a given leading electrolyte by running a visible compound and is valid for a specific current and the corresponding liquid flow rate. The time difference in a typical experiment is 3 to 6 seconds.

MATERIAL AND METHODS

Isotachophoresis

The isotachophoretic experiments were performed with the LKB 2127 Tachophor (LKB-Produkter AB, Bromma, Sweden) equipped with a counter-flow pump, Braun Perfusor IV (Braun, Melsungen, Fed. Rep. of Germany). The pump was used in position "5" giving a flow of 2.75 µl/min with a 500 µl Unimetrics syringe. It was connected to the Tachophor by a teflon capillary tube and a cannula through the counter-flow septum.

The usual type of capillary plate was substituted with a specially constructed plate with a 43 cm teflon capillary fitted with a T-connection immediately after the UV detector site. The connection consisted of a glass capillary tube (O.D. 0.25 mm, I.D. 0.04 mm) inserted through the teflon wall and reaching the inner surface of the separation tube. The center of the T-tube is positioned 0.5 mm behind (in the direction of the migration) the center of the UV-beam passing through the separation tube. To prevent the flow of leading electrolyte from the syringe pump from causing counter-flow in the whole separation tube, a semi-permeable membrane was mounted in the bottom of the terminating electrolyte reservoir and a valve in the terminating electrolyte syringe holder, thereby completely closing the terminator side of the capillary against liquid flow.

The balancing of pump flow against the isotachophoretic migration velocity (the current) was made by running ordinary counter-flow[4] with a visible sample. The pump was set at one position and the current varied until balance was reached, when the sample zones did not move in the capillary. The balancing current was then lowered by 5% and used for the sample collection phase in the preparative experiments.

The samples were collected as continuous streams on 3 mm wide strips of 0.11 mm thick cellulose acetate (Sartorius, Göttingen, Fed. Rep. of Germany). The strip was mechanically transported past the end of the glass tube and had a constant speed, adjustable between 40 and 55/min with a space between the tube end and the strip surface of 0.05 to 0.08 mm. The construction eliminated risks of contamination of the sample components collected on the strip. Each turn of the drive wheel (circumference 50 mm) marked the edge of the strip close to the T-tube end and simultaneously closed an electronic circuit producing a spike on the UV recording.

TRIS and β-alanine were bought from Sigma Chemical Co. (USA), histidin and barium hydroxide from Merck (Darmstadt, Fed. Rep. of Germany), hydroxypropylmethylcellulose (HPMC) from Dow Chemical Co. (Methocel 90 HG, 15000 cPs).

Radioactivity measurements

The radioactive nucleotide samples were bought from the Radiochemical Centre, Amersham, England. The uridine-4-C14-5'-monophosphate ammonium salt (batch 3) had a specific activity of 24 mCi/mmol and was dissolved in water to make a 5 mM solution. A sample of 1 μl therefore contained 5 mmoles or 120 nCi.

The adenosine-C14(U)-5'-monophosphate ammonium salt (batch 12) was supplied in an aqueous solution, containing 2% ethanol, with a specific activity of 570 mCi/mmol and a radioactive concentration of 50 μCi/ml. A sample of 1 μl therefore contained 88 pmoles or 50 nCi.

The radioactivity-containing cellulose acetate strips were mounted on a glass-plate covered with double-adhesive tape and evaluated by the LKB 2105 RadioChromatogram Camera, set for 20,000 counts, and a polarization of 1. The Polaroid Camera was set to f=11 (75 ASA). The sites of activity were measured on the photos with a special scale transforming ruler.

The scintillation countings were made in the LKB 81000 Liquid Scintillation Counter. The strips were cut along a ruler edge into 2.5 mm pieces, each piece was put into a vial, 1 ml distilled water was added and after a couple of minutes, including some slight swirling, 10 ml Instagel was added, and the vial shaken until it was clear. The counting times were 1 or 2 min/vial or 10 min with overnight counting.

Zymogram technique[5]

The sample used was cholinesterase (Sigma) with 8.6 units/mg protein and 0.70 mg protein/mg solid substance, type IV from horse serum, Lot No 120C-2240.

The zymogram reagent components were the following:
A. 1% α-naphtylacetate (Sigma) in acetone
B. Buffer, 0.12 M TRIS and 0.10 M HCl, pH 7.5
C. Fast Red TR salt (Sigma)

They were mixed in proportions A/B/C as 0.25 ml/9 ml/∿9 mg. The solution is unstable but lasts long enough for the soaking procedures, including the separation with sample collection.

RESULTS

Identification by means of radioactive tracer

The use of a radioactive detector has never been possible in capillary isotachophoresis because of the small sample amounts and other physical conditions encountered. However, the sample strip is easily investigated by radioactive methods.

The samples used in the experiments were nucleotide mixtures, containing the 5'-monophosphate of uridine, cytidine, guanosine, and adenosine and adenosine-3'-monophosphate, the A5'MP and U5'MP being marked with ^{14}C. Fig. 2 shows the UV absorbance pattern (278 nm) from an experiment using a leading electrolyte with a pH of 3.4. The corresponding cellulose acetate was analysed with the LKB RadioChromatogram Camera and the result is shown as the line at the top in Fig. 2, marked RCC. Liquid scintillation counting (LSC) of another strip from an identical experiment resulted in the lower picture in Fig. 2. The results are in full agreement with each other.

The separation between A5'MP and A3'MP is not absolute, but the overlap here is probably due to the normal zone boundary widths. The 88 pmol A5'MP sample will, in this "ordinary" concentration[3] experiment, occupy less than one mm in the capillary tube, and the mobilities of the two

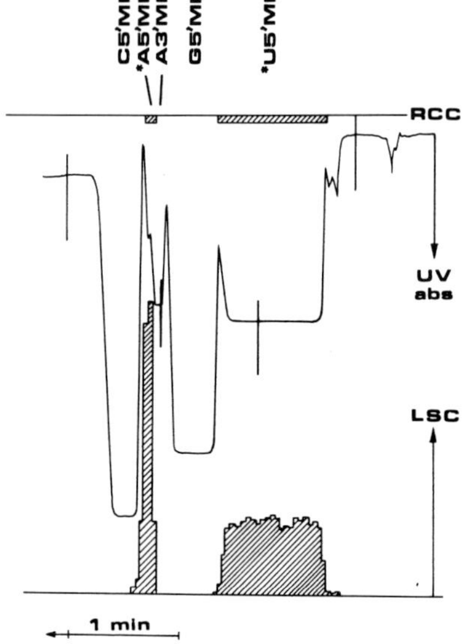

Fig. 2. Capillary isotachophoretic separation of C^{14}-labelled A5'MP (88 pmoles) and U5'MP (5nmoles) together with unlabelled A3'MP (0.77 nmoles), G5'MP and C5'MP (both 2.5 nmoles). The upper line, RCC, shows the activity as detected by the RadioChromatogram Camera. The lower tracing, LSC, shows liquid scintillation counting on pieces of the strip, cut to a length of 2.5 mm. Leading electrolyte was 5 mM HCl, 9 mM β-alanine, 0.25% HPMC at pH 3.4. Terminating electrolyte was 10 mM caproic acid. Capillary length 43 cm, current 80 µA and 30 µA during detection and sample collection; total time of experiment, 24 min; wavelength of recording, 270 nm; temperature, 20 ^{o}C.

AMP's will naturally be rather similar.

The LSC values show some spreading as can be seen in the UMP zone. This is certainly due to the author´s inability to cut the strip into pieces of exactly equal lengths with a scalpel at a right angle to a ruler, placed at the edge of the strip.

The RCC is much handier than the LSC for this type of radioactivity measurements. The RCC measurements take only a couple of minutes but have no quantitative accuracy. The intensity differences that can be seen on the photographs, however, will probably give sufficient information in most cases to distinguish between radionucleids with different activities.

Enzyme identification by zymogram

The possibility of specific analysis performed directly on the strip was demonstrated with the zymogram technique[5]. The cellulose acetate strip was treated with the reagent solution before sample collection. The sample chosen was cholinesterase, which was allowed to hydrolyse α-naphtylacetate, the product reacting with a diazo salt to form a reddish-brown colour. Because of the minute amounts of sample (about 20 µg or 0.17 units of enzyme) the reagent solution was concentrated (see Materials and Methods) more than usual[6]. The amount of reagents in the strip was further increased by repeated soaking of the strip in the reagent solution and drying under a hair dryer. This was repeated two times, for a total of only a few minutes. The impregnated length of strip was about 40 cm, for each experiment.

After sample collection on the pretreated strip (collection performed like on an untreated strip), it was again quickly dipped into the reagent solution and dryed under the dryer. The zymogram developed and was reinforced by one or two more soakings and dryings.

The cholinesterase was run in the electrolyte system given in Fig. 3, where in (a) the electrolytes are shown without any sample (electrolyte impurity pattern), in (b) with 20 µg cholinesterase. The zymogram colour intensities were visually estimated on a scale of 0 to 5 on a light table. The activity was found in a small part of the UV-absorbing protein region. In (c) an experiment with 1 µl of Ampholine pH 4-6, diluted 1:50, is shown with the same impurities as in Fig. (a) but spaced apart, giving no zymogram reaction. In (d) the enzyme and Ampholine samples were combined, giving an excellent separation of the enzymatic components with mobilities lower than those of the main part of the protein. This is more obvious in (e), where the amount of cholinesterase is almost doubled to 35 µg. The non-enzymatic protein part of the 35 µg sample

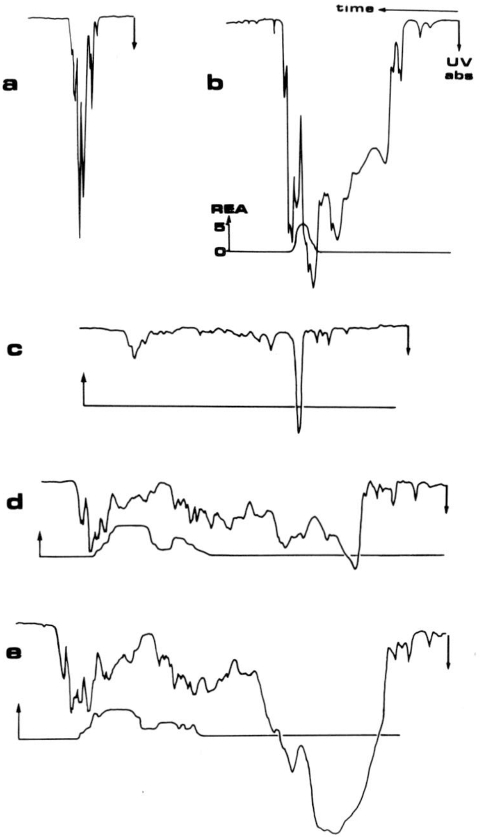

Fig. 3. Separation of a commercial cholinesterase preparation and zymogram detection of activity on pretreated sample collection strips. The axes are defined in (b), where REA is the relative enzyme activity, estimated from the zymogram color intensities on a scale of 0-5.
a) no sample, peak pattern is from electrolyte impurities. b) 20 μg cholinesterase. c) 1 μl Ampholine pH 4-6, diluted 1:50. d) 20 μg cholinesterase and Ampholine as in (c). e) 35 μg cholinesterase and Ampholine as in (c). Leading electrolyte: 5 mM HCl, 6 mM TRIS with 0.5% HPMC, pH 7.35. Terminating electrolyte: 10 mM glycine, $Ba(OH)_2$ added to pH 9.2. Capillary length 43 cm, current 45 μA, thermostat 20 °C, detection at 254 nm. Separation time until the first peaks were detected about 20 min.

was too large to completely reach isotachophoretic conditions, which is the reason for the difference in the shape of the zones directly after the leading ion. It is easily seen from the last two figures which zones contain activity. It would thus be an easy task to make another separation as in (e), collect the sample on an untreated strip, cut out the desired section and put it in a buffer solution to obtain an enzyme solution of high purity with a very high specific activity.

DISCUSSION

Experiments reported previously (2) indicate that the high resolution achieved with capillary isotachophoresis can be maintained during the preparative procedure. Another preliminary study of irreversible adsorption of two proteins on the cellulose acetate (2) did not show any such effect.

The radioactive and zymogram identification methods used here can of course be complemented with other methods. One obvious method is the immunological identification technique, which is shown to be easily combined (2,7) with preparative capillary isotachophoresis. The use of mass spectrometry is another possibility.

The usefulness of capillary isotachophoresis for the separation and analysis of complex biochemical molecules, such as proteins (8-12), will certainly be increasingly utilized. The preparative procedure will widen the applicability of capillary isotachophoresis because it offers totally new and powerful detection and identification possibilities, and also because it makes capillary isotachophoresis a high resolution purification technique for sample preparation.

REFERENCES

1. Arlinger, L.: Analytical Isotachophoresis. Principles of Separation and Detection. In Protides of the Biological Fluids (H. Peeters, ed.), 22, 661-667 (1974).

2. Arlinger, L.: Preparative Capillary Isotachophoresis. Principle and Some Applications. J. Chromatogr. 119, 9-24 (1976).

3. Arlinger, L.: Analytical Isotachophoresis. Resolution, Detection Limits and Separation Capacity in Capillary Columns. J. Chromatogr. 91, 785-794 (1974).

4. Everaerts, F.M., Vacik, J., Verheggen, T.P.E.M. and Zuska, J.: Displacement Electrophoresis. Experiments with Counterflow of Electrolyte. J. Chromatogr. 49, 262-268 (1970).

5. Wadström, T. and Smyth, C.J.: Zymogram Methods applied to Thin-layer Isoelectric Focusing in Polyacrylamide Gel. Sci. Tools 20, 17-21 (1973).

6. Young, C.W. and Bittar, E.S.: Isoelectric Focusing Comparison of Human Tissue Esterases with Those from Normal and Bacillus Calmette-Guèrin-treated Mice. Cancer Res. 34, 2675-2681 (1974).

7. Moberg, U., Hjalmarsson, S.G., Arlinger, L. and Lundin, H.: Preparative Capillary Isotachophoresis. Separation of μg Amounts of some Human Serum Proteins. This issue.

8. Arlinger, L.: Analytical Isotachophoresis in Capillary Tubes. Analysis of Hemoglobin, Hemiglobin Cyanide and Isoelectric Fractions of Hemiglobin Cyanide. Biochim. Biophys. Acta 393, 396-403 (1975).

9. Arlinger, L.: Analytical Isotachophoresis in Capillary Tubes: Analyses of Proteins in Spacer Gradients. In Proc. of the Electrofocusing and Isotachophoresis Symp., Milan 1974 (Righetti, P.G., ed.), North Holland Publ. Co., Amsterdam, 331-340 (1975).

10. Kjellin, K.G., Moberg, U. and Hallander, L.: Analytical Isotachophoresis of Cerebrospinal Fluid Proteins - A Preliminary Report. Sci. Tools 22, 3-7 (1975).

11. Kopwillem, A., Merriman, W.G., Cuddeback, R.M., Smolka, A.J.K. and Bier, M.: Serum Protein Fractionation by Isotachophoresis using Amino Acid Spacers. J. Chromatogr. 118, 35-46 (1976).

12. Kjellin, K.G., Hallander, L. and Moberg, U.: Analytical Isotachophoresis: A New Method for Analysis of Cerebrospinal Fluid Proteins. A Preliminary Report. J. Neurol. Sci., 26, 617-622 (1975).

Preparative Capillary Isotachophoresis: Separation of μg Amounts of Some Human Serum Proteins

U. Moberg, S.G. Hjalmarsson, L. Arlinger and H. Lundin

SUMMARY

A fraction collector device for isotachophoresis in capillary tubes has recently been developed. This makes it possible to analyse the highly resolved sample zones with further detectors for more complete characterization. The sample zones are transferred to a moving cellulose acetate strip without measurable loss of resolution. In this study μg amounts of human serum proteins were analysed isotachophoretically by using different Ampholine spacers and the sample collection strips were subsequently investigated by immunoelectrophoresis.

INTRODUCTION

Isotachophoresis (1-3) has been proved to have great potential both as an analytical technique in a capillary tube (4-7) and as a preparative method in polyacrylamide gel (8-11). The main advantages of analytical isotachophoresis in comparison with other analytical separation methods are the small sample amounts (μg), the short separation times (10-60 minutes) and the high resolution. One limitation in the use of capillary isotachophoresis has, however, been the impossibility of removing the highly resolved sample zones from the capillary. Therefore the introduction of a fraction collection capability to capillary tube isotachophoresis (12) implies a widened choice of methods for characterization and identification of the sample zones as well as giving a new micro preparative technique.
Serum was chosen as a model system in this investigation.

To check the homogeneity of protein zones and to detect proteins with the same mobility fusing into one peak in an isotachophoretic system, UV-detection has to be replaced by a more specific detection method such as immunological analysis, as will be reported in this work. Other described methods include radioactivity counting and zymogram techniques (12). To get good spacing of the serum proteins a study of the spacing mobility gradients created by Ampholine carrier ampholytes had first to be made.

PRINCIPLE OF SEPARATION WITH AMPHOLINE[R] AS SPACERS

In isotachophoresis the separated zones move in immediate contact with each other. In the analysis of proteins, detection difficulties are encountered because of the similar UV-absorbance and the difficulty therefore of distinguishing between narrow, though well-separated, isotachophoretic zones. In this case the use of spacing mobility gradients, consisting of non UV-absorbing spacers, is essential (5, 13, 14).

A spacing mobility gradient created by Ampholine[R] carrier ampholytes contains a large number of ion species with mobilites very close to each other. The ions in the gradient have to be distributed within the specific pH range suitable for spacing apart the protein zones of interest.

The different ampholyte molecules have different isoelectric points (pI). In isotachophoresis it is the different net mobilities of these ampholyte molecules at a specific pH in the leading electrolyte that is of importance. In a negative electrolyte system (for separation of negatively-charged sample ions), a stepwise increasing pH change occurs at each consecutive zone boundary, with the lowest pH in the leading electrolyte. It is shown (13, 14) that if proteins are to be spaced in the high-mobility region, the pH range of the Ampholine molecules must reach below the pH of the leading electrolyte. If proteins in the low-mobility region are to be spaced, the pH range need not exceed the isotachophoretic pH of the terminating electrolyte. In a positive electrolyte system it is the other way around.

Other types of spacers are low-molecular weight compounds, "discrete spacers", with defined mobilities and no UV-absorbance. Amino acids and peptides have mobilities similar to those of serum proteins, and could be used for this purpose (5, 8).

MATERIALS AND METHODS

Isotachophoresis

The isotachophoretic experiments were performed with an LKB 2127 Tachophor (LKB-Produkter AB, Bromma, Sweden) equipped with a fraction collector device. This apparatus with fraction collector arrangement is called Tachofrac and includes a counter flow pump, a capillary plate with T-tube, fraction collector and terminating reservoir block with membrane and valve attached to the standard Tachophor. See ref. 12 for detailed technical description.

The counter flow pump, Braun Perfusor IV (Braun Melsungen, Fed. Rep. of Germany), is connected to Tachophor via a 500 µl Unimetric syringe, teflon capillary tube and a cannula through the counter flow septum. The pump used in position "5" gives a flow of 2.75 µl/min with this syringe. The standard capillary is replaced by a specially constructed teflon capillary plate with a T-connection of glass immediately after the UV-detector site. The capillary lengths used in these experiments were 23 and 43 cm, thermostated at 10°C.

The sample zones were continuously transferred onto the cellulose acetate strip 0.15 mm thick and cut to 3 or 5 mm width.(Sartorius,Göttingen,G.F.R.)

30-50 cm of the strip (total length approx. 5 m) were used in each experiment. The strip is mechanically transported past the end of the glass T-tube at a constant adjustable speed. There is a distance of 0.05-0.08 mm between the end of the T-tube and the surface of the strip. The drive wheel (circumference 50 mm) marks the edge of the strip at each turn and simultaneously closes an electronic circuit to produce a spike on the UV-recording, which enables the identification of the sample zones of the strip from the UV-recording.

The leading electrolyte was 5 mM 2-N(morpholino)-ethanesulfonic acid (MES) from Sigma Chemical Co., USA adjusted to pH 9.0 with 2-amino-2-methyl-1,3-propanediol (ammediol) from Sigma. 0.5% hydroxypropylmethylcellulose (HPMC) Methocel, Standard 15000 CPS from Dow Chemical Co., USA was used for stabilization and to counteract electroendosmosis in the capillary. The terminating electrolyte was 5 mM ε-aminocaproic acid (EACA) (BDH Chemical Ltd., England) adjusted to pH 10.6 with $Ba(OH)_2$ (Merck, Germany). The wavelength of the UV-detector was 254 nm. The spacing mobility gradients were created by Ampholine^R carrier ampholytes (LKB-Produkter AB, Bromma, Sweden). The amino acids used as discrete spacers were β-alanine (BDH Chemicals Ltd, England) and γ-aminobutyric acid (GABA) from Sigma. The serum was prepared by defibrination of pooled human plasma (AB Kabi, Stockholm, Sweden). The human gammaglobulin, transferrin, albumin and ceruloplasmin were also from AB Kabi. The IgG and IgA samples were of unknown degree of purity.

Immunoelectrophoresis

The protein fractions collected on the cellulose acetate strip were analysed by immunoelectrophoresis (15) by simply laying the strip directly on the agarose gel containing either antibodies against total human serum or monospecific antisera. The strip was not soaked in advance. The current was applied immediately to minimize diffusion. The quantitative transfer of protein from the strip into the gel has previously been studied (12). The immunoelectrophoretic buffer was TRIS-barbital pH 8.6, analysis time was 15-20 h at 2.5 V/cm and the equipment used was LKB 2117 Multiphor. Glass slides 8x9 cm and 24x9 cm were covered with 12 and 36 ml respectively of 1% agarose gel (Litex, Olostrup, Denmark). The antibodies were rabbit anti-human type (Dakopatts, Copenhagen, Denmark). The concentration of antibodies varied from 50 μl to 400 μl/12 ml gel. The fixing and staining procedure was made according to ref. 15. The collected sample strips could be kept frozen before immunoelectrophoresis for at least two weeks (no longer time study made) without any sign of deterioration of the sample.

RESULTS

The pH of the leading electrolyte was chosen to be 9.0 and the terminating electrolyte was pH 10.6; the exactness of the latter figure is not important as long as it is between pH 10 and 11, because it will be adjusted in the capillary during the experiment to its isotachophoretic pH determined by the leading electrolyte. This alkaline electrolyte system implies negative charges for practically all the serum proteins and also for the Ampholine carrier ampholytes used in these experiments. MES was used as the leading ion because it has a mobility higher than albumin, but lower than that of the salt ions which are also present in serum but of no interest for this study. Therefore the curves shown are cut in front of the albumin peak. To get a record also containing serum components more mobile than albumin, one could use chloride as leading ion which has a higher mobility than the salt ions.

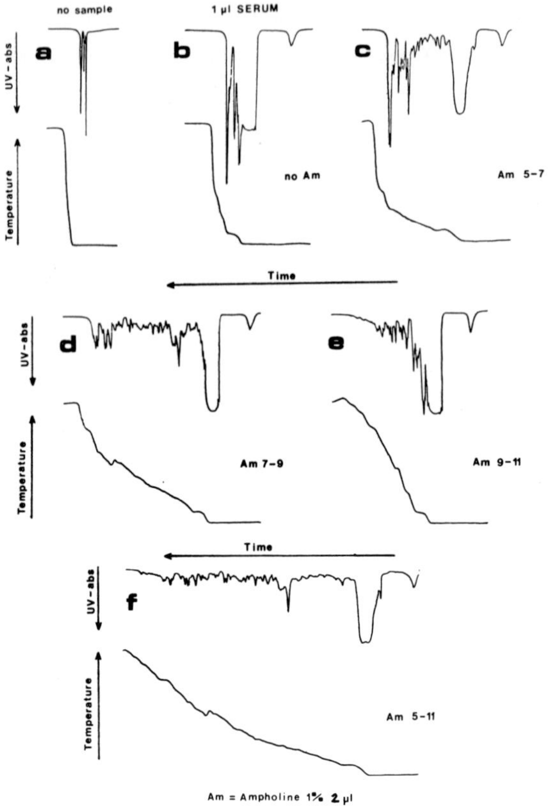

Fig. 1. Serum protein separation using spacing mobility gradients created by 3 different pH ranges of Ampholine® (Am) carrier ampholytes in the same electrolyte system. a) No sample. b) 1 µl serum without Am. c) 1 µl serum + 1 µl 1% Am pH 5-7. d) 1 µl serum + 1 µl 1% Am pH 7-9. e) 1 µl serum + 1 µl 1% Am pH 9-11. f) 1 µl serum + 1 µl each of 1% Am pH 5-7, 7-9, 9-11. Leading electrolyte (L): 5 mM MES + 10 mM ammediol + 0.5% HPMC, pH 9.0. Terminating electrolyte (T): 5 mM EACA adjusted with $Ba(OH)_2$ to pH 10.6. Length of the capillary 43 cm, thermostat temperature 10°C, current during start phase was 120 µA, during detection 70 µA, time for separation 30 minutes. UV detector set at 254 nm. Chart speed 2 cm/min.

Fig. 1 exemplifies the spacing of a complex protein mixture. The sample was human serum, with electrolytes and ampholyte pH ranges as described in the figure text. In (a) no sample was added. In (b) 1 µl serum was run without ampholytes. Only a few peaks can be distinguished. In (c) ampholytes pH 5-7 have been added and the high mobility region around albumin is well spaced. The ampholyte ions form a continuous spacer gradient in net mobility and consequently in temperature as can be seen from the thermal record. With ampholytes pH 7-9 (d) the medium mobility region after albumin is spaced, and finally (e) the low mobility region of the gammaglobulins is spaced by ampholytes pH 9-11. In (f) ampholytes of pH 5-7, 7-9, 9-11 were all mixed with 1 µl serum and analysed, producing a spacing effect over the whole range of protein mobilities.

In order to identify the different proteins in the pattern of a known serum the protein of interest could be added in a pure form (doping). The added protein will make an increase of one or several peaks in the known serum pattern which corresponds to its mobilities.

In Fig. 2 the doping technique is demonstrated. The sample was 1 µl serum, together with 1 µl each of 1% Ampholine pH 7-9 and pH 9-11 and 2 µl of 5 mM β-alanine and 5 mM GABA. β-alanine and GABA were used as discrete spacers and can be seen as the two non-UV-absorbing zones at the end of the UV record but were of little importance in this particular study as they are only spacing at the end of the gammaglobulin region. (a) shows an analysis of serum and spacers. In (b) transferrin was added and several peaks in the high to medium mobility region increased their heights. In (c) IgG was added which falls into the low mobility part of the pattern as would be expected. In (d) IgA was added which surprisingly appears in the high mobility region of the system. It is, however, difficult to know whether this is correct or due to large amounts of impurities in the sample. A specific antiserum is needed in order to fully establish the position of IgA.

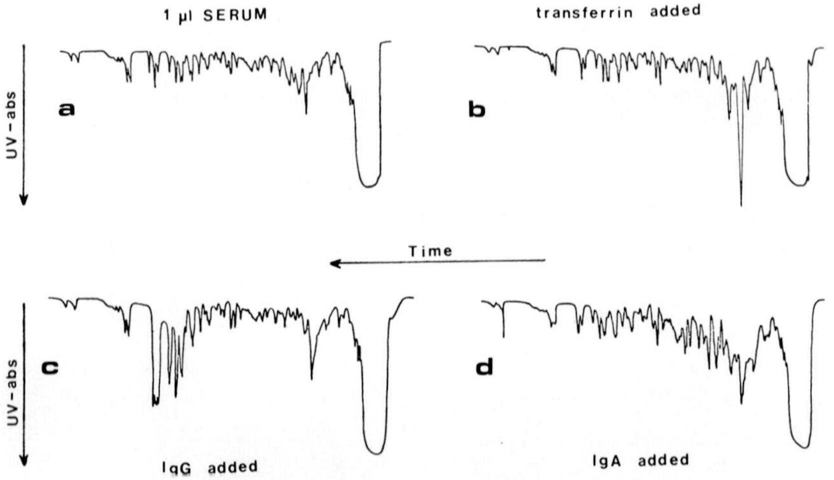

Fig. 2. Identification of different serum proteins by adding to a known serum pattern the protein of interest in a pure form (doping). Sample solutions: 1 µl serum; 1 µl 1% Ampholine pH 7-9, 1 µl 1% Ampholine pH 9-11, 2 µl 5 mM β-alanine, 2 µl 5 mM GABA. a) shows an analysis of serum and spacers. b) 10 µg transferrin was added. c) 10 µg IgG was added. d) 20 µg IgA was added. Current during start phase was 120 µA, during detection 70 µA, time for separation 30 minutes, other conditions as in Fig. 1.

Fig.3b and 4b show immunological characterizations of human serum proteins collected from isotachophoretic capillary experiments. The sample volume was 1.0 µl serum, with 0.5 µl 1% Ampholine pH 6-8 (experiment 3) and 0.5 µl 1% Ampholine pH 7-9 (experiment 4). The parts of the strips corresponding to the lines on top of the UV-records (Fig. 3a, 4a) were cut out and immunoelectrophoretically run against total anti human serum (1.0 ml anti bodies/ 36 ml gel). The vertical lines on the immuno patterns (fig. 3b, 4b) correspond directly to the spikes made electronically on the UV-records (Fig. 3a, 4a). The identification of the proteins were made by the doping procedure as described in Fig. 2 or better by running against monospecific antisera as will be described in connection with Fig. 5.

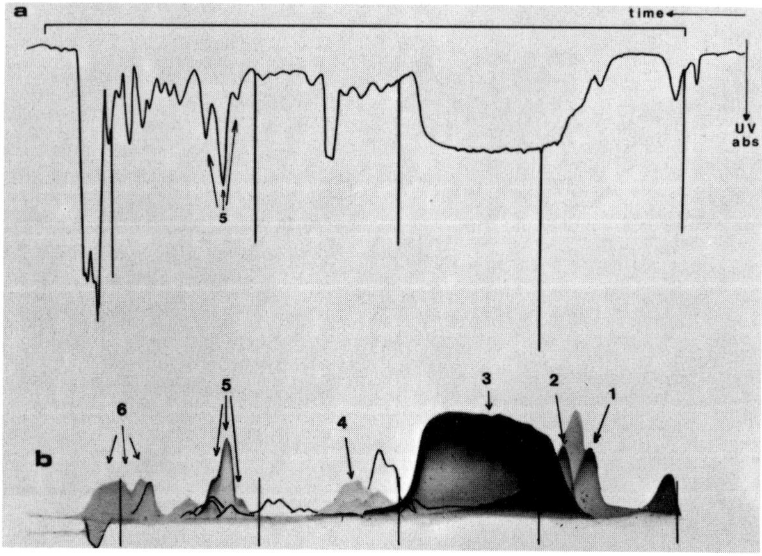

Fig. 3 and 4. Isotachophoretic and immunological analysis of human serum proteins. 1.0 µl serum and 0.5 µl 1% Ampholine pH 6-8 (Fig. 3) and 0.5 µl 1% Ampholine pH 7-9 (Fig. 4) were used for isotachophoresis. Analysis time 40 minutes, current during detection 50 µA. Chart speed 3 cm/min and speed of the sample collection strip was 2.6 cm/min. Capillary length was 43 cm, electrolytes and other conditions as in Fig. 1. The part of the cellulose acetate strip corresponding to the line above the UV-record (a) was cut out and applied to the agarose gel containing antibodies against whole human serum. The faintest immunoprecipitin lines were filled in by a filterpen. The different proteins marked in the figures were identified by the doping technique or by running against monospecific antisera described in Fig. 5 and 6. 1 = orosomucoid. 2 = α_1-antitrypsin. 3 = albumin. 4 = haptoglobulin. 5 = transferrin. 6 = IgG

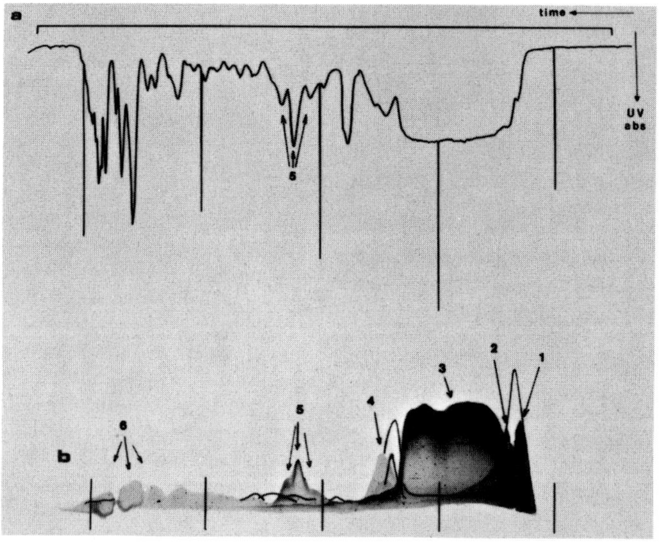

It is clearly shown from the UV-pattern that the Ampholine pH range 6-8 gives a spacing effect in the high mobility region (Fig. 3a, right part of the pattern). This is verified by the immunoelectrophoretic experiment where a great number of the high and medium-mobility proteins are spaced. In the low mobility region (left part of the pattern), however, the proteins are unresolved owing to lack of sufficient spacing. By using the spacing effect of Ampholine pH 7-9 the low mobility region will be spaced which is demonstrated in Fig. 4. The high-mobility proteins are now seen to move in immediate contact with or overlapping each other due to the lack of spacing (Fig. 4b).

Comparing the immunopatterns to the corresponding UV-records of experiments 3 and 4 shows no measurable loss of resolution when transferring the separated zones to the cellulose acetate strip. For example, transferrin (marked "5") shows up as three peaks in the immunopatterns which is in excellent accordance with the three peaks in the corresponding UV-record.

To make an immunological identification of the serum proteins, an intermediary gel with monospecific antisera can be used as demonstrated in Fig. 5b and 6b. The strip containing the serum protein part of interest was cut out and placed on the lower half of the agarose gel plate containing monospecific antisera. The upper part contained whole antisera 150 µl/6 ml gel. The sample solutions were 0.8 µl serum and 0.4 µl 1% Ampholine pH 6-8 in both experiments. In Fig. 5b monospecific antisera against α_1-antitrypsin (25 µl/6 ml gel was used) and in Fig. 6b against orosomucoid (100 µl/6 ml gel). Comparing Fig. 5b and 6b with the complete patterns of the whole serum in Fig. 3 and Fig. 4 peak No 2 can be identified as α_1-antitrypsin and No 1 as orosomucoid. The dark background colour in the upper part of Fig. 5b and 6b is due to protein impurities in the commercial monospecific antisera and as can be seen in Fig. 6b, the major part is albumin. The gel containing monospecific serum, could be run electrophoretically to let these impurities move into an antiserum-free agarose gel. When the impurity proteins have travelled into this gel it could be removed and replaced by a gel containing whole antiserum.

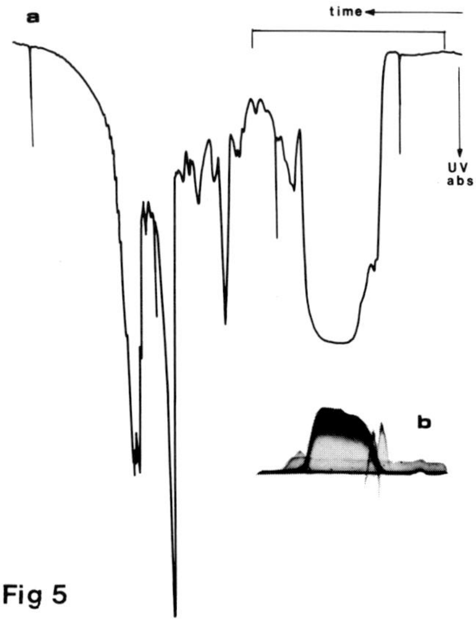

Fig 5

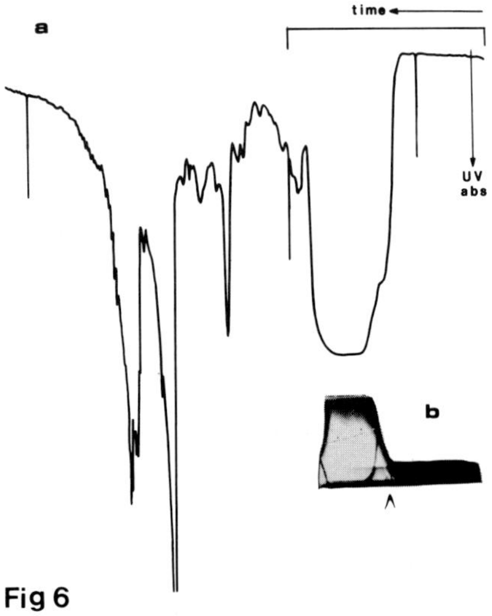

Fig 6

-Fig. 5 and 6. Immunological identification of α_1-antitrypsin (Fig.5) and orosomucoid (Fig. 6) using a monospecific antiserum in an intermediate gel. For isotachophoresis 0.8 µl of serum and 0.4 µl 1% Ampholine pH 6-8 were used. Separation time was 22 minutes, current during start phase 100 µA and during detection 50 µA. Speed of UV-recording chart and of collection strip was 4 cm/min. Capillary length was 23 cm. Electrolytes as in Fig. 1. The interesting part of the cellulose acetate strip was cut out as indicated by the line above the UV-records.

DISCUSSION

The proper choice of pH range for the spacing mobility gradient is most important in order to obtain a good spacing effect. An extended mobility gradient obtained by mixing different pH ranges gives a good overall picture of the serum proteins, but in special cases a narrow pH range in optimal ratio with the protein will suffice. The leading and terminating electrolytes can also be chosen so that the sample zones of no interest are eliminated. By selecting a leading ion with a mobility just above that of the protein to be studied and the terminating ion with a mobility lower, the isotachophoretic system will include just the part of interest. As many amino acids have mobilities similar to those of serum proteins they could be used for this discriminating purpose (5) or, of course, as discrete spacers as described by Kopwillem et al (8).

The fraction collector device to the LKB Tachophor opens new possibilities for studying the purity and homogeneity of protein samples. The microgram and microlitre amounts collected as a continuous flow on the cellulose acetate strip makes handling and storage very simple. No problems of adsorption to the strip has been found so far for the proteins investigated. Proteins can be purified by making one or two separations, cutting out the interesting part of the strip, and then eluting them with the proper buffer.

Only a few proteins in the serum sample were identified in this study. The aim, however, was to show the powerful combination of the specific immunological technique and the high resolving isotachophoretic separation.

REFERENCES

1. Haglund, H., A Principle for Analytical and Preparative Separation of Substances such as Proteins, Peptides, Nucleotides, Weak Acids and Metals, Science Tools 17,2 (1970).

2. Routs, R., Electrolyte Systems in Isotachophoresis and Their Applications to Some Serum Separations. Dissertation Eindhoven The Netherlands (1971).

3. Arlinger, L., Analytical Isotachophoresis - Principle of Separation and Detection, Protides of the Biological Fluids (H. Peeters, Ed.), 22,661-667 (1974).

4. Arlinger, L. Analytical Isotachophoresis in Capillary Tubes. Analysis of Hemoglobin, Hemoglobin Cyanide and Isoelectric Fractions of Hemoglobin Cyanide, Acta 393,396-403 (1975).

5. Kjellin, K.G., Moberg, U. and Hallander, L., Analytical Isotachophoresis of Cerebrospinal Fluid Proteins. A Preliminary Report, Science Tools 22,3-7 (1975).

6. Kopwillem, A., Moberg, U., Westin-Sjödahl, G., Lundin, R.and Sievertsson, H., Analytical Isotachophoresis in the Analysis of Synthetic Peptides, Anal. Biochem. 67,166-181 (1975).

7. Kopwillem, A., Analytical Isotachophoresis in Capillary Tubes. Transformation of Pyruvate to Succinate by Calf Heart Mitocondrial Enzymes, J. Chromatogr. 82,407-409 (1973).

8. Kopwillem, A., Merriman, W.G., Cuddeback, R.M., Smolka, A.J.K. and Bier, M., Serum Protein Fractionation by Isotachophoresis using Amino Acid Spacers, J. Chromatogr. 118,35-46 (1976).

9. Möllby, R., Hjalmarsson, S.G. and Wadström, T., Separation of E. Coli Heat-Labile Enterotoxin by Preparative Isotachophoresis, FEBS Letters 56 1,30 (1975).

10. Jefferis, R., and Butwell, A.J., Isolation of Human IgD Myeloma Protein by Isotachophoresis, Science Tools 22,1 (1975).

11. Svendsen, P.J., On the Procedure of Preparative Isotachophoresis, Science Tools 20,1 (1973).

12. Arlinger, L., Preparative Capillary Isotachophoresis. Principle and Some Applications, J. Chromatogr. 119,9-24 (1976).

13. Arlinger, L., Analytical Isotachophoresis in Capillary Tubes. Analysis of Proteins in Spacer Gradients, Progress in Isoelectrofocusing and Isotachophoresis Symp. Milano Ed. P.G. Righetti, North Holland Publ. Comp. p. 331 (1975).

14. Hjalmarsson, S.G., Preparative Isotachophoresis. The Effect of Using Ampholine of Different pH Ranges as Spacer Ions in the Fractionation of Serum Proteins, Science Tools 22,35 (1975).

15. A Manual of Quantitative Immunoelectrophoresis, edited by N.H. Axelsen, I. Kröll and B. Weeke, Universitetsforlaget, Oslo (1973).

Quantitative Determination of Picomoles ADP by Means of Steady State Mixed Zones in Isotachophoresis

J.P.M. Wielders and F.M. Everaerts

SUMMARY

For ionic components that can be detected specifically a method is described for analysing them via steady state mixed zones in an isotachophoretic analysis. Some practical remarks are given and the background is discussed. We recommend this method especially if the component of interest is present in very low concentrations, compared with other ions in the sample e.g. micromolar ADP concentrations in a centimolar buffer solution. The quantification of ADP in this type of sample by means of UV absorbance measurement of an ADP/acetate steady state mixed zone is shown.

INTRODUCTION

Isotachophoresis in capillaries is an electrophoretic analytical method by which ionic components are not only separated but also migrate in consecutive zones at equal velocity. Theory, instrumentation and applicability are discussed by Everaerts et al. (1). Within a certain operational system (consisting of leading electrolyte - LE - and terminating electrolyte - TE -) the concentration of a sample component A in the steady state of the experiment is determined. This means that the length of zone A is proportional to the absolute amount of A that has been injected. Of course this is only true, whenever there are no other ions than A and its counterion within this zone. At very low sample concentrations the zone length, that has to be measured for quantification of A, can even become less than the minimum necessary for accurate work, even when the sample volume is the maximum, that can be used for a given apparatus and a certain operational system. The following possibilities are left in this case:

- Lowering the concentration of the operational system.
- Changing the pH of the operational system. A higher pH can cause a higher charge of weak acid anions and this means a lowering of the concentration in the zones.
- "Dilution" of the component of interest in a steady state mixed zone, whenever this component can be detected specificially within that mixed zone.

This paper will only deal with that last possibility.

The detection principles that are applied in isotachophoresis are mainly temperature measurement, conductivity or field strength measurement and UV absorption measurement. The first three provide universal, the last one specific information. Until now the UV detector was mainly used to obtain qualitative information, although also quantitative work can be done with it (2). The application of the UV detector for the quantification of strongly UV absorbing components within steady state mixed zones will be described in this paper.

We had to measure nucleotide concentrations up to the micromolar level within a buffer of centimolar concentrations in studying the binding capacity of beef heart mitochondrial ATPase (3). Quantification of ADP and ATP in this sample by the normal isotachophoretic procedures is impossible, but since the principles of isotachophoresis allow us to force ions to form steady state mixed zones, we could fulfil these special demands.

DESCRIPTION AND THEORY

A mixed zone can be caused by impurities of the operational system. For example at an anion separation, it may contain other anions than the leading and terminating anion. Anionic impurities of the LE which have a lower effective mobility than sample anions, will be runned over by these sample ion zones, while also anionic impurities of the TE may migrate through sample component zones. Even when the operational system is free of impurities,mixed zones can occur. We can divide them into two groups, although a strict separation is not always possible. At first the detection may take place when the separation phase (a moving boundary process) has not ended yet. This happens when the separation compartment is too short for the amount of ionic material, that has been injected. Lowering of the sample volume or an increase of the capillary length (in reality or with the application of counter current) will lead to the complete resolution of the mixed zone. The second case arises when two (or more) components form a mixed zone because the difference of their mobilities under the local conditions (the effective mobility) is too small. In the most extreme case the result is a homogeneous mixed zone. Since these mixed zones occur in the steady state we will call them steady state mixed zones. Generally there is still a very small difference in the mobilities of steady state mixed zones components so there will be a small deviation from absolute homogeneity. These differences are often accentuated by resulting pH differences. Whenever these mobility differences are sufficient small, diffusion and convection will smooth concentration profiles within a mixed zone.

The concentration of components like nucleotides, which have a high molar absorption coefficient for UV light of a certain wavelength, can be measured accurately at a level far below the 10^{-2} till 10^{-3} greq*/liter, which is usual in isotachopho-

* greq = gramequivalent.

resis. In the 0.45 mm i.d. teflon capillary of the used apparatus ADP concentrations up till about 10^{-5} M can be measured easily. It needs no further explanation that other components should not interfere when for example ADP has to be measured within a steady state mixed zone. The reverse, quantification of low concentrations of not absorbing species within a mixed zone with an UV absorbing component, is not possible of course.

The analysis of an ADP/acetate mixture in a suitable operational system will result in a steady state mixed zone with the same length as the sum of the zone lengths of ADP and acetate, when the same amounts were analysed separately under the same conditions. The integrated UV absorbance signal of the mixed zone is proportional to the amount present in that zone. This dilution principle is illustrated in figure 1. Generally the formation of a steady state mixed zone requires the matching and influencing of effective mobilities of sample ions like ADP and carrier ions like acetate. The effective mobility (m_{eff}) is the migration velocity in a solution under a unity electrical field strength at the local conditions. At a certain temperature and in the absence of substances that will influence the m_{eff} via complex formation etc. we may write:

$$m_{eff} = \Sigma_i \alpha_i \beta_i \gamma_i m_{abs_i}$$

m_{abs_i} = absolute or ion mobility of ion with charge i in water
α_i = dissociation degree
β_i = solvent factor
γ_i = factor concerning the activity

Although the components must have an equal m_{eff} in order to form a steady state mixed zone, having an equal mobility in a certain operational system does not always mean that there will be a steady state mixed zone when a mix of them is analyzed.

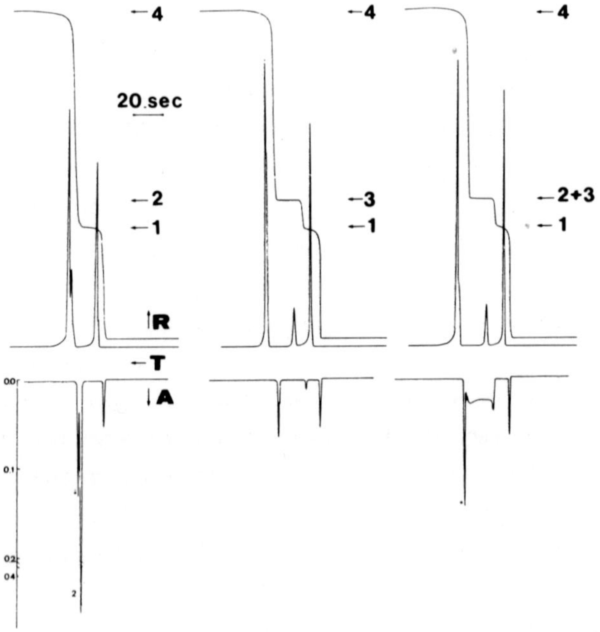

Figure 1:

Isotachopherograms showing the principles of quantitative analysis by means of steady state mixed zones.
R = increasing resistance; T = increasing time; A = increasing absorbance ($^{10}\log I_o/I$).
Operational conditions are presented in table III.
1 = bicarbonate; 2 = ADP; 3 = acetate; 4 = glucuronate; * = AMP.

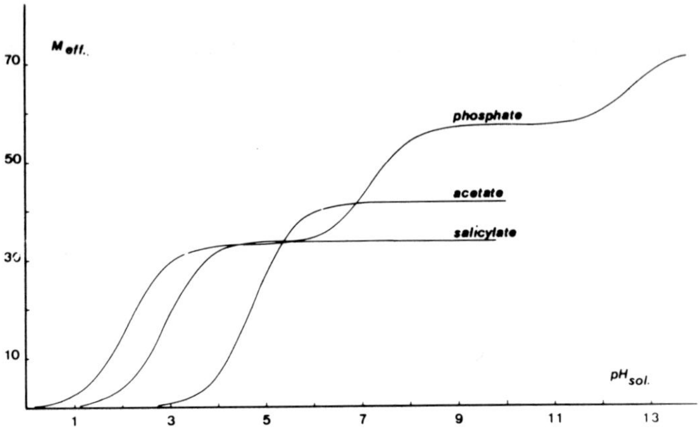

Figure 2:

Calculated $m_{eff}/pH_{solution}$ plots for phosphate, acetate and salicylate.

Next to this, experimentally it was found that in an operational system that causes a steady state mixed zone of two components A and B, the sequence of the A and B zone depends on the way in which A and B (injected separately) are layered in the injection system. This sequence was interchangeable.

Searching for equal m_{eff}'s as a function of the pH of the system is the opposite of what one has to do for the separation of components on pK values. In our case we are looking for the intersections of $m_{eff}/pH_{solution}$ plots. Those plots (fig. 2) show sections in which the m_{eff} is slightly or not influenced by small pH changes, while near the pK_a values the opposite takes place. If the conditions within a zone, during the steady state of an isotachophoretic experiment, were only determined by the LE and not also by the zone component properties, than every intersection of two $m_{eff}/pH_{solution}$ plots would lead to a steady state mixed zone.
We will show that the carrier ion should determine the pH of the mixed zone and that the sample ion should not possess a pK_a value within 1-1½ pH unit of the mixed zone pH. An ideal stable steady state mixed zone will result whenever the $m_{eff}/pH_{solution}$ plots of sample and carrier overlap during a long pH traject, especially when both ions migrate at their absolute mobility. To illustrate the occuring phenomena the steady state mixed zone systems acetate-phosphate and phosphate-salicylate have been calculated. Mixed zones in which nucleotides play a role can not be calculated yet, since no accurate mobility data are available.

Phosphate/acetate.

The theoretical $m_{eff}/pH_{solution}$ plots (figure 2) show two intersections for these components: at pH 5.35 and 6.93. So, in a solution at one of these pH's, the effective mobilities of acetate and phosphate are equal. Using pK values and mobility data from the literature, the pH of the zones ($pH_{acetate}$ and

$pH_{phosphate}$) and the mobilities of acetate and phosphate in their pure zones were calculated by a computer program developed by Beckers (1). The leading electrolyte in this calculation was 10^{-2} M chloride, histidine was the counterion, the pH_{LE} was varied from 6.5 till 6.85. The results were plotted in figure 3a and b.

These plots show that the mobilities are only equal at one pH_{LE} (6.80). More important however is the difference in the $pH_{acetate}$ and the $pH_{phosphate}$ when the mobilities of these components within their own zones are equal. Suppose we analyse an acetate/phosphate mixture in an operational system with pH_{LE} = 6.80. The concentration ratio acetate/phosphate determines the pH of the mixed zone when this arises. At a low ratio the $pH_{mixedzone}$ approaches the $pH_{phosphate}$. Although this $pH_{phosphate}$ is higher than the $pH_{acetate}$ in this operational system, the $m_{eff}/pH_{solution}$ plot of acetate is fairly flat in this region. This means that the change in m_{eff} of acetate due to the presence in the phosphate zone is very small, so the mixed zone will be stable.

The reverse, a high ratio acetate/phosphate causes such a change in the phosphate mobility, when phosphate is present in the acetate rich zone, that this mobility is lowered with at least one percent. So phosphate will gradually leave the mixed zone. For the intersection at pH 5.36 we find the reverse. Here only a acetate rich mixed zone can be a steady state mixed zone.

Phosphate/salicylate.

In figure 4a and b the results are shown for phosphate and salicylate at pH_{LE} 3.7 till 5.3. The leading electrolyte contained 10^{-2} M chloride, ε-aminocaproic acid was the counterion. These plots show clearly the large pH_{LE} region in which the resulting $pH_{phosphate}$ and $pH_{salicylate}$ are fully equal, while the m_{eff}'s are almost identical in that same region. These results can be understood as follows: The ionic mobili-

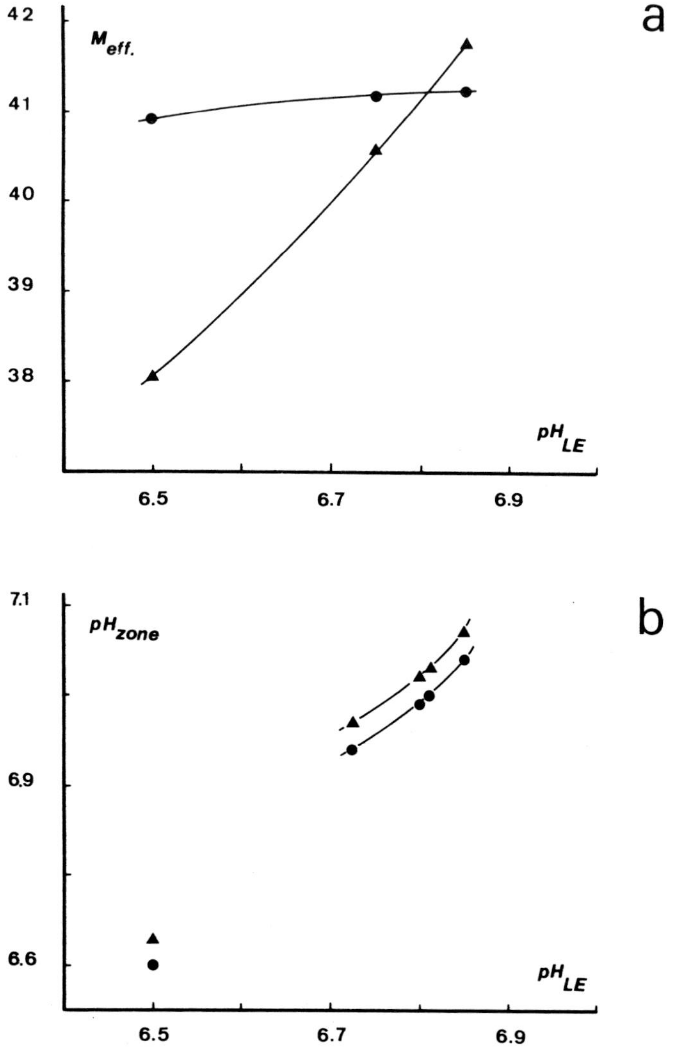

Figure 3a and 3b:

Calculated m_{eff}/pH_{LE} and pH_{zone}/pH_{LE} plots for phosphate (▲) and acetate (●).

Counterion in these calculations was histidine.

ties of phosphate and salicylate as a single charged anion are about equal and the first pK_a value of both components is at least 1½ pH unit apart from the pH of the mixed zone. Experiments showed that this capability to form steady state mixed zones has arisen at pH_{LE} 4 already. This decrepancy with the calculations will be due to inaccuracies in the used data and possible also to temperature and activity effects, for which no correction has been made in these calculations.

A second parameter that could influence the m_{eff} of ions was the electrolyte concentration. From the theories of Debye-Hückel and that of Onsager it follows that only when a steady state mixed zone has to be composed with ions of different charge (under the local conditions) the variation of concentration may be useful.
The 10^{-2} greq/liter leading ion concentrations we normally use for our operational systems, can be diluted practically until about 5×10^{-4} greq/liter. In this range the zone length of a certain amount of sample component is not inversely proportional to the leading ion concentration because of this activity effect.

Although conductivities and dissociation constants have been measured in solvent mixtures since long, the application of composed solvents to influence electrophoretic mobilities is rather new. The change in the solvents properties upon addition of other solvents can be very complex. For example Frank and Ives reviewed studies on the properties of mixtures of water with the lowest alcohols (4). The influence of composed solvents on the m_{eff}'s of the components that are dissolved in these, is difficult to explain some times. This does not mean at all, that solvent additives could not be useful to influence a m_{eff}. It can be a powerful tool to determine the m_{eff} of an ion as will be shown. Since data on mobilities, conductivities and pK_a values in composed solvents are only available for some model systems, the influence on effective mobilities by

solvent additives has to be studied experimentally.
When we are looking for a steady state mixed zone and we do not have the data to compose $m_{eff}/pH_{solution}$ plots, then we have to look for another plot that shows the influence of the pH on the components effective mobility. This purpose is full-filled by a conductivity plot. To obtain such a plot, we measure the conductivity preferably relative to an internal standard such as chlorate (1) and plot it as a function of the pH_{LE}.
Equal zone conductivities mean equal effective mobilities of those sample components. It must ve pointed out that these conductivity/pH_{LE} plots are different from $m_{eff}/pH_{solution}$ plots.

The applicability of this dilution method is shown in the next calculation. The minimum zone length for accurate quantification is about 0.5 mm when we use a conductivity, field strength or UV detector. By the way, this is about the detection limit for a thermocouple detector. Working with 0.45 mm i.d. capillaries at concentrations of 10^{-2} greq/liter means that we can quantify about 10^{-9} greq. So quantification if a triple charged anion B^{3-} (like ADP at high pH) at sample concentrations of 5 micro-molar, needs a sample volume of about 70 µl. For the normal isotachophoretic apparatus this is an enormous amount. Working at lower concentrations of the operational system may solve this problem, except when there is a large surplus of other anions in that sample. E.g. the sample contains 10^{-2} molar A^-. The 70 µl injection volume will cause a zone A^- that is 700 times as long as the B^{3-} zone. When A^- migrates faster than B^{3-}, we can estimate the minimal separation compartment length to reach the steady state phase by the formula:

$$L_{min} = \frac{Q_A \cdot q}{(1-m_B/m_A) \cdot p}$$

L_{min} = minimal separation compartment length
Q_A = total amount of A injected (greq)
p = concentration of A in the steady state (greq/ μl)
q = geometry factor; zone length per volume (mm/ μl)

The influence of the traces of B^{3-} on the A^- zone during the separation phase on the A^- zone conditions have been neglected. The sample is supposed to be an ideal cylinder in between LE and TE before the analysis starts. Let the ions migrate at their absolute mobilities under the experiment conditions and $m_A = 70 \times 10^{-5}$ $cm^2/Vsec$ and $m_B = 40 \times 10^{-5}$ $cm^2/Vsec$. Then for our capillary and 70 μl 10^{-2} molar A^- and 5×10^{-5} molar B^{3-} sample we need a minimal separation length of about 100 cm. The possibility of quantification of B within a steady state mixed zone changes this view completely.
For example to measure ADP concentrations of 5 micromolar, we need only 3-4 μl sample as will be shown. To compare with our example of 70μl, for a 0.5 mm steady state mixed zone we only need about 1 μl sample. The influence of this big decrease in sample volume on the whole system including the time of analysis, does not need any further explanation.

APPARATUS

The isotachophoresis apparatus used, is described extensively by Everaerts et al. (1). The separation compartment in these experiments was a PTFE capillary i.d. 0.45 mm o.d. 0.70 mm. Besides the UV detector also the conductivity detector was used. The UV source is a low pressure electrodeless mercury lamp, being excited in a high frequency electrical field. The UV light has to pass through a 0.3 mm slit to reach the capillary. Detection of transmitted light is done by a photo sensitive resistance. The transmission curve of the filterset used and the spectral lines of the mercury lamp make the absorption measurements take place at 254 nm. This set up is equal to absorption measurements at a rectangular cuvette of 0.39 mm cell length.

To obtain a detector signal proportional with the concentration the usual absorption signal (100-% transmittance) was transformed to an absorbance signal. This was done by an electronic device, that contained a logarithmic amplifier (Analogs Devices 755 N).

EXPERIMENTS

Looking at Perrins data (5) for the pK_a values of ADP it needs no further explanation that steady state mixed zones should be searched for at pH 7.8 or higher:

$pK_1 < 2$ $\quad$ $pK_3 = \pm 4.0$

$pK_2 < 2$ $\quad$ $pK_4 = \pm 6.3$

However the pK_4 of ADP (and the pK_5 pf ATP) are strongly influenced by the ionic strength and the presence of complex forming cations like Na^+, K^+ and Ca^{2+}. Using Phillips formula (6) we find $pK_4 = 6.98-7.00$ at an ionic strength of 0.008-0.01 (which is normal at 10^{-2} greq/liter leading ion concentration) and in the presence of non interfering cations. Tris (hydroxymethyl) amino methane (Tris) does not form complexes with the adenosine phosphates, so it can be used as a counterion in high pH operational systems.
A pK_3 of 4.0 and a pK_4 of 7.0 point to a possibility for mixed zones around pH 5.5 and at pH at least 8.5. Since working at high pH always meets difficulties due to the carbon dioxide dissolved in the electrolytes, we first screened the low pH region. A steady state mixed zone could be formed with α-hydroxybutyric acid. However the stability of this system was such that we switched to the high pH region. One disadvantage of the high pH operational systems is, that the difference between the ADP and ATP effective mobilities is much lower. So the analysis of minor ADP concentrations at high ATP concentrations and vice versa may be troubled.
At a high $pH_{mixed\ zone}$ we can search for carriers, that have

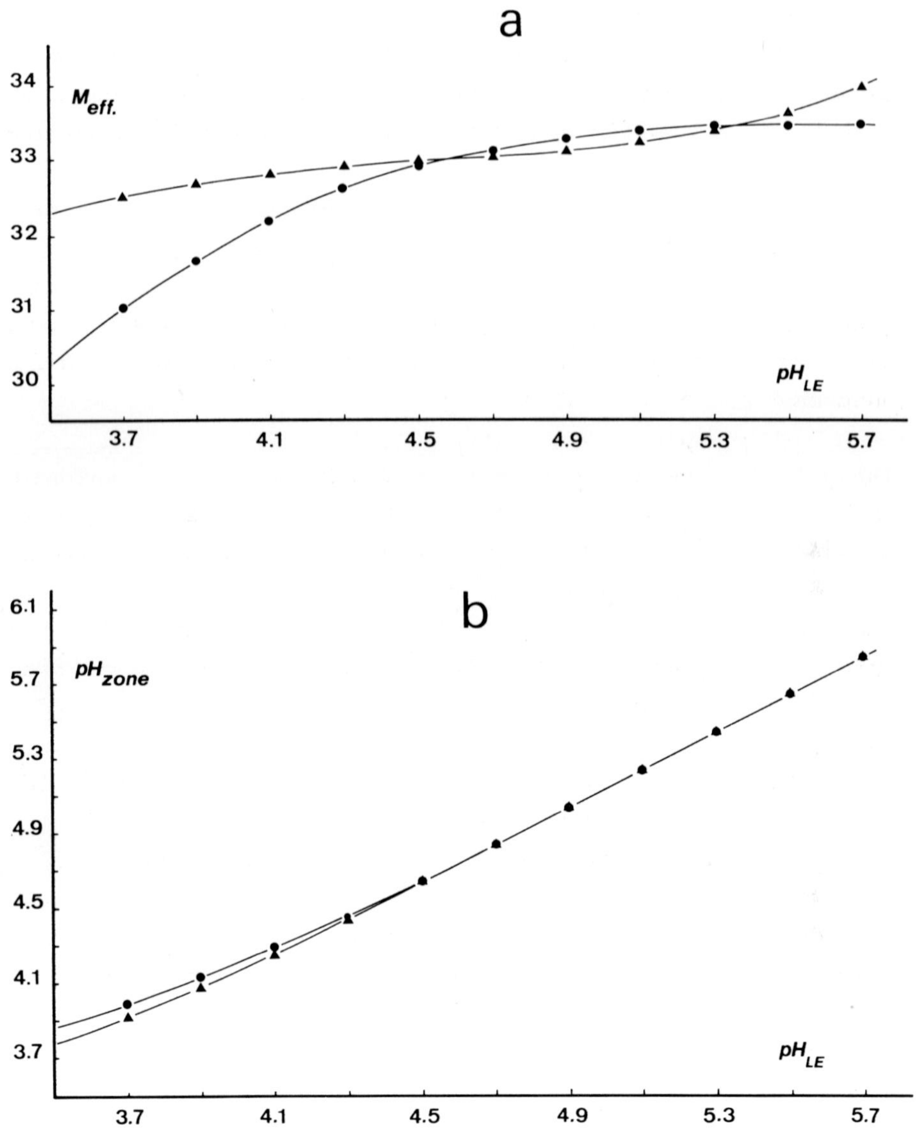

Figure 4a and 4b:

Calculated m_{eff}/pH_{LE} and pH_{zone}/pH_{LE} plots for phosphate (▲) and salicylate (●).
Counterion in these calculations was ε-aminocaproic acid.

a pK_a value at about 7-8 or 9-10. This carrier should have a relatively high mobility since ADP^{3-} and ATP^{4-} are quite fast at pH 8.5 (about 40-45 x 10^{-5} cm^2/Vsec). Components with pK_a values as indicated, such as amino acids or peptides are much to slow.

As mentioned above, a more ideal system will arise when both sample and carrier ion migrate at their absolute mobility. Solvent additives can be used to match their absolute mobilities. For this reason conductivity data of ADP, ATP and potential carrier zones were compared for the following operational systems.

Table I: Operational system at pH 8.5. Suitable for anionic separations.

Solvent: H_2O. Electric current: 80μA. Temperature 22^{0}C. I.D. capillary: 0.45 mm.

	Leading electrolyte	Terminating electrolyte
Anion	Chloride	Glucuronate
Concentration	0.01 M	Ca. 0.005 M
Counterion	Tris	Tris
pH	8.5	Ca. 8.5
Additive	0.05% polyvinyl-alcohol	None

Next to this, conductivities were measured in the same system but with 10 vol % methanol and also in the same system with 20 vol % methanol. No real conductivities were measured, but the differences in the resistances of the zones compared to the LE resistance. These stepheights (H) were compared with that of formiate. The results of these measurements are given in Table II.

	100 vol % H_2O			90/10 vol % H_2O/MeOH			80/20 vol % H_2O/MeOH		
	H_{anion}	$\frac{H_{anion}}{H_{ATP}}$	$\frac{H_{anion}}{H_{ADP}}$	H_{anion}	$\frac{H_{anion}}{H_{ATP}}$	$\frac{H_{anion}}{H_{ADP}}$	H_{anion}	$\frac{H_{anion}}{H_{ATP}}$	$\frac{H_{anion}}{H_{ADP}}$
Formiate	1			1			1		
ATP	1.84	1		2.10	1		2.55	1	
ADP	2.15		1	2.42		1	2.85		1
Acetate	2.17	1.18	1.01	2.08	0.99	0.85	2.27	0.89	0.80
Dichloroacetate	2.53	1.38	1.18	2.36	1.12	0.97	2.53	0.99	0.89
Pimelic acid	1.97	1.07	0.97	2.00	0.95	0.82	2.13	0.84	0.75
Bicarbonate	1.73			1.79			1.87		

Table II: Stepheights (H) and relative stepheigths of nucleotides and some potential carriers. The stepheight of formiate is choosen as unit.

The results of Table II show that we can expect suitable carrier zones for:

ADP and acetate at 0-1 vol % MeOH

ADP and dichloroacetate at about 8 vol % MeOH

ATP and pimelic acid at 4-5 vol % MeOH

ATP and acetate 9-10 vol % MeOH

Table II also shows the stepheight of bicarbonate which is faster tan the nucleotides at this pH. This means that only bicarbonate from the TE can be present in the nucleotide zones during the steady state.

For the quantification of ADP we choose acetate as carrier. The following system was selected for this purpose.

Table III: Operational system at pH 8.45 for formation of a steady state mixed zone acetate/ADP.

Solvent: H_2O. Electric current: 80μA. Temperature 22°C.
I.D. Capillary 0.45 mm

	Leading electrolyte	Terminating electrolyte
Anion	Chloride	Glucuronate
Concentration	0.01 M	Ca. 0.005 M
Counterion	Tris	Tris
pH	8.45	Ca. 8.5
Additives	0.005% polyvinylalcohol	None
	± 0.6 vol % methanol p.a.	

The leading electrolyte was prepared by weighing in of Tris and Tris-HCl. The use of freshly distilled water and storing the electrolytes under nitrogen gave much better results than the addition of $Ba(OH)_2$ to the terminating electrolyte.

Surrounding the capillary with nitrogen to prevent carbon dioxide diffusion through the capillary wall may also give an improvement.

Tests of this system proved that the total amount of ADP injected was proportional to the surface that is obtained by integration of the lineair UV absorbance signal of the mixed zone. In a series of experiments we used a sample that contained 165.1 μM ADP, 15 mM Na acetaat and Tris (added till pH 8.2). The results of analysis of this sample and dilutions of it (diluted with 15 mM NaAc) are shown in figure V together with some analysis of similar samples.

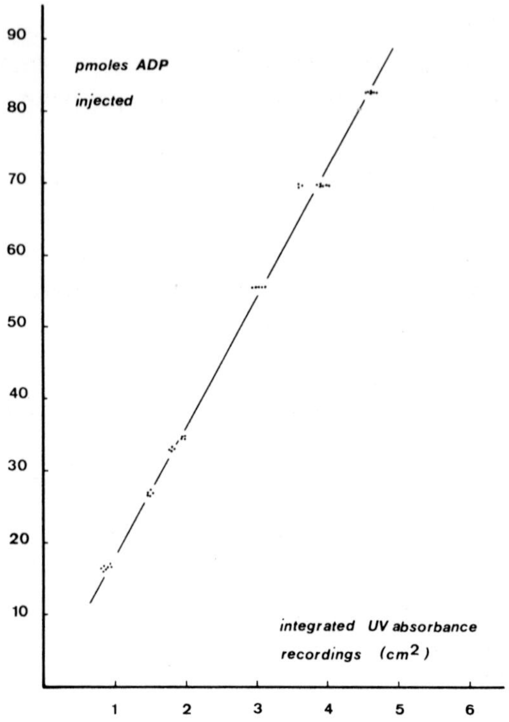

Figure 5:

Integrated UV absorbance recordings (one cm absorbance signal on the recorder was equal to A = 0.0179, chartspeed was 6 cm/min) as function of the amount ADP injected. Operational conditions are presented in table III.

ADP concentrations were determined enzymatically using phosphoenolpyruvate, pyruvate kinase and lactate dehydrogenase in a procedure similar to that described by Jaworek et al. (7).

If we neglect the influence of acetate on the UV absorbance of the mixed zone, then we find from figure 5 that 1 cm^2 UV surface corresponds on an average to 17.95 picomoles ADP injected.

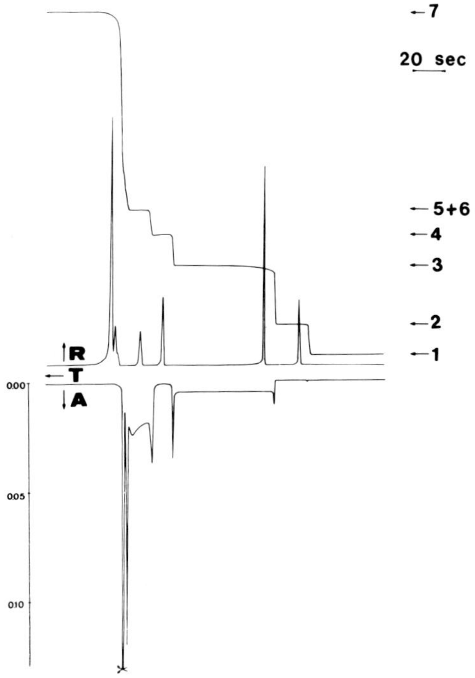

Figure 6:

Isotachopherogram of a sample, that contained 6.23 μM ADP, 1 mM acetate, 2 mM EDTA, 2 mM chlorate, 7.5 mM chloride and 15 mM tris.

R = increasing resistance; T = increasing time; A = increasing absorbance ($^{10}\log I_0/I$).

1 = chloride; 2 = chlorate; 3 = EDTA; 4 = bicarbonate; 5 = acetate; 6 = ADP; 7 = glucuronate.

This is very near 18.1 which can be calculated for the same apparatus and recorder settings when we use 14.3×10^3 as the molar absorption coefficient at a wavelength of 254 nm for ADP. Next to these simple ADP/acetate samples, analyses were done on a type of sample as that of figure 6. The actual ADP concentration in this sample was ± 6.23 μM (1 to 4 dilution of an 24.9 μM solution). From the isotachopherogram we calculated the ADP concentration to be 6.37 μM.
With this steady state mixed zone, concentrations up till 4 μM ADP have been measured already. Especially at these low concentrations a suitable intern standard should be added to the sample, so corrections for sampling errors can be made. The analysis of samples containing ADP and acetate in a molar ratio of 1/200 or even lower shows considerable contribution of the acetate to the total UV absorbance of the mixed zone: 10% or more. So for samples like 15 μM ADP and 4 mM acetate we have to correct for this by taking the absorbance level of a pure acetate zone or that of the terminating glucuronate (see fig. 1) as the zero level for the ADP absorbance.

DISCUSSION

Comparing this dilution method with the quantification by means of pure zones it will be clear, that only for components that can be detected specifically at very low concentrations a big advantage can be achieved. So the molar ratio carrier/sample will at least be 50-100. This means that the influence of the sample ions on the carrier zone conditions can be neglected. Next to UV absorbance measurements a radioactivity detector could be applicated for this dilution method.

To test the reproducibility of the method described, a number of 60 analysis were performed, spread out over 7 weeks. These experiments included different ADP/Ac^- ratios, injections with 1 μl and 10 μl Hamilton syringes and operational systems that were prepared freshly several times. For this group of analysis

we found 18.14 picomolen ADP to correspond with one cm^2 UV surface. ADP concentrations were from 4.1 till 165.1 μM. The standard deviation was 3.7 % so the 17.95 picomoles/cm^2 from figure 5 fits very well in this range. The spread of the results is owing to injection errors (1-2%), errors in enzymatic ADP concentration measurements (1-2%) and errors arising from the use of a ball and disc integrator (about 1%).
This quantification method can be used for ADP concentrations up till about one micromolar. The amount ADP injected has to be at least 10 picomoles.

ACKNOWLEDGMENT

This work was supported by a grant from the Netherlands Organisation for the Advancement of Pure Research (Z.W.O.).

REFERENCES

1. F.M. Everaerts, J.L. Beckers, Th.P.E.M. Verheggen, Isotachophoresis - Theory, instrumentation and applications (Journal of Chromatography Library, Vol. 6) Elsevier, Amsterdam, Oxford, London (1976).

2. L. Arlinger, Analytical Isotachophoresis. Resolution, detection limits and separation capacity in capillary columns. J. Chromatography 91 (1974) 785-794.

3. D.A. Harris, J. Rosing, R.J. van de Stadt, E.C. Slater, The tight binding of adenine nucleotides to beef-heart mitochondrial ATPase. Biochim. Biophys. Acta 314 (1973) 149-153.

4. F. Franks, D.J.G. Ives,
The structural properties of alcohol-water mixtures.
Quart. Rev. 20 (1966) 1-44.

5. D.D. Perrin,
Dissociation constants of organic bases in aqueous solutions. Butterworths, London (1965) p. 408.

6. R.C. Phillips, S.J.,Philip George, R.J. Rutman,
Thermodynamic studies of the formation and ionization of the magnesium (II) complexes of ADP and ATP over the pH range 5-9.
J. Am. Chem. Soc. 88 (1966) 2631-2640.

7. D. Jaworek, W. Graber, H.U. Bergmeyer in:
Methods of Enzymatic Analysis (ed. H.U. Bergmeyer)
Academic Press. (1974) p. 2127-2131.

Thin-Layer Polyacrylamide Gel Isotachophoresis, Crossed Immuno-Isotachophoresis and Crossed Isotachophoresis-Electrofocusing

C.H. Brogren

ABSTRACT

A simple analytical isotachophoretic method performed in a thin-layer polyacrylamide gel was developed with the purpose of combining this separation with transverse electromigration into an antibody containing gel or with thin-layer polyacrylamide gel isoelectric focusing. These methods provide a new tool for analysis of proteins, allow a study of the principles of protein separation by isotachophoresis and are an easy analytical way to predict the isotachophoretic separation of proteins on a preparative scale.

INTRODUCTION

A two dimensional analysis of proteins is most easily done when the separation in the first dimension is performed on a thin-layer supportive medium. Strips from this first dimension can then be cut out and transferred to the medium used for the second dimension. Söderholm and Smyth (1) have shown how easily IEF in a thin-layer PAG can be combined with immunochemical detection using the technique of crossed-immunoelectrofocusing. The problem of combining PAG with agarose gel was solved by using the so-called laying-on technique (1, 2). ITP in agarose gel layers has been performed by Uyttendaele and co-workers (3, 4), but because of the problems of electroendoosmosis, we have not found agarose a suitable supportive medium for ITP and have therefore chosen PAG.

ABBREVIATIONS: ITP, isotachophoresis; IEF, isoelectric focusing; CIITP, crossed immuno-isotachophoresis; CITPIEF, crossed isotachophoresis-electrofocusing;PAG, polyacrylamide gel.

+) Present address: Immunological Department, Pathological-Anatomical Institute, DK-1399 Copenhagen K, Denmark.

++) Sponsored by The Danish Cancer Society.

Electromigration of the ITP separated proteins into an antibody containing gel allows a specific, quantitative, and sensitive immunochemical detection of the proteins. When multispecific antibodies are used the method is multianalytic with a high resolution both in the 1st dimension ITP separation and the 2nd dimension immunochemical detection.

ITP followed by transverse IEF should allow an investigation of the differences between the separation principles of these two methods. In ITP, the degree of protein separation can be varied by changing the mobility of the leading ion (5,6) and in IEF, the separation of proteins occurs until the mobility is zero, a case of "ITP" at velocity 0 (7). A combined ITP-IEF separation may therefore allow separation of proteins of similar mobility under ITP separation (by IEF) and of proteins of similar isoelectric point under IEF separation (by ITP). The present studies provide experimental data for such separations.

MATERIALS AND METHODS

Human serum was obtained from local donors and was used as sample to be separated in some studies, and a bovine thymus extract as the sample in other studies. Ampholines (LKB-Produkter, Ltd., Bromma, Sweden) were applied as spacers.

1. Thin-layer PAG ITP. The buffer systems and PAG were the same as those used for preparative column experiments (8). The PAG was prepared in the leadin buffer (5), either between two glass plates (11 x 20.5 cm) or between a glass and a plexiglass plate, spaced by 1 mm using a rubber strip along the edges. After the polymerization one of the plates was removed. The gel containing plate was placed on a glass cooling plate (Wicklunds Glasindustri, Stockholm, Sweden) as used by Söderholm and Wadström (9,10). The temperature was maintained at 5° C. Buffer vessels were placed at the short borders of the cooling plate, with leading electrolyte as anode buffer and terminating electrolyte as cathode buffer. Each of the vessels contained at least 1 liter. A larger volume in the anode vessel is preferable to avoid any drastic fall in pH during the run. The gel and buffer vessels were connected with 8 layers of Whatman No. 1 paper as wicks. Care must be taken to avoid any contamination of the terminating electrolyte with leading buffer. The samples applied have been human serum (50-200 µl) with added Ampholine, 40% (50-400 µl) as spacers. In other experiments bovine thymic extract was applied without added spacers (since

the extract contained nucleic acids which gave a spacing effect) or with added Ampholines as spacers. Before the sample was applied, the terminating ion was migrated 1-2 cm into the gel by applying a current of 10 Watt for ½-1 hour. The sample, on a strip of Whatman No. 3 filter paper (normally 1 x 10 cm, but with a larger broadness in case of large sample volumes) was applied transversely onto that part of the gel now containing the terminating ion. This means that only one sample was applied to each plate. The electrophoretic migration was then continued at constant power (5-10 W preferred) until bromophenol blue added as marker to the sample, reached the anodic wick (after 4-8 hours at 5 W). The best results were obtained if the application filter paper was removed about 1 hour before the end of the run, since small amounts of proteins sticking to the paper can disturb the separation and can give a tailing effect. Depending on the kind of analysis, the plate can be stained directly, or a strip of 0.5 cm can be cut out and transferred to an agarose gel as described by Söderholm et al. (1 , 2) or to another PAG. The whole plate or the remainder of the gel on the plate was stained for proteins by the combined fixation-staining method bescribed by Vesterberg (10). Specific staining methods for enzymes (11) and specific detection of antibodies (12) have been tried successfully.

2. Crossed immuno-ITP. The procedure of Söderholm et al. (1, 2) for crossed immuno-IEF was followed. Whether the strip was placed upside-down or not on the agarose layer did not affect the results. A lengthwise strip (½-1 x 15 cm) starting from the anodic wick of the ITP gel was cut with a scalpel and transferred using a long razor blade to a part of the agarose gel free of antibodies. An electromigration was then performed at 10 Volt/cm (or 2 Volt/cm for thymic extract studies) at right angles to the strip for 4 hours (or 18 hours) into an antibody containing agarose gel layer. The high voltage was used to avoid loss of resolution due to diffusion. The PAG strip was removed and the agarose gel pressed, washed and stained as descibed for other quantitative immunoelectrophoretic techniques (13). The antibodies used in these studies were rabbit-anti human serum proteins (Code No. 100SF, Dakopatts, Ltd., Copenhagen, Denmark) and rabbit-anti bovine thymic extract antibodies delivered on contract immunization from Dakopatts Ltd. Agarose type, buffer and the esterase staining procedure were as described earlier (11).

3. Crossed ITP-IEF . This technique combines the high resolution of both ITP and IEF. A strip from the ITP PAG was transferred, using a long razor blade, to the cathodic side of a thin-layer (0.5 mm) PAG to be used for IEF separation (12). An Ampholine with a pH-gradient from 3.5-10 was used for the IEF separation. The procedure for focusing was as described by Peltre and Brogren (14). The staining of proteins was as described by Vesterberg (10).

RESULTS AND DISCUSSION

In the studies presented here, a method of thin-layer ITP is described. This method is essential for the development of the techniques of CIITP and CITPIEF, allows the specific detection of proteins separated by ITP, and will allow study of the pH gradient which forms during ITP using a surface electrode (15).

Fig. 1 shows a thin-layer ITP fractionation of a thymic extract without added spacers. The protein staining revealed clearly separated bands and these must be spaced by natural components in the sample. Since the protein bands seen are located near the terminating zone such natural "spacers" could be nucleic acids, which we know are present in high amounts in this thymic extract. A separation of the sample with Ampholine as spacers is seen in Fig. 4.

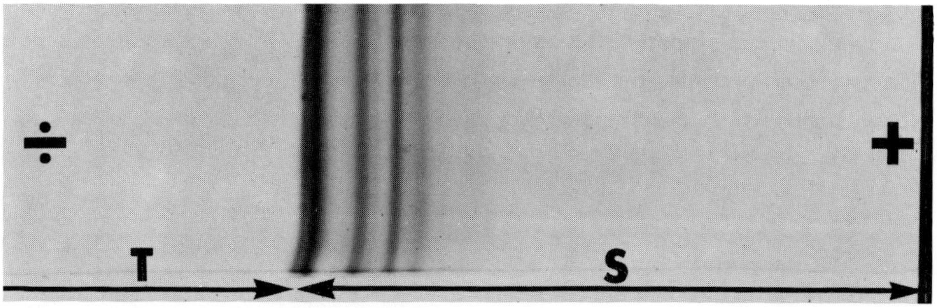

Fig. 1. Thin-layer PAG ITP analysis of bovine thymic extract without added spacers. The ITP separation was performed with 30 mM phosphate, 33 mM Tris, 2 mM TEMED, pH 6.25 (8) as leading buffer, and 230 mM ε-amino caproic acid, 12 mM Tris, pH 8.9 as terminating buffer. A 200 µl sample containing 10 mg dry weight of extract (this corresponded to 4 mg protein) was applied on a filter paper strip to the gel near the cathodic wick (left) and the separation performed by applying 10 W constant electric power for 4 hrs. In this experiment 10-fold concentrated leading buffer was used as anode buffer. The plate was stained for proteins. About half of the plate is shown. The leading/sample zone boundary is indicated to the right with a black line (position of the bromophenol blue marker). Note the spaced bands due to the presence of nucleic acids in the extract. Sample zone (S) and terminating zone (T) are indicated.

The leading/sample zone boundary was visualized using bromophenol blue as marker and the sample/terminating zone boundary was indicated by a yellow line (due to riboflavin or impurities in the gel). The ITP separation was considered to be at equilibrium when (i) the distance between the boundaries was constant and was no more than one half to two-thirds of the plate length, and (ii) when the profile obtained subsequently with CIITP showed separated and regular peaks without tailing. It was necessary to apply the spacers across the whole plate rather than localized with the sample on small pieces of filter paper, otherwise the proteins separated as "moon-shaped" bands. If several different samples were applied to the same plate, as done in IEF analysis, diffusion across the plate occurred during the run. This problem could be overcome if narrow strips of the large gel were cut away between the samples. Alternatively, different samples could be analysed on separate narrow plates placed on the same cooling plate.

Thin-layer ITP separation of human serum proteins is shown in Fig. 2 together with the CIITP analysis to illustrate the separation obtained with 50 ul Ampholine pH 5-8 as spacers in ITP. The peaks were more narrow than seen in a crossed immunoelectrophoretic analysis (13).

Fig. 3 illustrates the effect of increasing the amount of Ampholine spacers added to the sample. A pronounced spacing in the middle of the sample zone was seen. This indicates that the spacing was not equal in all zones.

Fig. 4 shows all the techniques described in this paper for an analysis of a thymic extract. The narrow strip in the middle (Fig. 4B) is the strip transferred to the second dimension analysis. CIITP (Fig. 4A) and CITPIEF (Fig. 4C) are shown. As seen in Fig. 4A the immunochemical analysis did not show all the proteins present in the sample, since there was no antibodies against some of them. The two pronounced basic proteins seen as "strong" bands in the ITP separation could therefore only be analysed by non-immunochemical methods such as e.g. IEF. The ITP-IEF two-dimensional analysis showed excellently the differences in the separation principles of ITP and IEF. In the Figure bands with nearly the same isoelectric points have been separated by isotachophoresis (vertical arrows) and bands with different isoelectric points have been separated from a single position in the ITP fractionation (horizontal arrows). This combined ITP-IEF analysis has also been applied to antibody fractionation (Fig. 7 in ref. 16). In this case the proteins are

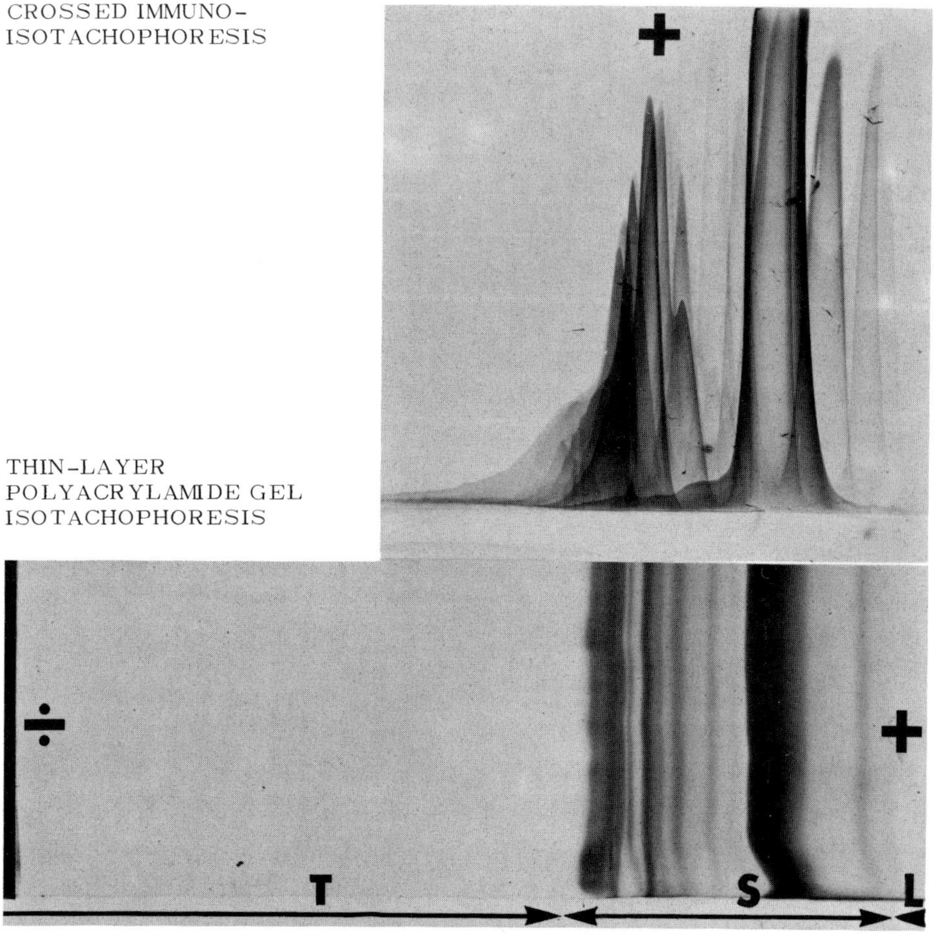

Fig. 2. Thin-layer PAG ITP and CIITP analysis of human serum proteins. A. The thin-layer PAG ITP was performed with 30 mM phosphate, 33 mM Tris. 2 mM TEMED, pH 6.25 as leading buffer, and 230 mM ε-amino caproic acid, 12 mM Tris, pH 8.9 as terminating buffer (8). A sample of 50 µl human serum with 50 µl Ampholine pH 5-8 was applied on a filter paper strip at the black line to the left. The migration was then performed for 6 hrs. at 5 W. The plate was fixed and stained for proteins. Note the sharp bands with albumin as the main band. Terminating zone is indicated with T and leading zone with L. At the bottom edge-disturdance is seen. Only half of the plate is shown. B. The CIITP was performed with a 1 cm broad strip (corresponsing to 5 µl human serum) from the plate in A using 20 $\mu l/cm^2$ in the second dimension of rabbit-anti human serum proteins antibodies (Code No. 100SF, Dakopatts, Ltd., Copenhagen, Denmark). The electromigration into the antibody-containing gel was done at 10 Volt/cm for 4 hrs.. The plate was stained for esterases and for proteins. Note the narrow peaks.

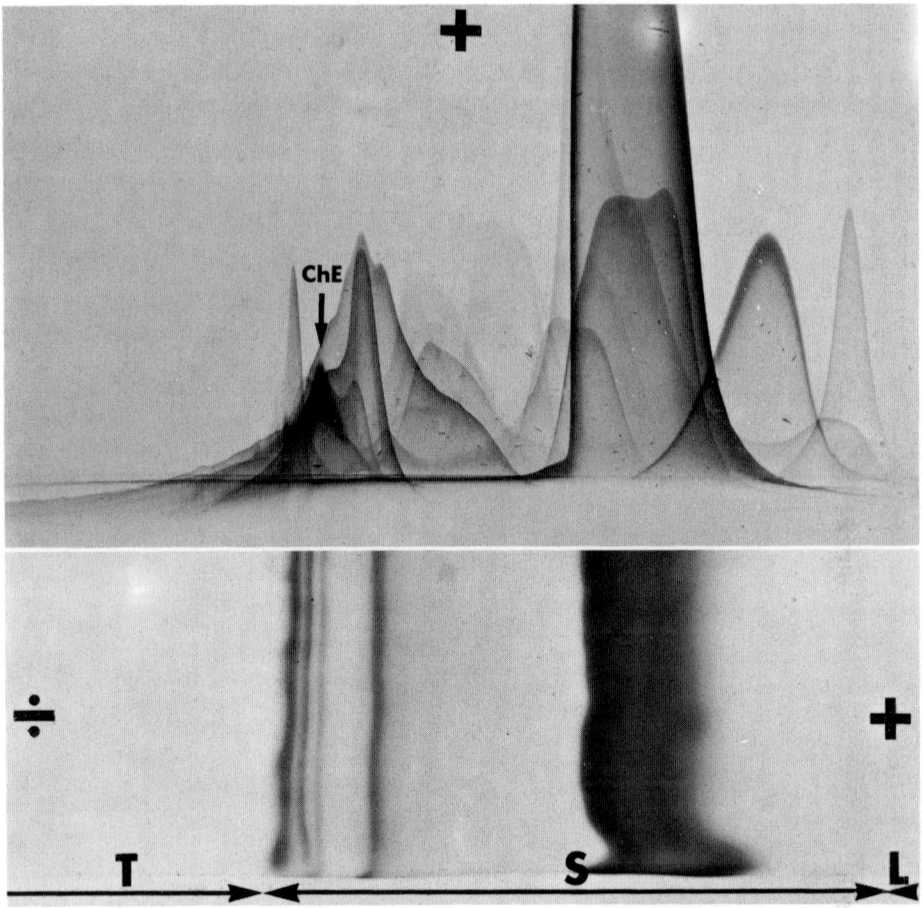

Fig. 3. Thin-layer PAG ITP and CIITP analysis of human serum proteins.
A. The thin-layer PAG ITP was done using the same buffer conditions as in Fig. 2. The sample was 50 µl human serum with 100 µl Ampholine pH 5-8 added as spacers. This was twice the amount of spacer used in the experiment shown in Fig. 2, and gives more spacing between the protein bands and gives broader bands (see the albumin band). The length of the sample zone was found to be twice the length of the sample zone in the experiment from Fig. 2. The sample zone fill up 2/3 of the plate length. The leading zone (L), sample zone (S), and terminating zone (T) are indicated. At the bottom the edge of the polyacrylamide gel is seen. Only one half of the plate is shown.
B. A CIITP analysis of the separation shown in Fig. 3A. The antibody concentration and other conditions were as for Fig. 2B. Compare the peaks with the bands seen in A. The cholinesterase peak is indicated with an arrow. Compare this analysis with the analysis seen in Fig. 2B.

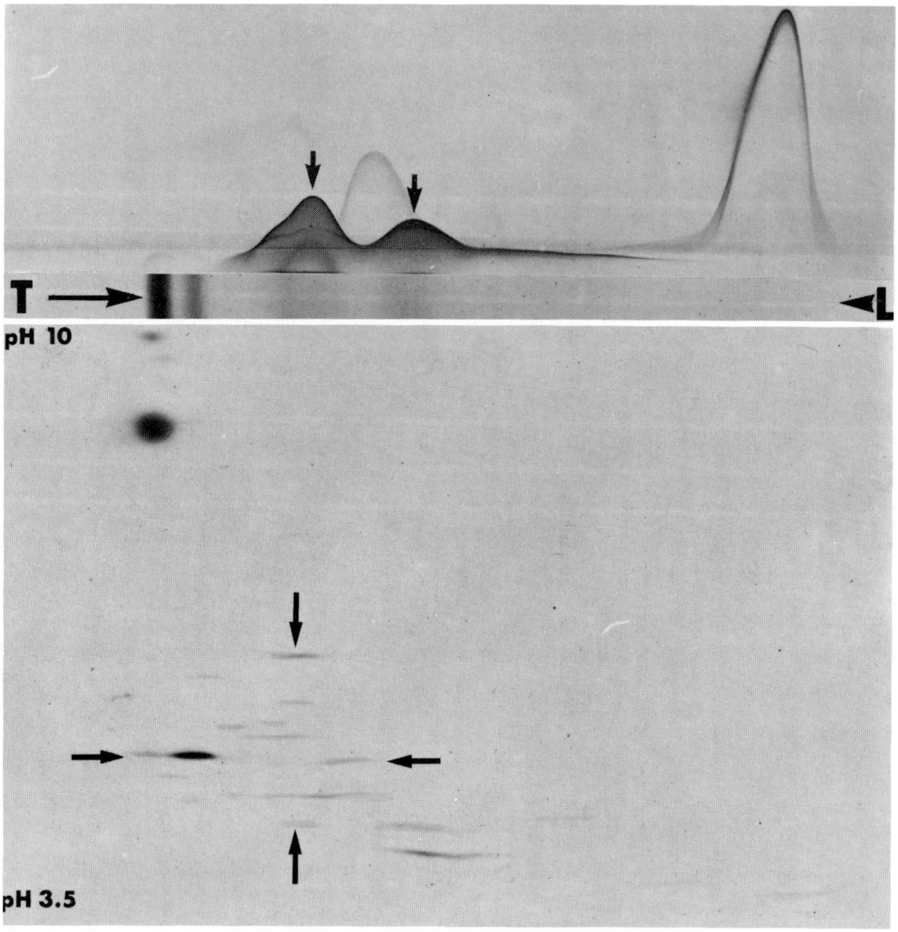

Fig. 4. Thin-layer PAG ITP, CIITP, and CITPIEF analysis of bovine thymic extract with Ampholine added as spacers for the ITP separation.
A. The ITP separation was performed under the same conditions as in Fig. 1, but with a sample of 5 mg extract in 100 µl with 100 µl Ampholine pH 5-8 added as spacers. A strip of the gel (0.5 x 10 cm) corresponding to 0.1 mg protein was applied to the CIITP and the CITPIEF plate. Such a strip is seen in the Figure. Leading (L) and terminating zones are indicated.
B. The CIITP analysis was performed with 28 µl/cm^2 of an rabbit-antiserum against thymic extract incorporated in the second dimension gel. The electromigration into the second dimension gel was performed with only 2 Volt/cm for 20 hrs., since higher voltage affected the precipitation. Note that the two bands seen to the left in the ITP strip did not form precipitates, and that a distinct peak is seen to the right in the CIITP plate, where no bands were visible in the ITP strip. Two proteins which separated in PAG ITP showed immunochemical identity in the CIITP analysis (indicated by arrows).
C. A CITPIEF analysis performed with an IEF pH-gradient from pH 3.5-10 (indicated). See the text for explanation of the picture.

not separated in discrete bands as seen for the thymic extract, but are separated in broader, closely associated bands presumably due to the similarity of the antibody molecules.

In conclusion, the thin-layer ITP was extremely easy to perform, inexpensive to work with, and does not need special and expensive equipment. Because of the problems of applying the sample on filter paper strips, other kinds of application methods will have to be investigated. Alternatively, if the thin-layer PAG ITP is performed in a vertical apparatus, the sample can be directly applied on top of the gel. We found it preferable to use the concentrated buffer systems, normally used for preparative separations,in this analytical method,since they gave better pH stability and better reproducibility in the separations. Particularly, a high buffer capacity is recommended for the anode buffer, which can be obtained by use of large volumes of anode buffer or more concentrated buffer than used for the leading electrolyte (cf. Fig. 1).

ACKNOWLEDGEMENTS

I want to thank Dr. J. Kieler and the whole staff of The Fibiger Laboratory for giving me excellent working facilities during these studies. I am indebted also to Dr. G. Peltre, Institut Pasteur, Paris, for initiating and discussing these methodological studies during my stay at the institute in 1975. Mrs. A. Hansen is acknowledged for her skillful technical assistance and independent performance in some of these experiments during my absence in summer 1976. I am most grateful to Dr. H. Warren for her patience in correcting the english in this manuscript, and Mrs. G. Samuelsen for her secretarial help. NOVO Industri A/S has supported part of this work.

REFERENCES

1. Söderholm, J. and Smyth, C.J.: Crossed immunoelectrofocusing for studies on protein microheterogeneity. In: Progress in Isoelectric Focusing and Isotachophoresis. Righetti, P.G., Ed. (ASP Biol. Med. Press B.V., Amsterdam), pp. 99-114, 1975.
2. Söderholm, J., Smyth, C.J., and Wadström, T.- A simple and reproduciple method for crossed immunoelectrofocusing. Scand. J. Immunol. 4, suppl. 2, 107-113 (1975).
3. Uyttendaele, K., De Groote, M., Blaton, V., and Peeters, H.: Detection of traces of proteins by isotachophoresis. Prot. Biol. Fluids 22, 743-747, (1975).
4. Uyttendaele, K., Blaton, V., Alexander, F., Peeters, H., De Groote, M., Vinaimont-Vandecasteele, N., and Chevalier, J.: Agarose isotachophoresis

of human sweat. In: Progress in Isoelectric Focusing and Isotachophoresis. Righetti, P.G., Ed. (ASP Biol. Med. Press B.V., Amsterdam), pp. 341-346, 1975.

5. Svendsen, P.J. and Rose, C.: Separation of proteins using Ampholine carrier ampholytes as buffer and spacer ions in an isotachophoresis system. Science Tools 17, 13-17 (1970).
6. Brogren, C.-H. and Svendsen, P.J.: Preparative isotachophoresis in combination with biospecific interaction and neuraminidase treatment in purification of human serum cholinesterase. Prot. Biol. Fluids 22, 685-689 (1975)
7. Chrambach, A and Nguyen, N.Y.: Electrofocusing as isotachophoresis. These proceedings.
8. Winter, A., Hjalmarsson, S.-G., and Karlsson, C.: Preparative isotachophoresis with the LKB 7900 Uniphor. I. Description of a standardized method. LKB Application Note 146 (1974).
9. Söderholm, J. and Wadström, T.: High voltage flat-bed isoelectric focusing in polyacrylamide gel with automatic constant wattage. In: Isoelectric Focusing. Arbuthnott, J.P. and Beeley, J.A., Eds. (Butterworth, London), pp.132, 1975.
10. Vesterberg, O.: Isoelectric focusing of proteins in polyacrylamide gels. Biochim. Biophys. Acta 257, 11-19 (1972).
11. Brogren, C.-H. and Bøg-Hansen, T.C. Enzyme characterization in quantitative immunoelectrophoresis. An enzymological study of human serum esterases. Scand. J. Immunol. 4, 37-51 (1975).
12. Peltre, G.: Immuno-isoelectric focusing study of the evolution of antibodies to pig lactate dehydrogenase produced in rabbits. In: Progress in Isoelectric Focusing and Isotachophoresis. Righetti, P.G., Ed. (ASP Biol. Med. Press B.V., Amsterdam), pp. 173-178, 1975.
13. Axelsen, N.H., Krøll, J., and Weeke, B.. Eds.: A manual of quantitative immunoelectrophoresis. Methods and applications. Scand. J. Immunol. 2, suppl. 1 (1973).
14. Peltre, G. and Brogren, C.-H.: Immuno-isoelectric focusing analysis of antibodies fractionated by isotachophoresis. These proceedings.
15. Gianazza, E., Righetti, P.G., Bordi, S., and Papeschi, G.: pH determination in isoelectric focusing with an iridium electrode. These proceedings.
16. Brogren, C.-H. and Peltre, G.: Analytical and preparative isotachophoresis in column and flat-bed gels for fractionation of antibodies. These proceedings.

Isotachophoresis of the Cerebrospinal Fluid (CSF) Proteins: Advantage of Prior Dialysis and its Practical Realization

P. Delmotte

ABSTRACT

In order to extract maximum profit out of the analysis of the proteins of CSF by isotachophoresis, the samples must be injected without prior concentration. All concentration techniques are time consuming and have an inherent risque of protein loss and denaturation. Unfortunately, due to its very low protein content, large sample volumes of CSF must be injected. However, the physiological concentrations of the small ions in the CSF cause an enormous loss of analysis time and, more important, a loss of resolution. A practical set-up, enabling the dialysis of CSF samples of 0.5 to 1 ml in about 7 minutes is presented. Of the small ions, Cl^-, Na^+ and K^+, the concentrations are reduced to about 4, 30 and 15 percent respectively. There is practically no loss of protein and a new sample can be dialyzed every 3 to 4 minutes. On isotachophoresis, the current needed to transport the small ions faster than the leading ion, is reduced to about 1/4. Total analysis time falls from more than 40 to about 25 minutes, with a clear cut gain in resolution.

INTRODUCTION

Isotachophoretic separation of complex protein mixtures on an analytical scale is a promising technique. Due to its very sensitive detection of separated protein bands by UV absorption, this technique is suitable for studying very diluted protein solutions. The importance of protein profile of the CSF in neurology needs not to be stressed again. Unfortunately, most of the methods used today to study this protein profile can only be applied to previously concentrated CSF. Although some concentrating techniques inflict minimal damage to the proteins, no absolute control over denaturation or loss exists and these procedures are all very time consuming.

In the light of these considerations, we have been experimenting with analytical isotachophoresis to study the protein profile of the CSF. It was soon evident that, due to the extremely low level of some individual protein fractions in the CSF, fairly large amounts of fluid had to be injected, so as to arrive at a minimal UV absorption.

However, these large volumes of CSF bring also into the isotachophoretic system large amounts of small ions and these have a deleterious effect on the isotachophoretic separation: Total separation time is considerably prolonged; the separation capability is greatly reduced.

Prior removal of the major part of these small ions is an absolute necessity. Ordinary dialysis, in the best conditions, is very difficult to apply to small volumes and due to its duration brings about a risk of denaturation of the proteins. This paper describes a dialysis technique which is very rapid and can be applied to small volumes of dilute protein solutions.

RESULTS AND DISCUSSION

As can be seen from Fig. 1, the system makes use of standard laboratory equipment: a Technicon Autoanalyzer dialysis unit and proportioning pump, combined with a LKB Uvicord II UV absorption meter. Layout and pumping tube values are evident from Fig. 1. As it was imperative that the protein concentration of the CSF remained at least at its original level, the stream of dialyzing fluid is continuously held at a constant positive pressure by keeping the fluid between two different pumping tube values. After passing through the dialyzer, the fluid stream is only partly debubbled, so that some of the air bubbles pass through the rest of the system. This arrangement keeps the samples better separated and exercise a cleaning effect on the tubings. Technicon Cuprophan dialyzing membranes are used and the dialyzing system is kept submerged in water at ordinary temperature. Due to the high speed counterflow, maximum dialysis during the short

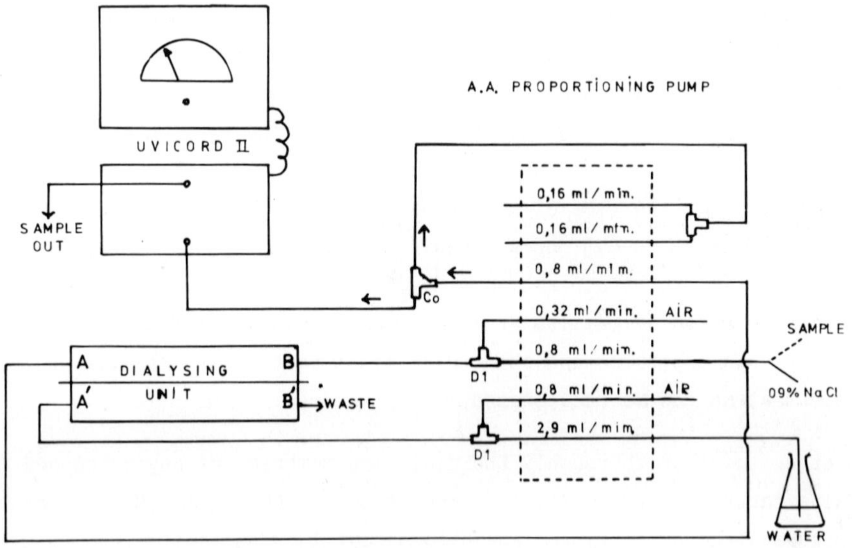

Fig. 1: Dialysis set-up including autoanalyzer dialysis unit and proportioning pump combined with LKB Uvicord II.

period of time is attained.

The system is equilibrated by continuously aspirating physiological salt solution and counterflow dialyzing water. At convenient moments, sample is aspirated during a short period of time. The passage of the fluid sec-

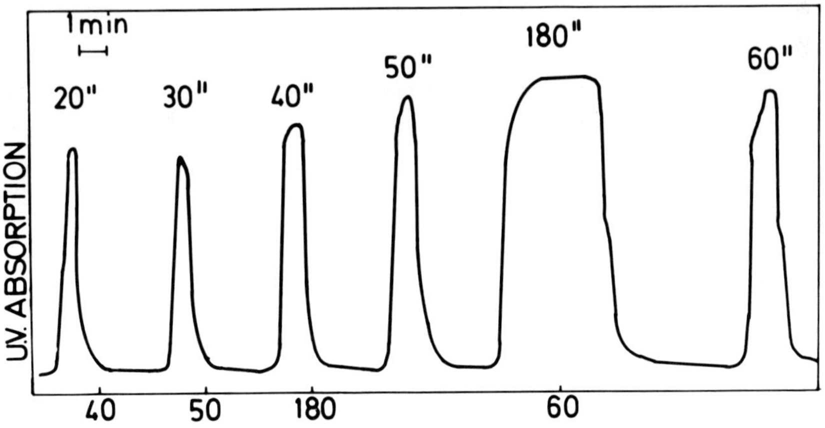

Fig. 2: Dialysis of CSF symple: Effect of sampling time.

tion corresponding to the sample is monitored with the Uvicord absorption meter. The output of the Uvicord is connected to a simple millivolt- or milliamperemeter. The occasional passage of an air bubble through the absorption cell is of no inconvenience to detect the passage of sample. Depending on the length of interconnecting tubing, which is best kept a strict minimum, the time of sample passage through the system is about 7 minutes. Every 3 to 4 minutes a new sample can be aspirated. As can be seen from Fig. 2, an aspiration time of 50 to 60 seconds is enough to attain a plateau of maximum UV absorption. During this time, about 0.75 ml of sample is aspirated and about 0.25 ml of sample with maximum protein concentration can be recuperated.

In order to clean thoroughly the dialyzing membrane of any entrapped material after the last sample, the connections at A - A' and B - B' are reversed and the pump started again for about 10 minutes.

	PROTEIN		Cl^-		Na^+		K^+		GLUCOSE	
	B.	A.	B.	A.	B.	A.	B.	A.	B.	A.
1	53	54	122	<4	155	34	2,95	0,4	126	68
2	40	39	119	<4	153	40	2,95	0,4	61	32
3	34	34	117	<4	150	34	2,9	0,45	39	20
4	41	42	120	<4	146	33	2,9	0,4	48	24

Table 1: Results of 4 different CSF samples. For details see the text.

Table 1 shows typical results obtained with 4 different samples of CSF. The mean and standard deviation of the percent of measured compounds remaining after dialysis is shown in table 2. There is no significant protein loss and for chlorine, sodium and potassium, there remains respectively less than 4, 28 and 15 percent of the original amounts.

Table 2: DIALYSIS OF 20 CSF SAMPLES

(Results as percent of conc. before dialysis)

Total protein:	99.7 (± 3.7)
Chlorine :	4
Sodium :	27.7 (± 3.8)
Potassium :	14.8 (± 1.9)

To evaluate any selective protein loss during dialysis, we concentrated samples of CSF without and after dialysis. The concentrated samples were submitted to quantitative cellulose acetate electrophoresis. Typical results are shown in table 3, and as can be seen there is no significant difference between the two samples.

Table 3: ELECTROPHORESIS OF DIALYZED CSF

	Before	After
Pre-alb.	3.7 %	3.8 %
Albumine	59.9 %	58.7 %
alpha-1-glob.	3.9 %	3.6 %
alpha-2-glob.	5.5 %	5.8 %
beta-glob.	10.8 %	10.7 %
gamma-glob.	16.2 %	17.4 %

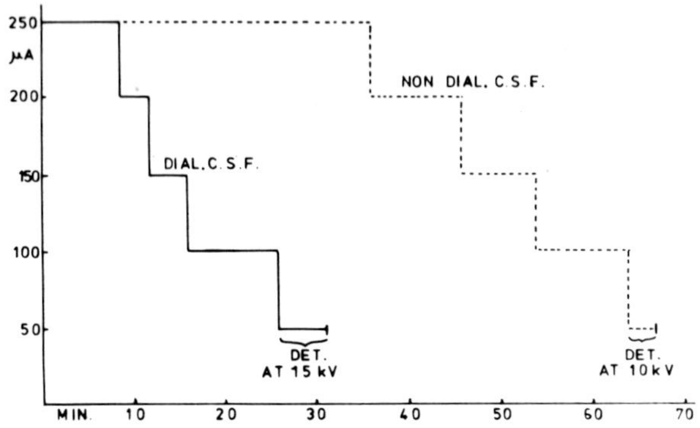

Fig. 3: Isotachophoresis of CFS current versus time. Comparison of dialyzed and non-dialyzed samples.

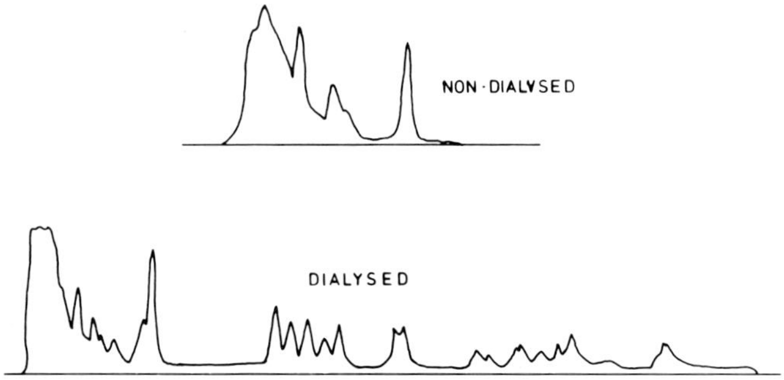

Fig. 4: Comparison of isotachophoresis patterns of cerebrospinal fluid before and after dialysis (sample application: 30 µl each).

The advantage of dialysis of CSF before isotachophoresis is evident from Fig. 3, which presents the plot of current versus time for native and dialysed CSF. Total separation time is roughly halved and the voltage at detection lies about 50 % higher with the dialyzed sample.

We cannot go into lenghty comparison between separation profiles obtained with native and dialyzed samples, but Fig. 4 shows typical results for one sample of CSF. It must be stressed that only with a dialyzed sample of CSF a protein profile can be obtained in which most of the protein fractions can be identified with the same fractions in the corresponding serum.

This work was supported by grant no 76/1 from the Belgian National Centre for Multiple Sclerosis.

2. Applications

Plasma Protein Fractionation in Sephadex by Isotachophoresis Using Discrete Spacers

M. Bier and A. Kopwillem

ABSTRACT

The use of Sephadex G-200 as stabilizing medium for preparative isotachophoresis has been investigated and the resolution was compared to that obtained in the customarily used polyacrylamide gels. Human plasma proteins were fractionated in the preparative LKB Uniphor column with bicine, asparagine and glutamine as discrete spacers. Using n spacers, n + 1 defined protein subgroups were obtained. The spacing characteristics of the ions mentioned were found to be dependent of the pH in the leading electrolyte, as clearly demonstrated by the immunoelectrophoretic analysis of ten specific proteins. It has also been shown that narrow mobility protein bands can be obtained using two spacers which isolate from human plasma 2 major and a few minor proteins. The immunoelectrophoresis shows that the majority of the proteins were sharply resolved indicating resolution compared to that obtained in polyacrylamide gels. Protein quantities ranging from 200 - 2000 mg have been applied in the fractionations.

INTRODUCTION

Isotachophoresis (ITP) is characterized by the use of a discontinuous buffer system where the sample is separated within zones formed between a high mobility leading ion electrolyte and a low mobility terminating electrolyte, using a common counterion (1-3). The sample components separate in order of net mobilities (i.e. degree of dissociation x ionic mobility, for small ions) and form zones of constant concentration, as first derived by Kohlrausch (4) in 1897). The concentrations and pH in the separated zones are regulated by the concentration and pH of the leading electrolyte. Once steady state is reached, the zone boundaries remain sharp due to self stabilization from the discontinuous electrical field created in the system (5). ITP has been applied extensively for the analysis of low molecular weight ions (1-3, 6-10) and the sharpness of separated zones and their high concentration makes it particularly appealing for preparative protein fractionation (11-17).

Because of large density gradients created by the sharp sample boundaries, stabilizing media are needed and polyacrylamide gels have been routinely used for preparative work (11-18). Unpredictable distortions of boundaries are frequently obtained, however, and their cause is not fully understood (18). A quest for alternate stabilizing media seemed desirable, therefore, and preliminary studies in this laboratory, using colored proteins, showed that good separations were obtained when using Sephadex for stabilization. While Sephadex G-10 and G-25 were unsuitable, G-50 to G-200 seemed to be comparable and G-200 was arbitrarily chosen for this study.

In an earlier study on polyacrylamide (18) we have shown the usefulness of amino acids as discrete spacers in ITP, in lieu of the customarily used Ampholine which provides a continuous mobility spectrum. It was found that using n discrete spacers, most serum proteins clearly separate into n + 1 mobility subgroups. Some proteins, however, were found to be present in more than one subgroup, presumably due to protein-protein interaction. The principal objective of the present study was to compare the resolution of proteins on Sephadex, to that previously obtained in polyacrylamide. It was anticipated that the different sieving characteristics of the two media would affect the relative position of spacers and proteins. A further objective of this study was to define the spacing characteristics of some individual spacers under different electrolyte conditions.

MATERIALS AND METHODS

Reagents: Cacodylic acid, tris (hydroxymethyl) aminomethane (TRIS), $Ba(OH)_2$, and sucrose were obtained from Fischer Scientific Company, Sephadex G-200 from Pharmacia Fine Chemicals, amino acids from Sigma Chemical Co., NN-bis-(2-hydroxyethyl)-glycine (bicine) from BDH Chemicals Ltd., agarose (Indubiose A45) from L'Industrie Biologique Francaise, S.A., and human antisera (rabbit) from Behring Diagnostics and Dakopatts A/S.

Sample: Either serum or outdated plasma was used to prepare an albumin poor protein fraction. This fraction was prepared by precipitation at 50% saturation of ammonium sulfate. The precipitate was dissolved in water and dialyzed against the leader buffer.

Preparative ITP in Sephadex: The separations were performed in the LKB Uniphor apparatus, with Sephadex G-200 as stabilizing medium. The 40 cm long column had a 2.5 cm I.D., and was kept at $12^{\circ}C$. Sephadex was equilibrated in the leading electrolyte and poured to an average column height of 20 cm. The spacing properties of asparagine, bicine and glutamine were investigated, these spacers covering the mobility range of plasma proteins. The leading electrolyte was 20 mM cacodylic acid, adjusted to pH 7.1 - 8.1 with TRIS (experiment 1-5a). In experiment 5b the leader was 20 mM glutamine adjusted to pH 8.6 with TRIS. Density stabilization of the leader was brought about by the addition of 10% sucrose. The terminator was in experiments 1-4 100 mM β-alanine and 100 mM ε-aminocaproic acid in experiment 5 brought to pH 9.2 with $Ba(OH)_2$. Barium hydroxide was added to precipitate bicarbonate ions whereas no TRIS counterion was necessary in the terminator.

Upon application of the sample, the initial voltage was set to 600V, (approximately 20 mA) to allow migration of the sample into the gel, and was increased to 1200 V after 4 hrs. The current decreased progressively to about 5 mA at 1200 V towards the end of the run. Time for separation was typically 20 hrs until the first protein boundary was eluted. Complete elution required about 7 to 15 hrs, depending on the zone lengths in the column, i.e. quantity of protein and spacers applied. In experiments 1-4 the leading electrolyte was used as elution buffer at a rate of 13-17 ml/hr, depending on the experiment. The eluant was monitored using the LKB 8300 Uvicord at 280 nm, and fractions were collected at 15 min intervals. The sample dilution due to elution was approximately 1:3. In experiment 5, the gel was extruded from the column and 1 cm long segments were collected. The double weight of water was added to the fractions in order to eluate the proteins and the pH and the optical density were measured.

Immunoelectrophoretic analysis: The protein fractions collected from the Uniphor were anlysed for ten specific proteins. (Experiment 1-2 and 5). A modified Laurell (20) rocket technique was used to analyse nine of them. A 5 mM HCl-TRIS buffer (pH 7.4) was used and only 0.1 ml of monospecific antisera/10 ml agarose was used, corresponding to a tenfold decrease of antisera as compared with the original method. IgG was analysed by immunodiffusion as outlined by Mancini (21) using the same buffer as above except that the antiserum concentration was 0.25 ml/10 ml. Calibration curves were established by analysing dilutions of the original plasma protein fraction and the data are reported in Fig. 1-2 as percentage of the original. The fractions from experiment 3 and 4 were analysed by the fused rocket immunoelectrophoresis as described by P.J. Svendsen (22). The veronal buffer contained 0.4 ml whole human antisera per 10 ml agarose.

RESULTS AND DISCUSSION

The spacing characteristics of a number of discrete spacers have been investigated in the Sephadex G-200 media (23), this report will, however, restrict itself to the use of bicine, asparagine and glutamine as spacers. The distribution of the plasma proteins in the presence of these spacers were investigated in two different electrolyte systems. The two first experiments (Fig. 1-2) shows the separations where the leading electrolyte was 20 mM cacodylic acid at pH 7.1 and 8.1 respectively, TRIS being the counterion and β-alanine the terminating ion. The UV recordings of the eluates are presented in the top tracings and the results from the quantitative immunoanalysis are in the bottom graphs.

Of the ten proteins analysed in the experiment reported in Fig. 1, five were sharply separated by the spacers into one subgroup each, albumin being confined to the first protein zone, transferrin, haptoglobin and hemopexin into the second zone and plasminogen into the last protein band. Of the remaining proteins, α_1-antitrypsin migrated mainly with the albumin but was also found as a second fraction in the next protein band, ceuroloplasmin and α_2-macroglobulin both showed tailing. The IgA component was essentially separated into two fractions i.e. into the second and third protein band and similarly the IgG was present in the two last

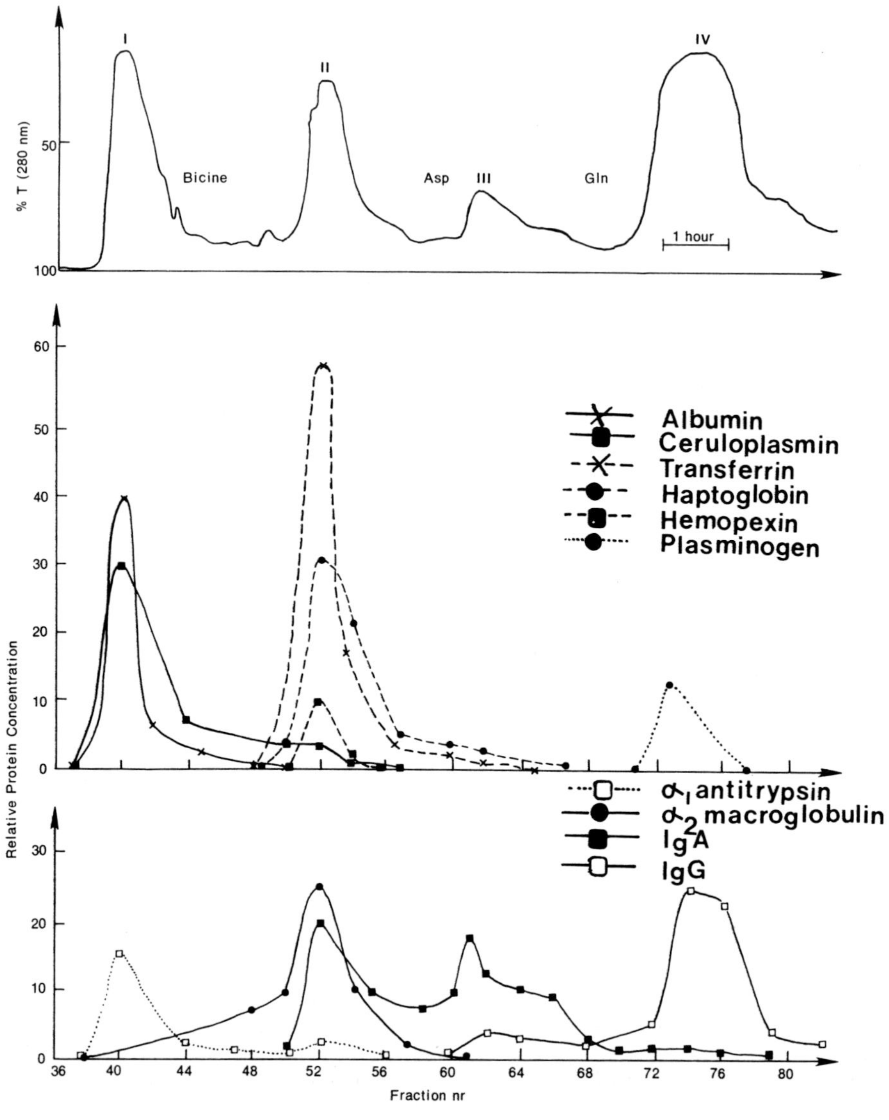

Fig. 1. Isotachophoretic separation of 180 mg of the plasma protein preparation using 240 μmol of bicine, asparagine and glutamine as spacers. The leading electrolyte was 20 mM cacodylic acid adjusted to pH 7.1 with TRIS, the terminator was β-alanine. The top graph reproduces the UV absorption of the column eluate by the Uvicord apparatus, while the two bottom graphs show the reuslts of immunoelectrophoretic analysis of selected fraction. The data are reported as relative protein concentration in the indicicual fractions, with reference to the original plasma protein preparation.

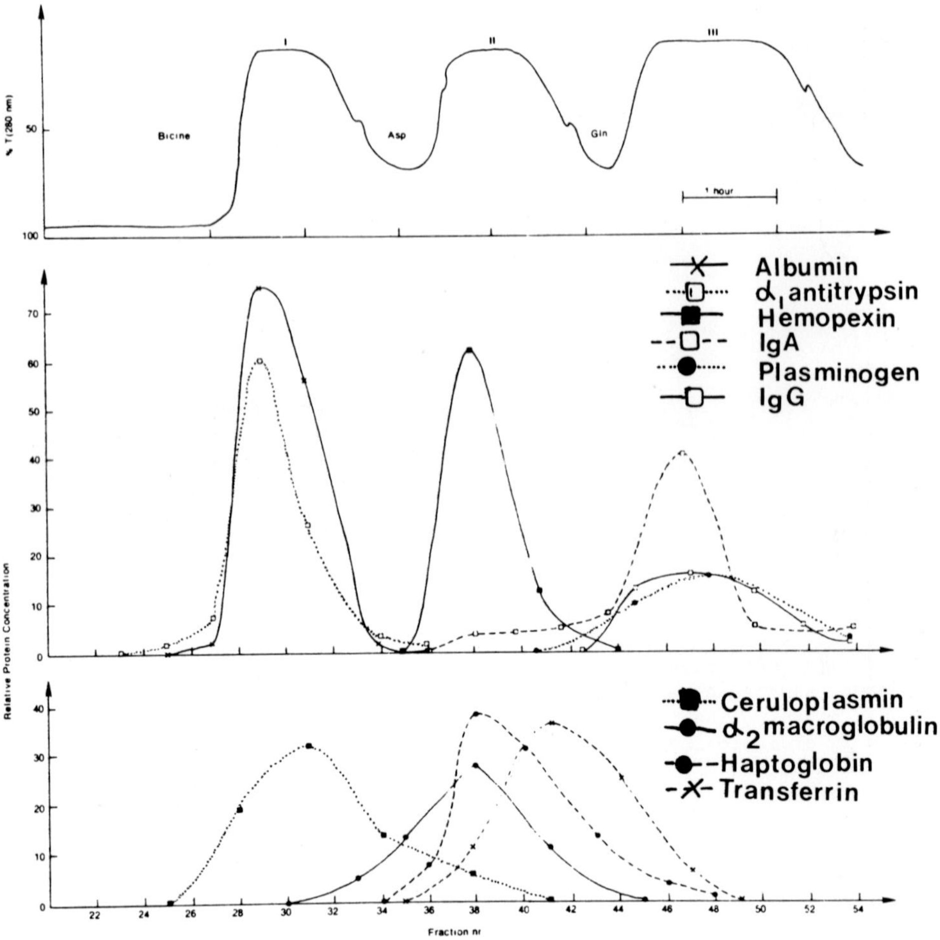

Fig. 2. Separation of 180 mg of the plasma protein fraction with 200 μmol of bicine, asparagine and glutamine as spacers. The leading electrolyte was 20 mM cacodylic acid, adjusted to pH 8.1 with TRIS, the terminator as in previous experiments. UV tracing (top) of the eluate and protein distribution (bottom) from analysis with individual antisera are shown.

protein bands.

A change of the leading electrolyte from pH 7.1 to 8.1 radically changed the characteristics of the system, as it increases the net mobility of the spacers more than the mobility of the proteins. This is shown in Fig. 2, using the same spacers as in experiment 1. Bicine now have higher mobilities than any of the protein components, and migrate adjacent to the cacodylic acid. The behaviour of the two spacers, asparagine and glutamine, at pH 8.1 now show that the six proteins shown in the center graph were well resolved, while there was substantial trailing for the four proteins shown in the bottom graph.

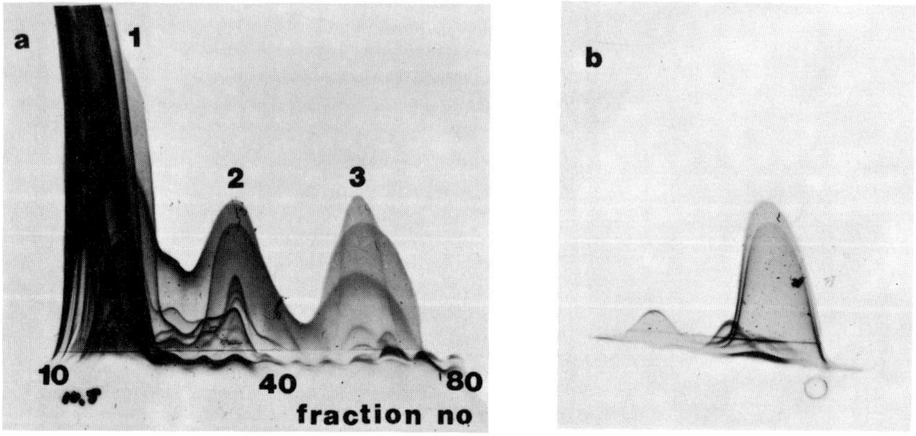

Fig. 3 a) Separation of 500 mg of the plasma protein fraction on the Uniphor column with asparagine and glutamine as spacers. The leading electrolyte was 20 mM cacodylic acid, adjusted to pH 7.1 with TRIS, the terminator was β-alanine. The fractions were analysed by fused rocket immunoelectrophoresis technique. b) The middle protein band in Fig. a) is analysed by crossed immunoelectrophoresis.

It seemed that using the electrolyte system from experiment 1 and the spacers asparagine and glutamine the protein band between these spacers would contain a narrow net mobility region. In order to verify this and also in order to investigate the separation capacity at an increased sample load the experiment was now performed using 500 mg of the plasma proteins and the fractions were analysed against whole human antisera (Fig. 3a). This experiment shows indeed that the bulk part of the plasma proteins migrate in front of the asparagine zone and that very few proteins are contained between the asparagine and glutamine spacers. Crossed immunoelectrophoresis (Fig. 3b) of the middle protein band shows two major proteins with identical mobilities (identified as IgA and IgG) and a few minor components.

The separation of the serum fraction containing two spacers (Fig. 4) at a leading electrolyte pH of 8.1 was analysed with the fused rocket immunoelectrophoresis, which showed that the majority of the proteins were present only in one of the protein bands. Analysis with individual antisera gave results which agreed with those obtained in experiment two. Albumin, α_1- antitrypsin and ceuroloplasmin were in the first protein band, the heavy precipitation line from the first to the second band is from α_2-macroglobulin. The IgA precipitation line reaches from band 2 to band 3, whereas the other analysed main proteins are present in only one of the bands. The length of the spacers zones were appr. 4 cm which explains the difference in apparent resolution compared to experiment two which had spacers zones of less than 2 cm.

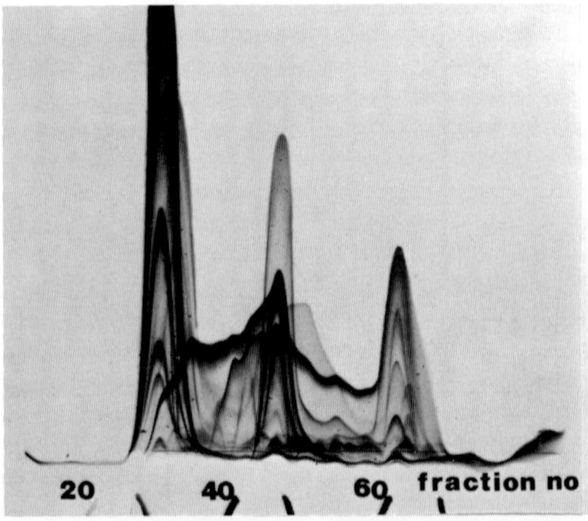

Fig. 4. Separation of 150 mg of the serum protein fraction on the Uniphor with asparagine and glutamine as spacers. The leading electrolyte was 20 mM cacodylic acid adjusted to pH 8.1 with TRIS, the terminator was β-alanine. The fractions were anlysed by fused rocket immunoelectrophoresis.

The consequence of a heavy sample load was investigated by applying 2 g of the plasma protein mixture (Fig. 5). Two different separations were performed, one with glutamine as spacer (left) and one with glutamine as the leading ion (right). The immunoanalysis showed that the last UV peak on the first run contained exclusively plasminogen, IgA and IgG of the analysed proteins. In the second run the proteins except for IgA and IgG should migrate zone-electrophoretically leaving only these proteins in the stack. Small deviations of the pH in the leading electrolyte or too short migration distances could be the cause to that also other proteins remained stacked.

In the present work we have studied the Sephadex G-200 as a replacement for the polyacrylamide gel, because of the greater ease of column preparation and the absence of boundary anomalies. It was hoped that the Sephadex would provide better resolution of all proteins than polyacrylamide due to the absence of the boundary anomalies. This was not the case. It seemed that at comparable operating conditions, (sample load, separation time and distance) comparable resolution was obtained. It has also been shown that pH changes radically changed the separation pattern of a protein mixture in the presence of the individual spacers. It is therefore possible to design a system where the leading electrolyte and two spacers are chosen such that proteins in the narrow mobility region are separated from the bulk protein mixture. Another indication of the applicability of ITP in Sephadex to preparative protein fractionation is that gram quantities of the plasma proteins can be applied to a 25 cm long gel (125 ml) and also at this load no boundary distortions occured.

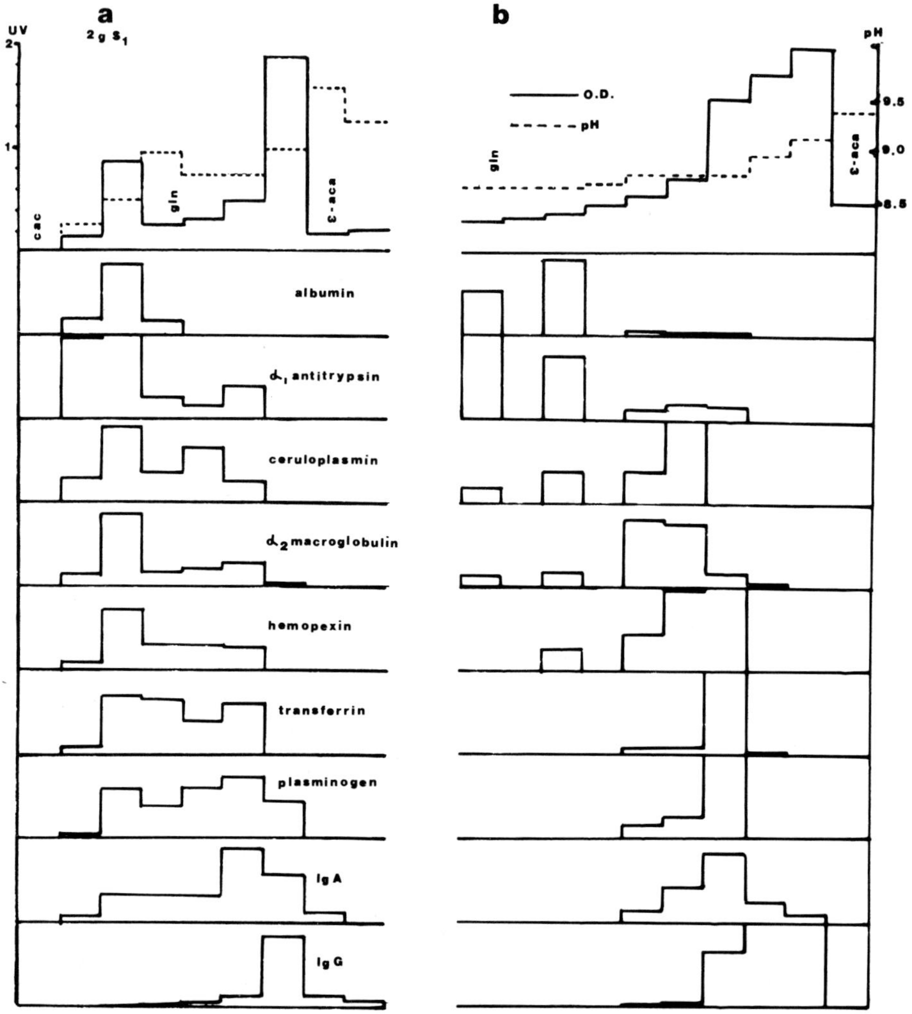

Fig. 5. Separation of 2 g of the plasma protein fraction on the Uniphor with a) 20 mM cacodylic acid at pH 8.1 as leading ion and using as spacer only glutamine (400 µmol) b) 20 mM glutamine at pH 8.6 as leading ion. The terminator was ε-aminocaproic acid (100 mM) for both experiments. The gel segments of 1 cm were extruded from the gel and the processed proteins from the gel segments were measured for optical density at 280 nm (after a dilution 1:20) and for pH (dotted line) (after a dilution of 1:2). The fractions were analysed for 9 individual proteins.

REFERENCES

1. Martin, A.J.P. and Everaerts, F.M.: Displacement Electrophoresis. Procl Roy. Soc. Lond. A., 316, 493 (1970).

2. Haglund, H.: Isotachophoresis - A Principle for Analytical and Prepa-

rative Separation of Substances such as Proteins, Peptides, Nucleotides, Weak Acids, Metals. Sci. Tools, 17, 1 (1970).

3. Routs, R.J.: Electrolyte Systems in Isotachophoresis and their Application to some Protein Separations. Thesis, University of Technology, Eindhoven, The Netherlands 1971.

4. Kohlrausch, F.: Ueber Concentrations - Verschiebungen durch Electrolyse im Inneren von Lösungen und Lösungsgemischen. Ann. Phys. Leipzig 62, 208 (1897).

5. Coxon, M. and Binder, M.J.: Radial Temperature Distribution in Isotachophoresis Columns of Circular Cross-Section. J. Chromatogr. 95, 133 (1974).

6. Vestermark, A.: Cons Electrophoresis - An experimental study. University of Stockholm, Sweden, 1966.

7. Beckers, J.L. and Everaerts, F.M.: Isotachophoresis. The Quantitative Separation of Anions. J. Chromatogr. 69, 165 (1972).

8. Everaerts, F.M. and Verheggen, T.P.E.M.: Isotachophoresis. Applications in the Biochemical Field. J. Chromatogr. 91, 837 (1974).

9. Hinckly, J.O.N.: Transphoresis and Isotachophoresis - Automatable fast Analysis of Electrolytes, Proteins and Cells with Suppression of Gravitational Effects. Clinical Chem., 20, 973 (1974).

10. Kopwillem, A.: Purification Control of Synthetic Peptides by means of Analytical Isotachophoresis. In H. Peeters, Protides of the Biological Fluids, 22, Pergamon Press, Oxford, 715 (1975).

11. Griffith, A. and Catsimpoolas, N.: General Aspects of Analytical Isotachophoresis in Polyacrylamide Gels. Anal. Biochem. 45, 192 (1972).

12. Svendsen, P.J.: On the Procedure of Preparative Isotachophoresis. Sci. Tools, 20, 1 (1973).

13. Catsimpoolas, N.: Isoelectric Focusing and Isotachophoresis of Proteins. Separation Science 8, 71 (1973).

14. Chrambach, A., Kapadia, G. and Cantz, M.: Isotachophoresis on Polyacrylamide Gel. Separation Science, 7, 785 (1972).

15. Jefferis, R. and Butwell, A.J.: Isolation of a Human IgD Myeloma Protein by Isotachophoresis. Sci. Tools, 22, 1 (1975).

16. Chrambach, A. and Skyler, J.S.: The Application of Steady-State Stacking to Macromolecular Fractionation by Polyacrylamide Gel Electrophoresis. In H. Peeters, Protides of the Biological Fluids, 22, Pergamon Press, Oxford, 701 (1975).

17. Bier, M., Hinckley, J.O.N., Smolka, A.J.K. and Snyder, R.S.: Potential Use of Isotachophoresis in Space. In H. Peeters, Protides of the Biological Fluids, 22, Pergamon Press, Oxford, 673 (1975).

18. Kopwillem, A., Merriman, W.G., Cuddeback, R.M. and Bier, M.: Serum Protein Fractionation by Isotachophoresis using Amino Acid Spacers. J. Chromatogr. 118, 35 (1976).

19. Arlinger, L.: Analytical Isotachophoresis in Capillary Tubes. Separation of Human Hemoglobin. In H. Peeters, Protides of the Biological Fluids, 22, Pergamon Press, Oxford, 691 (1975).

20. Laurell, C.B.: Quantitative Estimation of Proteins by Electrophoresis in Agarose Gels containing Antibodies. Anal. Biochem. 15, 45 (1966).

21. Mancini, G., Carbonara, A.O. and Heremans, J.F.: Immunochemical Quantities of Antigens by Single Radial Immunodiffusion. Immunochem. 2, 235 (1965).

22. Svendsen, P.J., in N.H. Axelsen, J.Kroll, B. Werke, Quantitative Immunoelectrophoresis, Universitetsforlaget, Oslo, 69-70 (1973).

23. Bier, M., Cuddeback, R.M. and Kopwillem, A.: Preparative Plasma Protein Fractionation by Isotachophoresis in Sephadex Columns. J. Chromatogr. in press.

Immuno-Isoelectric Focusing Analysis of Antibodies Fractionated by Isotachophoresis

G. Peltre and C.H. Brogren

ABSTRACT

We have studied the resolving power of preparative isotachophoresis in column and flat-bed gels for the fractionation of rabbit anti-pig lactate dehydrogenase antibodies. The heterogeneity of the specific antibodies in the fractions was analysed by immuno-electrofocusing - an analytical method where high resolution in the fractionation of antibodies is combined with sensitive detection of specific antibodies by precipitation with enzyme antigen.

INTRODUCTION

Our aim is to separate, in preparative scale, antibodies into fractions showing a restricted heterogeneity. We have selected here preparative ITP in PAG column to try to solve our problem. ITP is selected because it is, by principle, a preparative method of choice mainly due to the concentrating and the boundary sharpening effects. We have chosen the preparative ITP separation in columns as apparatus suited for such separations is available and already widely used for other applications.

In this work, the fractions obtained by preparative ITP of whole immune sera or globulin solutions containing antibodies are analysed by IEF in PAG flat slabs. The antibody activity of the subfractionated immunoglobulins is evidenced by an immune precipitation with the specific antigen, a pig enzyme (LDH). The presence of the enzyme which remains bound to the precipitated antibodies is visualized by specific LDH staining.

ABBREVIATIONS: ITP, isotachophoresis; IEF, isoelectric focusing; IIEF, immuno isoelectric focusing; PAG, polyacrylamide gel; Ip, isoelectric point; LDH, lactate dehydrogenase.

Different parameters of the isotachophoretic separation are tested (three leading buffers associated with different spacer systems: Ampholines (LKB, Bromma, Sweden) and/or amino acids) and their effect on the efficiency. Preparative ITP and IEF of immunoglobulins are compared.

MATERIALS AND METHODS

1 - Samples. Three immune sera obtained from two rabbits immunized twice with 1.5 mg of pig LDH V (Boehringer, Mannheim, Germany) are chosen for their high concentration of antibodies (approximately 1 mg/ml) and the great heterogeneity of the antibodies Ip (more than 30 bands detectable by analytical IIEF).
The γ-globulin fractions are obtained by ammonium sulphate precipitation (at 50% of saturation) from the preceeding sera.

2 - Analytical IEF in thin-layer PAG. The technique of Awdeh et al. (1) is used with some modifications (2). The PAG is made of 4.5% acrylamide, 2.6% cross-linking. The IEF plates are placed on a metallic cooling plate kept at 17°C. A constant electric power is applied between the carbon electrodes placed at the edges of the plates (5W per 10 x 20 cm plate, for $2\frac{1}{2}$ hours, at the beginning 300 V and 17 mA, at the end up to 1.2 kV, 5 to 4 mA). The detection of the antibody bands is made as described in (2).

3 - Preparative IEF in granulated gel. The technique of Radola (3, 4) is used. On a 20 x 20 cm glass plate, supporting 100 ml of pre-swollen Sephadex G-75 Superfine gel (Pharmacia Fine Chemicals, Uppsala, Sweden) containing 2% Ampholine pH 5-8, 0.6 ml of serum is deposited on a straight line at the anode. This glass plate is placed on a metallic cooling plate kept at 17°C. A direct current is applied to the electrodes and a constant power of 10 W is maintained during a 6 hours run (600 V and 17 mA at the beginning, approximately 1500 V and 6.6 mA at the end). After focusing, 40 fractions are sliced on the plate and their protein content eluted from the gel.

4 - Preparative ITP. This is done (as already described in (5, 6) on a Uniphor column (LKB, Bromma, Sweden) or a column prototype designed by P. Just Svendsen with an improved elution chamber (7). The leading or gel buffers used are numbered as in (6). Buffer No. 4: H_3PO_4 30 mM, Tris 33 mM, Temed 2 mM, pH 6.25. Buffer No. 6: H_3PO_4 30 mM, Tris 100 mM, Temed 2 mM,

pH 8.1. Buffer No. 7: HC1 60 mM, Tris 182 mM, Temed 2 mM, pH 8.5. The terminating buffer used is composed of ε-amino-caproic acid 230 mM and Tris 12 mM, pH 8.9. Preparative ITP in flat-bed granulated gel is performed as described in an other paper (these proceedings) with buffer No. 4 both as leading and anodic buffer. For further details consult this associated proceedings paper (8).

RESULTS AND DISCUSSION

This work describes seven of our separations by preparative ITP on PAG column and one preparative ITP in flat-bed granulated gel. They are arranged in 3 groups, each of them corresponding to the type of leading (or gel) buffer used in ITP (buffers No. 4, 6, 7). With buffer 6 five separations were performed using different spacers: Ampholine alone, Ampholines plus the rabbit serum proteins accompanying the specific antibodies against LDH in whole immuneserum, and amino acids together with the rabbit serum proteins.

ITP 1 (Fig. 1) was performed with leading buffer 4 and Ampholine pH 7 to 9 shows fractions of comparable heterogeneity in IIEF. Thirty-eight bands in total are visible. Each fraction contains 5 to 13 bands. Fractions 33 and 39 are inverted. This permits a good comparison of the pI of bands in these fractions with their neighbouring fractions.

ITP 2 (Fig. 2) was performed with buffer 6, Ampholines pH 7 to 9. The first fractions, up to fraction No. 49 are very homogeneous in IEF; the later (50-54) being more heterogeneous. Therefore, the spacing due to the same Ampholines as those added in ITP 1 is strongly efficient on the fast electrophoretically migrating antibodies with this slower moving leading ion.

This effect can be partly compensated by the use in ITP 3 (Fig. 3) of Ampholines whose pH ranges from 8 to 9.5. The end of the separation is slightly more spaced than in ITP 2.

In ITP 4 (Fig. 4) Ampholines from 7 to 9 and 8 to 9.5 are used. Smaller fractions (2.5 ml instead of 5 ml) are collected. The beginning and the end of the separation are well spaced. Some bands (arrows 2, 3 and 4) focus at the same pI as others (arrows without numbers) which have a higher electrophoretic mobility.

ITP

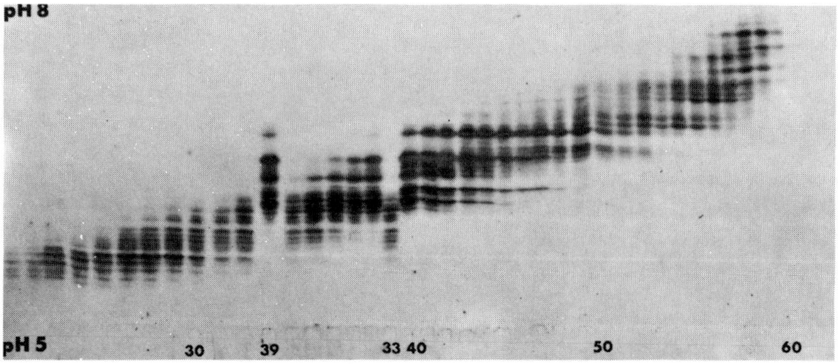

Fig.1: The IIEF analysis of the fractions from an ITP-PAG column experiment with buffer No.4 as leading electrolyte and 1.5 ml Ampholine pH 7-9 as spacer. Sample: 53 mg γ-globulin fraction from an anti-LDH V antiserum (153/20.7.74.) Two fractions are inverted. Anode and sample application at the bottom. Fraction numbers and pH-gradient inserted.

ITP 2

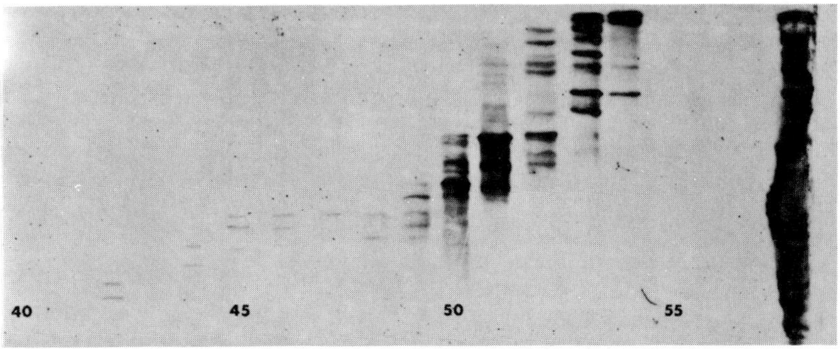

Fig.2: The IIEF analysis of the fractions from an ITP-PAG column experiment with buffer No.6 as leading electrolyte and 1.5 ml Ampholine pH 7-9 as spacer. Sample: 40 mg γ-globulin fraction from an anti-LDH V antiserum (151/1o.12.74). To the right an unseparated sample.

ITP 3

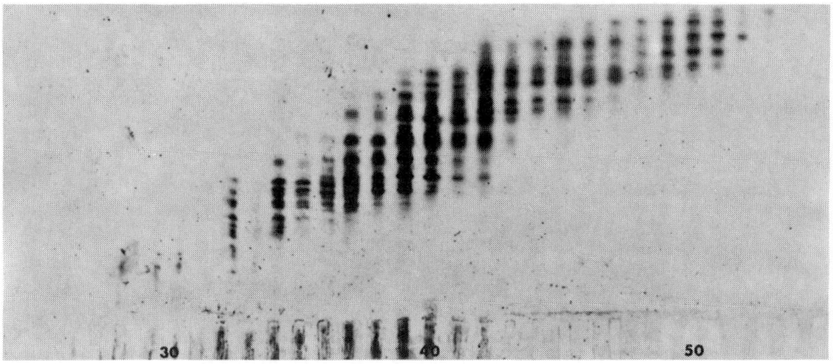

Fig.3: The IIEF analysis of the fractions from an ITP-PAG column experiment with buffer No.6 as leading electrolyte and 1.5 ml Ampholine pH 8-9.5 as spacer. Sample: 53 mg γ-globulin fraction from an anti-LDH V antiserum (153/2o.7.74).

ITP 4

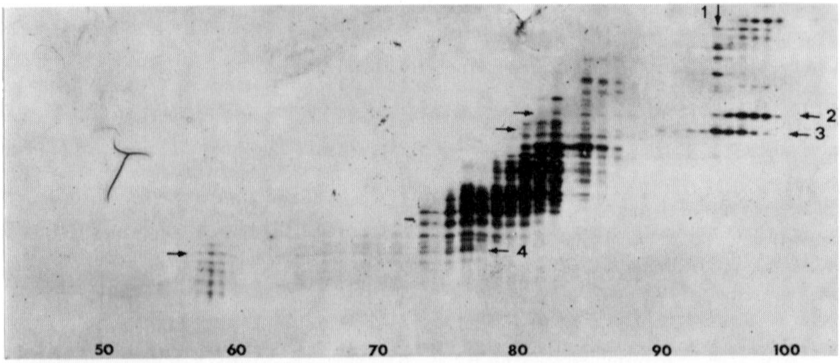

Fig. 4: The IIEF analysis of the fractions from an ITP-PAG column experiment with bufferNo.6 as leading electrolyte and 1 ml Ampholine pH 7-9 plus 1 ml Ampholine pH 8-9.5 as spacer. Sample : 56 mg γ-globulin fraction from an anti-LDH V antiserum (151/1o.12.74). Arrows explained in the text.

ITP 5

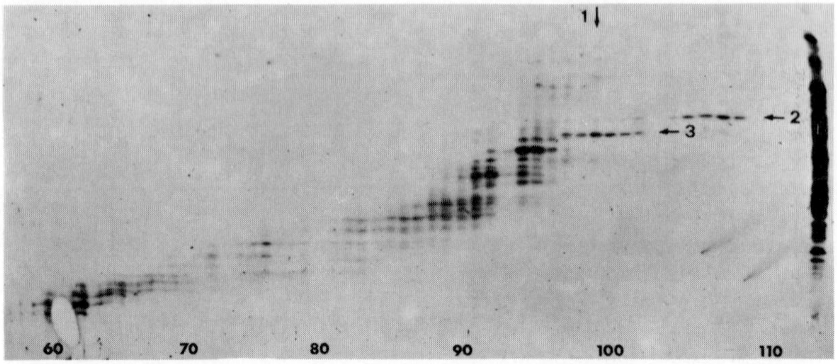

Fig. 5: The IIEF analysis of the fractions from an ITP-PAG column experiment with buffer No.6 as leading electrolyte and 1 ml Ampholine pH 7-9 plus 1 ml Ampholine pH 8-9.5 as spacer. Sample: 1.5 ml anti-LDH Vantiserum (151/1o.12.74) . See the text for explanation of inserted arrows.

ITP 6

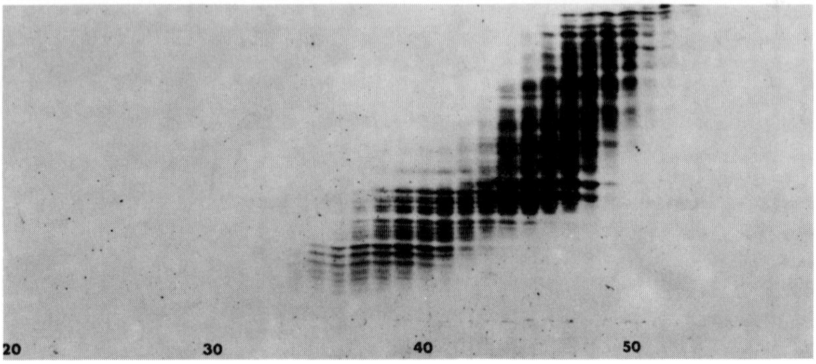

Fig.6: The IIEFanalysis of the fractions from an ITP-PAG column experiment with buffer No.6 leading electrolyte and 6 amino acids as spacer (only 3 of them spacing in the γ-globulin zone). Sample: 2 ml anti-LDH V antiserum (153/5.6.74). See also Fig.6 in ref. (8).

therefore, some unique IEF bands can be separated in this way by ITP into two different components.

The ITP 5 (Fig. 5) is performed under the same conditions as ITP 4 except that a whole serum globulin sample of the same serum is applied. The fractions are less heterogeneous than in ITP 4. The spacing added by the presence of serum proteins (perhaps non-specific γ-globulins,...) has modified the distribution of some bands especially those indicated by arrows 1 and 3 which are now well separated from the No. 2 (arrow) bands.

Another serum sample is separated in ITP 6 (Fig. 6) using 6 amino acids as spacers (asparagine, threonine, glutamine, glycine, alanine, β-alanine) each at 83 mM in 1 ml sample-spacer mixture. Other separations, which are not shown here, were performed combining amino acids and Ampholines as spacers. This combination permits good subdivisions into groups of bands. This spacing seems to be particularly promising for preparative fractionations.

The ITP 7 (Fig. 7) is performed with leading buffer 7. The sample is a globulin fraction. Spacers used are Ampholines pH 8 to 9.5. A rather good spacing is obtained in the beginning of the fractionation. Poor separation is observed at the end of the ITP. The comparison of ITP 7 and 3 (same spacing) illustrates the effect of a slower leading buffer on the separation.

The IIEF analysis of the fractions from preparative isotachophoresis in flat-bed granulated gel (Sephadex G-75 Superfine) is shown in Fig. 8. Here buffer No. 4 has been the leading and anodic buffer and Ampholine pH 8-9.5 the spacer. A good separation is observed of the faster electrophoretically migrating antibodies, but the separation is poor for the more basic antibodies. Compared to the column experiments we had expected a better separation of these more basic antibodies. Whether electroendoosmosis, sieving in the supporting Sephadex gel, or insufficient equilibrium of the isotachophoretic fractionation is the reason for the shift in spacing pattern have to be further investigated. We will try to improve this technique, since the device is much more simple than the column equipment, and extremely easy to handle. The use of a flat granulated gel bed as support medium for ITP may in a similar way as for preparative IEF greatly facilitate its utilisation.

The last figure (Fig. 9) shows the IIEF spectra of fractions obtained by pre-

ITP 7

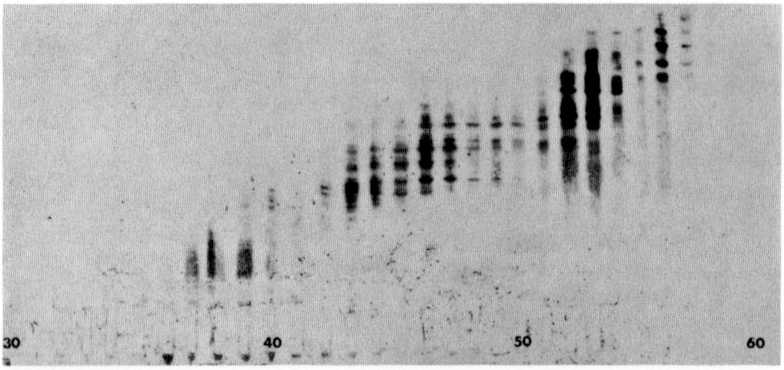

Fig. 7: The IIEF analysis of the fractions from an ITP-PAG column experiment with buffer No. 7 as leading electrolyte and 1.5 ml Ampholine pH 8-9.5 as spacer. Sample: 52 mg γ-globulin fraction from an anti-LDH V antiserum (153/20.7.74).

ITP 8

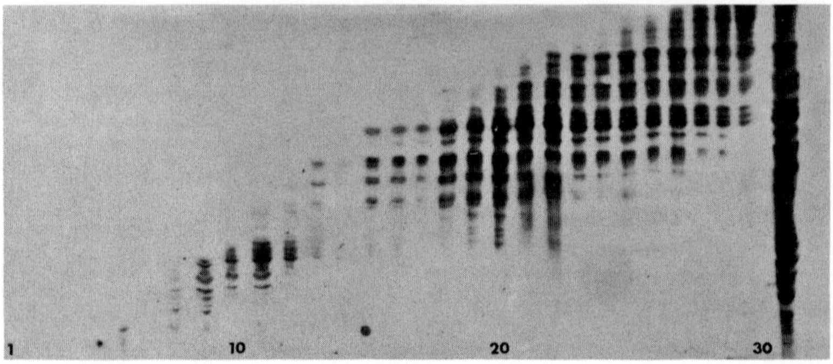

Fig. 8: The IIEF analysis of the fractions from a preparative ITP flat-bed experiment with Sephadex G-75 Superfine as support. Buffer No. 6 as leading electrolyte and anodic buffer. Ampholine pH 8-9.5 (2 ml) as spacer. Sample: 0.5 ml anti-LDH V antiserum (153/5.6.74).

IEF

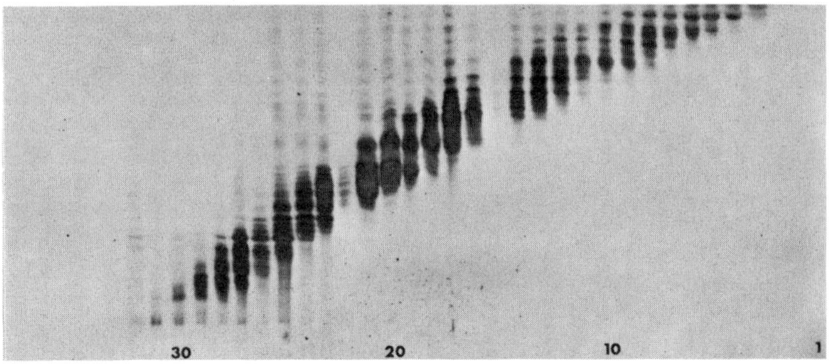

Fig. 9: The IIEF analysis of the fractions from a preparative IEF flat-bed experiment with Sephadex G-75 Superfine as support, and 2% Ampholine pH 5-8 to form the pH gradient. Sample: 0.6 ml anti-LDH V antiserum (153/5.6.74) to the right.

parative IEF on granulated gel (Radola's technique) of an antiserum to LDH. This method considerably shortens the preparation time (5 hours instead of 18 to 40 hours for ITP) using a simple device similar to the flat-bed isotachophoresis (cooled glass plate instead of the cumbersome and expensive preparative column and pump system). The fraction collection technique we used in this experiment needs to be improved and can be improved by using the LKB kit for granulated gel IEF. The contamination of the last fractions (left) by the first ones collected (right) as shown in this figure can be easily avoided. For some applications such a fractionation seems to be very competitive with ITP. When a leading buffer is coupled with a proper spacing system ITP remains a very powerful preparative method for protein fractionation.

ACKNOWLEDGEMENTS

The Danish Pasteur Society has supported C.-H. Brogren with a fellowship in 1975 and The Danish Medical Research Council an additional stay at Institut Pasteur in June 1976. We are thankful to Dr. Per Just Svendsen for the loan of his column and Dr. C. Bodard for the loan of the Uniphor column. Mrs. G. Bromblet is acknowledged for skillful technical assistance with the IIEF analyses and Mrs. Gerhild Samuelsen for secretarial help. We appreciated the inspirating and critical advice from professor Alain Bussard. The work was supported by CNRS (ER No. 119) and ATP INSERM (Nos. 714419 and 499). Part of this work has been presented at Réunion d'Immunologie (Societé d'Immunologie Francaise) in May 1975 in Toulouse, France (Ann. Immunol. 126 C, 363, 1975).

REFERENCES

1. Awdeh, Z.L., Willianson, A.R., Askonas, B.A.: Isoelectric focusing in polyacrylamide gel and its application to immunoglobulins. Nature 219, 66-67 (1968).
2. Peltre, G.: Immuno-isoelectric focusing study of the evolution of antibodies to pig LDH produced in rabbits. In: Progress in Isoelectric Focusing and Isotachophoresis. Righetti, P.G., Ed. (ASP Biol. Med. Press B.V., Amsterdam), pp. 173-178, 1975.
3. Radola, B.J.: Isoelectric focusing in layer of granulated gels: I. Thin-layer isoelectric focusing of proteins. Biochim. Biophys. Acta 295, 412-428 (1973).
4. Radola, B.J.: Isoelectric focusing in layer of granulated gels.: II. Preparative isoelectric focusing. Biochim. Biophys. Acta 386, 181-185 , (1974).

5. Brogren, C.-H., Svendsen, P.J., and Bøg-Hansen, T.C.: On the purification of proteins by neuraminidase treatment and preparative isotachophoresis. In: Progress in Isoelectric Focusing and Isotachophoresis. Righetti, P.G., Ed. (ASP Biol. Med. Press B.V., Amsterdam), pp. 359-369, 1975.

6. Winter, A., Hjalmarsson, S.-G., and Karlsson, C.: Preparative isotachophoresis with the LKB 7900 Uniphor. I. Description of a standardized method. LKB Application Note 146 (1974).

7. Svendsen, P.J.: On elution systems for column electrophoresis in gels. A universal elution system for column electrophoresis. Science Tools 19, 21-25 (1972).

8. Brogren, C.-H. and Peltre, G.: Analytical and preparative isotachophoresis in column and flat-bed gels for fractionation of antibodies. These proceedings.

Analytical and Preparative Isotachophoresis in Column and Flat-Bed Gels for Fractionation of Antibodies

C.H. Brogren and G. Peltre

ABSTRACT

Four different isotachophoretic methods have been applied to fractionation of rabbit anti-pig lactate dehydrogenase antibodies: analytical in capillary tube and in thin-layer polyacrylamide gel, and preparatively in polyacrylamide gel column and flat-bed granulated gel. The flat-bed techniques are a new and easy way to perform both analytical and preparative isotachophoresis shown here in a preliminary presentation. Different immunochemical methods have been used to test content and heterogeneity of specific antibodies besides the detection of total immunoglobulins in the fractions.

INTRODUCTION

The principle of ITP with spacer was applied to preparative separation of proteins by Svendsen and Rose in 1970 (1). They could obtain a good separation of individual proteins in human serum using Ampholines as spacers. The interest in this method has increased since large amounts of material can be separated with high resolution (2). Besides our own application of the method for purification of proteins (3, 4), the method has been used recently for purification of IgD immunoglobulin (5).

The purpose of the work presented was to apply the ITP principle to fractionation of antibodies in order to obtain homogeneous antibody fractions for immunological studies. If specific antibodies of restricted heterogeneity, corresponding to the product of a single clone (6), could be isolated, they could

---

ABBREVIATIONS: ITP, isotachophoresis;IEF, isoelectric focusing; LDH, lactate dehydrogenase.

+) Present address: Immunological Department, Pathological-Anatomical Institute, Kommunehospitalet, DK-1399 Copenhagen K, Denmark.

be used for studies of the regulation of the immune response. Furthermore, subfractionation of antibodies is a subject which has not been solved to satisfaction by other methods. Analytically, IEF reveals a great number of restricted bands, but preparatively in liquid phase in columns, precipitation problems disturb the separation. This problem should be managed by using ITP (3).

Our studies on antibody fractionation have led to some methodological development of ITP methods, of which preliminary results will be presented here. A thin-layer polyacrylamide gel ITP, which can be combined easily with immunochemical characterization or with a different separation in the second dimension will presented in more detail in another paper in these proceedings (7). A detailed study of the heterogeneity of the antibodies in the ITP fractions will be presented in a following paper (8). A further development of the capillary tube ITP (Tachofrac LKB), where detection of specific antibodies can be done is described in Arlinger and coworkers earlier (9) and in these proceedings (10). In our studies the capillary tube ITP has been used only in analytical pilot experiments for investigation of the different preparative buffer and spacer systems.

MATERIALS AND METHODS.

Three different antisera against pig lactate-dehydrogenase isoenzyme V (LDH V) (Boehringer, Mannheim, Germany) produced in rabbits by the procedure described earlier (11) were used. The immunoglobulin fraction was isolated by ammonium sulphate precipitation (50% saturation). Some of the capillary tube experiments were done with a more purified antiserum immunoglobulin fraction from Dakopatts Ltd., Copenhagen, Denmark (Code No. 100SF).

1. Liquid phase capillary tube ITP. Analytical ITP was performed in the LKB Tachophor with the buffer systems used for preparative experiments and with leading electrolyte as anode buffer (12, 13), or these buffers diluted 10-fold. Hydroxy-propyl-methyl cellulose (0.5%) was added as stabilizer to the leading buffer. Ampholines (40% or 1%) and several amino acids were investigated as spacers. Antisera and immunoglobulin fractions were used as samples.

2. Thin-layer polyacrylamide gel ITP. This analytical technique is described elsewhere in these proceedings (7). The ITP separation was performed with the buffer systems described for preparative experiments. Two combined modifications were tried. In crossed isotachophoresis-electrofocusing (7)

a strip from the ITP polyacrylamide gel was transferred to a thin-layer IEF polyacrylamide gel containing 2% Ampholine pH 5-8. The strip was placed on the anodic side with the laying-on technique (14). The focusing and staining procedure for specific anti-LDH antibodies are described in the following paper and earlier (8, 11). Furthermore, a reversed crossed immuno-isotachophoresis (7) was done with LDH V incorporated in the gel. Antisera were used as samples.

3. Polyacrylamide gel column ITP. The equipment for preparative ITP in columns was for some experiments the column designed by P. Just Svendsen, Copenhagen, and for other experiments the LKB Uniphor. The procedures, conditions, and buffers were as described earlier (2, 13) with a few modifications. The gel was increased in length and volume (to 50 ml). Four fractions of 5 ml were eluted per hour, in some experiments fractions of only 2.5 ml volume were collected. In one experiment the elution velocity was increased but without improved results. Both antisera and immunoglobulin fractions of antisera have been used as samples. Ampholines, amino acids or mixture of these were added as spacers.

4. Granulated flat-bed gel ITP. This new preparative method for ITP uses the same equipment as the analytical one (7) with a few modifications. An efficient cooling plate of glass (ad modum Söderholm modified after Vesterberg) together with two buffer vessels and the LKB granulated gel electrofocusing kit were used. The anode vessel should be able to contain 2-3 liters of leading buffer in order to avoid a drastic fall in pH under the separation. The separation was done in about 100 ml Sephadex G-75 superfine in leading buffer. The gel was prepared as described by Radola (15), and to the sedimented gel (100 ml) was added 10 ml of 10-fold concentrated leading buffer No. 4 (13)(H_3PO_4 0.3M, TRIS 0.33M, without TEMED). About 20% of the gel weight was removed by evaporation before use, giving a buffer concentration slightly higher than that used in the column experiments. The pH in the gel was 5.95. Eight layers of Whatmann No. 1 filterpaper were used as wicks and were placed at the short borders of the gel tray. A pre-run was done without sample at 20 W for one hour to force the leading/terminating boundary a few centimeters from the cathodic wick. 0.5-1 ml of antiserum sample was mixed together with 1-2 ml Ampholine and applied directly onto the surface of the gel behind the boundary either with a pipette or using the edge of a glass

plate, or was first mixed with some of the gel behind the boundary, and then reapplied. Bromophenol blue and hemoglobin were added to the sample to indicate the front of the sample zone and to evaluate the separation, respectively. The separation was performed at constant power (20 W) and the experiment was stopped when the bromophenol blue marker reached the anodic wick (12-16 hours). The fractionation grid from the LKB granulated gel electrofocusing kit was used for collection of the fractions.

5. Immunochemical analysis of the fractions. In addition to the UV elution profile, single radial immunodiffusion (SRID)(16) with LDH V incorporated in the gel and samples of the fractions applied in the wells was used for semi-quantitative estimation of the amount of specific antibodies. Furthermore, SRID with anti-rabbit Ig antibodies was used. Reversed rocket immunoelectrophoresis (17) with LDH V incorporated in the gel (no migration of LDH V at pH 8.6 in the agarose gel used (Litex, Batch No. 131A, Glostrup, Denmark)) has also been applied. Fused rocket immunoelectrophoresis (18) was used to evaluate all preparative fractionations. Detection of specific antibodies was performed by incubation of the cold dried plate with the enzyme antigen followed by enzymatic staining of antibody-bound enzyme. Immuno-isoelectric focusing was used to investigate the heterogeneity of the fractionated antibodies (see the following proceedings paper (8)).

RESULTS AND DISCUSSION

With the purpose of fractionating specific antibodies in more homogeneous fractions by ITP, we studied the fractionation of immunoglobulins by capillary tube ITP under different buffer and spacer systems using the analytical Tachophor. When we attempted to shorten the analysis time for these pilot experiments by diluting the preparative buffer systems, we obtained a different separation profile,so all pilot experiments were conducted using the concentrated buffer systems.

In Fig. 1 is shown the results from such a pilot experiment with 30 mM phosphate as leading ion, ε-amino caproic acid as terminating ion, and Tris as counterion (buffer No. 6 (13)). The sample applied contained 5 µl (180 ug) γ-globulin solution, and as spacers 1 µl (40% w/v) Ampholine pH 7-9. The relative ratio between amount of sample and spacer was about the same as normally used for preparative experiments. This was not the case for experiments done with

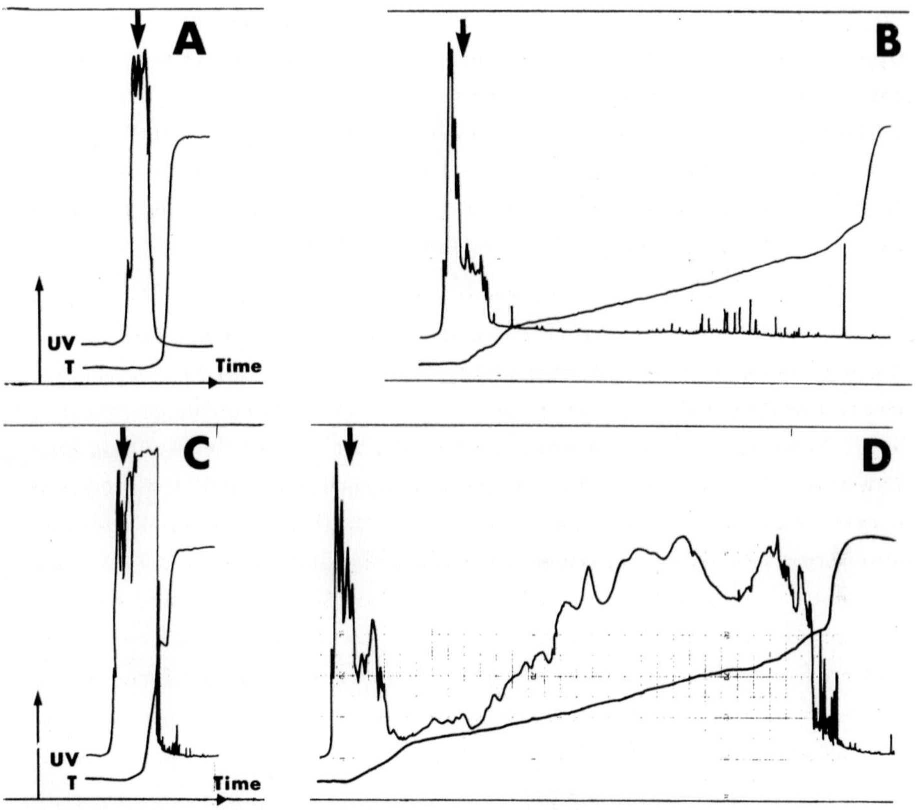

Fig.1: Isotachopherograms from analytical experiments on immunoglobulin fractionation with the LKB Tachphor (23 cm capillary). The buffer system was as applied to some preparative fractionations(30 mM phosphate, 100 mM Tris, 2 mM TEMED, pH 8.1 as leading electrolyte, and 240 mM ε-amino caproic acid, 12 mM Tris, pH 8.9 as terminating electrolyte). Hydroxy-propyl-methyl cellulose (0.5%) was added as stabilizer to the leading buffer.

Records from four experiments are shown: A. without sample, B. with 1 µl Ampholine pH 7-9, 40%, C. with 5 µl immunoglobulin solution (Code No. 100SF, Dakopatts, Ltd., Copenhagen, Denmark), and D. with 5 µl immunoglobulin solution plus 1 µl Ampholine pH 7-9. 40%.

The UV (280 nm) and thermal (T) detection are shown. Recording chart speed was 1 cm per min.. A constant current of 200 µA was applied at 5°C giving a voltage at the recording time of 6-12 kV. The single experiments took 1-2 hrs., with 5-30 minutes recording.

Note the high peak of UV absorbing impurities (arrow), and the separation profile in D showing insufficient spacing in the cathodic part (anode to the left).

For comparison with preparative fractionation in the same buffer system see Fig.2 and Fig.3, and Fig.2 (ITP 2) in ref. 8.

the more diluted buffers normally used in the Tachophor, where 1% solutions of Ampholines were used as spacers. We could increase the current to up to 200 µA corresponding to a maximum voltage at the recording time of 6-12 kV, but higher current gave rise to air bubble formation in the capillary. Under optimal conditions the time for a single analysis was 1-3 hours. The separation profile (UV-record) was in close agreement to the preparative elution profile (UV-record) under the same buffer and sample conditions and thus demonstrates the usefulness of pilot experiments in predicting the best conditions for immunoglobulin fractionation by ITP.

The immunochemical elution profile from a preparative experiment (ITP 2 in ref. 8) under the same buffer conditions as for the experiments shown in Fig. 1 is seen in Fig. 2. As seen in Fig. 2A, the immunoglobulin fraction of an antiserum applied as sample contained some impurities and a pronounced IgG peak. In Fig. 2B the plate was incubated with LDH V and stained enzymatically. Enzyme bound to the specific antibodies was distributed equally over the IgG peak, and was seen in co-precipitation with other proteins (19).

Fig. 3 shows a rocket immunoelectrophoretic analysis of the IgG containing fractions from the same experiment (ITP 2). The electromigration was done in a reversed modification where the antibodies are applied in the wells in an antigen (LDH V) containing gel. The rockets are stained enzymatically. Since the height of the peak is dependent both on the charge and amount of specific antibodies this analysis is not quantitative unless the samples have the same migration velocity. As expected, looking at the precipitate morphology, the analysis does show us that the antibodies were fractionated on the basis of mobility.

The fused rocket immunoelectrophoresis of fractions from a fractionation of 1.5 ml of an antiserum by column preparative ITP is shown in Fig. 4. The individual proteins from the serum can be seen as individual peaks and as for the plate shown in Fig. 2B this plate has been incubated with LDH V amd stained enzymatically before protein staining. The double staining can be done, if INT (2-(p-iodophenyl)-3-p-nitrophenyl-5-phenyltetrazolium salt) is used instead of MTT in the enzymatic staining, since INT in contrast to MTT is insoluble in acid ethanolic solution. The arrow indicates the red stained (LDH) and Coomassie Blue stained IgG peak. We know from Fig. 3

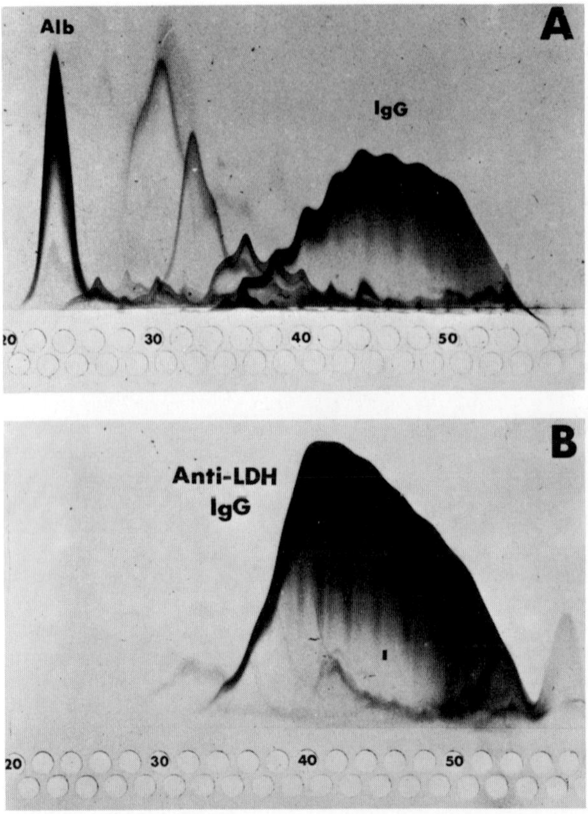

Fig.2: Fused rocket immunoelectrophoretic analysis of a preparative column ITP fractionation of antibodies. The ITP was performed with buffer No. 6 (13) as leading electrolyte, and ε-amino caproic acid as terminating ion. 40 mg γ-globulin fraction of an antiserum (151/10.12.74) was applied as sample added 1.5 ml Ampholine pH 7-9 as spacers.
Plate A: 5 µl of the fractions applied in the wells and 1 ml swine-anti rabbit serum proteins antibodies (Code No. 210SX, Dakopatts, Ltd., Copenhagen, Denmark) applied the gel above. The plate was stained for proteins by Coomassie Brilliant Blue R.
Plate B: 15 µl of the fractions applied in the wells, and 0.5 ml swine-anti rabbit serum proteins antibodies applied to the gel above. Antibodies against LDH V in the IgG peak were detected after incubation with LDH V followed by staining for LDH activity.
The size of the plates were 10 x 10 cm. Fraction numbers are indicated.

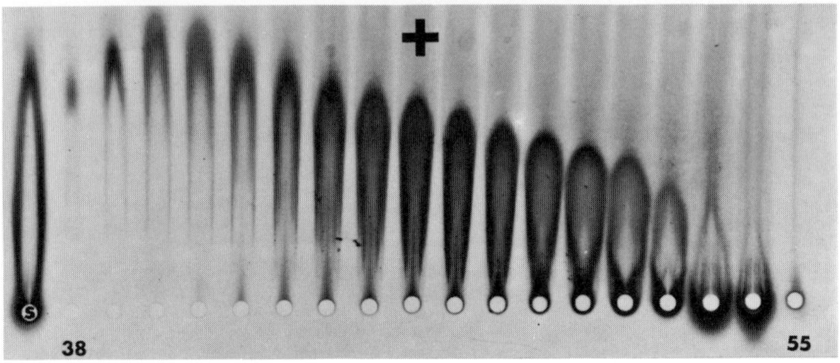

Fig.3: Reversed rocket immunoelectrophoresis of the IgG containing fractions (No. 38-55) seen in Fig.2. LDH V (50 µg) was incorporated in the gel and 20 µl of the fractions applied in the wells. An unfractionated antiserum sample was applied to the left (S). The electromigration was performed at 2 V/cm in 20 hrs. with the anode as indicated. The plate was stained enzymatically for LDH activity. Note the rocket morphology demonstrating high mobility antibodies in the fractions to the left and low mobility antibodies in the fractions to the right (see also Fig.2 in ref. 8). The size of the plate was 10 x 20 cm.

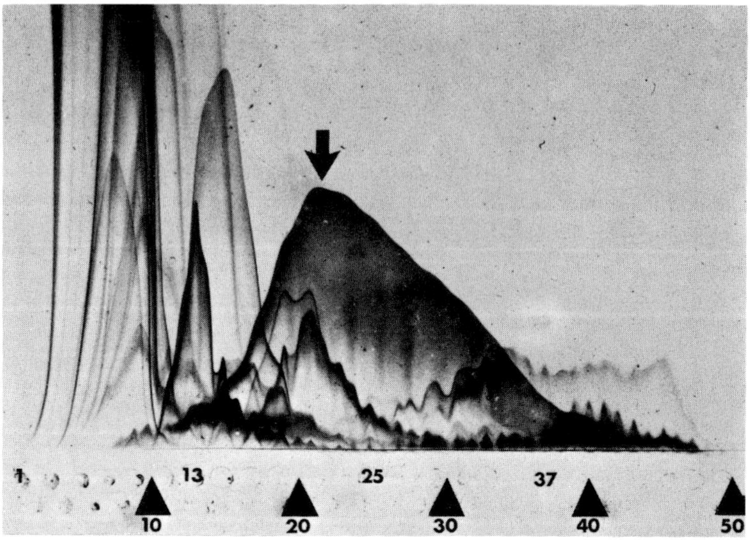

Fig.4: Fused rocket immunoelectrophoretic analysis of the fractions from a preparative column ITP experiment. The ITP was performed with buffers as in the experiment from Fig.2. Antiserum from a single bleeding (153/5.6.74) (2 ml) was applied as sample together with 1 ml Ampholine pH 8-9.5 as spacers. The immunochemical analysis was done with 10 µl of the fractions applied to the wells below, and 1 ml (5.6 µl/cm^2) swine-anti rabbit serum proteins antibodies (Code No. 210SX, Dakopatts, Ltd., Copenhagen, Denmark) added to the gel above. Note the individual proteins seen as peaks, and the IgG peak (arrow) containing specific antibodies. The plate was incubated with LDH V and enzymatically stained before protein staining. Fraction numbers indicated.

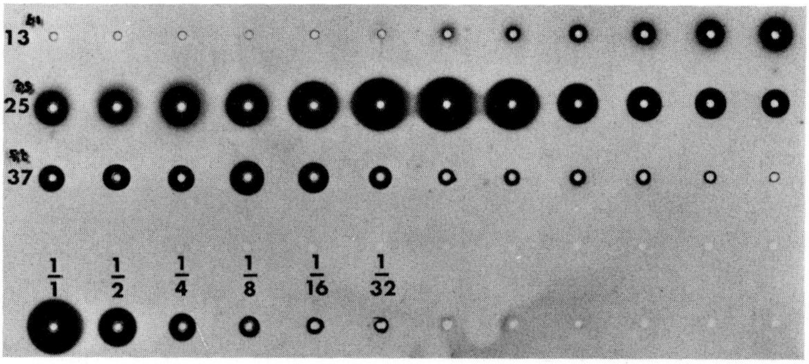

Fig.5: Single radial immunodiffusion analysis for specific antibodies of the IgG containing fractions identified in the analysis shown in Fig.4. The gel contained 50 µg LDH V, and 5 µl of the fractions were applied in the punched wells. As a reference standard diffusion pattern obtained with different dilutions of an antiserum are shown. Note that the highest amount of specific antibodies was seen in fraction No. 30-32 (cf. Fig.4). A similar analysis of the fractions studied in Fig.3 showed that the highest amount of specific antibodies in this experiment was found in fraction No.51.
Fraction numbers indicated. The size of the plate was 10 x 20 cm.

that the IgG profile does not represent the quantitative distribution of IgG in the fractions, but the analysis could be used for identification of IgG containing fractions.

Fig. 5 shows a radial immunodiffusion analysis of the IgG containing fractions identified in Fig. 4, performed in a LDH V containing gel. This method is independent to the charge of the antibodies and therefore can be used to semi-quantitate[1] the amount of the specific antibodies in the fractions.

The fractions from a preparative flat-bed gel experiment were analysed by fused rocket immunoelectrophoresis. The result is seen in Fig. 6. The leading buffer contained 30 mM phosphate and 100 mM Tris, pH 5.9, and the sample contained 500 µl of an antiserum and 2 ml Ampholine pH 8-9.5 as spacer. A pronounced spacing of the IgG is seen with nearly all other proteins unspaced in front of the sample zone. For evaluation of the subfractionation of the specific antibodies the immuno-isoelectric focusing method was used (see Fig. 8 in ref. 8). Probably, in this experiment the sample has been overloaded with spacer, since antibodies were also found in the collected fractions near the application point. From our experience the best results are obtained when the sample zone is from a half to two-thirds of the plate length.

In addition some studies have been done with fractionation of antibodies by thin-layer prolyacrylamide gel ITP followed by fixation of specific antibodies by LDH isoenzymes similar to the analysis by immuno-IEF (8). Such immuno-ITP analysis revealed no restricted bands as seen in immuno IEF, but rather a separation with several mixed zones as illustrated in a crossed ITP-IEF (7) (fig. 7). A reversed crossed immuno-ITP, done in a similar manner to crossed immuno-ITP (7), does not reveal restricted peaks of microheterogeneity as described by Söderholm et al. (14) for crossed immuno-IEF of proteins.

These studies demonstrate some of the limits of the ITP methods available. The mixed zone phenomenon is most pronounced, where the mobility of the separated ions are very close, and this is the case for separation of antibodies by ITP. However, compared to normal zone electrophoresis and disc-electrophoresis, ITP has some advantages in that large amounts of samples can

---

1) A critical study of methods for quantitation of specific antibodies is in preparation.

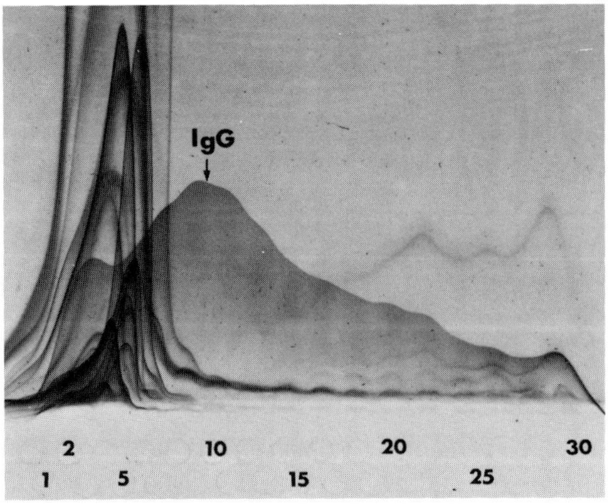

Fig.6: Fused rocket immunoelectrophoretic analysis of the fractions from a flat-bed granulated gel ITP. The ITP fractionation was performed by leading buffer No. 4 (13), and with ε-amino caproic acid as terminating ion. 2 ml Ampholine pH 8-9.5 was added as spacers to a 0.5 ml antiserum sample (156/5.6.74).
On the plate 15 µl of the fractions were applied in the wells below, and 1 ml swine-anti rabbit serum proteins antibodies (Code No. 210SX, Dakopatts, Ltd., Copenhagen, Denmark) added to the gel above. Note the pronounced spacing in the immunoglobulin region. The immuno-IEF analysis of this preparative ITP experiment is shown in Fig. 8 in ref. 8.

Fig.7: Thin-layer polyacrylamide gel ITP followed by transverse thin-layer polyacrylamide gel IEF (crossed ITP-IEF) of an antiserum sample (153/5.6.74). After IEF the plate was incubated with LDH V and then stained for LDH activity. This analysis showed the mixed zone phenomenon in ITP separation (cf. Fig.8 in ref. 8). The sample applied to the ITP plate (20.5 x 11 cm) was 100 µl antiserum plus 400 µl Ampholine pH 8-9.5. From this plate a strip of 0.5 x 15 cm was transferred to the IEF plate (10 x 20 cm) and IEF performed as described in ref. 8. The strip was placed at the bottom with the ITP anode (+) to the left. The pH gradient for the IEF separation is indicated.

be applied in large volumes due to the concentrating and sharpening effects of ITP. ITP is also a preferred method where proteins to be separated precipitate or are inactivated at the isoelectric point or have similar isoelectric points. The combination of ITP with other techniques such as IEF and zone electrophoresis should allow a hight resolution in protein fractionation.

The situation here with subfractionation of antibodies must be seen as one of the most difficult applications of ITP, and the results can probably be improved by development of special spacer systems for ITP, and by technical improvements. Our future plans are therefore to make further developments of flat-bed preparative ITP as a simple and convenient method for protein fractionation, avoiding loss of resolution due to an elution system.

Preparative ITP with spacers added for fractionation of proteins does not seem to follow the rules for ITP of restricted zones (cf. Fig. 7). However, preparative ITP as compared to other separation methods available is very powerful, and therefore justified as a tool for protein separation.

ACKNOWLEDGEMENTS

C.-H. Brogren thanks The Danish Pasteur Society for their support for a stay at Institut Pasteur in 1975, and the Danish Medical Research Council for a grant for a short stay in 1976, which allowed us to develop the new preparative ITP method. We are thankful to Dr. P. Just Svendsen, Copenhagen, for the loan of an ITP column and Dr. C. Bodard for the loan of the Uniphor. Furthermore, we thank LKB, France, for the possibility to work with the Tachophor in these studies. The skillful technical assistance of Mrs. G. Bromblet and Mrs. Å. Hansen, and secretarial help of Mrs. G. Samuelsen is gratefully acknowledged. We thank Dr. H.S. Warren for reading the manuscript. We appreciate the support and interest of Professor A. Bussard in this project. The work was supported by CNRS (ER No. 119) and ATP INSERM (Nos. 714419 and 499). Part of this work has been presented at Réunion d'Immunologie (Société Française d'Immunologie) in May 1975 in Toulouse (Ann. Immunol. 126 C, 363, 1975).

REFERENCES

1. Svendsen, P.J. and Rose, C.: Separation of proteins using Ampholine carrier ampholytes as buffer and spacer ions in an isotachophoresis system. Science Tools 17, 13-17 (1970).
2. Svendsen, P.J.: On the procedure of preparative isotachophoresis. Science Tools 20, 1-4 (1973).

3. Brogren, C.-H. and Svendsen, P.J.: Preparative isotachophoresis in combination with biospecific interaction and neuraminidase treatment in purification of human serum cholinesterase. Prot. Biol. Fluids 22, 685-689 (1975).

4. Brogren, C.-H., Svendsen, P.J. and Bøg-Hansen, T.C.: On the purification of proteins by neuraminidase treatment and preparative isotachophoresis. In: Progress in Isoelectric Focusing and Isotachophoresis. Righetti, P.G., Ed. (ASP Biol. Med. Press B.V., Amsterdam), pp. 359-369, 1975.

5. Jefferis, R. and Butwell, A.J.: Isolation of a human IgD Myeloma protein by isotachophoresis. Science Tools 22, 1-3 (1975).

6. Awdeh, Z.L., Williamson, A.R. and Askonas, B.A.: One cell - one immunoglobulin. Origin of limited heterogeneity of myelome proteins. Biochem. J. 116, 241-248 (1970).

7. Brogren, C.-H.: Thin-layer polyacrylamide gel isotachophoresis, crossed immuno-isotachophoresis, and crossed isotachophoresis-electrofocusing. These proceedings.

8. Peltre, G. and Brogren, C.-H.: Immuno-isoelectric focusing analysis of antibodies fractionated by isotachophoresis. These proceedings.

9. Arlinger, L.: Preparative capillary isotachophoresis. Principles and some applications. J. Chromatogr. 119, 9-24 (1976).

10. Arlinger, L.: The principle of preparative capillary isotachophoresis, and Moberg, U., Hjalmarsson, S.G., Arlinger, L., Lundin , H.: Preparative capillary isotachophoresis: Separation of µg amounts of some human serum proteins. These proceedings.

11. Peltre, G.: Immuno-isoelectric focusing study of the evolution of antibodies to pig lactate dehydrogenase produced in rabbits. In: Progress in Isoelectric Focusing and Isotachophoresis. Righetti, P.G., Ed. (ASP Bio. Med. Press B.V., Amsterdam), pp. 173-178, 1975.

12. Arlinger, L.: Analytical isotachophoresis in capillary tubes: Analyses of proteins in spacer gradients. In: Progress in Isoelectric Focusing and Isotachophoresis. Righetti, P.G., Ed. (ASP Biol. Med. Press B.V., Amsterdam), pp. 331-340, 1975.

13. Winter, A., Hjalmarsson, S.-G. and Karlsson, C.: Preparative isotachophoresis with the LKB 7900 Uniphor. I. Description of a standardized method. LKB Application Note 146 (1974).

14. Söderholm, J. and Smyth, C.J.: Crossed immunoelectrofocusing for stüdies on protein microheterogeneity. In: Progress in Isoelectric Focusing and Isotachophoresis. Righetti, P.G., Ed. (ASP Biol. Med. Press B.V., Amsterdam), pp. 99-114, 1975.

15. Radola, B.J.: Isoelectric focusing in layer of granulated gels. II. Preparative isoelectric focusing. Biochim. Biophys. Acta 386, 181-185 (1974).

16. Mancini, G., Carbonara, A.O. and Heremans, J.F.: Immunochemical quantitation of antigens by single radial immunodiffusion. Immunochemistry 2, 235-245 (1965).

17. Höyerall, H.M., Vandvik, B. and Mellbye, O.J.: Quantitation of Dipheria and tetanus antibodies by reversed rocket immunoelectrophoresis. J. Immunol. Methods 6, 385-398 (1975).

18. Svendsen, P.J.: Fused rocket immunoelectrophoresis. Scand. J. Immunol. 2, suppl 1, 69-70 (1973).

19. Brogren, C.-H., and Bøg-Hansen, T.C.: Enzyme characterization in quantitative immunoelectrophoresis. An enzymological study of human serum esterases. Scand. J. Immunol. 4, suppl. 2, 37-51 (1975).

Index